The University of North Carolina Monograph Series
in
PROBABILITY AND STATISTICS

Sponsored by the
Department of Statistics at
The University of North Carolina at Chapel Hill

Number 4

The University of North Carolina Monograph Series
in
PROBABILITY AND STATISTICS

Previously published:

NUMBER 1 RUSSIAN-ENGLISH DICTIONARY OF STATISTICAL TERMS AND EXPRESSIONS AND RUSSIAN READER IN STATISTICS, *by Samuel Kotz, with the collaboration of Wassily Hoeffding*

NUMBER 2 PROCEEDINGS OF THE SYMPOSIUM ON CONGESTION THEORY, *Edited by Walter L. Smith and William E. Wilkinson*

NUMBER 3 ESSAYS IN PROBABILITY AND STATISTICS *Edited by R. C. Bose* et al.

COMBINATORIAL MATHEMATICS AND ITS APPLICATIONS

The University of
North Carolina Press • Chapel Hill

COMBINATORIAL MATHEMATICS AND ITS APPLICATIONS

Proceedings of the Conference Held at the University of North Carolina at Chapel Hill April 10–14, 1967

Edited by
R. C. BOSE
and
T. A. DOWLING

Library of Congress Catalog Number
69-63006
Manufactured in the United States of America

General Editor's Preface

The University of North Carolina Monograph Series in Probability and Statistics is sponsored by the Department of Statistics at The University of North Carolina at Chapel Hill and provides an addition to the publication activities of the Institute of Statistics of the University.

Since the Department was established in Chapel Hill in 1946 by Professor Harold Hotelling, much of the work of the faculty, students, and visitors has been published in the Institute of Statistics Mimeograph Series. It is expected that original research similar in character to that published in the Mimeograph Series in the past will continue to be published as before. There is, however, work of value to research students and senior scholars that deserves publication in more permanent form, and it is this work that will be published in the Monograph Series.

Titles in the Series are approved by an editorial board from the Department of Statistics. Although it is expected that most of the titles will be the work of authors connected with the University, any suitable manuscripts will be considered for inclusion in the Series.

GEORGE E. NICHOLSON, JR.

Preface

Combinatorial mathematics concerns itself with the problems of operations on or arrangements and selections from a finite or discrete set. R. A. Fisher in the middle twenties showed that the design and analysis of experiments in which the response of the experimental units was subject to statistical variation required the solution of certain combinatorial problems viz., the construction of orthogonal Latin squares and incomplete block designs of certain types. Since the pioneering work of Fisher, the theory of statistically controlled experimental designs has had a great development requiring at every stage the solution of new combinatorial problems. More recently the advent of the electronic age and computer technology has opened up a vast new area for the application of combinatorial techniques. Thus combinatorial problems arise in communication theory, especially the construction of error correcting codes; transportation networks; systems for information retrieval; linear programming and a multitude of other areas. The basic mathematical tools for dealing with these problems include number theory, group theory, finite geometry, abstract algebra, matrix theory, graph theory and to a certain extent the theory of convex bodies.

It was therefore felt that an international conference on Combinatorial Mathematics and Its Applications, which brought together scholars from various disciplines—especially mathematics, statistics and electrical engineering—with a view to sharing their theoretical and practical insights, would be of great value.

As actually organized the Conference represented a joint meeting of the U.S.-Japan Seminar on Combinatorial Mathematics and Its Applications sponsored by the National Science Foundation, the Japan Society for Promotion of Science, and the U.S.-Japan Cooperative Sciences Program; and the Symposium on Combinatorial Mathematics and Its Applications sponsored by the U.S. Air Force Office of Scientific Research and the Department of Statistics of the University of North Carolina at Chapel Hill.

Professor George E. Nicholson, Jr., Chairman, Department of Statistics, The University of North Carolina at Chapel Hill, and Professor J. Ogawa, Chairman, Department of Statistics, Nihon University, were the organizers of the U.S.-Japan Seminar; Professors R. C. Bose and I. M. Chakravarti of the Department of Statistics, The University of North Carolina at Chapel Hill, organized the Symposium.

Encouragement and sustained enthusiastic support for the U.S.-Japan Seminar was encountered early from Dr. J. E. O'Connell when he was NSF representative in Tokyo and later when he returned to Washington as Program Director of the U.S.-Japan Cooperative Science Program. Grateful acknowledgment of his assistance is made here.

The U.S. Air Force Office of Scientific Research has supported basic research in probability and statistics in the Department of Statistics for many years; and the interest and generous assistance of Dr. Robert G. Pohrer, Chief, Mathematics Division, not only in providing support for the Symposium but in sustained interest in the research program which enabled the group in Chapel Hill to develop in this field was indispensable.

Twenty delegates participated in the U.S.-Japan Seminar, ten from each of the two countries. Thirty-three scholars active in the various branches of combinatorics participated in the Symposium. Among the fifty-three participants in the Conference were thirty-five speakers and eighteen discussants. Approximately sixty observers attended the Conference.

The work of the Conference was divided into five main areas: (i) General Problems of Combinatorial Mathematics (ii) Combinatorial Problems of Experimental Designs (iii) Error Correcting Codes and Other Problems of Information Theory (iv) Finite Geometries (v) Graphs. As far as practicable one day was devoted to each area.

The speakers invited to the Conference were asked to provide copies of their papers in advance so that these could be distributed during the various sessions. A discussant was assigned to each paper and was sent a copy of the paper in advance of the Conference. Following the presentation of each paper the discussant delivered his prepared remarks and the meeting was opened to general discussion. We have included in this volume those prepared discussions which were submitted to us for publication, but due to the size of both the audience and the meeting room it was considered impractical to attempt to tape record and subsequently to edit the general discussions. Thus the "discussions" appearing after many of the papers in this volume are remarks prepared in advance of the Conference.

The organizational details connected with a meeting of this size were immense in number, and we feel that proper acknowledgment should be made here to those whose assistance was directly responsible for its success. Mr. Bryant Dunlap and his staff in the Bureau of Residential Adult Education arranged for housing, catering, local transportation, registration, and provided numerous other services both in advance of and during the Conference. Mrs. Elsie Camp of the Department of Statistics was extremely helpful and cooperative in the preconference planning and arrangements.

The editors wish to acknowledge the valuable help they received from K. J. C. Smith, S. M. Heft and K. V. Suryanarayana, in preparing the volume for publication.

R. C. BOSE
T. A. DOWLING

Conference Participants

The following persons either presented papers at the conference or took an active part in the discussion.

E. F. ASSMUS, *Lehigh University and Sylvania Applied Research Laboratory*
M. L. BALINSKI, *City University of New York*
A. BARLOTTI, *University of Palermo, Italy*
C. BERGE, *International Computation Center, Italy*
E. R. BERLEKAMP, *Bell Telephone Laboratories*
B. B. BHATTACHARYYA, *North Carolina State University at Raleigh*
R. C. BOSE, *University of North Carolina at Chapel Hill*
A. T. BRAUER, *Wake Forest University*
R. H. BRUCK, *University of Wisconsin*
P. CAMION, *Toulouse University, France*
I. M. CHAKRAVARTI, *University of North Carolina at Chapel Hill*
H. CRAPO, *University of Waterloo, Canada*
T. A. DOWLING, *University of North Carolina at Chapel Hill*
D. R. FULKERSON, *The Rand Corporation*
S. W. GOLOMB, *University of Southern California*
M. HALL, *California Institute of Technology*
L. H. HARPER, *The Rockfeller University*
A. J. HOFFMAN, *IBM Watson Research Center*
S. IKEDA, *University of North Carolina at Chapel Hill*
M. IRI, *University of Tokyo, Japan*
T. KASAMI, *Osaka University, Japan*
T. KITAGAWA, *Kyushu University, Japan*
D. E. KNUTH, *California Institute of Technology*
G. KOCH, *University of North Carolina at Chapel Hill*
MRS. R. LASKAR, *University of North Carolina at Chapel Hill*
S. LIN, *University of Hawaii*
MRS. J. MACWILLIAMS, *Bell Telephone Laboratories*
H. B. MANN, *Mathematics Research Center, University of Wisconsin*
M. MASUYAMA, *The Catholic University of America and Meteorological College, Japan*
H. F. MATTSON, *Sylvania Applied Research Laboratory*

D. MESNER, *University of Nebraska*
J. OGAWA, *Nihon University, Japan*
T. G. OSTROM, *Washington State University and The University of Frankfurt, Germany*
W. W. PETERSON, *University of Hawaii*
C. R. RAO, *Indian Statistical Institute, Calcutta, India*
D. K. RAY-CHAUDHURI, *Ohio State University*
A. RÉNYI, *Mathematical Institute of the Hungarian Academy of Sciences*
J. RIORDAN, *Bell Telephone Laboratories*
R. T. ROCKAFELLAR, *University of Washington*
G. C. ROTA, *The Rockefeller University*
H. J. RYSER, *California Institute of Technology*
M. P. SCHÜTZENBERGER, *Institute Blaise Pascal, France*
S. S. SHRIKHANDE, *University of Bombay, India*
D. SMITH, *Duke University*
K. J. C. SMITH, *University of North Carolina at Chapel Hill*
J. N. SRIVASTAVA, *Colorado State University*
K. TAKEUCHI, *University of Tokyo, Japan*
W. T. TUTTE, *University of Waterloo, Canada*
M. WATKINS, *University of North Carolina at Chapel Hill*
E. J. WELDON, *University of Hawaii*
K. YAMAMOTO, *Florida Atlantic University*
S. YAMAMOTO, *Hiroshima University, Japan*

Contents

Part I

GENERAL PROBLEMS OF COMBINATORIAL MATHEMATICS

CHAPTER 1

Probabilistic Methods in Combinatorial Mathematics

ALFRED RÉNYI, *Mathematical Institute of the Hungarian Academy of Sciences*

1. THE RENAISSANCE OF COMBINATORIAL MATHEMATICS

It is beyond doubt that we are witnesses of a renaissance of combinatorial problems and methods in mathematics. This process started slowly and became evident only gradually in the last two decades, but its origins can be traced back to the 1920's. Without exaggeration one can say that all main branches of mathematics contributed to some extent to this development.

Probability theory and *statistics* was even at the time of stagnation of combinatorial mathematics in the nineteenth century and the beginning of our century the main source of problems and the main consumer of results of a combinatorial character. The rapid development of probability theory, which has been going on with increasing speed since the 1930's, meeting successfully the constantly growing challenge of its applications in practically every branch of human knowledge, was also the source of many problems of a combinatorial character. New types of combinatorial problems were raised in statistics, concerning the design of experiments and in the theory of order statistics. Another source of combinatorial problems was the arc-sine law, which led to the development of an important combinatorial theory[1]. Recently information theory became a rich source of combinatorial problems of a quite new type.

As a second source of the revival of interest in combinatorial mathematics, *graph theory* has to be mentioned. Before the second world war Hungary was one of the few countries where graph theory was taken seriously. The general opinion of mathematicians before the war is reflected in that some mathematicians called graph theory the "slums" of topology. However, in the years after the war it has been quickly realized all over the world that graph theory is a basic and independent chapter of combinatorial mathematics, having important applications, e.g., in operations research, in chemistry, in statistical physics, etc.. (Some such applications will be mentioned in what follows.)

A main cause of the revival of interest in combinatorial mathematics was the recent development of *numerical analysis*. The appearance of modern high speed computers led to a strong shift to finite mathematics in general, and thus to combinatorics, in particular.

Besides these main sources, stimulation came also from *algebra* (finite groups, Galois fields, matrices, lattices, etc.), *geometry* (finite geometries, discrete geometry), *number theory*, (difference-bases, combinatorial methods of P. Erdös), *set theory, topology* and *mathematical logic* (Sperner's theorem, Ramsey's therem) and also from *statistical mechanics* (Ising models), *genetics*, etc.. The wide spectrum of modern combinatorial mathematics and its various applications is exhibited by [1].

2. GENERAL METHODS AND SCOPE OF COMBINATORIAL MATHEMATICS

Characteristic for the present state of affairs of combinatorial mathematics is the wealth of particular problems and the lack of a systematic theory. Only a small number of basic methods are available. Among these the method of generating functions is to be mentioned first, which also plays a role in several other methods: in the counting method of G. Polya, generalized by N. G. de Bruijn, in symbolic methods, in the

[1] These problems were discussed in detail at a meeting held in Aarhus in 1962.

operator-method of G. C. Rota, etc.. The application of algebraic tools (finite fields, group theory, the Pfaffian, etc.) can also be considered as one of the general methods in combinatorial mathematics; in the analysis of networks, methods and concepts of Boolean algebra and mathematical logic are successfully applied. Another general method in combinatorial mathematics is the application of probability theory as a tool. We shall deal with this method in detail in the following sections of the present paper. In spite of the availability of these general methods of approach, if one is confronted with a particular problem of an unusual type, usually one has to attack it in an *ad hoc* way, and there is no general theory which would guide in the solution. The main reason for this seems to be that the field of combinatorial mathematics is still far from being systematized.

It is rather difficult to define at all what combinatorial mathematics really is, and to tell what its main chapters are, or what basic types of problems can be distinguished.

One possible and often used definition can be formulated as follows. Combinatorial mathematics is the theory of finite sets: it deals with relations and functions, and further, with sets of functions defined on finite sets, especially with problems of enumeration, construction and existence. It should be added that there are certainly some results concerning infinite sets (e.g., infinite graphs) which are undoubtedly of a combinatorial character. Thus in this respect the above definition seems to be a bit too restricted. However, in another respect the definition is not enough restricted, as clearly not every investigation concerning finite sets is of combinatorial character. For instance, though the study of the symmetric group of all permutations of a finite set is certainly a part of combinatorial mathematics, and in spite of the fact that every finite group is isomorphic to a subgroup of a symmetric group, nevertheless the main body of the theory of finite groups belongs to algebra and is not of combinatorial character.

Somewhat vaguely, problems concerning finite sets which are of a combinatorial character can be characterized by the fact that these problems are usually independent of the labelling of the elements of the basic set, i.e. invariant under any permutation of these elements. This is a common property of most (but not all) of those problems which we feel to be of combinatorial character. Problems in which the elements of the basic finite set have individual character and the set possesses an algebraic structure, belong usually to some other field of mathematics, e.g. algebra, number theory, etc.; nevertheless such problems may also exhibit combinatorial features.

3. PROBABILITY THEORY AS A TOOL OF COMBINATORIAL MATHEMATICS

Most classical results of combinatorial mathematics have been developed—since the time of Pascal and Fermat—with the aim of being applied in probability theory. However this relation between probability theory (as the field of application) and combinatorics (as a tool) can be reversed, and this led to interesting and even surprising results. Many important combinatorial problems can be attacked only in this way, or at least this is the easiest way of approach. For instance, in order to prove that combinatorial objects having certain properties exist, often the only available method is to show that by introducing a probability measure in a certain (finite) set of combinatorial objects, the subset of objects possessing the required properties has a positive probability and thus cannot be empty. This method of proof of course does not lead to an effective construction, only to *proofs of existence*. However, often such a proof shows not only that there exist objects of the required type but also that most of the objects of a certain set have the required properties, i.e., it is an exception if an object of the given type does not have the required properties.

As a matter of fact, the probabilistic approach is particularly useful if one wants to study *typical properties* of members of a class of combinatorial objects (like graphs, matrices, partitions, permutations, etc.), i.e., properties possesed by a majority of objects belonging to the class considered, as contrasted with properties which are possessed only by a negligibly small minority of these objects. In some cases in this way *problems of enumeration* can be solved also, at least asymptotically. In the following sections we shall illustrate these statements with some examples.

The aim of this paper is to present a few characteristic examples for the different ways in which probabilistic methods can be used in combinatorial mathematics. No attempt is made to achieve completeness in any respect and the selection of examples is of course strongly biased in favor of examples connected with my own work, done mainly in collaboration with P. Erdös.

4. PROBABILISTIC EXISTENCE PROOFS

One of the earliest examples of the application of probabilistic ideas in graph theory is the following theorem of P. Erdös [3]:

For every sufficiently large n there exist graphs G_n, having n vertices, such that neither G_n nor its complementary graph $\bar{G}$ contains a complete subgraph with more than $2 \log n / \log 2$ vertices.

The proof can be told in two different ways: its probabilistic nature can be disguised or be emphasized. We shall choose the second way, and prove slightly more than stated above, namely the following:

For every fixed p with $0 < p < 1$ and for every $n \geqq n_0(p)$ there exists a graph G_n having n vertices such that G_n does not contain a complete graph with more than $2 \log n / \log (1/p)$ vertices and $\bar{G}_n$ does not contain a complete graph of more than $2 \log n / \log (1/(1-p))$ vertices.

This can be proved as follows: Let H_n be a set having n elements. Let Γ_n be the random graph obtained by connecting any pair of points of H_n by an edge with probability p, independently for each pair. Let $\bar{\Gamma}_n$ denote the complementary graph of Γ_n (i.e., two points P and Q of H_n are connected by an edge in $\bar{\Gamma}_n$ iff they are not connected in Γ_n). Let $v(G, k)$ denote the number of complete sub-graphs of order k in the graph G and let $\mathbf{E}_p$ denote the expectation with respect to the probability measure introduced. We shall show that for sufficiently large n,

$$(1) \qquad \mathbf{E}_p\left(v\left(\Gamma_n, \left[\frac{2 \log n}{\log 1/p}\right] + 1\right) + v\left(\bar{\Gamma}_n, \left[\frac{2 \log n}{\log 1/(1-p)}\right] + 1\right)\right) < 1 .$$

As a matter of fact

$$\mathbf{E}_p(v(\Gamma_n, k)) = \binom{n}{k} p^{\binom{k}{2}} < n^k p^{\binom{k}{2}} / k!$$

and thus

$$(2) \qquad \mathbf{E}_p\left(v\left(\Gamma_n, \left[\frac{2\log n}{\log 1/p}\right] + 1\right)\right) = \sigma(1)\,.$$

Similarly we obtain

$$(3) \qquad \mathbf{E}_p\left(v\left(\bar{\Gamma}_n, \left[\frac{2\log n}{\log 1/(1-p)}\right] + 1\right)\right) = \sigma(1)\,,$$

because

$$\mathbf{E}_p(v(\Gamma_n, k)) = \mathbf{E}_{1-p}(v(\bar{\Gamma}_n, k))\,,$$

and thus (3) follows from (2), replacing p by $1-p$. Thus, for sufficiently large n, (1) holds, i.e. there exists at least one graph G_n having n vertices, for which

$$v\left(G_n, \left[\frac{2\log n}{\log 1/p}\right] + 1\right) + v\left(\bar{G}_n, \left[\frac{2\log n}{\log 1/(1-p)}\right] + 1\right) = 0$$

as was to be proved.

Clearly for $p = 1/2$ we obtain the theorem of Erdös mentioned above. It should be added that no method of construction is known for the graphs, the existence of which was shown above.

As another, more involved, example of the same type let us mention the probabilistic proof [8] of the fact that most of the graphs with n vertices are, for large values of n, almost as asymmetric as possible.

We consider only non-directed graphs without multiple edges and without loops. We call such a graph *symmetric*, if there exists a non-identical permutation of its vertices which leaves the graph invariant. In other words a graph is called symmetric if the group of its automorphisms has order greater than 1. A graph which is not symmetric will be called *asymmetric*. The degree of symmetry of a symmetric graph is evidently measured by the order of its group of automorphisms. The question which led us to the results mentioned is the following: how can we measure the degree of asymmetry of an asymmetric graph?

Evidently any asymmetric graph can be made symmetric by deleting certain of its edges and by adding certain new edges connecting its vertices. We shall call such a transformation of the graph its *symmetrization*. For each symmetrization of the graph let us take the sum of the number of deleted edges—say r—and the number of new edges—say s. It is reasonable to define the degree of asymmetry $A[G]$ of a graph G, as the minimum of $r+s$ where the minimum is taken over all possible symmetrizations of the graph G. Clearly the asymmetry of a symmetric graph is according to this definition equal to 0, while the asymmetry of any asymmetric graph is a positive integer.

The question arises: how large can the degree of asymmetry be for a graph of order n? We shall denote by $\mathbf{A}(n)$ the maximum of $A[G]$ for all graphs G of order n.

We have shown that the asymmetry of a graph of order n cannot exceed $(n-1)/2$ if n is odd, while if n is even the asymmetry cannot exceed $(n/2)-1$; further, that this estimate is asymptotically best possible, that is, for any $\varepsilon > 0$ there can be found an integer n_0 such that for any $n \geqq n_0$ there exists a graph G_n of order n for which $A[G_n] > (n/2)\cdot(1-\varepsilon)$. In other words,

$$\lim_{n\to\infty} \frac{\mathbf{A}(n)}{n} = \frac{1}{2}\,.$$

Our proof is not constructive, only a proof of existence. It uses probabilistic considerations. This method gives, however, more than stated above: it shows that for large values of n most graphs of order n are asymmetric, the degree of asymmetry of most of them being larger than $(n/2)\cdot(1-\varepsilon)$, where $\varepsilon > 0$ is arbitrary.

I would like to add that no method is known to me to construct effectively a graph G_n with degree of asymmetry k for any k, even if no restriction on the number n of vertices is made.

As a third example for a probabilistic proof of existence I mention the proof of the existence of codes with given properties by the method of "random codes" (see e.g. [26]). The basic idea of this method—due to Shannon—is similar to that of the proof of Theorem 1; an additional complication arises because one has to distinguish between the average error and the maximal error of a code.

5. PROBABILISTIC PROOF OF TYPICAL PROPERTIES AND ASYMPTOTIC ENUMERATION PROBLEMS

We start with the following simple example: in a tournament in which every player plays against every other exactly once and no game can end in a tie, the expected number of cyclic triples is exactly equal to one fourth of the total number of triples. (See e.g. [15]). This can be shown simply as follows: select any three players A, B and C. We can suppose that B has defeated A. In this case the triple A, B, C is cyclic if and only if C has defeated B and A has defeated C. If the chances of every game are equal (i.e. $1/2$) for each player, and independent from the outcome of every other game, then the probability of C defeating B and A defeating C is equal to $1/4$, which proves our assertion. It should be noted that it has been shown by Kendall and Smith that the maximal number of cyclic triples is asymptotically the same: if the number of players is denoted by n, the maximum is $(n^3 - n)/24$ if n is odd and $(n^3 - 4n)/24$ if n is even; thus the result can be stated as follows: "in most tournaments the number of cyclic triples is asymptotically maximal."

Many results of this type have been proved by P. Erdös and A. Rényi, in the theory of the evolution of random graphs ([4], [5], [6], [7], [12]). We mention here the first of these results: if n is large most graphs having n vertices and N edges are connected provided that $(2N - n \log n)/2n$ is a large positive number, while only a small minority of such graphs is connected if $(2N - n \log n)/2n$ is a large negative number. More recently we have shown that in case $(2N - n \log n)/2n$ is a large positive number the majority of graphs having n vertices and N edges contains a factor of degree 1 (see [12]). The theory of the evolution of random graphs has been recently applied also in chemistry, see e.g. [2].

Another recent result, due to A. Rényi and G. Szekeres [25], states that the order of magnitude of the diameter of most trees of order n is $\sqrt{n}$. Such results can usually be sharpened to results concering asymptotic distributions. For instance the result on random trees containing the above mentioned statement is as follows: let us consider the set $\mathcal{T}_n$ of all labelled trees of order n, with vertices $P_1, P_2, \ldots, P_n$ and let $h(T_n)$ denote the height of the tree $T_n \in \mathcal{T}_n$ over the point P_1 (i.e., the length of the longest path in T_n starting in P_1). Let $F_n(x)$ denote the probability that choosing at random a tree T_n

from the set $\mathcal{T}_n$ (with uniform distribution) one has $h(T_n) < x\sqrt{2n}$. Then the limit distribution

$$\lim_{n\to\infty} F_n(x) = F(x)$$

exists and is given by the formula (1.c [25])

$$F(x) = \frac{4\pi^{5/2}}{x^3} \sum_{n=1}^{\infty} n^2 e^{-n^2\pi^2/x^2} \qquad (0 < x < +\infty).$$

Hence the expectation of $h(T_n)$ is $\sim\sqrt{2\pi n}$.

Note that in this problem it is trivial that $h(T_n)$ can take on every value between 1 and $n-1$; the question is not one of existence, but one of (asymptotic) enumeration.

Let us mention in this direction a recent result of J. Komlós [18]: the great majority of all n by n zero-one matrices are non-singular if n is large, and the same holds for matrices with elements ± 1. (Note that though the two questions are closely related they are not identical: the probability of a random n by n matrix with elements ± 1 being non-singular is the same as the probability of a random n by n zero-one matrix being non-singular provided all elements of its first row are equal to 1).

A similar result (see [10]) which we obtained with Erdös is that most of the n by n zero-one matrices containing N ones and $n^2 - N$ zeros have a positive permanent provided that $c = (N - n\log n)/n$ is large. More exactly we have proved that if we select one among the $\binom{n^2}{N}$ such matrices at random (with uniform distribution), the probability of its permanent being positive is $\sim e^{-2e^{-c}}$. We have also proved that if $(N - n\log n - (k-1)n\log\log n)/n$ is large then the permanent of the majority of such matrices is $> k$ for $k = 1, 2, \ldots$.

Recently P. Erdös and P. Turán have studied random permutations (see [11]). They have shown that if $O(\Pi_n)$ denotes the order of a permutation Π_n of n elements, then except for $o(n!)$ permutations $\log O(\Pi_n)$ lies between the limits $-(1/2)\log^2 n \pm (\log n)^{3/2+\varepsilon}$ where $\varepsilon > 0$ is arbitrary small.

The proof of these statements consists of two parts. Let α_k denote the number of cycles of length k in the permutation Π_n. Then clearly $O(\Pi_n)$ is equal to the least common multiple of those numbers k for which $\alpha_k \geqq 1$. Now it is proved in [11] by a number theoretic argument (this is the hard part of the proof) that for the majority of the permutations $\log O(\Pi_n)$ is not too far from $\sum \alpha_k \log k = S_n$. On the other hand it is easy to show that the mean value of S_n is asymptotically equal to $-(1/2)\log^2 n$ and its variance to $(1/3)\log^3 n$; from this the result follows by Chebyschev's inequality. More exactly we have

$$\mathbf{E}(S_n) = \sum_{k \leq n} \frac{\log k}{k} \tag{4}$$

and

$$D^2(S_n) = \sum_{k \leq n} \frac{\log^2 k}{k} - \sum_{\substack{k+l>n \\ k \leq n,\, l \leq n}} \frac{\log k \log l}{kl}\,. \tag{5}$$

Formulas (4) and (5) can be deduced from the following remark, due to L. A. Shepp and S. P. Lloyd: if Π_n is chosen at random with uniform distribution on all $n!$ permutations of order n, the joint probability distribution of $(\alpha_1, \alpha_2, \ldots, \alpha_n)$ is the same as the conditional joint distribution of the independent random variables $\beta_1, \beta_2, \ldots, \beta_n$ such that β_k has Poisson distribution with parameter z^k/k $(0 < z < 1;\ k = 1, 2, \ldots, n)$ subject to the condition $\sum_{k=1}^{n} k\beta_k = n$.

Erdös and Turán have proved also in another paper (in print) that $\log O(\Pi_n)$ is asymptotically normally distributed with mean $\mathbf{E}(S_n)$ and variance $D^2(S_n)$. To prove this, besides the number theoretic estimations mentioned, a possible starting point is the following explicit formula for the characteristic function of the random variable S_n:

$$\mathbf{E}(e^{itS_n}) = (1 - z)\exp\left[\sum_{k=1}^{\infty} \frac{z^k}{k^{1-it}}\right]. \tag{6}$$

(6) follows also from the mentioned remark of Shepp and Lloyd.

6. OTHER USES OF PROBABILISTIC METHODS

Probabilistic methods can also be applied in the theory of search; it turns out that under certain conditions a random strategy of search may be asymptotically almost as good as the (usually much more complicated) best systematic strategy (see [21], [22], [23], [24]).

In other cases the number of operations to be carried out when applying a random strategy of search is larger by a constant factor only, compared with the corresponding number for the best systematic strategy. This is the case, for example, for the much discussed problem of separation of the fair and the counterfeit coins (see [9] and [19]) and for a related problem where the factor is $\log_2 3 = 1.58\ldots$.

We have proved that the random search algorithm needs $(\log_2 9)\cdot n/\log_2 n$ weighings, while Lindström has shown that the best systematic algorithm consists of $2n/\log_2 n$ steps only.

It should be mentioned that in the asymptotic evaluation of combinatorial results, the analytic methods developed under the influence and in close connection with probability theory may be successfully used even without a probabilistic interpretation. In this context I would like to call attention to the results of W. K. Hayman [17], who has shown that the moduli of the terms around the maximal term of the power series of an entire function are asymptotically normally distributed provided the function belogs to a certain rather wide class- called by Hayman the class of *admissible* functions. This class contains, e.g., the function e^{e^x}; in which case one gets as a particular case of Hayman's result an asymptotic formula for the number $T(n)$ of partitions of a set of n elements, namely the formula

$$T(n) \sim \frac{\exp\left(n\left(\tau_n + \frac{1}{\tau_n} - 1\right) - 1\right)}{\sqrt{\log n}}$$

where τ_n is defined as the positive root of the equation $\tau_n e^{\tau_n} = x$ (Compare [20]).

Another way to deduce this result is by the following probabilistic interpretation of the Bell-numbers $T(n)$: let $x_1, x_2, \ldots, x_n \ldots$ be independent random variables each having a Poisson distribution with mean value 1. Let ν be a random variable, independent from the x_ks, having a Poisson distribution with mean value e. Then the random variable $y = x_1 + x_2 + \ldots + x_\nu$ has the distribution

$$P(y = n) = \frac{T(n)}{n!} e^{1-e} \qquad (n = 0, 1, \ldots)\,.$$

The asymptotically normal distribution of many combinatorial functions has been proved by V. Gončarov [14] without a direct probabilistic interpretation. Those results where a probabilistic interpretation is found are more elegant, like that given by W. Feller [13] concerning the asymptotically normal distribution of the number of inversions and cycles of a random permutation and that given recently by L. H. Harper [16] on Stirling numbers.

We did not discuss in this paper the probabilistic proofs of combinatorial identities, because it has long been well known that this method of proof is available for a large number of identities.

References

1. Beckenbach, E. F. (ed.) *Applied Combinatorial Mathematics*, John Wiley and Sons, New York, 1964.
2. Bruneau, C. M. "Theorie des graphes Stochastiques appliquée à la synthèse et à la degradation aléatoires de composés macromoleculaires multifonctionnelles," Ph. C. Thesis, Paris, 1966, pp. 1-32.
3. Erdös, P. "Some Remarks on the Theory of Graphs," *Bull. Amer. Math. Soc.*, **53** (1947), 292-294.
4. Erdös, P. and Rényi, A. "On Random Graps I," *Publicationes Mathematicae* (*Debrecen*), **6** (1959), 290-297.
5. Erdös, P. and Rényi, A. "On the Evolution of Random Graphs," *Publ. Math. Inst. Hung. Acad. Sci.*, **5A** (1960), 17-61.
6. Erdös, P. and Rényi, A. "On the Strength of Connectedness of a Random Graph," *Acta Math. Acad. Sci. Hung.*, **12** (1961), 261-267.
7. Erdös, P. and Rényi, A. "On the Evolution of Random Graphs," *Bull. Inst. Internat. Statist.*, **38** (1961), 343-347.
8. Erdös, P. and Rényi, A. "Asymmetric Graphs," *Acta Math. Acad. Sci. Hung.*, **14** (1963), 297-315.
9. Erdös, P. and Rényi, A. "On Two Problems of Information Theory," *Publ. Math. Inst. Hung, Acad. Sci.*, **8** (1963), 241.
10. Erdös, P. and Rényi, A. "On Random Matrices," *Publ. Math. Inst. Hung. Acad. Sci.*, **8** (1963), 455-461.
11. Erdös, P. and Turán, P. "On some Problems of a Statistical Group Theory I," *Z. Wahrscheinlichkeitstheorie und Vers. Gebiete*, **4** (1965), 175-186.
12. Erdös, P. and Rényi, A. "On the Existence of a Factor of Degree One of a Connected Random Graph," *Acta Math. Acad. Sci. Hung.*, **17** (1966), 359-368.
13. Feller, W. *An Introduction to Probability Theory and Its Applications*, Vol. 1, John Wiley and Sons, New York, 1957, Ch. V, Sec. 6.
14. Gončarov, V. "On the Field of Combinatory Analysis," *Izvestia*

Akad. Nauk Ser. Math., 8 (1944), 3-48; see also Amer. Math. Soc. Translation Series, 1955.

15. Harary, F., Norman, R. Z. and Cartwright, D. *Structural Models*, John Wiley and Sons, New York, 1965.
16. Harper, L. H. "Stirling Behaviour is Asymptotically Normal," to be published.
17. Hayman, W. K. "A Generalization of Stirling's Formula," *J. reine und angew. Math.*, **196** (1956), 67-95.
18. Komlós, J. "On the Determinant of (0, 1) Matrices," to appear in *Studia Sci. Math. Hung.*
19. Lindström, B. "On a Combinatory Detection Problem," *Publ. Math. Inst. Hung. Acad. Sci.*, **9** (1964), 195-207.
20. Moser, L. and Wyman, M. "An Asymptotic Formula for the Bell Numbers," *Trans. Roy. Soc. Can.*, **49** (1955), 49-53.
21. Rényi, A. "On Random Generating Elements of a Finite Boolean Algebra," *Acta Sci. Math.* (*Szeged*), **22** (1961), 75-81.
22. Rényi, A. "On a Problem of Information Theory," *Publ. Math. Inst. Hung. Acad. Sci.*, **6** (1961), 505-516.
23. Rényi, A. "Statistical Laws of Accumulation of Information," *Bull. Inst. Internat. Statist.*, **39** (1962), 311-316.
24. Rényi, A. "On the Theory of Random Search," *Bull. Amer. Math. Soc.*, **71** (1965), 809-828.
25. Rényi, A. and Szekeres, G. "On the Height of Trees," to appear in *Austral. J. Math.*
26. Wolfowitz, J. *Coding Theorems of Information Theory*, 2nd ed., Springer, 1964, p. 96.

CHAPTER 2

Combinatorial Methods in the Distribution of kth Power Residues

ALFRED BRAUER, *Wake Forest University*

1. I want to consider some results on the distribution of kth power residues and some related problems. I do not want to give an encyclopedic survey of all known results, but mention only those which can be obtained by elementary combinatorial methods.

It must be stressed that the definitions of combinatorial mathematics in the books of P. A. MacMahon [31], E. Netto [37], J. Riordan [45], H. J. Ryser [46], and in the paper of D. H. Lehmer [30] in the Proceedings of the I. B. M. Scientific Computing Symposium on Combinatorial Problems are not exactly the same. Moreover, G. Pólya [40] finishes his paper " A Note of Welcome " in the first number of the Journal of Combinatorial Theory with the words " I have failed, I am afraid, in defining the scope or the methods or the applications of Combinatorial Theory." Therefore it could be said that some of the topics I want to consider here should have been omitted and others included in my talk. None of these results is studied in one of the mentioned books.

I believe that every mathematician will agree with Ryser [46] " that combinatorics and number theory are sister disciplines. They share a certain intersection of common knowledge, and each genuinely enriches the other. Combinatorial mathematics cuts across the many subdivisions of mathematics and this makes a formal definition difficult. But by and large it is concerned with the study of the arrangement of elements into sets."

Due to the last sentence every theorem on the distribution of the quadratic residues modulo a prime p could be considered a combinatorial result since we study the distribution of the integers $1, 2, \ldots, p-1$ into the two sets, the quadratic residues and the non-residues. But I would not call the reciprocity law a combinatorial theorem. Similarly, I would not call every proof combinatorial if in it the simplest fact of combinatorics is of importance, namely Dirichlet's principle of the drawers to the effect that if $n+1$ elements are distributed into n sets, at least one of the sets contains more than one element, while on the other hand deep analytical results are needed.

I always was interested to obtain in number theory simple combinatorial proofs even if they give less sharp results than deep analytical proofs. But often they have the advantage that they hold for every number n, or at least for every n which is greater than a fixed number C and not only for every sufficiently large number n where we are not able to find out for which n actually the theorem is correct.

2. Let p be a prime. We denote by RR the number of sequences of two quadratic residues, by RN the number of sequences of a residue and a non-residue. Similarly we define the number NR and NN. It was proved by N. S. Aladov [1] in 1896 that

$$4RR = p - 4 - \left(\frac{-1}{p}\right),$$

$$4NN = 4NR = p - 2 + \left(\frac{-1}{p}\right),$$

$$4RN = p - \left(\frac{-1}{p}\right).$$

A simpler proof was given by E. Jacobsthal [24], a student of G. Frobenius and I. Schur, in 1906 in his dissertation. Moreover, he determined the corresponding numbers of sequences of length 3 and obtained some results for sequences of length 4. A few of his results has been obtained earlier by R. Sterneck [52]. The corresponding formulas for sequences of length 2 of cubic and biquadratic residues and non-residues were obtained by W. Unger [54], another student of Schur, in 1921 in his dissertation.

3. Schur himself was highly interested in these problems and published a paper on the distribution on the quadratic residues and non-residues which we have to mention later [49]. For many years he conjectured that for every l and all sufficiently large primes there exist sequences of l quadratic residues and of l quadratic non-residues. For this problem he suggested to many mathematicians to prove the following conjecture. If the whole numbers $1, 2, \ldots, n$ are distributed arbitrarily in two classes, then at least one of these two classes must contain an arithmetic progression of l terms for all sufficiently large n. This conjecture remained unsolved for many years.

4. One day in September of 1927, my brother and I were visiting Schur when v. Neumann dropped in, having just returned from the annual meeting of the German Mathematical Society. He wanted to inform Schur that B. L. van der Waerden [59] had presented a proof of Schur's conjecture at the meeting in the slightly more general form that the numbers $1, 2, \ldots, n$ are distributed in k classes instead of two classes, by induction for 2 and every k, at the suggestion of E. Artin.

Schur was highly excited, but after a few minutes he was disappointed when he saw that this result did not prove his original conjecture. It follows directly from van der Waerden's Theorem only that there exist either a sequence of l residues or l non-residues. But two days later, I was able to send Schur a proof for the existence of sequences of quadratic residues [3].

For every l and all sufficiently large primes p there exists a sequence of l quadratic residues.

Proof. It is sufficient to consider those primes p only for which the smallest non-residue d is less than or equal to l since otherwise the integers $1, 2, \ldots, l$ form a sequence of l quadratic residues. We set

$$n = l(l-1) + 1 . \tag{1}$$

Then by the Theorem of van der Waerden there exists a number

W_n such that for all primes $p > W_n$ either the class of the residues or the class of the non-residues contains an arithmetic progression of n terms

$$r, r+s, r+2s, \ldots, r+(n-1)s .$$

Then the integers

$$(2) \qquad \frac{r}{s}, \frac{r+s}{s}, \frac{r+2s}{s}, \ldots, \frac{r+(n-1)s}{s} \pmod{p}$$

form a sequence of n quadratic residues or n non-residues according as r and s belong to the same class or not. We only have to consider those primes p for which we obtain a sequence of n non-residues. Let us denote these n integers (2) by

$$a, a+1, a+2, \ldots, a+n-1 .$$

They all are non-residues. Among them are the integers

$$(3) \qquad a, a+d, a+2d, \ldots, a+(l-1)d$$

since by (1) and $d \leqq l$ we have

$$(l-1)d \leqq (l-1)l - n = 1 .$$

Hence the numbers (3) form a progression of l non-residues whose difference d is a non-residue. Dividing the numbers (3) by $d \pmod{p}$ we obtain a sequence of l quadratic residues.

5. Schur [50] informed me that this method of proof can be used to improve the Theorem of van der Waerden and asked me to publish this result in my paper [3].

If the numbers $1, 2, \ldots, n$ *are distributed into* k *classes, then there exists a constant* $S(k, l)$ *such that for all* $n \geqq S(k, l)$ *at least one class contains an arithmetic progression of* l *terms whose difference belongs to the same class as the elements of the progression.*

Proof. For $k = 1$, this is obvious; we choose $S(1, l) = l$. We assume that the theorem is proved for k classes and we want to prove it for $k+1$ classes. Then it is also correct if for any positive integer c the numbers $1c, 2c, 3c, \ldots, nc$ are distributed into k classes since this represents a distribution of the numbers $1, 2, \ldots, n$ into k classes.

Let $W(k+1, z)$ be the constant of van der Waerden for progressions of length z in the distribution of the numbers $1, 2, \ldots, n$ into $k+1$ classes. We set

$$(4) \qquad S(k+1, l) = W\{k+1, (l-1)S(k, l)+1\} .$$

Then for $n > S(k+1, l)$ there exists at least one class C which contains a progression

$$(5) \qquad a, a+d, a+2d, a+(l-1)S(k, l)d$$

of $(l-1)S(k, l)+1$ terms. Among the numbers (5) are the progressions

$$(6) \qquad a, a+xd, a+2xd, \ldots, a+(l-1)xd$$

for each $x = 1, 2, \ldots, S(k, l)$ by (4). If for one of these values x the number xd belongs to the class C, then this progression (6) and its difference belong to C.

Otherwise the numbers

$$d, 2d, \ldots, S(k, l)d$$

are distributed into the k classes different from C; hence contain an arithmetic progression of l terms whose difference belongs to the same class as the elements of the progression by the induction assumption. Therefore the theorem is proved.

6. In the following we shall consider kth power residues module a prime p. We always will assume that $p \equiv 1 \pmod{k}$.

Theorem. *For every k and l and all sufficiently large primes p there exists at least one sequence of kth power residues of length l.*

Proof. Since there exists an arithmetic progression of length l whose elements belong all to the same class of kth power residues or non-residues and since the difference of this progression belongs to the same class by Schur's extension of the Theorem of van der Waerden, we only have to divide the elements of this progression by its difference to obtain a sequence of length l of kth power residues.

7. More efforts were necessary to find a proof for the

existence of long sequences of quadratic non-residues, but the proof itself is simple. I shall prove somewhat more.

Theorem [3]. *For all sufficiently large primes p with $p \equiv 1$* (mod k) *there exist sequences of any given length l of kth power non-residues belonging to the class $1/d$ where d is the smallest positive kth power .non-residue.*

Proof. We set

$$n = l!\,(l-1) + 1\,.$$

It follows from the preceeding theorem that there exists a number $z = z(k, l)$ such that for all primes $p > z$ there exists a sequence of kth power residues of length l

$$a, a+1, a+2, \ldots, a+n-1\,.$$

If here $d < l!$, then

$$a, a+d, a+2d, \ldots, a+(l-1)d$$

form a progression of kth power residues of length l whose difference is the kth power non-residue d. Dividing by d (mod p), we obtain a sequence of l non-residues which belong to the class of $1/d$.

Since d must be a prime, the case $d = l!$ is impossible.

Now we assume that $d > l!$. We devide d by $l!$

$$d = l!\,m + b \qquad (0 < b < l!)\,. \tag{7}$$

The numbers

$$(b-d)+d, (b-d)+2d, \ldots, (b-d)+ld \tag{8}$$

are positive. It follows from (7) that

$$b - d + \lambda d \equiv 0 \pmod{\lambda} \qquad (\lambda = 2, 3, \ldots, l)$$

and

$$0 < \frac{b-d}{\lambda} + d < d\,.$$

Hence $\dfrac{b-d}{\lambda} + d$ and $\left(\dfrac{b-d}{\lambda} + d\right)\lambda$ are kth power residues since $\lambda \leqslant l < d$.

It follows that the numbers (8) form an arithmetic progression of kth power residues of length l whose difference is a non-residue. Dividing them by d we obtain a sequence of kth power non-residues of length l which belong to the class of $1/d$. This proves the theorem.

Four years later, I proved that actually each of the $k-1$ classes of kth power non-residues contains sequences of length l for all sufficiently large primes. But the proof is too long to be given here [4].

8. It seems that the title of van der Waerden's paper "Beweis einer Baudetschen Vermutung" [59] is not justified. Certainly van der Waerden heard about the conjecture from Baudet, a student at Goettingen. But at that time, E. Landau, a close friend of Schur since 1900, was one of the four professors at Goettingen. Both saw each other every year and corresponded regularly. Both were interested in the distribution of the quadratic residues. Schur's paper [49] "Einige Bemerkungen zu der vorstehenden Arbeit des Herrn G. Pólya: Ueber die Verteilung der quadratischen Reste und Nichtreste" was introduced by Landau at the meeting of the Goettinger Gelehrten Gesellschaft and Landau improved some of Schur's results in the same volume of the *Goettinger Nachrichten* [28]. Hence it is obvious that Landau knew Schur's conjecture. Landau used to tell every conjecture he could not prove himself to every mathematician he met. Therefore it can be assumed that Baudet heard about Schur's conjecture directly or indirectly.

9. It is very interesting that A. Y. Khintchin in his book *Three Pearls of Number Theory* [27] picks the Theorem of van der Waerden as one of the three pearls. But the only known application of this theorem is the number theoretical application just discussed, and this is not even mentioned.

10. *R. Rado* [41] generalized the Theorem of van der Waerden as follows.

Theorem. *Let k, l, r be positive integers. There exists a number $N = N(k, l, r)$ with the following property. We consider systems of r numbers $x_1, x_2, \ldots, x_r$ where each x_ρ takes all the values $1, 2, \ldots, N$. If we distribute all these systems into k classes, then there exist $2r$ numbers $x_1, x_2, \ldots, x_r$ and $d_1, d_2, \ldots, d_r$ such* *that all the systems* $a_1 + \lambda_1 d_1, a_2 + \lambda_2 d_2, \ldots, a_r + \lambda_r d_r$ *belong to the same class where the* λ_ρ *take all the values*

$$0 \leqq \lambda_\rho \leqq l \qquad \text{for} \quad \rho = 1, 2, \ldots, r .$$

Using this result, Rado obtains some number theoretical results, for instance the following theorems.

(1) *For every* k, l, r *there exists a constant* $N = N(k, l, r)$ *such that for every given* r *different primes* $p_1, p_2, \ldots, p_r$ *which are all greater than* N, *there exist* l *consecutive integers* $a, a+1, \ldots, a+l-1$ *which are all kth power residues for each of the primes* $p_1, p_2, \ldots, p_r$.

(2) *For each* l *and* r *there exists a constant* $C = C(l, r)$ *such that if* $p_1, p_2, \ldots, p_r$ *and* $q_1, q_2, \ldots, q_r$ *are all different primes greater than* C, *there exist* l *consecutive integers which are all quadratic residues for the primes* $p_1, p_2, \ldots, p_r$ *and all quadratic non-residues for the primes* $q_1, q_2, \ldots, q_r$.

Actually these applications can be obtained simpler from our preceeding theorems without using the Theorem of Rado.

Let $G(k, l)$ be a constant such that there exists a sequence of length l of kth power residues for all primes $p > G(k, l)$. Assume that for the different primes $p_1, p_2, \ldots, p_r$ we obtain the following sequences of kth power residues

$$a_1, a_1 + 1, \ldots, a_1 + l - 1 \pmod{p_1} ,$$

$$a_2, a_2 + 1, \ldots, a_2 + l - 1 \pmod{p_2} ,$$

$$\cdots\cdots\cdots\cdots\cdots\cdots$$

$$a_r, a_r + 1, \ldots, a_r + l - 1 \pmod{p_r} .$$

We solve the simultaneous congruences

$$x \equiv a_\rho \pmod{p_\rho} \qquad (\rho = 1, 2, \ldots, r) .$$

Then the numbers $x, x+1, \ldots, x+l-1$ are kth power residues for all the primes $p_1, p_2, \ldots, p_r$.

This proves the first application. In the same way the second application and similar results can be obtained.

11. Estimates for the number of sequences of residues and non-residues of a given length l have been obtained in a number of papers of K. Dörge [16], H. Hopf [21], H. Davenport [13], [14], H. Salié [47], H. Hasse [19], [20] and A. Weil [60]. But the proofs are not purely combinatorial.

12. O. Perron [38] proved the following theorem.

Let p be a prime of the form $4n-1$ and $r_1, r_2, \ldots, r_{2r}$ the quadratic residues (mod p), including the number 0. Let $\mathfrak{K}$ be the matrix which is obtained from the matrix

$$\begin{bmatrix} r_1 & r_2 & \ldots & r_{2n} \\ r_1+1 & r_2+1 & \ldots & r_{2n}+1 \\ \cdots & \cdots & \cdots & \cdots \\ r_1+p-1 & r_2+p-1 & \ldots & r_{2n}+p-1 \end{bmatrix}$$

by reducing the elements (mod p). Then this matrix of $4n-1$ rows and $2n$ columns has the following properties:

I. Each row has exactly n elements in common with each other row.

II. The corresponding elements of each pair of rows have the same difference (mod $4n-1$).

Hence such matrices exist if $4n-1$ is a prime. Perron raised the interesting question: For which other numbers m of form $4n-1$ do such matrices exist, and how can they be obtained? He gives an example for $m = 15$.

I saw that this example could be obtained by using the Jacobian symbols for $m = 15$. Therefore I determined the number of sequences of length 2 for the Jacobian symbols for odd square free numbers m [10]. Since the symbol takes the values 1, -1, and 0, eight such numbers had to be determined. It followed that matrices satisfying Perron's conditions exist if m is the product of a pair of twin primes. Using this result I could prove that Hadamard matrices of order $p(p+2)+1$ exist where p and $p+2$ are twins [9].

My student R. Vause [55], [56] determined in his doctoral dissertation the number of sequences of Jacobian symbols of length 3 and proved that there exist Hadamard matrices of order p^2q+1 where p and q are odd primes and $p^2 = q+2$.

After the publication of my paper [9], I received a letter from H. Hadwiger informing me that he had obtained the result of my paper [9], but not that of [10] and that it was published in the paper of W. Gruner [18] under the title "Einlagerung des regulaeren n-Simplex in den n-dimensionalen

Wuerfel." Neither this title, nor the review in the *Jahrbuch fuer die Fortschritte der Mathematik* indicated that results on Hadamard matrices were obtained. Gruner actually proved more, namely that Hadamard matrices of order $p^i q^j + 1$ exist where p and q are odd primes and where $q^j - p^i = 2$. My proof is more elementary than that of Gruner.

13. We consider now the inverse problem. Given a prime p, what is the greatest length l of the sequences of quadratic residues or non-residues? It follows from a paper of Pólya [39], one of Schur [49] and one of Landau [28] that

$$l = O(\sqrt{p} \log p) .$$

H. Davenport and P. Erdös [15] state in a paper published in 1952: "We draw attention to the problem of estimating the maximum number, say l, of consecutive quadratic residues or non-residues. All we are able to prove is $l = O(p^{1/2})$." L. Rédei begins his paper [44] published in 1953 with the words "Let p be an odd prime and $[x]$ the greatest integer less than or equal to x. One of the most interesting results on the distribution of the quadratic residues and non-residues is the following Theorem of Vinogradov [58]: There exists at least one quadratic non-residue in each set of $3[\sqrt{p}] - 1$ consecutive integers mod p."

But already in 1932 I had obtained a sharper result [6] not mentioned in [15] nor [44].

Theorem. *Let k be any integer and p a prime with $p \equiv 1 \pmod{k}$. If l is the length of the greatest sequence in any of the k classes of kth power residues or non-residues, then*

$$l < \sqrt{2p} + 2 \tag{9}$$

Only in one of the k classes can lie a sequence whose length is greater than $\sqrt{p}$.

Without mentioning this result D. A. Burgess [12] proved that for $k = 2$, we have

$$l = O(p^{1/4+\delta}) \tag{10}$$

for every positive δ.

In the proof of (10) very deep theorems are used. The result is better for sufficiently large primes p while (9) holds for every k and all primes, and this is obtained by a very simple combinatorial method.

In order to obtain these results we first prove the following lemma.

Lemma. *If for $p \equiv 1 \pmod k$ one of the k classes of kth power residues or non-residues contains a sequence*

$$(11) \qquad a, a+1, \ldots, a+l-1$$

of length $l > \sqrt{p}$, then all the numbers

$$(12) \qquad \left[\frac{p}{l}\right]+1, \left[\frac{p}{l}\right]+2, \ldots, l$$

are kth power residues.

Proof. Let z be any integer satisfying $[p/l]+1 \leqq z \leqq l$. We consider the numbers

$$(13) \qquad az, (a+1)z, \ldots, (a+l-1)z$$

and reduce them mod p. Then these l reduced numbers divide the interval $\{0 \ldots p\}$ in subintervals. The length of each of these $l+1$ subintervals is less than or equal to z since

$$lz \geqq l\left\{\left[\frac{p}{l}\right]+1\right\} > l\frac{p}{l} = p\,.$$

Hence at least one of each z consecutive numbers is congruent to one of the numbers (12). But $l \geqq z$. Therefore one of the numbers (11) is congruent to one of the numbers (13). Hence the numbers (11) and (13) belong to the same residue class. It follows that z is a kth power residue.

Corollary. *If for any number $c > \sqrt{p}$, there is a kth power non-residue d in the closed interval*

$$(14) \qquad \left\{\left[\frac{p}{c}\right]+1 \ldots c\right\},$$

then the greatest length l of the sequences in all the k classes is less than c.

Proof. If this is not correct, then $l \geqq c$. Hence (14) is a subinterval of (12) and would only contain kth power residues. This gives a contradiction.

Theorem. *Only one of the k classes of kth power residues and non-residues can contain a sequence of length greater than $\sqrt{p}$.*

Proof. Assume that the numbers (11) form a sequence of length $l > \sqrt{p}$ of the class K. Then the number $z = [p/l] + 1$ is a kth power residue by (12), and the numbers (13) reduced mod p belong to K. Hence of each z consecutive integers at least one belongs to K. It follows that each of the $k - 1$ classes different from K can only contain sequences of maximal length λ where

$$\lambda \leqq z - 1 = \left[\frac{p}{l}\right] \leqq \left[\frac{p}{\sqrt{p}}\right] < \sqrt{p}\,.$$

Corollary. *If p is a prime of form $4n - 1$, then the greatest length of any sequence of quadratic residues or non-residues is less than or equal to $[\sqrt{p}]$.*

Proof. If $a, a + 1, \ldots, a + l - 1$ is a sequence of quadratic residues, then $p - (a + l - 1), p - (a + l - 2), \ldots, p - a$ is a sequence of non-residues. Therefore the maximal length of the sequences is the same in both classes, hence $l \leqq [\sqrt{p}]$.

We may have equality here as the examples $p = 3, 7, 19, 23$ show. For $p = 4n + 1$, we may have $l > [\sqrt{p}]$ as in the case $p = 13$ where 5, 6, 7, 8 are quadratic non-residues.

Now we want to get bounds for the maximal length of the sequences for the primes of form $4n + 1$ and arbitrary k and for the primes of form $4n - 1$ and odd k. It is sufficient to find a number $c > \sqrt{p}$ such that there is at least one kth power non-residue in the interval (14).

Assume first that 2 is a kth power non-residue. We want to prove that we can choose $c = \sqrt{2p} + 2$.

The two integers $[\frac{1}{2}\sqrt{2p}] + 1$ and $2[\frac{1}{2}\sqrt{2p}] + 2$ belong to different classes, hence at least one of them is a kth power non-residue. Moreover, they lie both in the interval

$$\left\{\left[\frac{p}{\sqrt{2p} + 2}\right] + 1 \ldots \sqrt{2p} + 2\right\}$$

since

$$\left[\frac{p}{\sqrt{2p} + 2}\right] + 1 \leqq \left[\frac{p}{\sqrt{2p}}\right] + 1 = \left[\frac{1}{2}\sqrt{2p}\right] + 1$$

and

$$2\left[\frac{1}{2}\sqrt{2p}\right] + 2 < \sqrt{2p} + 2\,.$$

Hence in this case

$$l < \sqrt{2p} + 2\,.$$

Now we consider the case that 2 is a kth power residue. Here we use the fact that the smallest kth power non-residue satisfies

(15) $$d < \sqrt{2p}$$

which will be proved soon. We choose $c = \sqrt{2p}$ and consider the interval

(16) $$\left\{\frac{1}{2}c \ldots c\right\} = \left\{\frac{1}{2}\sqrt{2p} \;\ldots\; \sqrt{2p}\right\}.$$

This interval contains a number of the form $2^t d$ with integral positive t, hence a kth power non-residue. But the smallest integer of the interval (16) is the number $[\frac{1}{2}\sqrt{2p}] + 1$. Hence the interval contains a non-residue, and it follows that $l < \sqrt{2p}$ in this case.

14. We want to consider the length of the special sequence of kth power residues $1, 2, \ldots, d-1$, that is to find estimates for the least kth power non-residue. For this purpose we first prove the important Theorem of L. Aubry [2] and A. Thue [53].

The congruence

(17) $$ax + by \equiv 0 \pmod{m}$$

has at least one non-trivial solution x, y for which

(18) $$|x| \leqq \sqrt{m} \quad \text{and} \quad |y| \leqq \sqrt{m}\,.$$

Proof. We set $[\sqrt{m}] = q$. In $ax + by$ we substitute the values $0, 1, 2, \ldots, q$ for x and y. Then we obtain $(q+1)^2$ integers. Since $(q+1)^2 > m$, at least 2 of these integers are congruent (mod m), say

$$ax_1 + by_1 \equiv ax_2 + by_2 \pmod{m}\,.$$

Then

$$X = x_1 - x_2\,, \qquad Y = y_1 - y_2$$

is a solution of (17). Here X and Y cannot be both zero. They satisfy (18) since the difference of two non-negative numbers which are less than or equal to $\sqrt{m}$ has an absolute value which is less than or equal to $\sqrt{m}$. Vinogradov [57] showed that (17) also has a non-trivial solution where

$$0 < x \leqq k \quad \text{and} \quad |y| \leqq \frac{m}{k}\,.$$

Using geometry of numbers, L. Rédei [43] improved the Theorem of Aubry-Thue.

Finally this theorem was generalized by me as follows [8].

Let r and s be rational integers with $r < s$. The system of r linear homogeneous congruences in s unknowns

$$\sum_{\sigma=1}^{s} a_{\rho\sigma} x_\sigma \equiv 0 \pmod{m} \qquad (\rho = 1, 2, \ldots, r)$$

always has a non-trivial solution for which $|x_\sigma| \leqq m^{r/s}$ for $\sigma = 1, 2, \ldots, s$.

15. The most difficult part of the first proof of the law of reciprocity, as Gauss [17] himself states, was the proof of the following theorem.

If p is a prime of the form $8n + 1$, then there exists an odd prime $q < 2\sqrt{p} + 1$ for which p is a quadratic non-residue.

Using now the reciprocity law we obtain the following theorem.

For every prime p of the form $8n + 1$ there exists an odd prime q which is a quadratic non-residue (mod p).

That the latter theorem is correct for all primes $p < 3$ was proved by T. Nagell [32]. A much better result was given by I. Schur (see [5]) as a problem in his classes on number theory since 1920.

The least quadratic non-residue d (mod p) *is less than $\sqrt{p}$ except for $p = 3, 7$, and 23.*

For primes of the form $8n \pm 3$ this is trivial since 2 is a non-residue. For primes of the form $8n + 1$ this follows at once from the Theorem of Thue by considering the congruence

$$x \equiv ny \pmod{p}$$

where n is an arbitrary quadratic non-residue. One of the numbers x and y must be a non-residue. We can assume that both are of absolute value less than $\sqrt{p}$, and with x, $-x$ is also a quadratic non-residue.

For primes of the form $4n-1$ we have proved already that $d < \sqrt{p}+1$, since there do not exist sequences of non-residues of length greater than $\sqrt{p}$. To prove that $d < \sqrt{p}$ is a little more tedious. We can omit this proof since we shall get a much better bound soon.

The corresponding result holds for kth power non-residues [8]. This is obvious for $k = 2m$ since every quadratic non-residue is a $2m$th power non-residue. If k is odd, then -1 is a kth power residue, and the result follows from the Theorem of Thue and the congruence $x \equiv ny \pmod{p}$ where n is a kth power non-residue.

For $k = 2$ and primes of the form $8n+1$ a little better result can be obtained. Each of the two sets of d integers $p-d+1, p-d+2, \ldots, p$ and $p, p-1, \ldots, p-d+1$ contains a multiple of d, say qd and $(q+1)d$. These numbers are quadratic residues and q and $q+1$ are non-residues. Since 2 is a residue, either $\frac{1}{2}q$ or $\frac{1}{2}(q+1)$ is a whole number and a non-residue. It follows that $d \leqq \frac{1}{2}(q+1)$. On the other hand $qd < p$; hence

$$2d^2 \leqq qd + d < p + d\,,$$

$$d < \frac{1}{4}(1 + \sqrt{1+8p})\,.$$

These and similar results were also obtained recently in papers of H.-J. Kanold [25], T. Nagell [33, 34, 35, 36], L. Rédei [44], and Th. Skolem [51].

16. But already by 1931, I was able to improve these results for all primes except those of the form $8n+1$. [5].

Theorem. *Let p be a prime of the form $8n+7$. The smallest quadratic non-residue d satisfies*

$$d < (2p)^{2/5} + 3(2p)^{1/5} + 1\,.$$

I shall only outline the proof.

Let r be an integer satisfying $1 < r < d$ which we shall

choose later in a suitable way. Then r is a residue and the numbers

$$p - d + 1, p - d + 2, \dots, p - 1 \tag{18}$$

are non-residues. Let

$$kr, (k + 1)r, \dots, (k + l - 1)r \tag{19}$$

be the multiples of r in (18). It follows that

$$k, k + 1, \dots, k + l - 1 \tag{20}$$

form a sequence of l non-residues which must lie between two consecutive squares, say a^2 and $(a + 1)^2$. We divide the interval $\{(a + 1)^2 \dots a(a + 1)\}$ by the points

$$(a + 1)^2 - \nu^2 = (a + 1 + \nu)(a + 1 - \nu) \qquad (\nu = 1, 2, \dots, [\sqrt{a}\,]) \tag{21}$$

and the interval $\{a(a + 1) \dots a^2\}$ by the points

$$a(a + 1) - \nu(\nu + 1) = (a + 1 + \nu)(a - \nu) \qquad (\nu = 1, 2, \dots, [\sqrt{a}\,]) \tag{22}$$

into subintervals. It is easy to see that the maximal length of these subintervals is

$$s \leqq 2[\sqrt{a}\,] + 2 \,. \tag{23}$$

For an indirect proof we now assume that

$$d \geqq \max \{a + 2 + [\sqrt{a}\,], rs\} \,. \tag{24}$$

It follows now from (21), (22), and (24), that these end-points are quadratic residues. Since the maximal length of the subintervals is s, the interval $\{a^2 \dots (a + 1)^2\}$ cannot contain any sequence of s non-residues.

On the other hand, it follows from (24) that $d \geqq rs$; hence the interval (18) contains at least s multiples of r. It follows from (19) that the exact number of such multiples equals l; hence $l \geqq s$. The interval $\{a^2 \dots (a + 1)^2\}$ would contain the sequence (20) of l non-residues. This gives a contradiction.

Hence by (24) and (23)

$$d < \max \{a + 2 + [\sqrt{a}\,], 2r([\sqrt{a}\,] + 1)\} \,,$$

$$d \leqq \max \{a + 1 + [\sqrt{a}], 2r([\sqrt{a}] + 1) - 1\} .$$

It follows from (20) and (19) that

$$a^2 < k + l - 1 < \frac{p}{r} .$$

Hence

$$d < \max \left\{ \sqrt{\frac{p}{r}} + \sqrt[4]{\frac{p}{r}} + 1, 2r\left(\sqrt[4]{\frac{p}{r}} + 1\right) - 1 \right\} . \tag{25}$$

If we assume now that the theorem is not correct, then we have

$$d \geqq (2p)^{2/5} + 3(2p)^{1/5} + 1 , \tag{26}$$

and we are permitted to choose

$$r = \left[\left(\frac{p}{16}\right)^{1/5}\right] + 1 \tag{27}$$

since the only condition was $r < d$. If we substitute (27) in (25), then we obtain

$$d < (2p)^{2/5} + 3(2p)^{1/5} + 1 .$$

This contradicts (26) and the theorem is proved.

By a similar, but more difficult, proof it was shown [5] that the smallest odd quadratic non-residue u for the primes of form $8n \pm 3$ satisfies

$$u < \{(4p)^{2/5} + (4p)^{1/5}\} + 1 . \tag{28}$$

In a later paper [7], it was possible to replace the coefficients of (28) by smaller ones. This was of importance for a theoretical application. Due to this refinement only the primes below 3,300,000 had to be checked directly while otherwise this bound would have been far beyond the existing tables, and no computer existed at that time.

Extending this method my student C. T. Whyburn [61], [62] obtained bounds for the second smallest quadratic non-residue for all primes except those of form $24n + 1$.

By analytic methods better bounds have been obtained by J. M. Vinogradov [57], Davenport and Erdös [15], and D. A. Burgess [12]. These results hold for sufficiently large primes.

L. K. Hua [22] found bounds for the last three prime quadratic non-residues for primes $p > e^{250}$. H. Salié [48] proved that there exists a constant c such that for infinitely many primes p the least quadratic non-residue is greater than $c \log p$.

I want to mention a few other papers whose proofs are not combinatorial. H.-J. Kanold [26] and L. Rédei [42] obtained estimates for the number of kth power residues less than $\sqrt{p}$ and T. Nagell [33] and K. Inkeri [23] found bounds for the least prime quadratic residue.

17. Finally we consider the related problem of finding bounds for the least primitive root mod p. Such bounds were found with analytic methods by Vinogradov [57] and Landau [29]. But very little has been obtained by elementary methods, except for special classes of primes.

Using the generalization of the Theorem of Aubry-Thue [8] which we already mentioned, we prove two simple theorems. [11].

Theorem. *Let p be a prime of form $4n+1$ and k the number of different prime divisors of $p-1$.*

We set $r = 2^k$. Then the smallest positive primitive root (mod p) *is less than $p^{(r-1)/r}$.*

Proof. Let g be any primitive root. Assume that $-g$ belongs to the exponent z (mod p). Then $(-g)^z \equiv 1 \pmod p$ and $g^{2z} \equiv 1 \pmod p$. Hence $2z$ is divisible by $p-1$ and z is even. It follows that $(-g)^z \equiv g^z \equiv 1 \pmod p$, hence $z = p-1$ and $-g$ is a primitive root, too.

Let $l_1 = 1$ and let $l_2, l_3, \ldots, l_{r-1}$ be the products of different prime divisors of $p-1$, except the product of all the different prime divisors. We consider the system of $r-1$ congruences

$$\xi \equiv g^{l_\rho} y_\rho \qquad (\rho = 1, 2, \ldots, r-1)\,. \tag{29}$$

It follows from the generalization of the Theorem of Aubry-Thue that this system has a solution $x, y_1, y_2, \ldots, y_{r-1}$ for which

$$|x| < p^{(r-1)/r},\ |y_1| < p^{(r-1)/r}, \ldots, |y_{r-1}| < p^{(r-1)/r}\,. \tag{30}$$

If x is a primitive root, the proof is finished. Otherwise we set $x \equiv g^m \pmod p$. Since x is not a primitive root, at least one of the prime divisors of $p-1$ divides m. We consider the product of those different prime divisors of $p-1$ which are relatively prime to m. This product must be one of the numbers $l_1, l_2, \ldots, l_{r-1}$, say l_ν. It follows from the νth of the congruences (29) that

$$y_\nu \equiv g^{m-l_\nu} \pmod p .$$

Since m and l_ν are relatively prime and since each prime divisor of $p-1$ divides either m or l_ν, it follows that $m-l_\nu$ and $p-1$ are relatively prime. Hence y_ν and $-y_\nu$ are primitive roots satisfying (30).

Sometimes the following theorem is better [11].

Theorem. *Let p be a prime of the form $4n+1$. Let s be the maximal length of the sequences of consecutive integers which are all not relatively prime to $p-1$. Then the smallest positive primitive root* (mod p) *is less that $p^{s/(s+1)}$.*

Proof. Let g be an arbitrary primitive root. The system of s congruences

$$\xi \equiv g^\sigma \eta_\sigma \qquad (\sigma = 1, 2, \ldots, s)$$

has a solution $x, y_1, y_2, \ldots, y_s$ such that

$$|x| < p^{s/(s+1)}, |y_1| < p^{s/(s+1)}, \ldots, |y_s| < p^{s/(s+1)} .$$

If we set

$$x \equiv g^m \pmod p ,$$

then

$$y_\sigma \equiv g^{m-\sigma} \pmod p ,$$

and the $s+1$ consecutive powers of g, namely $g^m, g^{m-1}, \ldots, g^{m-s}$ are congruent to the numbers $x, y_1, y_2, \ldots, y_s$ whose absolute values are less than $p^{s/(s+1)}$. At least one of these exponents of g must be relatively prime to $p-1$ and the corresponding power of g is a primitive root.

References

1. Aladov, N. S. (1896). "Sur la distribution des résidues quadratiques et non-quadratiques d'un nombre premier p dans la suite $1, 2, \ldots, p-1$," *Matematicheskii Sbornik* 18, 61–75.

2. Aubry, L. (1913). "Un théorème d'arithméthique," *Mathesis* (sér. 4) **3**, 33-35.
3. Brauer, Alfred (1928). "Ueber Sequenzen von Potenzresten," *Sitzungsber. Preuss. Akad. Wiss. Phys. Math. Kl.*, 1928, 9-16.
4. Brauer, Alfred (1931). "Ueber Sequenzen von Potenzresten. II," *Sitzungsber. Preuss. Akad. Wiss. Phys. Math. Kl.*, 1931, 329-341.
5. Brauer, Alfred (1931). "Ueber den kleinsten quadratischen Nichtrest," *Math. Zeitschr.*, **33**, 161-176.
6. Brauer, Alfred (1932). "Ueber die Verteilung der Potenzreste," *Math. Zeitschr.*, **32**, 39-50.
7. Brauer, Alfred (194). "On the Non-Existence of the Euclidean Algorithm in Certain Quadratic Number Fields," *Amer. J. Math.* **62**, 697-716.
8. Brauer, Alfred and Reynolds, T. L. (1951). "On a Theorem of Aubry-Thue," *Canad. J. Math.*, **3**, 367-374.
9. Brauer, Alfred (1953). "On a New Class of Hadamard Determinants," *Math. Zeitschr.*, **58**, 219-225.
10. Brauer, Alfred (1953). "On the Distribution of the Jacobian Symbols," *Math. Zeitschr.*, **58**, 226-231.
11. Brauer, Alfred (1954). "Elementary Estimates for the Least Primitive Root," *Studies in mathematics and mechanics presented to Richard von Mises*, pp. 20-29, Academic Press, New York.
12. Burgess, D. A. (1957). "The Distribution of the Quadratic Residues and Non-Residues," *Mathematika*, **4**, 106-112.
13. Davenport, H. (1931, 1933). "On the Distribution of Quadratic Residues (mod p)," *J. London Math. Soc.*, **6**, 49-54; **8**, 46-52.
14. Davenport, H. (1932). "On the Distribution of the l-th Power Residues (mod p)," *J. London Math. Soc.*, **7**, 117-121.
15. Davenport, H. and Erdös, P. (1952). "The Distribution of Quadratic and Higher Residues, *Publ. Math. Debrecen*, **2**, 252-265.
16. Dörge, K. (1929). "Zur Verteilung der quadratischen Reste," *Jahresbericht Deutsche Math. Verein.*, **38**, 41-49.
17. Gauss, C. F. (1870). *Werke I. Disquisitiones Arithmeticae*, art. 125 and 129, Gesellschaft d. Wiss. Göttingen 2nd ed.—See also Dirichlet-Dedekind (1894), Vorlesungen ueber Zahlentheorie, 4th ed., 116, Vieweg, Braunschweig.
18. Gruner, W. (1940). "Einlagerung des regulaeren n-Simplex in den n-dimensionalen Wuerfel," *Commentarii Math. Helv.*, **12**, 149-152.
19. Hasse, Helmut (1934). "Abstrakte Begruendung der komplexen Multiplikation und Riemannsche Vermutung in Funktionenkoerpern," *Abhandl. Math. Sem. Hamburg*, **10**, 325-348.
20. Hasse, Helmut (1964). *Vorlesungen ueber Zahlentheorie*, 2nd ed., Springer, Berlin, 145-167.
21. Hopf, Heinz (1930). "Ueber die Verteilung der quadratischen Reste," *Math. Zeitschr.*, **32**, 222-231.
22. Hua, L.-K. (1944). "On the Distribution of Quadratic Non-Resi-

dues and the Euclidean Algorithm in Real Quadratic Fields, I," *Trans. Amer. Math. Soc.*, **56**, 537-546.

23. Inkeri, K. (1950). "On the Least Prime Quadratic Residue," *Ann. Acad. Sc. Fenn. Ser. A*, **73**, 1-10.
24. Jacobsthal, Ernst (1906). *Anwendungen einer Formel aus der Theorie der quadratischen Reste*, Diss. Univ. Berlin, Kästner Göttingen, 1-39.
25. Kanold, Hans-Joachim (1950). "Saetze ueber Kreisteilungspolynome und ihre Anwendungen auf einige zahlentheoretische Probleme, I," *J. Reine Angew. Math.*, **187**, 169-182.
26. Kanold, Hans-Joachim (1950). "Eine Bemerkung zur Verteilung der r-ten Potenznichtreste einer ungerader Primzahl," *J. Reine Angew. Math.* **188**, 74-77.
27. Khinchin, A. Y. (1952). *Three Pearls of Number Theory*, Graylock, Rochester, 11-17.
28. Landau, Edmund (1918). "Abschaetzungen von Charaktersummen, Einheiten und Klassenzahlen," *Nachr. Ges. Wiss. Göttingen Phys. Math. Kl.*, 1918, 79-97.
29. Landau, Edmund (1927). *Vorlesungen ueber Zahlentheorie II*, Hirzel, Leipzig, 178-180.
30. Lehmer, D. H. (1964). "Combinatorial Types in Number-Theory Calculations," Proc. IBM Sc. Comp. Symp., Yorktown Heights.
31. MacMahon, P. A. (1915/16). *Combinatory Analysis*, Cambridge Univ. Press.
32. Nagell, T. (1923). Zahlentheoretische Notizen. Vidensk. selsk. Skrifter, *Mat. Mat. Kl.* 1923, No. 13, II.
33. Nagell, T. (1950). "Sur les restes et les non-restes quadratiques suivant un module premier," *Arkiv Mat.* **1**, No. 16, 185-193.
34. Nagell, T. (1951). "Sur un théorème d'Axel Thue," *Arkiv Mat.* **1**, No. 33, 489-496.
35. Nagell, T. (1951). "Sur le plus petit non-reste quadratique impair," *Arkiv Mat.* **1**, No. 38, 573-578.
36. Nagell, T. (1952). "Den minste positive n^{te} ikke-potensrest modulo p," *Norsk Mat. Tidsskr.*, **34**, 13.
37. Netto, Eugen (1901). *Lehrbuch der Kombinatorik*. 2nd ed., reprint 1958, Chelsea, N. Y.
38. Perron, Oskar (1952). "Bemerkungen über die Verteilung der quadratischen Reste," *Math. Zeitschr.*, **56**, 122-130.
39. Pólya, Georg (1918). "Ueber die Verteilung der quadratischen Reste und Nichtreste," *Nachr. Ges. Wiss. Göttingen Phys. Math. Kl.*, 1918, 21-29.
40. Pólya, Georg (1966). "A Note of Welcome," *J. Combinatorial Theory* **1**, 1-2.
41. Rado, Richard (1933). "Verallgemeinerung eines Satzes von van der Waerden mit Anwendungen auf ein Problem der Zahlentheorie," *Sitzungsber. Preuss. Akad. Wiss., Phys. Math. Kl.*, 1933, 589-596.

42. Rédei, L. (1950). "Ueber die Anzahl der Potenzreste mod p im Interval 1, $\sqrt{p}$," *Nieuw Archief*, **23**, 150-162.
43. Rédei, L. (1951). "Ueber eine Verschaerfung eines zahlentheoretischen Satzes von Thue, *Acta Math.* (*Budapest*), **2**, 75-82.
44. Rédei, L. (1953). "Die Existenz eines ungeraden quadratischen Nichtrestes mod p im Intervall 1, $\sqrt{p}$," *Acta Math. Sc. Szeged*, **15**, 12-19.
45. Riordan, John (1958). *An Introduction to Combinatorial Analysis*, Wiley, New York.
46. Ryser, H. J. (1963). *Combinatorial Mathematics*, Carus Math. Monograph No. 14, John Wiley and Sons, New York.
47. Salié, Hans (1933). "Ueber die Verteilung der quadratischen Reste," *Math. Zeitschr.*, **37**, 594-602.
48. Salié, Hans (1949). "Ueber den kleinsten positiven quadratischen Nichtrest nach einer Primsahl," *Math. Nachr.*, **3**, 7-8.
49. Schur, I. (1918). "Einige Bemerkungen zu der vorstehenden Arbeit des Herrn Pólya," *Nachr. Ges. Wiss. Göttingen. Phys. Math. Kl.*, 1918, 30-36.
50. Schur, I. (1928) see [3].
51. Skolem, Th. (1951). "Eksistensen av en n^{te} ikke-potensrest (mod p) mindre enn $\sqrt{p}$," *Norsk Mat. Tidsskrift*, **33**, 123-126.
52. Sterneck, R. von (1897-99). "Sur la distribution des résidues et non-résidues quadratiques d'un nombre premier," *Mat. Sbornik*, **20**, 269-284.
53. Thue, A. (1917). "Et bevis for at lignigen $A^3 + B^3 = C^3$ er remulig i hele fra nul forsk jellige tal A, B og B," *Archiv Math. Naturvid*, **34** No. 15.
54. Unger, Walter (1921). *Ueber einige Summen cubischer und biquadratischer Charaktere*, Dissertation Bonn, Kaestner, Göttingen, pp. 1-25.
55. Vause, R. Z. (1954). *On the Distribution of the Jacobian Symbols*, Dissertation Univ. North Carolina, pp. 1-44.
56. Vause, R. Z. (1956). "On the Distribution of the Jacobian Symbols," *J. Elisha Mitchell Sci. Soc.*, **72**, 15-24.
57. Vinogradov, J. M. (1927). "On a General Theorem Concerning the Distribution of the Residues and Non-Residues of Powers," Trans. *Amer. Math. Soc.*, **29**, 209-217; "On the bound of the least non-residue of nth powers," Trans. *Amer. Math. Soc.*, **29**, 218-226.
58. Vinogradov, J. M. (1954). *Elements of Number Theory*, Dover Publications New York.
59. Waerden, B. L. van der (1927). "Beweis einer Baudetschen Vermutung," *Nieuw Arch. Wisk.*, **15**, 212-216.
60. Weil, André (1948). *Sur les courbes algébriques et les variétés qui s'en déduisent*, Actualités Scientifiques 1041. Hermann, Paris.
61. Whyburn, C. T. (1964). *On the Second Smallest Quadratic Non-Residue*, Dissertation, Univ. North Carolina, pp. 1-56.

62. Whyburn, C. T. (1965). " The Second Smallest Quadratic Non-Residue," *Duke Math. J.*, **32**, 519-528.

Discussion on Professor Brauer's Paper

PROFESSOR D. A. SMITH : Mathematical history is a sadly neglected subject. Most of this history belongs to the twentieth century, and a good deal of it is in the memories of mathematicians still living. The younger generation of mathematicians has been trained to consider the product, mathematics, as the most important thing, and to think of the people who produced it only as names attached to theorems. This frequently makes for a rather dry subject matter, and it is a delightful relief when a participant in some of the developments of 30 or 40 years ago sets down his recollections of the people involved in these developments, as well as the mathematical content. I want to express my personal thanks to Professor Brauer for this very readable and enjoyable historical account of an important topic in the interaction between combinatorial theory and number theory, and also for the excellent bibliography he has provided.

If one were to draw a very narrow definition of combinatorial theory, it might be said that the combinatorial content of Sections 1–11 of the paper is contained in the Theorem of van der Waerden and its extensions by Schur (Section 5) and Rado (Section 10), while the rest of the material applies these theorems to number theory. Section 12, on the existence of Hadamard matrices, clearly involves an application of number theory to combinatorial theory. The remainder of the paper, which naturally complements the first 11 sections, would have to be described as purely number-theoretic and not combinatorial at all, by this definition. However, the distinction is not really this neat. As Professor Brauer notes in his introduction, we have no satisfactory definition of combinatorial theory, and it is probably better that we do not. Recent developments in this rapidly growing field have made it quite clear that there are interconnections with practically all of the traditional " branches " of mathematics, and these interconnections will continue to produce substantial benefits in both directions as discrete mathematics continue to develop.

PROFESSOR D. E. KNUTH : We thank Professor Brauer for

this informative paper, and especially for the historial remarks and the excellent bibliography he has included. The application of combinatorial methods in number theory is well illustrated here and these applications form a nice companion to the applications of number theoretical methods (in particular the use of kth power residues) to combinatorial construction problems with which most of us are familiar.

Although the methods of this paper guarantee the existence of l consecutive residues or nonresidues for all large enough primes, they do not seem to say anything about any of the "non-pure" sequences. For example, we can apparently not use the methods described here to conclude that there will be sequences of 10 residues followed by 10 nonresidues, for all large primes, nor to prove that sequences of l elements which are alternately residues and nonresidues exist, etc..

CHAPTER 3

On Canonical Bases for Subgroups of an Abelian Group

HENRY B. MANN[1], *Mathematics Research Center, University of Wisconsin*

Notation. The letter G will denote a finitely generated Abelian group, written additively. The elements $\omega_1, \ldots, \omega_n$ of a basis of G will always be chosen so that their order is either infinite or a prime power. The number n is then an invariant of G.

If A and B are sets of elements of G then $A + B$ denotes the Schnirelmann sum of A and B.

$$A + B = \{a + b\,;\ a \in A, b \in B\}\,.$$

If H_1 and H_2 are subgroups of G then $H_1 + H_2$ is the smallest subgroup containing H_1 and H_2. We shall also write

$$(A \cap B) + C = A \cap B + C\,.$$

The elements $\eta_1, \ldots, \eta_s$ of G will be called *independent* if

$$x_1\eta_1 + \ldots + x_s\eta_s = 0$$

[1] Sponsored by the Mathematics Research Center, United States Army, Madison, Wisconsin, under Contract No.: DA-31-124-ARO-D-462.
Author's Note: **While this paper was in print, it has come to my attention that the main theorem and its proof are correct only in the case that G is torsion-free, and this assumption must be made throughout the paper.**

implies $x_i\eta_i = 0$. If $\eta_1, \ldots, \eta_s$ are independent elements of G then

$$[\eta_1, \ldots, \eta_s]$$

will denote the subgroup generated by $\eta_1, \ldots, \eta_s$. The GCD of a and b will be denoted by (a, b) and the letters p, q will always denote primes.

Summary of results. Let H be a subgroup of G. It is well known that there exists a basis $\omega_1, \ldots, \omega_n$ of G such that

$$(1) \qquad H = [d_1\omega_1, \ldots, d_n\omega_n]$$

where the d_i are non-negative integers depending on H and on the choice of $\omega_1, \ldots, \omega_n$. One can in fact find $\omega_1, \ldots, \omega_n$ so that d_i / d_{i+1} but we shall not require this in what follows.

A basis $\omega_1, \ldots, \omega_n$ of G which satisfies (1) will be called a *canonical basis* for the subgroup H. In [2] it was shown that a basis canonical for the subgroups $H_1, \ldots, H_s$ always exists if $H_1, \ldots, H_s$ are ideals in an algebraic numberfield F and G is the unit ideal of F. An example was also given of two subgroups H_1, H_2 such that no basis exists which is canonical for both H_1 and H_2.

In this paper we shall find necessary and sufficient conditions that a basis of G exists which is canonical for the subgroups $H_1, \ldots, H_s$. Using these conditions the result on ideals in algebraic numberfields mentioned above follows easily and it will also be easy to construct examples of subgroups H_1, H_2 such that no basis exists which is canonical for H_1 as well as for H_2. We shall also be able to formulate and prove a converse to the result on ideals in algebraic numberfields proved in [2].

1. NECESSARY AND SUFFICIENT CONDITIONS FOR CANONICAL BASES

Let H be a subgroup of G. The elements $\eta_1, \eta_2, \ldots$ are called a *system of residues* mod H *if*

$$G = (H + \eta_1) \cup (H + \eta_2) \cup \ldots$$

is the coset decomposition of $G \bmod H$.

Definition 1. *The elements $\omega_1, \ldots, \omega_t$ of G will be called a basis for the residues* mod H *if for each $\eta \in G$*

$$(2) \qquad \eta \equiv x_1\omega_1 + \ldots + x_t\omega_t \pmod{H},$$

is solvable in integers x_i and

$$x_1\omega_1 + \ldots + x_t\omega_t \equiv 0 \pmod{H}$$

implies

$$x_i\omega_i \equiv 0 \pmod{H}.$$

If for some positive integer x we have $x\omega_\alpha \equiv 0 (\bmod H)$, let d_α be the smallest such integer. If no such integer exists let $d_\alpha = 0$. It is not difficult to see that the x_i in (2) are uniquely determined mod d_α by η.

Now let $H_1, \ldots, H_s$ be subgroups of G. Let $H_1 \cap H_2 \cap \ldots \cap H_s = H$. We shall prove that the following three statements are equivalent.

I. *There exists a basis of G which is canonical for $H_1, \ldots, H_s$.*

II. *There exists a basis $\omega_1, \ldots, \omega_t$ for the residues* mod H *such that $\omega_1, \ldots, \omega_t$ is a basis for the residues* mod H_i *for each i.*

III. *There exist subgroups $\kappa_1, \ldots, \kappa_t$ with cyclic factor groups G/κ_i such that*

(i) *for $1 \leqslant j \leqslant s$ the group H_j is the intersection of some of the κ_i.*

(ii) *For each choice $\alpha_1, \ldots, \alpha_u$ and α either $\kappa_\alpha \subseteq \kappa_{\alpha_j}$ or $\kappa_{\alpha_j} \subset \kappa_\alpha$ for some j or $\kappa_{\alpha_1} \cap \ldots \cap \kappa_{\alpha_u} + \kappa_\alpha = G$.*

We shall show the equivalence of these three conditions by showing that

$$\mathrm{I} \rightarrow \mathrm{II} \rightarrow \mathrm{III} \rightarrow \mathrm{I}.$$

Thus II as well as III are necessary and sufficient conditions for the existence of such a basis.

(a) We first show that I $\rightarrow$ II. Let $\omega_1, \ldots, \omega_n$ be a canonical basis for $H_1, \ldots, H_s$

$$H_i = [d_{i1}\omega_1, \ldots, d_{in}\omega_n].$$

Then

$$H_1 \cap \ldots \cap H_s = H = [d_1\omega_1, \ldots, d_n\omega_n]$$

where $d_\alpha = \mathrm{LCM}\,(d_{1\alpha}, \ldots, d_{s\alpha})$. It is easy to see that $\omega_1, \ldots, \omega_n$ satisfy for $H_1, \ldots, H_s$ and $H = H_1 \cap \ldots \cap H_s$ the conditions of Definition 1.

One can of course simplify the basis for the residue system and delete ω_α whenever $d_\alpha = 1$.

(b) We next show that II → III.

Let $\omega_1, \ldots, \omega_t$ be a basis for the residues mod H_i, $i = 1, \ldots, s$ and mod H, and let d_α and $d_{i\alpha}$ be defined for H and H_i respectively as in the paragraph following Definition 1.

We have

$$(3) \qquad \begin{aligned} g &= h + x_1\omega_1 + \ldots + x_t\omega_t \\ &= h_i + y_1\omega_1 + \ldots + y_t\omega_t \end{aligned}$$

where $x_1, \ldots, x_t, y_1, \ldots, y_t$ are uniquely determined mod d_α and mod $d_{i\alpha}$ respectively by g and $d_{i\alpha}\omega_\alpha \in H_i$. It follows that $d_\alpha \equiv 0 (\mathrm{mod}\, d_{i\alpha})$ and that for $g \in H_i$, we must have $x_i \equiv 0(\mathrm{mod}\, d_{i\alpha})$. Hence for every $h_i \in H_i$ we have

$$(4) \qquad h_i = h + x_1 d_{i1}\omega_1 + \ldots + x_t d_{it}\omega_t$$

where the x_α are uniquely determined by $h_i \bmod d_\alpha / d_{i\alpha}$ if $d_{i\alpha} \neq 0$.

If $d_{i\alpha} = 0$ we put

$$\kappa_{i\alpha 1} = \{H + x_1\omega_1 + \ldots + x_{\alpha-1}\omega_{\alpha-1} + x_{\alpha+1}\omega_{\alpha+1} + \ldots + x_t\omega_t\}\,.$$

If $d_{i\alpha} = 1$ we put $\kappa_{i\alpha 1} = G$. If $d_{i\alpha} > 1$ let

$$d_{i\alpha} = p_1^{l_1} \ldots p_u^{l_u}\,.$$

We put

$$\begin{aligned} \kappa_{i\alpha v} = \{H + x_1\omega_1 + \ldots + x_{\alpha-1}\omega_{\alpha-1} + p_v^{l_v} x_\alpha\omega_\alpha + x_{\alpha+1}\omega_{\alpha+1} \\ + \ldots + x_t\omega_t\} \qquad 1 \leqslant v \leqslant u\,. \end{aligned}$$

Because of the uniqueness of the x_i in (3) we must have

$$\begin{aligned} \kappa_{i\alpha} = \bigcap_v \kappa_{i\alpha v} = \{H + x_1\omega_1 + \ldots \\ + x_{\alpha-1}\omega_{\alpha-1} + d_{i\alpha}x_\alpha\omega_\alpha + x_{\alpha+1}\omega_{\alpha+1} + \ldots + x_t\omega_t\}\,. \end{aligned}$$

By (3) we then have

$$H_i = \bigcap_{\alpha, v} \kappa_{i\alpha v} .$$

It is also easy to verify that the $\kappa_{i\alpha v}$ satisfy the condition (ii) of III. This completes the proof that II $\rightarrow$ III.

We next prove three lemmas.

Lemma 1. *Let G/H be cyclic. Let $\omega_1, \ldots, \omega_n$ be a basis of G such that $\omega_2, \ldots, \omega_n \in H$. Then*

$$H = [d_1\omega_1, \omega_2, \ldots, \omega_n] .$$

Proof. For $\eta \in G$ we have

$$\eta \equiv x\omega_1 \pmod{H} .$$

If $x\omega_1 \notin H$ for any value of x, then

$$H = [0\omega_1, \omega_2, \ldots, \omega_n] .$$

Otherwise let d_1 be the smallest positive x such that $x\omega_1 \in H$. Then

$$H = [d_1\omega_1, \omega_2, \ldots, \omega_n] .$$

This proves Lemma 1.

Lemma 2. *Let H_1, H_2 be subgroups of G with cyclic factor group either of prime power order or of infinite order. If $H_1 \subseteq H_2$ and $\omega_1, \ldots, \omega_n$ is a canonical basis for H_1 then $\omega_1, \ldots, \omega_n$ is a canonical basis for H_2.*

Proof. Arrange the notation so that

$$H_1 = [d_1\omega_1, \omega_2, \ldots, \omega_n] .$$

Then $\omega_2, \ldots, \omega_n \in H_1 \subseteq H_2$ and Lemma 2 follows from Lemma 1.

Lemma 3. *Let $\omega_1, \ldots, \omega_n$ be a basis of G. Let G/H be cyclic of order p^r. Let $(p, b) = 1$. Then we can find a basis $\omega_1^*, \ldots, \omega_n^*$ canonical for H such that $\omega_i^* \equiv \omega_i \pmod{b}$ and*

$$H = [p^r\omega_1^*, \omega_2^*, \ldots, \omega_n^*] .$$

Proof. The order mod H of each element of G is a power of

p. Let p^u be the highest among the orders of $\omega_1, \ldots, \omega_n$. Since every $\eta \in G$ is a linear combination of $\omega_1, \ldots, \omega_n$, it follows that η has order at most p^u; and since G/H is cyclic of order p^r, we must have $u = r$. This shows that at least one of the elements $\omega_1, \ldots, \omega_n$ must generate $G \bmod H$. Let t be the smallest index such that ω_t generates $G \bmod H$. If $t = 1$, then we put $\omega_1^* = \omega_1$. If $t > 1$, then $\omega_1 + b\omega_t$ generates $G \bmod H$; and we put $\omega_1^* = \omega_1 + b\omega_t$. For each i we determine x_i so that

$$\omega_i - x_i b\omega_1^* \equiv 0(H), \qquad 2 \leqslant i \leqslant n.$$

(This is possible because $(b, p) = 1$ and $b\omega_1^*$ therefore generates $G \bmod H$.)

We now put

$$\omega_i^* = \omega_i - x_i b\omega_1^*, \qquad 2 \leqslant i \leqslant n.$$

It follows from Lemma 1 that $\omega_1^*, \ldots, \omega_n^*$ satisfy the conditions of Lemma 3.

(c) We are now prepared to prove that III $\to$ I. It is not difficult to see that it will be sufficient to construct a basis canonical for $\kappa_1, \ldots, \kappa_t$. We can without loss of generality delete G from $\kappa_1, \ldots, \kappa_t$ if it occurs among them. Furthermore, if κ_i has finite index we can represent each κ_i as an intersection of subgroups with cyclic factor groups of prime power order and these subgroups themselves satisfy III. We may therefore assume that G/κ_i is either infinite or cyclic of prime power order for $i = 1, \ldots, t$.

We shall give the proof by induction on n. If there is among the κ_i a subgroup of infinite index under G, let κ_1 be such a group and let $\omega_1, \ldots, \omega_n$ be a basis for G canonical for κ_1 such that

$$\kappa_1 = [0\omega_1, d_2\omega_2, \ldots, d_n\omega_n].$$

Since G/κ_1 is cyclic of infinite order one easily sees that $d_i = 1$ for $i \geqslant 2$ and that ω_1 must be of infinite order.

Hence

$$\kappa_1 = [\omega_2, \ldots, \omega_n].$$

If $\kappa_1 \supseteq \kappa_i$ then κ_i must be cyclic of infinite index under G and this implies $\kappa_1 = \kappa_i$.

On the other hand if $\kappa_1 \subseteq \kappa_i$ then $\kappa_i \supseteq [\omega_2, \ldots, \omega_n]$ and by Lemma 1

$$\kappa_i = [d_{i1}\omega_1, \omega_2, \ldots, \omega_n]$$

and $\kappa_i \supset \kappa_1$ implies $d_{i1} > 0$.

Now let $\kappa'_1, \ldots, \kappa'_{t'}$ be those of the subgroups $\kappa_1, \ldots, \kappa_t$ which do not contain κ_1 and let

$$\kappa = \kappa'_1 \cap \ldots \cap \kappa'_{t'} .$$

By condition (ii) of III we have

$$\kappa + \kappa_1 = G .$$

In particular this means that κ must contain an element ω'_1 of the form

$$\omega'_1 = \omega_1 + x_2\omega_2 + \ldots + x_n\omega_n .$$

We replace ω_1 by ω'_1 and arrive at a basis $\omega'_1, \omega_2, \ldots, \omega_n$ canonical for κ_1 and such that either $\kappa_1 \subseteq \kappa_i$ or $\omega'_1 \in \kappa_i$. Assume that $\omega_1, \ldots, \omega_n$ already have this property so that $x_1\omega_1 + \ldots + x_n\omega_n \in \kappa_i \rightarrow x_2\omega_2 + \ldots + x_n\omega_n \in \kappa_i$. We put

$$\kappa_i^* = \kappa'_i \cap [\omega_2, \ldots, \omega_n] = \kappa'_i \cap \kappa_1 , \qquad i=1, \ldots, t' .$$

Let

$$\kappa_i^* = [d_2^*\omega'_2, \ldots, d_n^*\omega'_n]$$

where $\omega'_2, \ldots, \omega'_n \in \kappa_1$. Then

$$\kappa'_i = [\omega_1, d_2^*\omega'_2, \ldots, d_n^*\omega'_n] .$$

Hence all d_i^* except one must be unity, κ_i^* is cyclic under κ_1, and

$$G/\kappa'_i = \kappa_1/\kappa_i^* .$$

It is also not difficult to see that the κ_i^* satisfy the condition (ii) of III with respect to κ_1. Hence by induction on n we can construct $\omega_2^*, \ldots, \omega_n^*$ where $\omega_i^* = \sum_{j=2}^{n} c_{ij}\omega_j$, $i = 2, \ldots, n$, is a basis of κ_1 canonical for $\kappa_1^*, \ldots, \kappa_{t'}^*$. Since $\omega_2^*, \ldots, \omega_n^* \in \kappa_1$, it is clear that $\omega_1, \omega_2^*, \ldots, \omega_n^*$ is a canonical basis for $\kappa_1, \ldots, \kappa_t$.

We now consider the case that G/κ_j is of finite order for $j = 1, \ldots, t$.

In constructing a basis canonical for $\kappa_1, \ldots, \kappa_t$ we may by Lemma 2 delete κ_j if $\kappa_i \subseteq \kappa_j$ for some $j \neq i$. It is then sufficient to show that there is a basis canonical for $\kappa_1, \ldots, \kappa_\tau$ if $\kappa_i \nsubseteq \kappa_j$ for any pair i, j, $i \neq j$, and if $\kappa_1, \ldots, \kappa_\tau$ satisfy condition (ii) of III.

We arrange the notation so that

$$|G/\kappa_1| = p_1^{r_{11}}, \ldots, |G/\kappa_{t_1}| = p_1^{r_{1t_1}}$$
$$\vdots$$
$$|G/\kappa_{t_{i-1}+1}| = p_i^{r_{i1}}, \ldots, |G/\kappa_\tau| = p_i^{r_{it_i}}$$

where $p_1, \ldots, p_i$ are distinct primes and $r_{j1} \geq r_{j2} \geq \ldots \geq r_{jt_j} > 0$.

Suppose that we have constructed a basis $\omega_1, \ldots, \omega_n$ which is canonical for $\kappa_1, \ldots, \kappa_{t_{i-1}+j}$, $0 \leq j < t_i - t_{i-1}$. If $j = 0$ then we can construct by Lemma 3 a basis $\omega_1^*, \ldots, \omega_n^*$ such that

$$\kappa_{t_{i-1}+1} = [p_i^{r_{i1}}\omega_1^*, \omega_2^*, \ldots, \omega_n^*]$$

and such that

$$\omega_i^* \equiv \omega_i \pmod{b}$$

where $b = p_1^{r_{11}} p_2^{r_{21}}, \ldots, p_{i-1}^{r_{i-1,1}}$.

If $j > 0$ put $\kappa_{t_{i-1}+u} = L_u$ for $u = 1, \ldots, t_i - t_{i-1}$ and assume that

$$(5)\quad \begin{aligned} L_\alpha &= [\omega_1, \ldots, \omega_{\alpha-1}, d_\alpha\omega_\alpha, \omega_{\alpha+1}, \ldots, \omega_n]\,, \qquad 1 \leqslant \alpha \leqslant j\,, \\ L &= L_1 \cap \ldots \cap L_j = [d_1\omega_1, \ldots, d_j\omega_j, \omega_{j+1}, \ldots, \omega_n]\,, \end{aligned}$$

where $d_\alpha = p_i^{r_{i\alpha}}$.

By (ii) of III we must have $L + L_{j+1} = G$. Moreover since $d_\alpha \geq |G/L_{j+1}|$ for $1 \leq \alpha \leq j$ we must have $d_\alpha\omega_\alpha \in L_{j+1}$ for $1 \leqslant \alpha \leqslant j$. Hence

$$\omega_1 = l + x_{j+1}\omega_{j+1} + \ldots + x_n\omega_n$$

with $l \in L_{j+1}$. Now since $(b, p_i) = 1$ we can solve

$$p_i^{r_{i1}}x + by = 1\,, \qquad x_\beta(p_i^{r_{i1}}x + by) = x_\beta$$

and obtain since $p_i^{r_{i1}}\omega_\alpha \in L_{j+1}$

$$\omega_1 = \omega_1^* + byx_{j+1}\omega_{j+1} + \ldots + byx_n\omega_n$$

with $\omega_1^* \in L_{j+1}$. Equation (5) shows that $\omega_{j+1}, \ldots, \omega_n \in L$. Hence $\omega_1^*, \omega_2, \ldots, \omega_n$ is still a canonical basis for $\kappa_1, \ldots, \kappa_{t_{i-1}+j}$ such that (5) holds. We transform $\omega_2, \ldots, \omega_j$ similarly to arrive at the basis

$$\omega_1^*, \ldots, \omega_j^*, \omega_{j+1}, \ldots, \omega_n$$

canonical for $\kappa_1, \ldots, \kappa_{t_{i-1}+j}$ for which

$$\omega_i^* \in L_{j+1} \qquad 1 \leqslant i \leqslant j .$$

We now apply the construction of the proof of Lemma 3 to L_{j+1} and $\omega_{j+1}, \omega_1^*, \ldots, \omega_j^*, \omega_{j+2}, \ldots, \omega_n$ to obtain ω_{j+1}^* such that

$$\omega_1^*, \ldots, \omega_{j+1}^*, \omega_{j+2}, \ldots, \omega_n$$

is a canonical basis for L_{j+1}, while

$$\omega_i^* \equiv \omega_i \pmod b \qquad 1 \leqslant i \leqslant j+1 .$$

Note that in our construction the basis elements $\omega_1^*, \ldots, \omega_j^*$ are not involved at all since they are already in L_{j+1}. Hence $\omega_1^*, \ldots, \omega_{j+1}^*, \omega_{j+2} \ldots, \omega_n$ is also a canonical basis for $L_1, \ldots, L_j$; hence also for $\kappa_1, \ldots, \kappa_{t_{i-1}+j+1}$. Moreover

$$L_{j+1} = [\omega_1, \ldots, \omega_j, d_{j+1}\omega_{j+1}, \omega_{j+2}, \ldots, \omega_n] ,$$

$$L_1 \cap \ldots \cap L_{j+1} = [d_1\omega_1, \ldots, d_{j+1}\omega_{j+1}, \omega_{j+2}, \ldots, \omega_n]$$

with $d_\alpha = p_i^{r_{i\alpha}}$, so that also (5) is satisfied with j replaced by $j+1$. This completes the proof that III → I. We have proved

Theorem 1. *For any set* $H_1, \ldots, H_s$ *of subgroups of* G *it is true that each of the conditions* I, II, III *implies the other two.*

2. APPLICATIONS

Let H be a subgroup of G. Let $\alpha_1, \ldots, \alpha_n$ be a basis of H. Then

$$\alpha_i = \sum_{j=1}^{n} a_{ij}\omega_j , \qquad i = 1, \ldots, n , \tag{6}$$

where $\omega_1, \ldots, \omega_n$ is a basis of G. We may write (6) in matrix form as

$$(7) \qquad \boldsymbol{\alpha} = A\boldsymbol{\omega}$$

where $\boldsymbol{\alpha}$ and $\boldsymbol{\omega}$ are column vectors and $A = (a_{ij})$ is a matrix with integral coefficients. Conversely every integral matrix A corresponds to a subgroup.

A change in the basis $\boldsymbol{\alpha}$ of H corresponds to left multiplication of A by a unimodular matrix; a change in the basis $\boldsymbol{\omega}$ corresponds to right multiplication. Let G be torsion free and let $A_1, \ldots, A_s$ be matrices defining the subgroups $H_1, \ldots, H_s$. Then there exists a basis canonical for $H_1, \ldots, H_s$ if and only if there exist unimodular matrices $X_1, \ldots, X_s$ and a matrix Y such that $X_i A_i Y$ are diagonal matrices.

Let H_1, H be subgroups of G and let A_1, A be matrices defining H_1 and H respectively. We have

$$(8) \qquad A_1 = XA$$

where X is an integral matrix if and only if $H_1 \subseteq H$.

Now let H_1, H_2 be defined by A_1, A_2 and let $H = H_1 + H_2$. Then $H \supseteq H_1$ and $H \supseteq H_2$. Hence

$$(9) \qquad A_1 = X_1 A\,, \qquad A_2 = X_2 A\,,$$

where A corresponds to H. Moreover let

$$\alpha_i^{(1)} = \sum a_{ij}^{(1)} \omega_j\,, \qquad \alpha_i^{(2)} = \sum a_{ij}^{(2)} \omega_j\,, \qquad \alpha_i = \sum a_{ij} \omega_j$$

where $A_1 = (a_{ij}^{(1)})$, $A_2 = (a_{ij}^{(2)})$, $A = (a_{ij})$. Then since $\alpha_i \in H_1 + H_2$, we can solve the equations

$$\sum_i (x_{ti}\alpha_i^{(1)} + y_{ti}\alpha_i^{(2)}) = \alpha_t\,, \qquad t = 1, \ldots, n\,.$$

Hence with $X = (x_{ti})$, $Y = (y_{ti})$ we have

$$(10) \qquad XA_1 + YA_2 = A\,.$$

We shall write

$$A = (A_1, A_2)_r$$

and shall call A the greatest common right divisor (GCRD) of

A_1 and A_2. Our argument can be extended to any set of matrices and shows that in the ring of matrices with integral coefficients every ideal is principal.

Similarly let $H = H_1 \cap H_2$. Let A correspond to H and A_1, A_2 to H_1, H_2 respectively. Since $H \subset H_1$, $H \subset H_2$ we have

$$A = X_1A_1 = X_2A_2 . \tag{11}$$

Moreover, if $H' \subseteq H_1$, $H' \subseteq H_2$; then $H' \subseteq H$. Hence if

$$A' = X_1'A_1 = X_2'A_2 , \tag{12}$$

then

$$A' = XA . \tag{13}$$

We shall call the class of matrices A which satisfy (11) and (13) the least common right multiple of A_1 and A_2 and shall write

$$A = \mathrm{LCRM}\,(A_1, A_2) .$$

As before A is determined up to unimodular left factors. Again the definition extends easily to any set of matrices and the existence of the LCRM for any set of matrices follows from the fact that $H = H_1 \cap H_2 \cap \ldots$ is defined for any set $H_1, H_2 \ldots$.

Let H be a subgroup of G. We can find a basis $\omega_1, \ldots, \omega_n$ of G such that

$$H = [d_1\omega_1, \ldots, d_n\omega_n] .$$

If G/H is cyclic, then we can determine $\omega_1, \ldots, \omega_n$ so that all but one of the d_j are 1. A matrix A will be called cyclic if A defines a subgroup with cyclic factor group. If A is cyclic then there exist unimodular matrices X, Y such that

$$XAY = \mathrm{diag}\,[d_1, 1, \ldots, 1] .$$

Theorem 1 now gives the following corollary.

Corollary 1. *Let $A_1, \ldots, A_s$ be s matrices with integral coefficients. There exist matrices $X_1, \ldots, X_s$ and a matrix Y such that*

$$X_i A_i Y = D_i\,, \qquad i = 1, \ldots, s\,,$$

are diagonal matrices if and only if there exist cyclic matrices $B_1, \ldots, B_t$ *such that each* A_i *is a LCRM of some of the matrices* $B_1, \ldots, B_t$, *and such that for each choice* $\alpha_1, \ldots, \alpha_u$ *and* α, *either* $B_{\alpha_i} = C_i B_\alpha$ *where* C_i *or* C_i^{-1} *is an integral matrix or*

$$(\text{LCRM}\,(B_{\alpha_1}, \ldots, B_{\alpha_u}), B_\alpha)_r = I.$$

(I represents the class of unimodular matrices).

If in Theorem 1 all the factor groups $G/H_1, \ldots, G/H_s$ are cyclic of prime power order or of infinite order then the groups H_i must themselves occur among the subgroups $\kappa_1, \ldots, \kappa_t$. Hence in this case we may take $s = t$ and $\kappa_i = H_i$. This gives

Corollary 2. *If* $H_1, \ldots, H_s$ *are subgroups of* G *with cyclic factor group of prime power order or if infinite order then* I *and* II *hold if and only if for any choice* $\alpha_1, \ldots, \alpha_u$ *and* α *either* $H_{\alpha_i} \subset H_\alpha$ *or* $H_\alpha \supseteq H_{\alpha_i}$ *for some* i *or*

$$H_{\alpha_1} \cap \ldots \cap H_{\alpha_i} + H_\alpha = G\,.$$

Corollary 2 makes it easy to find examples of subgroups $H_1, H_2, \ldots, H_s$ for which no canonical basis exists. The example given in [2] is of this type.

It is well known that the integers of an algebraic numberfield F form a finite module over the rational integers. Let $\mathcal{A}_1, \ldots, \mathcal{A}_s$ be s ideals in F and let $\mathcal{A}$ be their LCM. It is well known that there exists a basis $\eta_1, \ldots, \eta_t$ for the residues mod $\mathcal{A}_i$ such that $\eta_1, \ldots, \eta_t$ is a basis for the residues mod $\mathcal{A}_i$ for each i. Hence $\mathcal{A}_1, \ldots, \mathcal{A}_s$ satisfy the condition II and we therefore have

Corollary 3. *Let* $\mathcal{A}_1, \ldots, \mathcal{A}_s$ *be ideals of an algebraic numberfield* F. *There exists an integral basis* $\omega_1, \ldots, \omega_n$ *of* F *such that*

$$\mathcal{A}_i = [d_{i1}\omega_1, \ldots, d_{in}\omega_n]\,, \qquad i = 1, \ldots, n\,.$$

Consider a lattice L in the n-dimensional Euclidean space R_n given by all points $x = (x_1, \ldots, x_n)$ where

$$x_i = \sum_{i=1}^{n} a_{ij} y_j\,, \qquad |a_{ij}| \neq 0\,,$$

and the y_j are integers. There exists a basis of L canonical for the sublattices $L_1, \ldots, L_s$ if there exists a fundamental parallelepiped of L spanned by the vectors $e_1, \ldots, e_n$ such that $d_{i1}e_1, \ldots, d_{in}e_n$ span L_i. We shall call such a parallelepiped canonical for L_i. We can state

Corollary 4. *Let $L_1, \ldots, L_s$ be sublattices of the n-dimensional lattice L. There exists a fundamental parallelepiped for L canonical for $L_1, \ldots, L_s$ if and only if there exist cyclic sublattices $\kappa_1, \ldots, \kappa_t$ of prime power index such that L_i is for each i the intersection of some of the κ_i, and such that for each choice $\alpha_1, \ldots, \alpha_i$ and α, we have either $\kappa_{\alpha_i} \supseteq \kappa_\alpha$ or $\kappa_{\alpha_i} \subset \kappa_\alpha$ or*

$$\kappa_{\alpha_1} \cap \ldots \cap \kappa_{\alpha_i} + \kappa_\alpha = L\,.$$

Let $\omega_1, \ldots, \omega_n$ be an integral basis for the algebraic numberfield F. Let α be an integer of F then

$$\alpha = x_1\omega_1 + \ldots + x_n\omega_n\,, \qquad x_i \text{ integral}\,.$$

We call $x_1, \ldots, x_n$ the *coordinates* of α. The lattice of an ideal $\mathcal{A}$ of F is the totality of all points $\mathbf{x} = (x_1, \ldots, x_n)$ whose coodinates are the coordinates of an integer in $\mathcal{A}$. We have

Corollary 5. *Let $L_1, \ldots, L_s$ be the lattices of the ideals $\mathcal{A}_1, \ldots, \mathcal{A}_s$ of an algebraic numberfield F. Then there exists a fundamental parallelepiped of the lattice L of points with integral coordinates which is canonical for $L_1, \ldots, L_s$.*

3. THE CONVERSE OF COROLLARY 3 OF THEOREM 1

The converse to Corollary 3 of Theorem 1 also holds in the sense that if a canonical basis for a set of subgroups of finite index exists in a torsion free finitely generated Abelian group G then these subgroups can in fact be represented as ideals in some algebraic numberfield. We shall prove

Theorem 2. *Let G be a finitely generated torsion free Abelian group and let $H_1, \ldots, H_s$ be subgroups of G of finite index given by the matrices $A_1, \ldots, A_s$ and assume that there is a basis in G canonical for $H_1, \ldots, H_s$. Then there exists an algebraic*

numberfield F *with an integral basis* $\omega_1, \ldots, \omega_n$ *and ideals* $\mathcal{A}_1, \ldots, \mathcal{A}_s$ *of* F *such that* $\boldsymbol{\alpha}^{(i)} = \boldsymbol{A}_i\boldsymbol{\omega}$ *is a basis for* $\mathcal{A}_i$ *where*

$$\boldsymbol{\alpha}^{(i)} = \begin{pmatrix} \alpha_1^{(i)} \\ \vdots \\ \alpha_n^{(i)} \end{pmatrix}, \qquad \boldsymbol{\omega} = \begin{pmatrix} \omega_1 \\ \vdots \\ \omega_n \end{pmatrix}.$$

We first prove a lemma.

Lemma 4. *Let* $p_1, \ldots, p_s$ *be prime numbers. There exists an algebraic numberfield* F *of degree* n *such that each of the primes* $p_1, \ldots, p_s$ *splits in* F *into* n *prime ideals of degree and order* 1.

Proof. Let $p_i^{\alpha_i} > n$. Let $a_1^{(i)}, \ldots, a_n^{(i)}$ be distinct mod $(p_i^{\alpha_i})$ for $i = 1, \ldots, s$. Let $p_i^{\delta_i}$ be the p_i component of the discriminant of

$$f_i(x) = (x - a_1^{(i)}) \ldots (x - a_n^{(i)}).$$

Let $m_i \equiv 1(\text{mod } p_i^{\delta_i+1})$, $m_i \equiv 0(p_j^{\delta_j+1})$ for $j \neq i$, and set

$$f(x) = x^n + \sum_{i=1}^{s} m_i(f_i(x) - x^n).$$

Let q be a prime distinct from $p_1, \ldots, p_s$ and set $m = p_1^{\delta_1+1} \ldots p_s^{\delta_s+1}$. Determine $g(x)$ of degree $< n$ so that all coefficients of

$$F(x) = f(x) + mg(x)$$

except the first are divisible by q and the constant term not divisible by q^2. Then $F(x)$ is a monic polynomial of degree n, irreducible by Eisenstein's criterion and factors mod $p_i^{\delta_i+1}$ into linear factors. It follows from a theorem of Oystein Ore [1, p. 77] that p_i splits into n factors in $R(\theta)$ where R is the field of rational numbers and θ a root of $F(x)$.

We now proceed to the proof of Theorem 2. Under the hypothesis of Theorem 2 there are unimodular matrices $\boldsymbol{X}_1, \ldots, \boldsymbol{X}_s$ and $\boldsymbol{Y}$ such that

$$\boldsymbol{X}_i\boldsymbol{A}_i\boldsymbol{Y} = \text{diag.}\,(d_{i1}, \ldots, d_{in}), \qquad i = 1, \ldots, s.$$

The subgroups H_i then have the basis $\boldsymbol{X}_i\boldsymbol{A}_i\boldsymbol{\omega}$ and G has the basis $\boldsymbol{Y\omega}$. Suppose we can find an algebraic numberfield F that contains ideals $\mathcal{A}_1, \ldots, \mathcal{A}_s$ and a basis $\omega_1, \ldots, \omega_n$ such that

$$\mathcal{A}_i = [d_{i1}\omega_1, \ldots, d_{in}\omega_n]\boldsymbol{D}_i\boldsymbol{\omega}\,. \tag{11}$$

We then have

$$\boldsymbol{A}_i = \boldsymbol{X}_i^{-1}\,\text{diag.}\,(d_{i1}, \ldots, d_{in})\boldsymbol{Y}^{-1}\,.$$

Moreover $\boldsymbol{X}_i^{-1}\boldsymbol{D}_i\boldsymbol{Y}^{-1}(\boldsymbol{Y\omega})$ is a module basis for $\mathcal{A}_i$ and $\boldsymbol{\omega}' = \boldsymbol{Y\omega}$ is an integral basis for F so that

$$\boldsymbol{\alpha}^{(i)} = \boldsymbol{A}_i\boldsymbol{\omega}'$$

is a module basis for the ideal $\mathcal{A}_i$. Hence Theorem 2 will be proved if we can construct an algebraic numberfield F in which there are ideals $\mathcal{A}_1, \ldots, \mathcal{A}_s$ such that

$$\mathcal{A}_i = \boldsymbol{D}_i\boldsymbol{\omega}$$

where $\boldsymbol{D}_1, \ldots, \boldsymbol{D}_s$ are given diagonal matrices and $\omega_1, \ldots, \omega_n$ is an integral basis of F.

Let

$$d_{i\alpha} = p_1^{l_{1i\alpha}} \ldots p_u^{l_{ui\alpha}}\,, \qquad i = 1, \ldots, s\,;\ \alpha = 1, \ldots, n,\ l_{ji\alpha} \geq 0\,.$$

By Lemma 4 we can construct a field F in which all the ideals p_i split completely into prime ideals of degree and order 1. Hence in F we have the decomposition

$$p_j = \mathfrak{p}_{j1}\mathfrak{p}_{j2} \ldots \mathfrak{p}_{jn}\,, \qquad j = 1, \ldots, u\,.$$

Let $N = \max\limits_{j,i,\alpha}(l_{ji\alpha})$. By Lemma 1 of [2] we can construct an integral basis $\omega_1, \ldots, \omega_n$ such that

$$\omega_h \equiv 0 \;(\text{mod}\ \mathfrak{p}_{jt}^N) \quad \text{for} \quad h \neq t\,.$$

We put

$$\mathcal{A}_i = \bigcap_j \bigcap_\alpha \mathfrak{p}_{j\alpha}^{l_{ji\alpha}}\,,$$

then by [2]

$$\mathcal{A}_i = [d_{i1}\omega_1, \ldots, d_{in}\omega_n]\,.$$

This completes the proof of Theorem 2.

Together with Corollary 3 of Theorem 1 we have

Corollary. *Let $H_1, H_2, \ldots, H_s$ be subgroups of finite index of G given by matrices $\boldsymbol{A}_1, \ldots, \boldsymbol{A}_s$. There exists a basis canonical for $H_1, \ldots, H_s$ if and only if there exists an algebraic numberfield F of degree n and an integral basis $\omega_1, \ldots, \omega_n$ of F such that*

$$\mathcal{A}_i = [\alpha_1^{(i)}, \ldots, \alpha_n^{(i)}]$$

are ideals of F, where

$$\boldsymbol{\alpha}^{(i)} = \boldsymbol{A}_i \boldsymbol{\omega} .$$

The proof of Theorem 1 and its corollaries can be generalized to the case that G is a *finitely generated module over an integral domain $\mathcal{J}$ in which every ideal is principal.* One can show that $\mathcal{J}$ must have a unit element which we denote by 1 and we shall assume that $1\omega = \omega$ for all $\omega \in G$.

We introduce the following terminology.

The *elements of $\mathcal{J}$ are called integers.*

If $\omega \in G$ let (a) be the ideal of all integers x such that $x\omega = 0$. *The ideal* (a) *is called the order of ω.* (We shall use the symbol 0 for the unit element of G and also for the 0 of $\mathcal{J}$.)

G is called cyclic if it is generated by one element ω over $\mathcal{J}$. The order of ω is then also called the order of G.

If G is an $\mathcal{J}$ module, H an $\mathcal{J}$ submodule of G, and x any integer then $xH\omega = Hx\omega$. Hence G/H is an $\mathcal{J}$ module. *If G/H is cyclic then the order of G/H is called the index of H under G.*

In order to generalize Theorem 1 and its corollaries to the case that G is a finitely generated $\mathcal{J}$ module read *order* $\neq (0)$ for *finite order* and *order* $= 0$ for *infinite order*. Moreover in Corollary 3 let Σ be the quotient field of $\mathcal{J}$ and read *finite separable algebraic extension of Σ for algebraic numberfield.* (Local separability is not necessary). More generally Corollary 3 carries over to any algebraic extension F of Σ such that an integral basis exists for the integers of F over the integers of Σ. This is for instance always the case if $\mathcal{J}$ is the domain of polynomials over a field [3, p. 398].

A careful examination of the proof of the theorem of Ore in [1] shows that the theorem carries over to any principal ideal ring and in fact to any Dedekind ring. The proof of

Lemma 4 requires only a prime q which does not divide any of the $d_{i\alpha}$. Hence the converse to Corollary 3 does hold whenever such a prime is available. In particular it will always hold if $\mathcal{J}$ contains infinitely many primes.

A special case of Corollary 3 and its converse was proved by Olga Taussky [4].

References

1. Fricke, Robert. *Lehrbuch der Algebra*, Vol. 3, Vieweg and Sohn Akt.-Ges., Braunschweig, 1928.
2. Mann, H. B. and Koichi Yamamoto. "On Canonical Bases of Ideals," *J. Combinatorial Theory*, **2** (1967), 71–76.
3. Roquette, Peter. "Über den Riemann-Rochschen Satz in Funktionenkörpern vom Transzendenzgrad I," *Math. Nachr.*, **19** (1958), 375–404.
4. Taussky, Olga. "Ideal Matrices II," *Math. Ann.*, **150** (1963), 218–225.

Discussion on Professor Mann's Paper

PROFFESSOR H. H. CRAPO: We find in H. B. Mann's talk a wealth of detailed information about the lattice of subgroups of an Abelian group. By proving the equivalence of three conditions for the existence of a basis canonical for several subgroups, he illuminates the relationship between a product structure in the lattice, a relativized basis notion, and the spanning capacity of independent sets of elements near the top of the lattice. We inquire whether an equivalent condition might be phrased in terms of the lattice substructure (distributive ?) generated by the several subgroups.

These equivalent conditions are then shown to be necessary and sufficient for the existence of an algebraic number field with a related ideal structure.

CHAPTER 4

Permanents and Systems of Distinct Representatives

H. J. RYSER[1] *California Institute of Technology*

1. INTRODUCTION

The purpose of this paper is to survey some of the recent literature dealing with the combinatorial aspects of the permanent function of a matrix. In Section 2 we discuss the König-Egerváry theorem on the term rank of a matrix and the P. Hall theorem on systems of distinct representatives for subsets. Section 3 is devoted to evaluation techniques for permanents. Then in Section 4 we summarize some of the main results that center around van der Waerden's well-known conjecture on the permanent of a doubly stochastic matrix. In our concluding Section 5 we deal with upper and lower estimates for permanents of (0, 1)-matrices and the applications of these results to Latin rectangles. Throughout our discussion we place the main emphasis on unsettled issues.

2. THE KÖNIG-EGERVÁRY AND P. HALL THEOREMS

Let A be a matrix of m rows and n columns and let the

[1] This research was supported in part by U.S. Army Research Office (Durham) and carried out at Syracuse University.

elements of A be the integers 0 and 1. We call such a matrix A a (0, 1)-*matrix* of *size* m by n. These (0, 1)-matrices play an essential role in the development of many combinatorial problems. One of the main reasons for this is the following. Let S be an n-set of elements $a_1, a_2, \ldots, a_n$ and let $S_1, S_2, \ldots, S_m$ be subsets of S. Now let $a_{ij} = 1$ if a_j is a member of S_i and let $a_{ij} = 0$ if a_j is not a member of S_i. Then

$$(2.1) \qquad A = [a_{ij}] \quad (i = 1, 2, \ldots, m;\ j = 1, 2, \ldots, n)$$

is a (0, 1)-matrix of size m by n called the *incidence matrix* for the subsets $S_1, S_2, \ldots, S_m$ of the n-set S. The 1's in row i of A specify the elements that belong to set S_i and the 1's in column j of A specify the sets that contain element a_j. Thus it follows that A contains a complete description of the subsets $S_1, S_2, \ldots, S_m$ of S. Moreover, if a (0, 1)-matrix A of size m by n is given and if S is an arbitrary n-set, then there exist subsets $S_1, S_2, \ldots, S_m$ of S such that A is the incidence matrix for these subsets.

From the preceding remarks it is clear that theorems on (0, 1)-matrices have direct combinatorial consequences. One of the central results dealing with (0, 1)-matrices is the theorem of König and Egerváry. A *line* of a matrix designates either a row or a column of the matrix.

Theorem 2.1. (König-Egerváry) *Let A be a* (0, 1)-*matrix of size m by n. The minimal number of lines in A that contain all of the* 1's *in A is equal to the maximal number of* 1's *in A with no two* 1's *on a line.*

We have here an example of a minimax theorem and, indeed, the König-Egerváry theorem may be derived directly from the max-flow min-cut theorem of Ford and Fulkerson [3]. It is customary to call the maximal number of 1's in A with no two 1's on a line the *term rank* of A.

Let $S_1, S_2, \ldots, S_m$ be subsets of a set S and suppose that $a_1, a_2, \ldots, a_m$ are m *distinct* elements such that

$$(2.2) \qquad a_i \in S_i \qquad (i=1, 2, \ldots, m).$$

Then the subsets $S_1, S_2, \ldots, S_m$ are said to have a *system of distinct representatives* (abbreviated SDR). An obvious necessary condition in order that the subsets $S_1, S_2, \ldots, S_m$ have an

SDR is that every k of the sets contain in their union at least k distinct elements. The fact that this necessary condition is also sufficient is the essential content of the following theorem of P. Hall.

Theorem 2.2. (P. Hall) *The subsets* $S_1, S_2, \ldots, S_m$ *of* S *have an SDR if and only if the set* $S_{i_1} \cup S_{i_2} \cup \ldots \cup S_{i_k}$ *contains at least* k *elements for* $k = 1, 2, \ldots, m$ *and for all* k*-combinations* $\{i_1, i_2, \ldots, i_k\}$ *of the integers* $1, 2, \ldots, m$.

It is easy to rephrase the P. Hall theorem in the terminology of matrix theory. We now restrict S to be an n-set and we let A be the incidence matrix of size m by n for the subsets $S_1, S_2, \ldots, S_m$ of S. Then the P. Hall theorem asserts that the matrix A has term rank m if and only if every k rows of A contain 1's in at least k columns for $k = 1, 2, \ldots, m$. The König-Egerváry theorem and the P. Hall theorem are in fact equivalent and it is not difficult to derive one from the other [7]. An impressive combinatorial literature centers around these theorems and their various ramifications and extensions (see, for example, the recent very extensive survey paper by Mirsky and Perfect [18]). At this point we remark that combinatorics would be greatly enriched by the development of other general combinatorial theorems unrelated to the above but with equally far-reaching consequences.

3. THE PERMANENT

The literature on combinatorial enumeration and the closely allied topic of combinatorial identities is enormous. At the outset we mention the well-known book by Riordan [20] and an important recent paper by Rota [21] dealing with a very general principle of enumeration. In what follows we confine ourselves to a tiny facet of the subject involving permanents.

Let $A = [a_{ij}]$ be a matrix of size m by n with $m \leqq n$ and let the elements of A be real (or complex) numbers. Then the *permanent* of A is defined by

$$\text{(3.1)} \qquad \operatorname{per}(A) = \sum a_{1i_1} a_{2i_2} \ldots a_{mi_m},$$

where the summation in (3.1) extends over all m-permutations $(i_1, i_2, \ldots, i_m)$ of the integers $1, 2, \ldots, n$. This scalar function of the matrix A occurs throughout the literature of combinatorics and is especially important in problems dealing with enumeration. The following theorem displays the quantitative role played by the permanent in SDR theory. Its proof is an immediate consequence of the terminology [22].

Theorem 3.1. *Let $S_1, S_2, \ldots, S_m$ be subsets of an n-set S and let $m \leqq n$. Let A be the incidence matrix for these subsets. Then the number of SDR's for $S_1, S_2, \ldots, S_m$ is per(A).*

In the case of square matrices the permanent is the same as the determinant apart from a factor ± 1 preceding each product on the right side of equation (3.1). But most of the highly developed techniques used in the evaluation of determinants are not available for permanents, and consequently there are many square matrices with easily computed determinants but undetermined permanents. The following helpful rule for the evaluation of permanents is a consequence of the principle of inclusion and exclusion [22].

Theorem 3.2. *Let A be a matrix of size m by n with $m \leqq n$. Let A_r denote a matrix obtained from A by replacing r columns of A by zeros. Let $S(A_r)$ denote the product of the row sums of A_r and let $\sum S(A_r)$ denote the sums of the $S(A_r)$ over all of the choices for A_r. Then*

$$\text{(3.2)} \quad \operatorname{per}(A) = \sum S(A_{n-m}) - \binom{n-m+1}{1} \sum S(A_{n-m+1}) + \binom{n-m+2}{2} \sum S(A_{n-m+2}) - \cdots + (-1)^{m-1}\binom{n-1}{m-1} \sum S(A_{n-1}).$$

In particular, let A be a square matrix of order n. Then

$$\text{(3.3)} \quad \operatorname{per}(A) = S(A) - \sum S(A_1) + \sum S(A_2) - \cdots + (-1)^{n-1} \sum S(A_{n-1}).$$

We remark in passing that a formula closely related but distinct from (3.3) has been investigated recently by Wilf [23].

It turns out that Theorem 3.2 is the source of some rather remarkable identities. We illustrate this point with two examples. For our first example we let I denote the identity matrix of order n and we let J denote the matrix of order n with every entry equal to 1. Then it is easy to verify directly that

$$\text{per}\,(J - I) = D_n\,, \tag{3.4}$$

where

$$D_n = n!\left(1 - \frac{1}{1!} + \frac{1}{2!} - \dots + (-1)^n\frac{1}{n!}\right) \tag{3.5}$$

is the derangement number. On the other hand an application of (3.3) gives us the relationship

$$D_n = \sum_{r=0}^{n-1} (-1)^r \binom{n}{r} (n - r)^r (n - r - 1)^{n-r}\,. \tag{3.6}$$

For our second example we let A denote the matrix of order n with x's in the main diagonal, y's in the positions $(1, 2), (2, 3), \dots, (n - 1, n), (n, 1)$, and 0's in all other positions. Then by applying the Laplace expansion for permanents to the first column of A we obtain

$$\text{per}\,(A) = x^n + y^n\,. \tag{3.7}$$

Let $g(n, k)$ denote the number of ways of selecting k objects, no two consecutive, from n objects arranged in a circle. Kaplansky [11] in his well-known derivation of the formula for ménage numbers shows that

$$g(n, k) = \frac{n}{n - k}\binom{n - k}{k} \qquad (n > k)\,. \tag{3.8}$$

We may now use (3.3) and (3.8) to evaluate per(A). But then by equating the evaluation with (3.7) we obtain the identity

$$x^n + y^n = \sum_{r=0}^{[n/2]} (-1)^r \frac{n}{n - r}\binom{n - r}{r} (xy)^r (x + y)^{n-2r}\,, \tag{3.9}$$

where the bracket denotes the integral part of $n/2$.

Next we remark that Theorem 3.2 is also useful in the derivation of certain inequalities for permanents. A matrix A of order n is called *row substochastic* provided that its elements are nonnegative real numbers and its row sums are each no more than 1. Let A denote a row substochastic matrix of order n and let I denote the identity matrix of order

n. Brualdi and Newman [2] have shown that

$$\text{per}(\boldsymbol{I} - \boldsymbol{A}) \geqq 0 . \tag{3.10}$$

Gibson [4] has recently pointed out that (3.10) is also a rather direct consequence of (3.3) because it is not difficult to verify that

$$(-1)^r S((\boldsymbol{I} - \boldsymbol{A})_r) \geqq 0 . \tag{3.11}$$

The argument here is highly suggestive and indicates the possibility of the derivation of other interesting inequalities by way of special identities for permanents.

In a recent paper Jurkat and Ryser [9] have introduced an entirely different technique for the evaluation of the permanent of a matrix $\boldsymbol{A}$ of order n. The method is also valid for determinants and is illustrated in what follows for matrices of orders 2, 3, and 4:

$$[a_1 \quad a_2]\begin{bmatrix} b_2 \\ \pm b_1 \end{bmatrix} = a_1 b_2 \pm a_2 b_1 = \begin{cases}\text{per}(\boldsymbol{A}) \\ \det(\boldsymbol{A})\end{cases}, \quad \boldsymbol{A} = \begin{bmatrix} a_1 & a_2 \\ b_1 & b_2 \end{bmatrix},$$

$$[a_1 \quad a_2 \quad a_3]\begin{bmatrix} b_2 & b_3 & 0 \\ \pm b_1 & 0 & b_3 \\ 0 & \pm b_1 & \pm b_2 \end{bmatrix}\begin{bmatrix} c_3 \\ \pm c_2 \\ c_1 \end{bmatrix} = \begin{cases}\text{per}(\boldsymbol{A}) \\ \det(\boldsymbol{A})\end{cases},$$

$$\boldsymbol{A} = \begin{bmatrix} a_1 & a_2 & a_3 \\ b_1 & b_2 & b_3 \\ c_1 & c_2 & c_3 \end{bmatrix},$$

$$[a_1 \quad a_2 \quad a_3 \quad a_4]\begin{bmatrix} b_2 & b_3 & b_4 & 0 & 0 & 0 \\ \pm b_1 & 0 & 0 & b_3 & b_4 & 0 \\ 0 & \pm b_1 & 0 & \pm b_2 & 0 & b_4 \\ 0 & 0 & \pm b_1 & 0 & \pm b_2 & \pm b_3 \end{bmatrix}$$

$$\times \begin{bmatrix} c_3 & c_4 & 0 & 0 \\ \pm c_2 & 0 & c_4 & 0 \\ 0 & \pm c_2 & \pm c_3 & 0 \\ c_1 & 0 & 0 & c_4 \\ 0 & c_1 & 0 & \pm c_3 \\ 0 & 0 & c_1 & c_2 \end{bmatrix}\begin{bmatrix} d_4 \\ \pm d_3 \\ d_2 \\ \pm d_1 \end{bmatrix} = \begin{cases}\text{per}(\boldsymbol{A}) \\ \det(\boldsymbol{A})\end{cases}.$$

Here the plus sign applies to per(A) and the minus sign applies to det(A). Notice that in each of the preceding equations the product of the first two factors is a vector whose components are the minors of order two of the first two rows of A, etc.

For general n per(A) and det(A) regarded as a 1 by 1 matrix may be written as a product of n rectangular matrices, each formed from only one row of A. The factors themselves of this matrix equation have an interesting combinatorial structure and one may apply the powerful techniques of matrix theory to them. Actually this device yields a highly effective procedure for establishing certain matrix inequalities. The preceding remarks suggest the possibility of expressing a wide variety of combinatorial and algebraic quantities as products of rectangular matrices. In certain instances such representations may be far more revealing than the more conventional type of algebraic equation. This subject holds many possibilities and has scarcely been touched upon.

4. THE VAN DER WAERDEN CONJECTURE

A matrix A of order n is called *doubly stochastic* provided that its elements are nonnegative real numbers and its row and column sums are each equal to 1 (for a general reference on doubly stochastic matrices, see, for example [17]). The following classical theorem of G. Birkhoff is an easy consequence of the König-Egerváry theorem.

Theorem 4.1. (G. Birkhoff) *Let A be a doubly stochastic matrix of order n. Then*

$$A = c_1P_1 + c_2P_2 + \dots + c_tP_t , \tag{4.1}$$

where the P_i are permutation matrices and the c_j are positive reals such that

$$c_1 + c_2 + \dots + c_t = 1 . \tag{4.2}$$

The preceding theorem implies that every doubly stochastic matrix A has per$(A) > 0$. But for A a doubly stochastic matrix of order n the determination of the minimal value of per(A) remains a difficult unsolved problem. A celebrated conjecture of van der Waerden asserts that every doubly stochastic matrix A of order n satisfies

$$\text{per}(A) \geqq \frac{n!}{n^n} . \tag{4.3}$$

Equality holds in (4.3) for $A = n^{-1}J$, where J is the matrix of 1's of order n. This may actually be the only case of equality. The validity of van der Waerden's conjecture is known only for the first few values of n.

Recently a great many authors have studied one aspect or another of the van der Waerden conjecture. It is clear that (4.3) implies that a doubly stochastic matrix $A = [a_{ij}]$ of order n must have a permutation σ such that

$$\prod_{j=1}^{n} a_{j\sigma(j)} \geqq \frac{1}{n^n} . \tag{4.4}$$

This consequence of the van der Waerden conjecture is in fact the following theorem of Marcus and Minc [12].

Theorem 4.2. *Let $A = [a_{ij}]$ be a doubly stochastic matrix of order n. Then A has a permutation σ such that*

$$\prod_{j=1}^{n} a_{j\sigma(j)} \geqq \frac{1}{n^n} . \tag{4.5}$$

Also, the following theorem of Marcus and Newman [14] tells us that the van der Waerden conjecture is valid, but under certain additional stipulations on the matrix A.

Theorem 4.3. *Let A be a symmetric positive semidefinite doubly stochastic matrix of order n. Then*

$$\text{per}(A) \geqq \frac{n!}{n^n} , \tag{4.6}$$

and equality holds in (4.6) *if and only if $A = n^{-1}J$.*

The preceding results tend to confirm the validity of the van der Waerden conjecture. But certain conjectured inequalities more general than (4.3) are now known to be false. It has been conjectured that if A and B are doubly stochastic matrices of order n then

$$\text{per}(AB) \leqq \{\text{per}(A), \text{per}(B)\} . \tag{4.7}$$

Note that (4.7) with $B = n^{-1}J$ implies (4.3). The following counter-example of (4.7) is by W. B. Jurkat.

$$A = \frac{1}{24}\begin{bmatrix} 11 & 5 & 8 \\ 13 & 11 & 0 \\ 0 & 8 & 16 \end{bmatrix} \qquad B = \frac{1}{2}\begin{bmatrix} 1 & 1 & 0 \\ 1 & 1 & 0 \\ 0 & 0 & 2 \end{bmatrix}$$

$$\operatorname{per}(A) = \frac{3808}{13824} < \operatorname{per}(AB) = \frac{3840}{13824}.$$

Also, it has been conjectured that if A is a doubly stochastic matrix of order n then

$$\operatorname{per}(AA^T) \leqq \operatorname{per}(A), \tag{4.8}$$

where A^T denotes the transpose of A. Note that (4.8) implies (4.3) because of Theorem 4.3. The following counter-example of (4.8) is by Morris Newman.

$$A = \frac{1}{2}\begin{bmatrix} 1 & 1 & 0 & 0 \\ 0 & 1 & 1 & 0 \\ 0 & 0 & 1 & 1 \\ 1 & 0 & 0 & 1 \end{bmatrix} \qquad AA^T = \frac{1}{4}\begin{bmatrix} 2 & 1 & 0 & 1 \\ 1 & 2 & 1 & 0 \\ 0 & 1 & 2 & 1 \\ 1 & 0 & 1 & 2 \end{bmatrix}$$

$$\operatorname{per}(AA^T) = \frac{9}{64} > \operatorname{per}(A) = \frac{8}{64}.$$

These counter-examples are disturbing because they cannot help but make us somewhat less confident of the actual validity of the van der Waerden conjecture itself.

Let $A = [a_{ij}]$ be a matrix of size m by n with nonnegative real elements. Let the sum of row i of A be r_i ($i = 1, 2, \ldots, m$) and let the sum of column j of A be s_j ($j = 1, 2, \ldots, n$). We call $R = (r_1, r_2, \ldots, r_m)$ the *row sum vector* and $S = (s_1, s_2, \ldots, s_n)$ the *column sum vector* of A. It follows at once that the components of R and S must be compatible in the sense that

$$r_1 + r_2 + \ldots + r_m = s_1 + s_2 + \ldots + s_n. \tag{4.9}$$

Now let R and S denote nonnegative vectors whose components satisfy (4.9). Then we denote by

$$\mathfrak{A} = \mathfrak{A}(R, S) \tag{4.10}$$

the class of all nonnegative matrices of size m by n with row sum vector $\boldsymbol{R}$ and column sum vector $\boldsymbol{S}$.

The extremal matrices in the class $\mathfrak{A}(\boldsymbol{R}, \boldsymbol{S})$ regarded as a convex set have been investigated in some detail recently [1; 10]. In what follows we merely point out that the class of doubly stochastic matrices with $\boldsymbol{R} = \boldsymbol{S} = (1, 1, \ldots, 1)$ is by no means the only class in which every matrix has a positive permanent. Actually there are available simple necessary and sufficient conditions on the components of $\boldsymbol{R}$ and $\boldsymbol{S}$ in order for this requirement to hold, and this fact enables us to place van der Waerden's problem on minimal permanents within a much broader setting. But the evaluation of such a minimal positive permanent over a class with $m = n$ has not been carried out successfully for a single specified $\boldsymbol{R}$ and $\boldsymbol{S}$ and general n. For example, let $\mathfrak{A}(\boldsymbol{R}, \boldsymbol{S})$ be the class of all nonnegative matrices of order n with

$$\boldsymbol{R} = (1, n+1, n+1, \ldots, n+1), \qquad (4.11)$$

$$\boldsymbol{S} = (n+1, n+1, \ldots, n+1, 1).$$

Each matrix in this class has a positive permanent. The integral matrix A with 1's on the main diagonal, n's in positions $(2, 1), (3, 2), \ldots, (n, n-1)$, and 0's elsewhere belongs to this class and has $\operatorname{per}(A) = 1$. But is this the minimal permanent over the class?

The van der Waerden conjecture has inspired a variety of inequalities and conjectures on permanents and we refer the reader to the review article by Marcus and Minc [13] for additional material. We conclude this section with the statement of a theorem by Jurkat and Ryser [10] concerning an upper estimate for the permanent of a matrix of order n in the class $\mathfrak{A}(\boldsymbol{R}, \boldsymbol{S})$.

Theorem 4.4. *Let A be a matrix of order n in the class $\mathfrak{A}(\boldsymbol{R}, \boldsymbol{S})$ and let the components of $\boldsymbol{R}$ and $\boldsymbol{S}$ be arranged in increasing order. Then*

$$\operatorname{per}(A) \leq \prod_{j=1}^{n} \min(r_j, s_j). \qquad (4.12)$$

The case of equality in (4.12) poses some difficulties. But the inequality is sharp in the sense that the class always contains extremal matrices for which equality is attained.

5. PERMANENTS OF (0, 1)-MATRICES.

A number of years ago M. Hall [6] established the following quantitative refinement of the P. Hall theorem.

Theorem 5.1. *Let the subsets $S_1, S_2, \ldots, S_m$ of S satisfy the necessary condition for the existence of an SDR and let each of these subsets contain at least t elements. Then the subsets $S_1, S_2, \ldots, S_m$ have at least the following number of SDR's:*

$$\prod_{j=1}^{\min(t,m)} (t+1-j) = \begin{cases} t! & (t \leqq m), \\ \dfrac{t!}{(t-m)!} & (t > m). \end{cases} \tag{5.1}$$

Only recently R. Rado [18; 19] communicated the following extension of Theorem 5.1.

Theorem 5.2. *Let the subsets $S_1, S_2, \ldots, S_m$ of S satisfy the necessary condition for the existence of an SDR. Furthermore, let each S_j be an r_j-subset of S and let the subsets be ordered so that $r_1 = t \leqq r_2 \leqq \ldots \leqq r_m$. Then the subsets $S_1, S_2, \ldots, S_m$ have at least the following number of SDR's:*

$$\prod_{j=1}^{\min(t,m)} (r_j + 1 - j). \tag{5.2}$$

Let A be the incidence matrix for the subsets $S_1, S_2, \ldots, S_m$ of an n-set S. Then it is clear that the preceding theorems yield lower estimates for $\operatorname{per}(A)$ provided that A is of term rank m. In what follows we describe some upper estimates for the permanent of a (0, 1)-matrix A of order n with row sum vector $R = (r_1, r_2, \ldots, r_n)$. Minc [15] has shown that

$$\operatorname{per}(A) \leqq \prod_{j=1}^{n} \left(\frac{r_j + 1}{2}\right) \tag{5.3}$$

and he has conjectured that

$$\operatorname{per}(A) \leqq \prod_{j=1}^{n} (r_j!)^{1/r_j}. \tag{5.4}$$

Jurkat and Ryser [9] using the techniques of matrix factorization described in Section 3 have shown that

$$\operatorname{per}(A) \leqq \prod_{j=1}^{n} (r_j!)^{1/n} \left(\frac{r_j + 1}{2}\right)^{(n - r_j)/n} \tag{5.5}$$

Notice that the general factor $r!^{1/n}((r+1)/2)^{(n-r)/n}$ in (5.5) is a weighted geometric mean of $(r!)^{1/r}$ and $(r+1)/2$, the corresponding factors in (5.4) and (5.3). Furthermore, one may prove that [9]

$$(5.6)\qquad \prod_{j=1}^{n} \max(r_j + 1 - j, 0) \leqq \operatorname{per}(A) \leqq \prod_{j=1}^{n} \min(n + 1 - j, r_j)$$

and (5.3) may be refined to yield [16]

$$(5.7)\qquad \operatorname{per}(A) \leqq \prod_{j=1}^{n} \left(\frac{r_j + \sqrt{2}}{1 + \sqrt{2}}\right).$$

We conclude with a discussion of an application of the preceding inequalities to Latin rectangles. We recall that a *Latin rectangle* based on the n-set S is an r by s rectangular array

$$(5.8)\qquad A = [a_{ij}] \quad (i = 1, 2, \ldots, r;\ j = 1, 2, \ldots, s)$$

with the requirement that each row of (5.8) is an s-permutation of elements of S and each column of (5.8) is an r-permutation of elements of S. In particular, if $r = s = n$ we call (5.8) a *Latin square* of order n. We let $L(r, s)$ denote the number of r by s Latin rectangles based on the n-set S.

Now let $r < n$ and let A denote an r by n Latin rectangle based on the n-set S. Let S_j denote the set of all elements of S that do not appear in column j of A and let A' denote the incidence matrix for the subsets $S_1, S_2, \ldots, S_n$ of S. Then $\operatorname{per}(A')$ denotes the number of ways in which A may be extended to an $(r+1)$ by n Latin rectangle based on the n-set S. It is not difficult to verify that A' is a $(0, 1)$-matrix of order n with exactly $k = n - r$ 1's in each row and column. It follows that $k^{-1}A'$ is a doubly stochastic matrix and hence A' is of term rank n. Thus by Theorem 5.1 (or Theorem 5.2) we have

$$(5.9)\qquad \operatorname{per}(A') \geqq k!$$

and by repeated applications of (5.9) we obtain the following inequality of M. Hall [5; 6] concerning Latin rectangles:

$$(5.10)\qquad L(r, n) \geqq \prod_{j=1}^{r} (n + 1 - j)!\,.$$

In particular, (5.10) tells us that

(5.11)
$$L(n, n) \geqq \prod_{j=1}^{n} (n+1-j)! .$$

We may apply the van der Waerden conjecture and the Minc conjecture to the matrix A' of (5.9) and thereby obtain

(5.12)
$$\operatorname{per}(A') \geqq k^n \frac{n!}{n^n} ,$$

(5.13)
$$\operatorname{per}(A') \leqq k!^{n/k} .$$

But unfortunately the validity of both of these inequalities is also undetermined. Using (5.12) and (5.13) we may formulate the following conjectures concerning $L(r, n)$:

(5.14)
$$L(r, n) \geqq \left(\frac{n!}{n^n}\right)^r \prod_{j=1}^{r} (n+1-j)^n ,$$

(5.15)
$$L(r, n) \leqq \prod_{j=1}^{r} (n+1-j)!^{n/(n+1-j)} .$$

References

1. Brualdi, R. A. (1968). "Convex Sets of Nonnegative Matrices," *Canad. J. Math.*, **20**, 144–157.
2. Brualdi, R. A. and Newman, M. (1966). "Proof of a Permanental Inequality," *Quart. J. Math.*, **17**, 234–238.
3. Ford, L. R. Jr. and Fulkerson, D. R. (1962). *Flows in networks*, Princeton Univ. Press.
4. Gibson, P. M. (1966). "A Short Proof of an Inequality for the Permanent Function," *Proc. Amer. Math. Soc.*, **17**, 535-536.
5. Hall, M. Jr. (1945). "An Existence Theorem for Latin Squares," *Bull. Amer. Math. Soc.*, **51**, 387-388.
6. Hall, M. Jr. (1948). "Distinct Representatives of Subsets," *Bull. Amer. Math. Soc.* **54**, 922-926.
7. Hall, M. Jr. (1958). *Some aspects of analysis and probability*, John Wiley and Sons, New York, pp. 35–104.
8. Hall, M. Jr. (1967). *Combinatorial theory*, Blaisdell, Waltham, Massachusetts.
9. Jurkat, W. B. and Ryser, H. J. (1966). "Matrix Factorizations of Determinants and Permanents," *J. Algebra*, **3**, 1-27.
10. Jurkat, W. B. and Ryser, H. J. (1967). "Term Ranks and Permanents of Nonnegative Matrices," *J. Algebra*, **5**, 342–357.
11. Kaplansky, I. (1943). "Solution of the 'Problème des Ménages'," *Bull. Amer. Math. Soc.*, **49**, 784–785.

12. Marcus, M. and Minc, H. (1962). "Some Results on Doubly Stochastic Matrices," *Proc. Amer. Math. Soc.*, **13**, 571-579.
13. Marcus, M. and Minc, H. (1965). "Permanents," *Amer. Math. Monthly*, **72**, 577-591.
14. Marcus, M. and Newman, M. (1962). "Inequalities for the Permanent Function," *Ann. Math.*, **75**, 47-62.
15. Minc, H. (1963). "Upper Bounds for Permanents of (0, 1)-Matrices," *Bull. Amer. Math. Soc.*, **69**, 789-791.
16. Minc, H. (1967). "An Inequality for Permanents of (0, 1)-Matrices," *J. Combinatorial Theory*, **2**, 321-326.
17. Mirsky, L. (1963). "Results and Problems in the Theory of Doubly Stochastic Matrices," *Z. Wahrscheinlichkeitstheorie*, **1**, 319-334.
18. Mirsky, L. and Perfect, H. (1966). "Systems of Representatives," *J. Math. Anal. Appl.*, **15**, 520-568.
19. Rado, R. "On the Number of Systems of Distinct Representatives of Sets," (To be published).
20. Riordan, J. (1958). *An introduction to combinatorial analysis*, John Wiley and Sons, New York.
21. Rota, G.-C. (1964). "On the Foundations of Combinatorial Theory I, Theory of Möbius Functions," *Z. Wahrscheinlichkeitstheorie*, **2**, 340-368.
22. Ryser, H. J. (1963). *Combinatorial mathematics*, Carus Math. Monograph No. 14, Math. Assoc. Amer., John Wiley and Sons, New York.
23. Wilf, H. S. "Permanental Methods in the 'Problème des Ménages'," (To be published).

Discussion on Professor Ryser's Paper

PROFESSOR M. IRI: As a non-mathematician, but merely a mathematical-minded engineer who is not a specialist in Combinatorics, I do not consider myself to be well qualified for leading the discussion on Professor Ryser's highly valuable and instructive paper. But I dare take this opportunity to expound what I usually have in mind about the subjects dealt with by Professor Ryser.

The first point that I am concerned with is the connection of Combinatorics with the problems in other fields. I take it for granted that Combinatorics has its own esthetic *raison d'être*, but the non-specialists would be much profited if the relations of combinatorial problems to the problems familiar to them were clarified wherever possible.

In this respect, I think, the network-flow formulation of

the SDR problems is a splendid gift, as was mentioned by Professor Ryser. But, as far as I know from Ford and Fulkerson's book on network flows and Berge's book on graph theory, the network-flow techniques are ordinarily used only for proving the König-Egerváry theorm or one of the Hall-type theorems. Other theorems are made to follow from the basic theorem thus proved. Couldn't they be proved more directly and intuitively by the network-flow technique? For example, I will take up the boy-friend girl-friend problem (as is treated in Berge's book) and the Birkhoff-von Neumann theorem (Theorem 4.1 of Ryser's paper). The boy-friend girl-friend theorem states that, if there are n boys and n girls and if each boy has m girl friends and each girl has m boy friends, then it is possible to establish n one-to-one partnerships such that each boy is coupled with one of his m girl friends and each girl with one of her m boy friends. The simplest way to prove this theorem would not be to derive it from the Hall theorem, but to show the existence of the polygamy-polyandry relation such that every boy is coupled with all of his girl friends, with each by $1/m$, and every girl with all of her boy friends, each by $1/m$. The existence of such a relation is trivially obvious. But, the existence of a monogamy-monoandry relation, which we require, is a direct consequence of that of a polygamy-polyandry relation in view of the fundamental theorem concerning the existence of integer solutions in a maximum-flow problem with integer capacities.

The situation is quite similar for the decomposition of a doubly stochastic matrix $\boldsymbol{A}$ into a weighted mean of permutation matrices. The central point in question is the existence of a permutation matrix $\boldsymbol{P}$ such that

$$\text{if} \quad a_{ij} = 0 \quad \text{then} \quad p_{ij} = 0 .$$

Apart from this condition and the integrity of elements, the conditions to be satisfied by $\boldsymbol{P}$ are

$$p_{ij} \geqq 0, \quad \sum_j p_{ij} = 1, \quad \sum_i p_{ij} = 1.$$

But all these conditions except for the integrity of elements are already satisfied by $\boldsymbol{A}$ itself, as is evident, and so we may put $\boldsymbol{P}$ equal to $\boldsymbol{A}$. Therefore, from this point of view, the existence of a $\boldsymbol{P}$ with integer elements is a direct consequence of the double stochasticness of $\boldsymbol{A}$.

I was much impressed by Professor Ryser's novel technique

for expressing the determinant or permanent of a square matrix A of order n as the product of n rectangular matrices, each consisting of elements of one row or one column of A, and I perfectly agree with Professor Ryser in that, in many instances, such representations are far more revealing than those in ordinary textbooks on matrices and determinants. I can say so based on my own experiences in developing the topological theory of electric networks using tensor notations for determinants.

Finally, I would like to point out the difference between determinants and permanents, from the standpoint of numerical calculation. The amount of computational labor required in calculating the determinant of a matrix of order n is proportional to n^3 according to the Gauss elimination method, as is well known. It would be prohibitively large if we would calculate it according to the very definition of determinant or to the Laplace expansion or even to Jurkat and Ryser's novel expansion. The amount of computations required in obtaining an SDR for n members and m subsets is proportional to nm^3 according to the standard method. The same for determining the term-rank of an $n \times m$ matrix. How about the permanent? When the permanents find useful practical applications, we, practical engineers, shall be eager to demand an efficient way of calculating them.

CHAPTER 5

Abel Identities and Inverse Relations

JOHN RIORDAN, *Bell Telephone Laboratories*

1. INTRODUCTION

N. H. Abel's celebrated generalization of the binomial formula (given in his *Oevres Complètes*, Christiania, C. Groendahl, 1839), in one form, that was preferred by A. Hurwitz [2], is as follows

$$(1) \quad x^{-1}(x+y+na)^n = \sum_{k=0}^{n} \binom{n}{k} (x+ka)^{k-1}(y+(n-k)a)^{n-k} .$$

Replacing x by ax, y by ay, this is the same as

$$(1a) \quad x^{-1}(x+y+n)^n = \sum_{k=0}^{n} \binom{n}{k} (x+k)^{k-1}(y+n-k)^{n-k} ;$$

so the parameter a is disposable.

To prove (1a), and to derive additional results, I consider the class of Abel sums defined by

$$(2) \quad A_n(x, y; p, q) = \sum_{k=0}^{n} \binom{n}{k} (x+k)^{k+p}(y+n-k)^{n-k+q} .$$

In this notation Abel's result is $A_n(x, y; -1, 0) = x^{-1}(x+y+n)^n$.

The symmetry and recurrence relations enjoyed by (2) lead to a simple proof of (1a) and the evaluation of a number of its companions. Moreover they are readily extended to the corresponding multinomial sums defined by

$$(3) \qquad A_n(x_1, \ldots, x_m; p_1, \ldots, p_m) = \sum [n; k_1, \ldots, k_m] \prod_{j=1}^{m} (x_j + k_j)^{k_j + p_j}$$

with $[n; k_1, \ldots, k_m] = n!/k_1! \ldots k_m!$, $k_1 + \ldots + k_m = n$, the multinomial coefficient, and summation over all nonnegative integral values of the k_i, $i = 1(1)m$, subject to $k_1 + \ldots + k_m = n$. The three evaluations of such multinomial sums due to A. Hurwitz [2] are not as well known as they should be.

Finally, a number of inverse relations, for the binomial case, which are the natural companions of the Abel identities, are developed.

2. SYMMETRY AND RECURRENCE RELATIONS FOR ABEL BINOMIAL SUMS

It is clear from (2) that

$$(4) \qquad A(x, y; p, q) = A(y, x; q, p)\,.$$

Next, by the basic recurrence for binomial coefficients,

$$(5) \qquad A_n(x, y; p, q) = A_{n-1}(x+1, y; p+1, q) + A_{n-1}(x, y+1; p, q+1)\,.$$

Also, writing $(x+k)^{k+p}$ as $(x+k)(x+k)^{k+p-1}$ leads to

$$(6) \qquad A_n(x, y; p, q) = xA_n(x, y; p-1, q) + nA_{n-1}(x+1, y; p, q)$$

or

$$(6a) \qquad xA_n(x, y; p-1, q) = A_n(x, y; p, q) - nA_{n-1}(x+1, y; p, q)\,.$$

The two forms on iteration lead to different kinds of sums.

The iteration of (6) is immediate; the result is

$$(7)\qquad A_n(x, y; p, q) = \sum_{k=0}^{n} \binom{n}{k} k!(x+k)A_{n-k}(x+k, y; p-1, q).$$

That of (6a) is more intricate. In the first place

$$\begin{aligned} & x^2(x+1)A_n(x, y; p-2, q) \\ &= x(x+1)A_n(x, y; p-1, q) - nx(x+1)A_{n-1}(x+1, y; p-1, q) \\ &= (x+1)A_n(x, y; p, q) - n(2x+1)A_{n-1}(x+1, y; p, q) \\ &\quad + n(n-1)xA_{n-2}(x+2, y; p, q). \end{aligned}$$

This and (6a) suggest writing

$$(8)\qquad \begin{aligned} & x^k(x+1)^{k-1}\dots(x+k-1)A_n(x, y; p-k, q) \\ &= \sum_{j=0}^{k} (-1)^j (n)_j a_{k,j}(x) A_{n-j}(x+j, y; p, q) \end{aligned}$$

with $(n)_j = n(n-1)\dots(n-j+1)$, the falling factorial, and the a_{kj} undetermined polynomials in x. Use of (6a) in (8) shows that

$$\begin{aligned} a_{k0}(x) &= (x+1)\dots(x+k-1)a_{k-1,0}(x) \\ &= (x+1)^{k-1}(x+2)^{k-2}\dots(x+k-1), \end{aligned}$$

$$a_{kk}(x) = x^{k-1}a_{k-1,k-1}(x) = x^{k-1}(x+1)^{k-2}\dots(x+k-2),$$

and

$$a_{kj}(x) = (x+1)\dots(x+k-1)a_{k-1,j}(x) + x^{k-1}a_{k-1,j-1}(x), \qquad j = 1(1)k-1.$$

Writing $A(x; k) = x^k(x+1)^{k-1}\dots(x+k-1)$, it turns out that for $k > j$,

$$(9)\qquad \begin{aligned} a_{kj}(x) &= \frac{a_{jj}(x)}{j!} A(x+j+1; k-j-1) \\ &\quad \cdot \sum_{i=0}^{j} (-1)^i \binom{j}{i} [(x+j)_{j-i}(x+i-1)_i]^{k-j+1}. \end{aligned}$$

Of course $(x)_j$ is the falling factorial and $(x)_0 = 1$.

3. THE ABEL BINOMIAL IDENTITY AND COMPANIONS

Combining (5) and (6) gives

$$(10) \qquad A_n(x, y; p, q) = xA_{n-1}(x, y+1; p-1, q+1) + (x+n)A_{n-1}(x+1, y; p, q)\,.$$

The instance of this for $p = 0$, $q = -1$ is

$$A_n(x, y; 0, -1) = xA_{n-1}(x, y+1; -1, 0) + (x+n)A_{n-1}(x+1, y; 0, -1)$$

or, using the symmetry relation and interchanging x and y,

$$(11) \qquad A_n(x, y; -1, 0) = yA_{n-1}(y, x+1; -1, 0) + (y+n)A_{n-1}(x, y+1; -1, 0)\,,$$

a recurrence relation for Abel's sum (1a). Directly from its definition

$$A_0(x, y; -1, 0) = x^{-1}\,, \qquad A_1(x, y; -1, 0) = x^{-1}(x+y+1)\,,$$

and if $A_k = x^{-1}(x+y+k)^k$, $k = 0(1)n-1$, it follows from (11) that

$$\begin{aligned} A_n(x, y; -1, 0) &= yy^{-1}(x+y+n)^{n-1} + (y+n)x^{-1}(x+y+n)^{n-1} \\ &= x^{-1}(x+y+n)^n \end{aligned}$$

as in (1a).

The first companion $A_n(x, y; -1, -1)$ follows from (5) and symmetry; thus

$$\begin{aligned} (12) \qquad & A_n(x, y; -1, -1) \\ &= A_{n-1}(x+1, y; 0, -1) + A_{n-1}(x, y+1; -1, 0) \\ &= A_{n-1}(y, x+1; -1, 0) + A_{n-1}(x, y+1; -1, 0) \\ &= (x^{-1} + y^{-1})(x+y+n)^{n-1} \\ &= (xy)^{-1}(x+y)(x+y+n)^{n-1}\,. \end{aligned}$$

Also, by (6a)

$$xA_n(x, y; -2, 0) = A_n(x, y; -1, 0) - nA_{n-1}(x+1, y; -1, 0)$$

$$= x^{-1}(x+y+n)^n - n(x+1)^{-1}(x+y+n)^{n-1},$$

or

$$\text{(13)} \qquad x^2(x+1)A_n(x, y; -2, 0) \\ = (x+1)(x+y+n)^n - nx(x+y+n)^{n-1}.$$

Again, by the first iteration of (6a)

$$x^2(x+1)A_n(x, y; -3, 0) \\ = (x+1)A_n(x, y; -1, 0) - n(2x+1)A_{n-1}(x+1, y; -1, 0) \\ + n(n-1)xA_{n-2}(x+2, y; -1, 0)$$

or

$$\text{(14)} \qquad x^3(x+1)^2(x+2)A_n(x, y; -3, 0) \\ = (x+1)^2(x+2)(x+y+n)^n \\ - nx(x+2)(2x+1)(x+y+n)^{n-1} \\ + n(n-1)x^2(x+1)(x+y+n)^{n-2}.$$

Next, by (7)

$$\text{(15)} \qquad A_n(x, y; 0, 0) = \sum_{k=0}^{n} \binom{n}{k} k!(x+k)A_{n-k}(x+k, y; -1, 0) \\ = \sum_{k=0}^{n} \binom{n}{k} k!(x+y+n)^{n-k} \\ = (x+y+n+\alpha)^n, \qquad \alpha^k \equiv \alpha_k = k!.$$

This is usually called Cauchy's formula. Continuing the use of (7),

$$\text{(16)} \qquad A_n(x, y; 1, 0) = \sum_{k=0}^{n} \binom{n}{k} k!(x+k)A_{n-k}(x+k, y; 0, 0) \\ = \sum_{k=0}^{n} \binom{n}{k} k!(x+k)(x+y+n+\alpha)^{n-k} \\ = (x+y+n+\alpha+\beta(x))^n, \\ \beta^k(x) \equiv \beta_k(x) = k!(x+k).$$

Noting that

$$\exp t\alpha = \sum_{n=0}^{\infty} t^n\alpha_n/n! = (1-t)^{-1},$$

$$\exp t\beta(x) = \sum_{n=0}^{\infty} t^n \beta_n(x)/n! = \sum_{n=0}^{\infty} t^n(x+n)$$
$$= (1-t)^{-2}(t + x(1-t)),$$

it follows that

$$\exp t(\alpha + \beta(x)) = (1-t)^{-3}(t + x(1-t))$$
$$= \sum_{n=0}^{\infty} t^n \left[\binom{n+1}{2} + x(n+1)\right],$$

and (16) may be rewritten as

(16a) $$A_n(x, y; 1, 0)$$
$$= \sum_{k=0}^{n} \binom{n}{k} k! \left[\binom{k+1}{2} + x(k+1)\right](x+y+n)^{n-k}.$$

Again, by (7)

(17) $$A_n(x, y; 1, -1) = \sum \binom{n}{k} \beta_k(x) A_{n-k}(x+k, y; 0, -1)$$
$$= \sum \binom{n}{k} \beta_k(x) y^{-1}(x+y+n)^{n-k}$$
$$= y^{-1}(x+y+n+\beta(x))^n,$$
$$\beta^k(x) \equiv \beta_k(x) = k!(x+k)$$

and

(18) $$A_n(x, y; 1, 1) = \sum \binom{n}{k} \beta_k(x) A_{n-k}(x+k, y; 0, 1)$$
$$= \sum \binom{n}{k} \beta_k(x)(x+y+n+\alpha+\beta(y))^{n-k}$$
$$= (x+y+n+\alpha+\beta(x)+\beta(y))^n.$$

An alternate form of (18) follows from the generating function

$$\exp t(\alpha + \beta(x) + \beta(y)) = (1-t)^{-5}[t + x(1-t)][t + y(1-t)].$$

The case of $A_n(x, y; 2, 0)$ is worth noting because of an added complication. First, by (7)

$$A_n(x, y; 2, 0) = \sum_{k=0}^{n} \binom{n}{k} \beta_k(x) A_{n-k}(x+k, y; 1, 0)$$

and, using (16)

$$\begin{aligned}
&A_{n-k}(x+k, y; 1, 0)\\
&= (x+y+n+\alpha+\beta(x+k))^{n-k}\\
&= \sum_{j=0}^{n-k}\binom{n-k}{j}(x+y+n+\alpha)^{n-k-j}j!(x+k+j)\\
&= (x+y+n+\alpha+\beta(x))^{n-k}\\
&\quad + k(x+y+n+\alpha+\alpha)^{n-k}\,.
\end{aligned}$$

Hence

$$\begin{aligned}
(19)\qquad A_n(x, y; 2, 0) &= (x+y+n+\alpha+\beta(x; 2))^n\\
&\quad + (x+y+n+\alpha(2)+\gamma(x))^n
\end{aligned}$$

where $\exp t\alpha(m) = (\exp t\alpha)^m$, $\exp t\beta(x; m) = (\exp t\beta(x))^m$, and $\gamma^k(x) \equiv \gamma_k(x) = k\beta_k(x)$. Since

$$\exp t\gamma(x) = t\frac{d}{dt}\exp t\beta(x) = (1-t)^{-3}(2t^2 + (1+x)(1-t))\,,$$

it follows that

$$\begin{aligned}
\exp t(\alpha(2)+\gamma(x)) &= (1-t)^{-5}(2t^2 + (1+x)(t-t^2))\\
&= \sum_{n=0}^{\infty}\left[2\binom{n+2}{4} + (1+x)\binom{n+2}{3}\right]t^n\,,
\end{aligned}$$

while

$$\begin{aligned}
\exp t(\alpha+\beta(x; 2)) &= (1-t)^{-5}[t + x(1-t)]^2\\
&= \sum_{n=0}^{\infty}\left[\binom{n+2}{4} + 2x\binom{n+2}{3} + x^2\binom{n+2}{2}\right]t^n\,.
\end{aligned}$$

Thus an alternate form for (19) is

$$\begin{aligned}
(19a)\qquad &A_n(x, y; 2, 0)\\
&= \sum_{k=0}^{n}\binom{n}{k}(x+y+n)^{n-k}\\
&\quad \times k!\left[3\binom{k+2}{4} + (1+3x)\binom{k+2}{3} + x^2\binom{k+2}{2}\right]\\
&= (x+y+n+\delta(x))^n\,,
\end{aligned}$$

with $\delta^k(x) \equiv \delta_k(x) = k!\left[3\binom{k+2}{4} + (1+3x)\binom{k+2}{3} + x^2\binom{k+2}{2}\right]$.

Note that

$$\exp t\delta(x) = (1-t)^{-5}[3t^2 + (1+3x)t(1-t) + x^2(1-t)^2] .$$

Table 1 summarizes these and a few additional results.

Table 1. ABELIAN BINOMIAL IDENTITIES

$$A_n(x, y; p, q) = \sum_{k=0}^{n} \binom{n}{k} (x+k)^{k+p}(y+n-k)^{n-k+q}$$

p	q	$A_n(x, y; p, q)$
-3	0	$x^{-3}(x+1)^{-2}(x+2)^{-1}[(x+1)^2(x+2)(x+y+n)^n$ $- nx(x+2)(2x+1)(x+y+n)^{n-1} + (n-1)x^2(x+1)(x+y+n)^{n-2}]$
-2	0	$x^{-2}(x+1)^{-1}[(x+1)(x+y+n)^n - nx(x+y+n)^{n-1}]$
-1	0	$x^{-1}(x+y+n)^n$
0	0	$(x+y+n+\alpha)^n$
1	0	$(x+y+n+\alpha+\beta(x))^n$
2	0	$(x+y+n+\alpha+\beta(x;2))^n + (x+y+n+\alpha(2)+\gamma(x))^n$
-1	-1	$(xy)^{-1}(x+y)(x+y+n)^{n-1}$
-1	1	$x^{-1}(x+y+n+\beta(y))^n$
-1	2	$x^{-1}(x+y+n+\beta(y;2))^n + x^{-1}(x+y+n+\alpha+\gamma(y))^n$
1	1	$(x+y+n+\alpha+\beta(x)+\beta(y))^n$
1	2	$(x+y+n+\alpha+\beta(x)+\beta(y;2))^n + (x+y+n+\alpha(2)+\gamma(y))^n$
2	2	$(x+y+n+\alpha+\beta(x;2)+\beta(y;2))^n$ $+ (x+y+n+\alpha(2)+\gamma(x)+\beta(y;2))^n$ $+ (x+y+n+\alpha(2)+\beta(x;2)+\gamma(y))^n$ $+ (x+y+n+\alpha(3)+\gamma(x)+\gamma(y))^n$

Notation: $\alpha^k \equiv \alpha_k = k!$

$[\alpha(j)]^k \equiv \alpha_k(j) = (\alpha + \ldots + \alpha)^k (j \text{ terms}) = \binom{k+j-1}{k} k!$

$\beta^k(x) \equiv \beta_k(x) = k!(x+k)$

$[\beta(x;j)]^k \equiv \beta_k(x;j) = (\beta(x) + \ldots + \beta(x))^k (j \text{ terms})$

$\gamma^k(x) \equiv \gamma_k(x) = k!k(x+k)$

4. MULTINOMIAL IDENTITIES

The symmetry properties of the multinomial sums defined by (3) are apparent in the definition. The basic recurrence for multinomial coefficients, namely

$$\begin{aligned}[n; k_1, \ldots, k_m] = [n-1; k_1 - 1, k_2 \ldots, k_m] + \ldots \\ + [n-1; k_1, \ldots, k_{j-1}, k_j - 1, k_{j+1}, \ldots, k_m] + \ldots \\ + [n-1; k_1, \ldots, k_{m-1}, k_m - 1]\end{aligned}$$

implies the extension of (5), that is

$$\begin{aligned}(20)\quad & A_n(x_1, \ldots, x_m; p_1, \ldots, p_m) \\ &= A_{n-1}(x_1 + 1, x_2, \ldots, x_m; p_1 + 1, p_2, \ldots, p_m) \\ &\quad + A_{n-1}(x_1, x_2 + 1, \ldots, x_p; p_1, p_2 + 1, \ldots, p_m) + \ldots \\ &\quad + A_{n-1}(x_1, \ldots, x_m + 1; p_1, \ldots, p_m + 1).\end{aligned}$$

The extension of (6) is almost immediate and reads

$$\begin{aligned}(21)\quad & A_n(x_1, \ldots, x_m; p_1, \ldots, p_m) \\ &= x_1 A_n(x_1, \ldots, x_m; p_1 - 1, p_2, \ldots, p_m) \\ &\quad + n A_{n-1}(x_1 + 1, x_2, \ldots, x_m; p_1, \ldots, p_m).\end{aligned}$$

This implies the extension of (7), which is

$$\begin{aligned}(22)\quad & A_n(x_1, \ldots, x_m; p_1, \ldots, p_m) \\ &= \sum_{k=0}^{n} \binom{n}{k} k!(x_1 + k) \\ &\quad \cdot A_{n-k}(x_1 + k, x_2, \ldots, x_m; p_1 - 1, p_2, \ldots, p_m).\end{aligned}$$

The extension of (8) in the same way is

$$\begin{aligned}(23)\quad & x_1^k(x_1 + 1)^{k-1} \ldots (x_1 + k - 1) \\ &\quad \cdot A_n(x_1, \ldots, x_m; p_1 - k, p_2, \ldots, p_m) \\ &= \sum_{j=0}^{k} (-1)^j (n)_j a_{k,j}(x_1) A_{n-j}(x_1 + j, x_2, \ldots, x_m; p_1, \ldots, p_m).\end{aligned}$$

The first of the Hurwitz extensions, mentioned in the Introduction, is that of (1a). A derivation by iteration, quite different from the Hurwitz derivation, is as follows. First

$$\begin{aligned}& y^{-1} A_n(x, y + z; -1, 0) \\ &= (xy)^{-1}(x + y + z + n)^n \\ &= \sum \binom{n}{k} (x + k)^{k-1} (y + z + n - k)^{n-k} y^{-1} \\ &= \sum \binom{n}{k} (x + k)^{k-1} \sum \binom{n-k}{j} (y + j)^{j-1} (z + n - k - j)^{n-k-j} \\ &= A_n(x, y, z; -1, -1, 0),\end{aligned}$$

since

$$\binom{n}{k}\binom{n-k}{j} = [n; k, j, n-k-j]\,.$$

Repetitions of the iteration lead to Hurwitz's result

$$(24)\quad A_n(x_1, \ldots, x_m; -1, \ldots, -1, 0) = (x_1 \ldots x_m)^{-1} x_m (x+n)^n$$

with $x = x_1 + \ldots + x_m$.

The second Hurwitz extension is that of $A_n(x, y; 0, 0)$. A derivation like that above is as follows. First

$$\begin{aligned} &A_n(x, y+z+\alpha; 0, 0) \\ &= (x+y+z+n+\alpha(2))^n \\ &= \Sigma \binom{n}{k}(x+k)^k(y+z+\alpha+n-k)^{n-k} \\ &= \Sigma \binom{n}{k}(x+k)^k \, \Sigma \binom{n-k}{j}(y+j)^j(z+n-k-j)^{n-k-j} \\ &= A_n(x, y, z; 0, 0, 0)\,. \end{aligned}$$

The general result is clearly

$$(25)\quad \begin{aligned} &A_n(x_1, \ldots, x_m; 0, \ldots, 0) \\ &= [x+n+\alpha(m-1)]^n\,, \qquad (x = x_1 + \ldots + x_m) \\ &= \sum_{k=0}^{n} \binom{n}{k}(x+n)^{n-k} k! \binom{m+k-2}{k}\,, \end{aligned}$$

which is Hurwitz's result. The evalution

$$\alpha_k(m-1) = k!\binom{m+k-2}{m}$$

which also appears in Table 1, of course follows from

$$\exp t\alpha(m) = (\exp t\alpha)^m = (1-t)^{-m}\,.$$

The third Hurwitz extension is that of $A_n(x, y; -1, -1)$. The result follows at once from recurrence (20), (24) and symmetry; it is

$$(26)\quad \begin{aligned} &A_n(x_1, \ldots, x_m; -1, \ldots, -1) \\ &= (x_1 x_2 \ldots x_m)^{-1}(x+n)^{n-1}\,, \qquad (x = x_1 + \ldots + x_m)\,. \end{aligned}$$

Many other extensions are possible; indeed, it is impossible to be exhaustive. It is possible to exhaust the extensions of the chief Abel binomial identity, $A_n(x, y; -1, 0)$. The other extreme to Hurwitz's extension, that is $A_n(x_1, \ldots, x_m; -1, -1, \ldots, -1, 0)$, is $A_n(x_1, \ldots, x_m; -1, 0, \ldots, 0)$. This may be obtained as follows. First

$$\begin{aligned} &A_n(x, y + z + \alpha; -1, 0) \\ &= x^{-1}(x + y + z + \alpha + n)^n \\ &= \sum \binom{n}{k}(x + k)^{k-1}(y + z + \alpha + n - k)^{n-k} \\ &= \sum \binom{n}{k}(x + k)^{k-1} \sum \binom{n-k}{j}(y + j)^j(z + n - k - j)^{n-k-j} \\ &= A_n(x, y, z; -1, 0, 0)\,. \end{aligned}$$

Repetitions lead to

$$\begin{aligned} (27) \qquad &A_n(x_1, \ldots, x_m; -1, 0, \ldots, 0) \\ &= x_1^{-1}(x + n + \alpha(m - 2))^n\,, \qquad (x = x_1 + \ldots + x_m)\,. \end{aligned}$$

Note that the binomial case $m = 2$ requires the interpretation $\alpha_k(0) = \delta_{k0}$. An alternate derivation, useful below, is as follows. By (21) and (25)

$$\begin{aligned} &x_1 A_n(x_1, \ldots, x_m; -1, 0, \ldots, 0) \\ &= A_n(x_1, \ldots, x_m; 0, \ldots, 0) - nA_{n-1}(x_1 + 1, x_2, \ldots, x_m; 0, \ldots, 0) \\ &= (x + n + \alpha(m - 1))^n - n(x + n + \alpha(m - 1))^{n-1}\,. \end{aligned}$$

Replacing the umbral variable $\alpha(m - 1)$ by its equivalent $\alpha(m - 2) + \alpha$ leads to

$$\begin{aligned} &xA_n(x_1, \ldots, x_m; -1, 0, \ldots, 0) \\ &= (x + n + \alpha(m - 2))^n \\ &\quad + \sum_{k=1}^{n}\left[\binom{n}{k}\alpha_k - n\binom{n-1}{k-1}\alpha_{k-1}\right](x + n + \alpha(m - 2))^{n-k}\,. \end{aligned}$$

and each term of the sum vanishes. Note that the identity

$$\begin{aligned} (x + n + \alpha(m - 2))^n = (x + n + \alpha(m - 1))^n \\ - n(x + n + \alpha(m - 1))^{n-1} \end{aligned}$$

holds for $m = 2, 3, \ldots$ (with $\alpha_k(0) = \delta_{k0}$, as already noted) while for $m = 1$, it is replaced by

$$x(x+n)^{n-1} = (x+n)^n - n(x+n)^{n-1} = (x+n+\alpha(-1))^n,$$

the last of which is consistent with $\exp t\alpha(-1) = (\exp t\alpha)^{-1} = 1 - t$.

Now, by (21) and (27)

$$\begin{aligned} &x_2 A_n(x_1, \ldots, x_m; -1, -1, 0, \ldots, 0) \\ &= A_n(x_1, \ldots, x_m; -1, 0, \ldots, 0) \\ &\quad - nA_{n-1}(x_1, x_2+1, \ldots, x_m; -1, 0, \ldots, 0) \\ &= x_1^{-1}[(x+n+\alpha(m-2))^n - n(x+n+\alpha(m-2))^{n-1}] \end{aligned}$$

or, using the identity above,

$$\begin{aligned} &A_n(x_1, \ldots, x_m; -1, -1, 0, \ldots, 0) \\ &= (x_1 x_2)^{-1}(x+n+\alpha(m-3))^n, \qquad (x = x_1 + \ldots + x_m). \end{aligned}$$

The general result is now evident; it is, with $p_j = -1$, $j = 1(1)k$; $p_j = 0$, $j = k+1(1)m$

$$\begin{aligned} &A_n(x_1, \ldots, x_m; -1, \ldots, -1, 0, \ldots, 0) \\ &= (x_1 \ldots x_k)^{-1}(x+n+\alpha(m-k-1))^n. \end{aligned} \tag{28}$$

Note that (26) is included with the interpretation of $(x+n+\alpha(-1))^n$ given above.

Next, turn to the case where p_j is 0 or 1 for $j = 1(1)m$. First, by (22) and (25)

$$\begin{aligned} &A_n(x_1, \ldots, x_m; 1, 0, \ldots, 0) \\ &= \sum \binom{n}{k} \beta_k(x_1) A_{n-k}(x_1+k, x_2, \ldots, x_m; 0, \ldots, 0) \\ &= \sum \binom{n}{k} \beta_k(x_1)(x+n+\alpha(m-1))^{n-k} \\ &= (x+n+\alpha(m-1)+\beta(x_1))^n, \qquad (x = x_1 + \ldots + x_m). \end{aligned} \tag{29}$$

Repetitions give the general case (the first k p's are 1, the rest zero), namely

$$\begin{aligned} &A_n(x_1, \ldots, x_m; 1, \ldots, 1, 0, \ldots, 0) \\ &= (x+n+\alpha(m-1)+\beta(x_1)+\ldots+\beta(x_k))^n. \end{aligned} \tag{30}$$

Finally, the generalization of both (28) and (30) is (the specification of numbers p_1 to p_m is $p_i = -1$, $i = 1(1)k$; $p_i = 0$, $i = \overline{k+1}(1)\overline{k+j}$, $p_i = 1$, $i = \overline{k+j+1}(1)m$

(31) $$A_n(x_1, \ldots, x_m; -1, \ldots, -1, 0, \ldots, 0, 1, \ldots, 1) \\ = (x_1 \ldots x_k)^{-1}[x + n + \alpha(m - k - 1) \\ + \beta(x_{k+j+1}) + \ldots + \beta(x_m)]^n .$$

One example must serve to show that these results have combinatorial aspects. Consider the rooted trees with n distinct (labeled) points (including the root) and m lines at the root; if $R_n(m)$ is the number of such trees, then by a known formula (cf. [4], p. 128)

$$m!R_n(m) = n(R + \ldots + R)^{n-1} \ (m \text{ terms}) , \qquad R^k \equiv R_k ,$$

and $R_0 = 0$, $R_k = k^{k-1}$. This is the same as

$$\begin{aligned} m!R_n(m) &= n \sum [n-1; k_1, \ldots, k_m] k_1^{k_1-1} \ldots k_m^{k_m-1} \\ &\qquad\qquad (k_i \neq 0,\ i = 1(1)m) \\ &= n(n-1)_m \sum [n-1-m; k_1, -1, \ldots, k_m, -1] \\ &\quad \cdot k_1^{k_1-2} \ldots k_m^{k_m-2} \qquad (k_i \neq 0,\ i = 1(1)m) \\ &= n(n-1)_m \sum [n-1-m; k_1, \ldots, k_m] \\ &\quad \cdot (k_1+1)^{k_1-1} \ldots (k_m+1)^{k_m-1} \\ &= (n)_{m+1} A_{n-1-m}(1, \ldots, 1; -1, \ldots, -1) \end{aligned}$$

with A_{n-1-m} multinomial in m variables. Using (26)

$$m!R_n(m) = (n)_{m+1} m(m + n - 1 - m)^{n-m-2}$$

or

$$R_n(m) = n\binom{n-2}{m-1}(n-1)^{n-1-m} .$$

Let $T_n(m)$ be the number of free trees with n distinct points and a given point of degree m (that is, with m lines incident to it). Then clearly

$$T_n(m) = n^{-1}R_n(m) = \binom{n-2}{m-1}(n-1)^{n-1-m}$$

in agreement with L. E. Clark [1].

5. INVERSE RELATIONS

Consider first the Abel identity (1a), namely

$$(x + y + n)^n = \sum_{k=0}^{n} \binom{n}{k} x(x + k)^{k-1}(y + n - k)^{n-k} .$$

Writing $a_n \equiv a_n(x, y) = (x + y + n)^n$, $b_n \equiv b_n(y) = (y + n)^n$, this becomes

(32a)
$$\begin{aligned} a_n &= \sum_{k=0}^{n} \binom{n}{k} x(x + k)^{k-1} b_{n-k} \\ &= \sum_{k=0}^{n} \binom{n}{k} x(x + n - k)^{n-k-1} b_k . \end{aligned}$$

But also

(32b)
$$\begin{aligned} b_n &= (-x + x + y + n)^n \\ &= \sum_{k=0}^{n} \binom{n}{k} (-x)(-x + k)^{k-1}(x + y + n - k)^{n-k} \\ &= \sum_{k=0}^{n} (-1)^k \binom{n}{k} x(x - k)^{k-1} a_{n-k} \\ &= \sum_{k=0}^{n} (-1)^{n+k} \binom{n}{k} x(x - n + k)^{n-k-1} a_k . \end{aligned}$$

Each of these two relations implies the other, as is evident in the orthogonal relation obtained by substituting one into the other, namely

$$\begin{aligned} \delta_{n,m} &= \sum_{k=m}^{n} (-1)^{k+m} \binom{n}{k} \binom{k}{m} x^2(x + n - k)^{n-k-1}(x - k + m)^{k-m-1} \\ &= \binom{n}{m} \sum_{k=0}^{n-m} \binom{n-m}{k} (-x)(-x + k)^{k-1} \\ &\quad \cdot x(x + n - m - k)^{n-m-k-1} \end{aligned}$$

with $\delta_{n,m}$ the Kronecker delta. This relation is readily proved by noticing that it is equivalent to

$$\delta_{n0} = -x^2 A_n(-x, x; -1, -1)$$

and

$$A_n(x, y; -1, -1) = (xy)^{-1}(x + y)(x + y + n)^{n-1} .$$

The first few instances of the inverse relations (32a) and (32b) are $a_0 = b_0$,

$$a_1 = xb_0 + b_1, \qquad a_2 = x(x+2)b_0 + 2xb_1 + b_2,$$

$$b_1 = -xa_0 + a_1, \quad b_2 = x(x-2)a_0 - 2xa_1 + a_2,$$

and

$$a_3 = x(x+3)^2 b_0 + 3x(x+2)b_1 + 3xb_2 + b_3,$$

$$b_3 = -x(x-3)^2 a_0 + 3x(x-2)a_1 - 3xa_2 + a_3.$$

Another pair of inverse relations arising from (1a) is derived as follows. First replacing x by $-x$, y by $-y$ leads to

$$(x+y-n)^n = \sum_{k=0}^{n} \binom{n}{k} x(x-k)^{k-1}(y-n+k)^{n-k}$$

or

$$(x+y)^n = \sum_{k=0}^{n} \binom{n}{k} x(x-k)^{k-1}(y+k)^{n-k},$$

and by symmetry

$$(x+y)^n = \sum_{k=0}^{n} \binom{n}{k} (x+k)^{n-k} y(y-k)^{k-1}.$$

Writing $a_n = (x+y)^n$, $b_n = y(y-n)^{n-1}$, this is the same as

$$a_n = \sum_{k=0}^{n} \binom{n}{k} (x+k)^{n-k} b_k. \tag{33a}$$

The inverse is found from

$$\begin{aligned} b_n = y(y-n)^{n-1} &= \sum_{k=0}^{n-1} \binom{n-1}{k} (-n)^k y^{n-k} \\ &= \sum_{k=0}^{n-1} \binom{n-1}{k} (-n)^k (x+y-x)^{n-k} \\ &= \sum_{k=0}^{n-1} \binom{n-1}{k} (-n)^k \sum_{j=0}^{n-k} \binom{n-k}{j} (x+y)^{n-k-j} (-x)^j \\ &= \sum_{k=0}^{n} (-1)^k (x+y)^{n-k} \sum_{j=0}^{k} n^j \binom{n-1}{j} \binom{n-j}{k-j} x^{k-j}. \end{aligned}$$

Write $b_{n,k}$ for the inner sum; then

$$\begin{aligned} b_{n,k} &= \binom{n}{k} \sum_{j=0}^{k} \binom{k}{j} (n-j) n^{j-1} x^{k-j} \\ &= \binom{n}{k} [(x+n)^k - k(x+n)^{k-1}] \\ &= \binom{n}{k} (x+n-k)(x+n)^{k-1} . \end{aligned}$$

Hence

$$b_n = \sum_{k=0}^{n} (-1)^{n+k} \binom{n}{k} (x+k)(x+n)^{n-1-k} a_k . \tag{33b}$$

The pair (33a) and (33b), that is

$$a_n = \sum_{k=0}^{n} \binom{n}{k} (x+k)^{n-k} b_k ,$$

$$b_n = \sum_{k=0}^{n} (-1)^{n+k} \binom{n}{k} (x+k)(x+n)^{n-1-k} a_k ,$$

is equivalent to a result in H. W. Gould [3]. Note that (33b) may be rendered more concisely by

$$b_n = (a-x)(a-x-n)^{n-1} , \qquad (a^k \equiv a_k) .$$

Also multiplying each of (33a) and (33b) by $(x+n)^m$ and replacing $(x+n)^m a_n$ by a_n, $(x+n)^m b_n$ by b_n, yields the alternate

$$\begin{aligned} a_n &= \sum_{k=0}^{n} \binom{n}{k} (x+n)^m (x+k)^{n-m-k} b_k , \\ b_n &= \sum_{k=0}^{n} (-1)^{n+k} \binom{n}{k} (x+n)^{n+m-1-k} (x+k)^{-m+1} a_k . \end{aligned} \tag{34}$$

The instance $m = 1$, $x = 0$ of (34) appears in Jacques Touchard [7] and also, without noticing this prior result, in my paper [5].

Next consider

$$\begin{aligned} A_n(x, y; 0, 0) &= \sum_{k=0}^{n} \binom{n}{k} (x+k)^k (y+n-k)^{n-k} \\ &= (x+y+n+\alpha)^n , \qquad (\alpha^k \equiv \alpha_k = k!) \end{aligned}$$

and write $a_n = (x+y+n+\alpha)^n$, $b_n = (y+n)^n$, so that

(35a) $$a_n = \sum_{k=0}^{n} \binom{n}{k}(x+k)^k b_{n-k} = \sum_{k=0}^{n} \binom{n}{k}(x+n-k)^{n-k} b_n\,.$$

To find the inverse, first write α' as the inverse umbral variable to α; that is

$$\exp t\alpha' = (\exp t\alpha)^{-1} = 1 - t$$

and

$$\alpha'_0 = 1 = -\alpha'_1\,, \quad \alpha'_n = 0\,, \qquad n = 2, 3, \ldots\,.$$

Then

$$\begin{aligned} b_n = (y+n)^n &= (\alpha' - x + x + y + n + \alpha)^n \\ &= (\alpha' - x)A_n(\alpha' - x, x + y + \alpha; -1, 0) \\ &= \sum_{k=0}^{n} \binom{n}{k}(\alpha' - x)(a' - x + k)^{k-1} a_{n-k}\,, \end{aligned}$$

and

$$\begin{aligned} &(\alpha' - x)(\alpha' - x + k)^{k-1} \\ &= \sum_{j=0}^{k-1} \binom{k-1}{j}(\alpha' - x)^{k-j} k^j \\ &= \sum_{j=0}^{k-1} \binom{k-1}{j} k^j[(-x)^{k-j} - (k-j)(-x)^{k-j-1}] \\ &= \sum_{j=0}^{k-1} \binom{k-1}{j} k^j(-x)^{k-j-1}(-x-k+j) \\ &= (-x-k)(-x+k)^{k-1} + k(k-1)\sum_{j=0}^{k-2} \binom{k-2}{j} k^j(-x)^{k-2-j} \\ &= [(-x-k)(-x+k) + k^2 - k](-x+k)^{k+2} \\ &= (x^2 - k)(-x+k)^{k-2}\,. \end{aligned}$$

Hence

(35b) $$\begin{aligned} b_n &= \sum_{k=0}^{n} (-1)^k \binom{n}{k}(x^2 - k)(x - k)^{k-2} a_{n-k} \\ &= \sum_{k=0}^{n} (-1)^{n+k} \binom{n}{k}(x^2 - n + k)(x - n + k)^{n-k-2} a_k\,. \end{aligned}$$

Two further pairs of somewhat different character are derived as follows. First

$$(-1)^{n+1}xA_n(-x, -y; -1, 0) = (x + y - n)^n ,$$

so that

$$(-1)^{n+1}(x + 2n)A_n(-x - 2n, x; -1, 0)$$
$$= n^n = \sum_{k=0}^{n} \binom{n}{k}(x + 2n)(x + n + k)^{n-k-1}(-1)^k(x + k)^k$$

and

$$(-1)^{n+1}(x + 2n)A_n(-x - 2n, 0; -1, 0)$$
$$= (x + n)^n = \sum_{k=0}^{n} \binom{n}{k}(x + 2n)(x + n + k)^{n-k-1}(-1)^k k^k .$$

Writing $a_n = n^n$, $b_n = (x + n)^n$, these are the same as

(36)
$$a_n = \sum_{k=0}^{n} \binom{n}{k}(x + 2n)(x + n + k)^{n-k-1}(-1)^k b_k ,$$
$$b_n = \sum_{k=0}^{n} \binom{n}{k}(x + 2n)(x + n + k)^{n-k-1}(-1)^k a_k .$$

Replace a_n by $(x + 2n)a_n$, b_n by $(x + 2n)b_n$ to get the alternate form

(36a)
$$a_n = \sum_{k=0}^{n} \binom{n}{k}(x + 2k)(x + n + k)^{n-k-1}(-1)^k b_k ,$$
$$b_n = \sum_{k=0}^{n} \binom{n}{k}(x + 2k)(x + n + k)^{n-k-1}(-1)^k a_k .$$

For the second pair, first

$$xA_n(x, -2n; -1, 0)$$
$$= (x - n)^n = \sum \binom{n}{k}(n + k)^{n-k}(-1)^{n+k}x(x + k)^{k-1}$$

or, with $a_n = (x - n)^n$, $b_n = x(x + n)^{n-1}$,

$$a_n = \sum_{k=0}^{n} \binom{n}{k}(n + k)^{n-k}(-1)^{n+k}b_k .$$

Now $b_n = (x + n)^n - n(x + n)^{n-1}$, and

$$(x + n)^n = (-1)^{n+1}2nA_n(-2n, -x; -1, 0)$$

$$= \sum_{k=0}^{n} \binom{n-1}{k} (2n-1)(n+k)^{n-k-2}(x-k)^k ,$$

so that

$$b_n = \sum_{k=0}^{n} \binom{n}{k} [2n(n+k) - (2n-1)(n-k)](n+k)^{n-k-2} a_k$$
$$= \sum_{k=0}^{n} \binom{n}{k} [n + k(4n-1)](n+k)^{n-k-2} a_k .$$

Hence the inverse pair is

$$a_n = \sum_{k=0}^{n} \binom{n}{k} (n+k)^{n-k} (-1)^{n+k} b_k ,$$

(37)

$$b_n = \sum_{k=0}^{n} \binom{n}{k} (n+k)^{n-k-2} [n + k(4n-1)] a_k .$$

These results are summarized in Table 2; undoubtedly they are a small sample of the possible results and not necessarily the most signal. It should be remembered that sign factors may be moved about in the pairs by changing either a_n to $(-1)^n a_n$, b_n to $(-1)^n b_n$, or both. Also it is worth noting that the pairs remain true when the binomial coefficient $\binom{n}{k}$ is replaced by $\binom{n+p}{k+p}$.

To show that at least one of the relations has combinatorial interest, consider the instance $x = 0$ of (33a) and (33b), that is,

Table 2. ABEL INVERSE RELATIONS

$$a_n = \sum_{k=0}^{n} \binom{n}{k} A_{n,k}\, b_k \qquad b_n = \sum_{k=0}^{n} (-1)^{n+k} \binom{n}{k} B_{n,k}\, a_k$$

No.	$A_{n,k}$	$B_{n,k}$
1	$x(x+n-k)^{n-k-1}$	$x(x-n+k)^{n-k-1}$
2	$(x+n-k)^{n-k}$	$(x^2-n+k)(x-n+k)^{n-k-2}$
3	$(x+k)^{n-k}$	$(x+k)(x+n)^{n-k-1}$
3a	$(x+n)(x+k)^{n-k-1}$	$(x+n)^{n-k}$
4	$(x+2n)(x+n+k)^{n-k-1}$	$(x+2n)(x+n+k)^{n-k-1}$
4a	$(x+2k)(x+n+k)^{n-k-1}$	$(x+2k)(x+n+k)^{n-k-1}$
5	$(n+k)^{n-k}$	$(n+k)^{n-k-2}[n+k(4n-1)]$

Reminders: Sign factors may be moved by redefining a_n, b_n, or both. Relations remain true when $\binom{n}{k}$ is replaced by $\binom{n+p}{k+p}$.

the pair

$$a_n = \sum \binom{n}{k} k^{n-k} b_k \qquad b_n = \sum (-1)^{n+k} \binom{n}{k} k n^{n-1-k} a_k .$$

Then, the relation

$$\Delta^n 0^n = \sum_{k=0}^{n} (-1)^{k+n} \binom{n}{k} k^n = n! S(n, n) = n!$$

($\Delta^n 0^n$ is of course an abbreviation for $\Delta^n x^n$ evaluated at $x = 0$, with $\Delta x^n = (x+1)^n - x^n = (E-1)x^n$ and $Ex^n = (x+1)^n$; $S(n, n)$ is a Stirling number of the second kind). Then the first of the pair is satisfied with $a_n = (-1)^n n!$, $b_n = (-1)^n n^n$; the inverse is

$$n^n = \sum_{k=0}^{n} \binom{n}{k} k n^{n-1-k} k! = \sum_{k=1}^{n} \binom{n-1}{k-1} n^{n-k} k!$$
$$= \sum_{k=0}^{n-1} \binom{n-1}{k} n^{n-1-k} (k+1)! .$$

This is the formula appearing in my paper [6] for the total number of mappings of a finite set of n elements into itself, as classified by the number of map components. Moreover, replacing a_n by $(-1)^n b_n$, b_n by $(-1)^n a_n$, the pair may be rewritten

$$a_n = \sum \binom{n}{k} k n^{n-1-k} b_k , \qquad b_n = \sum (-1)^{k+n} \binom{n}{k} k^{n-k} a_k .$$

The first is the same as

$$a_n = \sum \binom{n-1}{k} n^{n-1-k} b_{k+1} = b(b+n)^{n-1} , \qquad (b^k \equiv b_k)$$

and if $f(z) = \exp zb$, $b^n \equiv b_n$, $D = d/dx$,

$$a_n = D^{n-1}[f'(x)e^{nx}]_{x=0} .$$

By the Lagrange theorem in the form

$$f(z) = f(0) + \sum_{n=1}^{\infty} \frac{(ze^{-z})^n}{n!} D^{n-1}[f'(x)e^{nx}]_{x=0} ,$$

it follows that (since $a_0 = b_0$)

$$\exp zb = b_0 + \sum_{n=1}^{\infty} \frac{(ze^{-z})^n}{n!} a_n = \exp(ze^{-z}a), \qquad (a^n \equiv a_n, \ b^n \equiv b_n).$$

The presence of ze^{-z} suggests labeled rooted trees, since $ze^{-z} = w$ implies

$$z = R(w) = \sum_{n=1}^{\infty} n^{n-1} \frac{w^n}{n!},$$

with $R(w)$ the enumerator of rooted trees with distinct points by number of points. Setting $b_n = x^n$ and $ze^{-z} = w$ the equation above becomes

$$\exp wa = \exp xR(w), \qquad (a^n \equiv a_n(x)).$$

Hence, by Equation (1) of [6],

$$a_n(x) = \sum_{k=0}^{n} \binom{n}{k} kn^{n-1-k} x^k$$

is the enumerator of forests of rooted trees with distinct (labeled) points by number of trees.

References

1. Clarke, L. E. "On Cayley's Formula for Counting Trees," *J. London Math. Soc.*, **33** (1958), 471–475.
2. Hurwitz, A. "Über Abel's Verallgemeinerung der binomischen Formel," *Acta Math.*, **26** (1902), 199–203.
3. Gould, H. W. "A Series Transformation for Finding Convolution Identities," *Duke Math. J.*, **28** (1961), 193–202.
4. Riordan, John. *An Introduction to Combinatorial Analysis*, John Wiley and Sons, New York, 1958.
5. Riordan, John. "Inverse Relations and Combinatorial Identities," *Amer. Math. Monsly*, **71** (1964), 485–498.
6. Riordan, John. "Enumeration of Linear Graphs for Mappings of Finite Sets," *Ann. Math. Statist.*, **33** (1962), 178–185.
7. Touchard, Jacques. "Sur la Théorie des Différences," *Proc. Int. Math. Congress 1924*, Vol. 1, pp. 623–629.

Discussion on Mr. Riordan's Paper

PROFESSOR D. E. KNUTH: I would like to begin this discussion by making some presumptuous comments about notation, in the fond hope that the distinguished audience gathered here may help to overcome a rather unfortunate situation that exists today. Mr. Riordan has used the symbolism "$(x)_n$" to stand for the falling factorial power $x(x-1)\ldots(x-n+1)$. This is one of several commonly used notations for this function, which is also often written "$x^{(n)}$", "$x_{(n)}$", "$x^{[n]}$", etc.. The problem is that the same notations are also commonly used for the *rising* factorial powers $x(x+1)\ldots(x+n-1)$, specially in the literature concerning hypergeometric functions. Since the history of mathematics clearly indicates the importance of good notations, I believe it necessary for me to speak up in behalf of a notational convention which offers a good solution to this dilemma: Let us agree to write

$$x^{\underline{n}} = x(x-1)\ldots(x-n+1)\,,$$

$$x^{\overline{n}} = x(x+1)\ldots(x+n-1)\,, \qquad \text{when } n \geq 0\,.$$

Such a symbolism appears to possess all of the properties of a good notation. For example it is mnemonic, unambiguous, brief, relatively easy to set in type, and it strongly suggests analogies with the usual notation for non-factorial powers, as in the Abel-like sums

$$(x+y+na)^{\underline{n}} = \sum_k \binom{n}{k} x(x+ka-1)^{\underline{k-1}}(y+(n-k)a)^{\underline{n-k}}\,;$$

$$(x+y+na)^{\overline{n}} = \sum_k \binom{n}{k} x(x+ka+1)^{\overline{k-1}}(y+(n-k)a)^{\overline{n-k}}\,.$$

Note that another notational convention, the use of "umbral variables", is the key reason why Mr. Riordan has been able to compile the formulas shown in Table 1 in such a suggestive form.

Abel published his short paper on the extended binomial formula in the first volume of Crelle's Journal, together with two of his most famous other memoirs dealing with the unsolvability of the quintic and the beginnings of a rigorous theoretical treatment of infinite series in connection with the convergence of $\sum \binom{n}{k} x^k$ when n is not an integer. In this connection it is interesting to raise a problem that is still apparently unsolved: for what complex values of x, y, and n

does the sum $\sum_{k\geq 0}\binom{n}{k}x(x+k)^{k-1}(y+n-k)^{n-k}$ converge to $(x+ +y+n)^n$?

I believe Mr. Riordan's references to Adolf Hurwitz's paper give a misleading supression of the contents of that paper. The main point of Hurwitz's work here was not to give a multinomial generalization of Abel's identity; that follows easily from the binomial case, and it is discussed (in a much more general setting) only in the last paragraphs. The point was to exhibit and prove a much more interesting generalization of Abel's formula, namely

$$(*) \qquad (x+y)(x+y+z_1+z_2+\dots+z_n)^{n-1}$$
$$=\sum x(x+\varepsilon_1 z_1+\dots+\varepsilon_n z_n)^{\varepsilon_1+\dots+\varepsilon_n-1}$$
$$\cdot\, y(y+(1-\varepsilon_1)z_1+\dots+(1-\varepsilon_n)z_n)^{(1-\varepsilon_1)+\dots+(1-\varepsilon_n)-1}$$

summed over all 2^n choices of $\varepsilon_1,\dots,\varepsilon_n$ independently taking the values 0 and 1. This is an identy in $2n+2$ variables, and Abel's binomial formula is the special case $z_1=z_2=\dots=z_n$. Hurwitz proved this identity and some related results by using techniques quite similar to those of the present paper, inserting an arbitrary p and q in place of the -1's which occur in the exponents here and deriving appropriate recurrence relations.

I have mentioned Hurwitz's general identity (*) here primarily because it has an interesting combinatorial significance. To see this connection, let us first mention an auxiliary result about cycle-free directed graphs; Let $r\geq 1$ and $1\leq p\leq q$, and consider a directed graph with $q+r$ vertices $U_1,\dots,U_q$, $V_1,\dots,V_r$, and $q-p$ arcs from U_t to $V_{f(t)}$, for $p<t\leq q$, where f is any function from $\{p+1,\dots q\}$ into $\{1,2,\dots,r\}$. Then the number of ways to add r additional arcs, from V_t to $U_{g(t)}$ for $1\leq t\leq r$ where g is a function from $\{1,\dots,r\}$ into $\{1,\dots,q\}$, such that the resulting directed graph contains no cycles, is exactly $q^{r-1}p$. In fact there is a one-to-one correspondence between such functions g and the sequences of r integers $a_1,a_2,\dots,a_r$ where $1\leq a_j\leq q$ and $a_r\leq p$; this correspondence may be obtained by generalizing a construction of Prüfer [*Arch. Math. u. Phys.*, 27 (1918), 142–144] in a straight-forward manner.

Now consider the directed graph with $x+y+n+z_1+z_2+\dots+z_n$ vertices, $U_1,U_2,\dots,U_{x+y}$, $V_1,V_2,\dots,V_n$, W_{ij} for $1\leq i\leq n$, $1\leq j\leq z_i$, and $z_1+z_2+\dots+z_n$ arcs, from W_{ij} to V_i for $1\leq i\leq n$, $1\leq j\leq z_i$. By the above argument the number of ways to add n additional arcs, from V_j to one of the U's or W's for $1 \leq j \leq n$, is exactly $(x + y + z_1 + z_2 + \ldots + z_n)^{n-1}(x + y)$. In any graph which results from this construction, there is a unique path from each of the vertices $V_1, \ldots, V_n$ to one of the U's; if we set $\varepsilon_j = 1$ if and only if there is a path from V_j to U_k for some $k \leq x$, and apply the formula just derived, we obtain Hurwitz's equation (*) whenever $x_1, y_1, z_1, \ldots, z_n$ are positive integers. Since both sides of (*) are polynomials, the identity must hold in general.

Finally I would like to point out that the relations between trees and Abel's identities which appear in this paper have been discussed in a much more general setting in two important papers by George N. Raney [*Trans. Amer. Math. Soc.*, **94** (1960), 441–451; *Canadian J. Math.*, **16** (1964), 755–762].

PROFESSOR D. A. SMITH: The work of G.-C. Rota (see his earlier paper in *Z. Wahrscheinlichkeitstheorie*, (1964)) has given new impetus to the study of algebraic structures with discrete convolution-type products. The identities obtained by Mr. Riordan can be interpreted in the context of such a structure, many of them in an algebra of arithmetic functions familiar to number theorists.

An *incidence function* on the ordered set N of natural numbers is a function f of two variables with values in a field K such that $f(k, n) = 0$ unless $k \leq n$. The *incidence algebra* of N is the set of all incidence functions with the obvious vector space operations and the convolution product

$$f * g(h, n) = \sum_{j=k}^{n} f(k, j)g(j, n) .$$

The functions f whose values depend only on the difference of the arguments form a commutative subalgebra. By abuse of notation, we will write $f(k, n) = f(n - k)$ for such functions, which gives a natural identification with the algebra of arithmetic functions $f: N \to K$ with the "Cauchy product:"

$$f * g(n) = \sum_{j=0}^{n} f(j)g(n - j) .$$

For the most part, we can restrict our attention to this subalgebra. However, some of Mr. Riordan's results can only be interpreted in the larger algebra. In what follows, it is convenient to think of K as being a field of rational functions in two or more variables with coefficients in the real or complex numbers.

For $p \in N$, $z \in K$, let $F_{p,z}$ be the function defined by $F_{p,z}(n) = (z+n)^{n+p}/n!$. These functions are related to the Abel sums A_n by

$$A_n(x, y; p, q) = n! F_{p,x} * F_{q,y}(n) .$$

Thus the Abel binomial identities may be thought of as evaluations of the products $F_{p,x} * F_{q,y}$ as arithmetic functions. In particular (with reference numbers matching those of Mr. Riordan's paper), the simplest of these formulas are

$$\text{(1a)} \qquad xF_{-1,x} * F_{0,y} = F_{0,x+y} ,$$

$$\text{(12)} \qquad (xF_{-1,x}) * (yF_{-1,y}) = (x+y)F_{-1,x+y} .$$

Somewhat more complicated, and not expressible directly in terms of the F-functions themselves, are

$$\text{(13)} \qquad F_{-2,x} * F_{0,y}(n) = \left(\frac{1}{x^2}F_{0,x+y} - \frac{n}{x(x+1)}F_{-1,x+y}\right)(n) ,$$

and Cauchy's formula

$$\text{(15)} \qquad F_{0,x} * F_{0,y}(n) = \sum_{k=0}^{n} F_{0,x+y+k}(n-k) = \sum_{k=0}^{n} (n)_k F_{-k,x+y}(n) .$$

(The special case of (15) with $x = y = 0$ gives $F_{0,0}^2(n) = \sum_{k=0}^{n} n^k/k!$ = the sum of first $n+1$ terms in the expansion of e^n.)

The evaluations of the binomial sums are not made any easier by this interpretation in terms of arithmetic functions. However, the extensions to multinomial identities and the inverse relations are simpler in this context.

For example, iteration of (12) gives immediately

$$\text{(26)} \qquad \prod_1^m (x_i F_{-1,x_i}) = xF_{-1,x} , \qquad x = \sum_1^m x_i ,$$

and combining this with (1a) we get

$$\text{(24)} \qquad \prod_1^{m-1} (x_i F_{-1,x_i}) * F_{0,x_m} = F_{0,x} , \qquad x = \sum_1^m x_i .$$

With a little more effort and frequent changes of order of summation, iteration of (15) yields

$$(25)\qquad \left(\prod_{1}^{m} F_{0,x_i}\right)(n) = \sum_{k=0}^{n} \binom{m+k-2}{k} F_{0,x+k}(n-k)\,, \qquad x = \sum_{1}^{m} x_i\,.$$

The derivation of inverse relations in this context makes it clearer why relations derived with particular functions a_n, b_n are valid for arbitrary functions a_n, b_n. For example, from (1a) we have

$$(32b)\qquad F_{0,y} = F_{0,-x+(x+y)} = (-xF_{-1,-x}) * F_{0,x+y}\,.$$

Comparison of this with (1a) shows that $xF_{-1,x}$ and $-xF_{-1,-x}$ are inverses of each other in the function algebra, which is also evident from (12). (The last two sentences are merely a condensation of the first paragraph of Section 5 of Mr. Riordan's paper.)

Similarly, line 2 of Table 2 states that the inverse of $F_{0,x}$ is $G(n) = (x^2 - n)F_{-2,-x}(n)$. Line 3 of Table 2 is a good example of an inverse relationship involving functions of two variables, not merely dependent on their difference. Even though this relation is derived from (1a), the two-variable situation arises from the replacement of y for $y - n$ in the third paragraph of Section 5. The inverse functions here are $A(k, n) = (x + k)^{n-k}/(n - k)!$ and $B(k, n) = (-1)^{n-k}(x + k)(x + n)^{n-k-1}/(n - k)!$. The fact that these are inverses is equivalent to the formula

$$\sum_{k=0}^{n} \frac{(-1)^k x(x+k)^{n-1}}{k!(n-k)!} = \delta_{0,n}\,,$$

since the Kronecker delta function is the identity element of the incidence algebra. The inverse relationship of line 3a reduces to essentially the same formula (different on the left by a factor of $1 + n/x$), which is not surprising, since 3a is a variant of 3.

DR. V. R. RAO UPPULURI: First, I wish to point out that Abel identities have been used in mathematical statistics as illustrated by the work of Z. W. Birnbaum and R. Pyke, and

Bernard Harris. Next I wish to ask Mr. Riordan whether it is true that the only solutions of the difference equation (10) are indeed given by (2)?

MR. RIORDAN replied as follows:

The answer is no; (2) is the solution of (10) with the initial condition $A_0(x, y; p, q) = x^p y^q$.

CHAPTER 6

On the Automorphism Groups of Paley-Hadamard Matrices

E. F. ASSMUS, JR.[1], *Lehigh University and Sylvania Applied Research Laboratory*
H. F. MATTSON, JR., *Sylvania Applied Research Laboratory*

For a prime[2] of the form $l = 4N - 1$, the Paley-Hadamard matrix of order $l + 1$ is defined as the $(l + 1) \times (l + 1)$ matrix of $+1$'s and -1's with the first row and first column all $+1$'s; the second row is defined to be -1 at 0 and at the quadratic nonresidues modulo l and $+1$ elsewhere. The columns are indexed as $(\infty; 0, 1, 2, \ldots, l - 1)$. The remaining $l - 1$ rows are defined to be the cyclic shifts of the " finite part " of the second row.

The automorphism group of a Hadamard matrix is defined as the group of $(l + 1) \times (l + 1)$ monomial matrices (with entries $0, \pm 1$) modulo $\{\pm \mathbf{I}\}$, $\mathbf{I}$ being the identity, which act on the right on the Hadamard matrix in such a way that the result is the original Hadamard matrix except for a permutation of rows and a possible change of sign of some rows.

A monomial matrix is of course the product of a diagonal matrix and a permutation matrix. The mapping which sends

[1] The research reported in this paper was partly sponsored by the Air Force Cambridge Research Laboratories, Office of Aerospace Research under contract AF19 (628)-5998.

[2] In fact l could be a prime power.

each element of the above automorphism group to the associated permutation matrix is an isomorphism, as one can easily verify. Because we are concerned with the permutation group which is the image of this isomorphism, we shall speak of the automorphism group of the matrix as being, or being contained in, this or that permutation group.

It is known ([5], [2]) that when $l = 11$, the automorphism group is the Mathieu group M_{12}. What we prove here is the following.

Theorem. *When l is a prime of the form $12N - 1$ with $6N - 1$ also prime and $23 \leq l \leq 4{,}079$, then the automorphism group G of the Paley-Hadamard matrix of order $l + 1$ is the projective unimodular group $PSL_2(l)$.*

$PSL_2(l)$ is the group of all 2×2 matrices with determinant 1, modulo $\pm\begin{pmatrix}1 & 0\\ 0 & 1\end{pmatrix}$, over $GF(l)$. It is a 2-fold transitive permutation group on the projective line.

Proof of the Theorem. Hall [5] has proved that the automorphism group of a Paley-Hadamard matrix contains $PSL_2(l)$. Consider the rows of the matrix to be vectors over $GF(3)$. Their linear span is contained in[3] a so-called extended $(l + 1, (l + 1)/2)$ quadratic-residue code over $GF(3)$, of dimension $(l + 1)/2$, of which there are two, here called A and B. Let A be the row-space of the matrix. Then G leaves A invariant. The group $PSL_2(l)$ acts on both codes together, and any element of $PGL_2(l)$ not in $PSL_2(l)$ maps A onto B and B onto A. Also, $A \cap B = 0$. ([4], [9]) This motivates our choice of l so that $(3/l) = +1$; otherwise the row-space has dimension l.

Case 1. $23 < l \leq 4{,}079$. The subgroup of $PSL_2(l)$ which fixes the "infinite" coordinate of the code (or column of the Paley-Hadamard matrix) is the group sending x to $a^2x + b$ for $a, b \in GF(l)$ with $a \neq 0$. It is transitive on the "finite" coordinates (columns). Nikolai and Parker [8] show, however, that there are no transitive nonsolvable groups on l letters. Let G_∞ be the subgroup of G fixing the column ∞. Then G_∞ is solvable, and by elementary arguments one sees that G_∞ is contained in the "$ax + b$" group, which sends x to $ax + b$ for all $a, b \in GF(l)$ with $a \neq 0$. But G_∞ contains the "$a^2x + b$" group, and if G_∞ were equal to the $ax + b$ group, then G

[3] The relation is equality but we only need the inclusion.

would not leave A invariant but would contain elements mapping A onto B. Therefore G_∞ is the $a^2x + b$ group and $G = PSL_2(l)$.

Case 2. $l = 23$. Here there is of course a nonsolvable group, namely the Mathieu group M_{23}, and hence there is the possibility that M_{24} could be the automorphism group of the Paley-Hadamard matrix of order 24 since M_{23} is contained in the 5-fold transitive group M_{24} as the stability subgroup of a point. Furthermore, M_{24} is the only 5-fold transitive group on 24 letters [6, p. 80].

Ito [7] proves that if l and $(l - 1)/2$ are primes with $l > 11$, then a nonsolvable, transitive permutation group on l letters is 4-fold transitive. It follows from this result that if G for $l = 23$ is larger than $PSL_2(23)$, then it is 5-fold transitive and therefore is M_{24}. We now sketch a proof that M_{24} is not the automorphism group.

M_{24} is best regarded as the automorphism group of a certain tactical configuration, namely, a Steiner system of type 5-8-24. This configuration is a collection D of 8-subsets of a set of 24 points such that every 5-subset of the 24-set is contained in exactly one element of D. Witt [11] proved that such a configuration is unique (up to action by the symmetric group Σ_{24} on the 24 points) and that the subgroup of Σ_{24} which permutes the elements of D among themselves is a 5-fold transitive group of order $48 \cdot 24 \cdot 23 \cdot 22 \cdot 21 \cdot 20$. This group is, by definition, the automorphism group of the 5-8-24 Steiner system; and, since Witt, this is the most commonly used definition of M_{24}.

For our proof we need some simple properties of M_{24}. From the 5-fold transitivity it follows that the subgroup G_0 of M_{24} which fixes each of 5 points has order 48. We need to know the action of G_0 on the remaining 19 points.

Lemma. *G_0 has two orbits—one of length* 3 *and one of length* 16*—on the remaining* 19 *points.*

Proof. The 5 fixed points of G_0 are contained in a unique 8-set belonging to D. Therefore G_0 acts on the remaining 3 points X of the 8-set and on the other 16 points Y. To see that these actions are transitive one can refer to [11], where an explicit description of G_0 as 3×3 matrices over $GF(4)$ acting on the projective space of 21 points over $GF(4)$ is given.

It is also possible to prove transitivity from the definition and properties of M_{24} listed above. One proves that only the identity element of M_{24} can fix 7 points not contained in an 8-set belonging to D. (One sees this by picking pairs of 5-subsets of the 7 points which meet in 3 points. These give rise to pairs of 3-subsets outside the 7 meeting in 1 point, which must be also fixed. Continuing, one gets all points fixed.) Then consider the stability subgroup G_1 in G_0 of a point from the 16-set Y. G_1 must have order at least 3, but no non-trivial element σ of G_1 can fix any more points. This means in particular that σ is a 3-cycle on the 3 points of X, and thus G_0 is transitive on X. Moreover, σ^3 is the identity on 9 points and therefore $\sigma^3 = 1$. Since now every nontrivial element of G_1 has order 3, G_1 must have order 3, since $|G_0| = 3 \cdot 16$. Therefore G_0 is transitive on Y also.

We use this lemma in the situation arising from the code generated by the rows of the Paley-Hadamand matrix over $GF(3)$. This code is the whole extended quadratic-residue code A of type (24, 12) over $GF(3)$, because the latter is the sum of $\alpha(1\ 1\ 1 \ldots 1; 1)$, $\alpha = 0, \pm 1$, and the irreducible (24, 11) extended quadratic-residue code. We have proved [3] that A has the following properties.

Proposition. *The minimum distance in A is* 9; *the* 9*-subsets of coordinate-places of A holding minimum-weight code-vectors form a tactical configuration of type* 6; 5-9-24. (*Thus every* 5*-subset is in precisely* 6 *of the given subsets.*)

We use our lemma to show that M_{24} cannot act on this design, as it would certainly do if it were the automorphism group of the matrix. Consider a given 5-subset of the 24-set. There are the two subsets X and Y, of cardinalities 3 and 16, respectively, on which G_0 of the lemma is transitive. The 6 different 9-subsets, containing the given 5-subset, of the new design have $6(9-5) = 24$ points distributed with multiplicities among the remaining 19 points. Arrange them in a 6×19 incidence matrix. If M_{24} acts, then G_0 acting on the columns must permute the 6 rows. By the lemma, there are the same number of these incidences, say x, in each of the three columns determined by X and there are y in each column determined by Y. This means $3x + 16y = 24$, but since $x \leq 6$ this is not solvable in integers. Therefore M_{24} does not act on the 6; 5-9-24 design. Hence the automorphism group of the code is $PSL_2(23)$, and the same for the Paley-Hadamard matrix.

As we showed in [2], the row space generated over $GF(3)$ by the rows of the Paley-Hadamard matrix of order 12 yields the 5-6-12 Steiner system (a "5-design") of which M_{12} is the automorphism group. Thus, for order 24 the 5-design remains but the group is no longer large. We do not yet know whether the order 48 Paley-Hadamard matrix yields a 5-design, but there is a possibility that it will.

Note. On the above result for $l = 23$, M. Hall has informed us that Gordon Keller has shown that M_{24} cannot act on any Hadamard matrix of order 24.

In [10] is indicated in effect that the subgroup of G_∞ (for $l = 23$) which acts without any signs, thus fixing row ∞ as well as column ∞, is the "$a^2x + b$" group; but it is not clear how to proceed from this sub-group to G_∞.

Acknowledgment

Our interest and what knowledge we have of these matters owes much to continued conversations with A. M. Gleason and Richard Turyn.

References

1. Assmus, E. F., Jr. and Mattson, H. F., Jr. "Cyclic Codes," Scientific Report AFCRL-65-322, Air Force Cambridge Research Laboratory, Bedford, Mass., 1965.
2. Assmus, E. F., Jr. and Mattson, H. F., Jr. "Perfect Codes and the Mathieu Groups," *Arch. Math.*, **17** (1966), 121–135.
3. Assmus, E. F., Jr., and Mattson, H. F., Jr. "New 5-Designs," *J. Comb. Theory*, to appear.
4. Gleason, A. M. Private Communication (presented in [1]).
5. Hall, Marshall, Jr. "Note on the Mathieu Group M_{12}," *Arch. Math.*, **13** (1962), 334–340.
6. Hall, Marshall, Jr. *The Theory of Groups*, Macmillan, New York, 1959.
7. Ito, N. "Transitive Permutation Groups of Degree $p = 2q + 1$, p and q being Prime Numbers, III," *Trans. Amer. Math. Soc.*, **116** (1965), 151–166.
8. Parker, E. T. and Nikolai, Paul J. "A Search for Analogues of the Mathieu Groups," *Math. Tables Aids Comput.*, **12** (1958), 38–43.
9. Prange, Eugene A. "Codes Equivalent Under the Projective Group, III," unpublished memorandum (presented in [1]).

10. Todd, J. A. "A Combinatorial Problem," *J. Mathematical Phys.*, **12** (1933), 321–333.
11. Witt, E. "Die 5-fach transitiven Gruppen von Mathieu," *Abh. Math. Sem. Hans. Univ.*, **12** (1938), 256–264.

CHAPTER 7

The Elementary Vectors of a Subspace of R^N

R. T. ROCKAFELLAR[1] *University of Washington*

1. INTRODUCTION

This paper concerns some connections between convex analysis, network flows and matroid theory.

Let K be an arbitrary subspace of R^N, where R is the real number system. Regarding K as a chain group in the sense of Tutte, one can pass to the corresponding matroid. Combinatorial facts deduced from general matroid theory may then be reinterpreted in terms of the original vectors in K. The results so obtained reflect the fact that K is not just a real vector space, but has further structure because of its particular disposition within R^N. Specifically, the matroid analysis of K deals with the way K intersects the special subspaces of R^N spanned by the canonical coordinate axes. Now, the natural ordering of R allows one to enlarge the finite category of intersections under scrutiny to include closed "orthants", and in general all the polyhedral convex cones generated by the various positive and negative halves of coordinate axes. The

[1] The research was supported in part by the Air Force Office of Scientific Research through Grant No. AFOSR-1202-67 at the University of Washington, Seattle.

combinatorial theory of such intersections is what we want to consider here.

One motivation is the question of "orientation" in matroids. In the elementary cycles and cocycles of a directed linear graph, there is a natural sense in which two elements have the same or the opposite orientation. There are certain combinatorial results, such as Minty's "colored arc lemma", which involve orientations but are otherwise really assertions about matroids. Minty has recently presented in [8] an interesting development of matroid theory, in which certain orientations are introduced axiomatically in terms of "digraphoids", so that abstract generalizations of the graph results are true. A "digraphoid", he has shown, corresponds to a dual pair of matroids which are "regular" in Tutte's terminology. There does exist, then, an abstract theory of "oriented" or "signed" matroids which implies that the matroids involved are regular.

A much broader theory of orientation ought to be possible, in our opinion. Regular matroids arise from subspaces of R^N, but only subspaces of an extremely special type. For any subspace K of R^N, however, there is a natural way of using the signs of the coordinates of the vectors to introduce orientations into the corresponding matroid. The study of the signed matroid amounts to the generalized intersection problem posed above. We shall demonstrate that, for such signed matroids, several theorems are valid which are far from obvious, and which even have important well-known non-matroid theorems as consequences.

No attempt is made here to develop a theory of signed matroids axiomatically. We are concerned, rather, with showing that there are interesting and significant examples which any such theory ought to encompass.

The paper is partly expository, in that we also aim to describe a certain bridge between results in convex analysis and graph theory. Well-known theorems about systems of linear inequalities can sometimes be reformulated as seemingly much simpler combinatorial theorems about the way a subspace K intersects some orthant. This is true of the duality theorem for linear programs, as has been pointed out by Tucker. We want to show that, in this form, the theorems about inequalities correspond to other well-known theorems of a combinatorial character about graphs, which have been arrived at by an entirely different route. The idea is to specialize K to a space of network flows. It turns out, for instance, that Minty's "colored arc lemma" is essentially a special case of the classical lemma of Farkas.

The result to which we would most like to draw the reader's attention is Theorem 3, a versatile existence theorem which extends Minty's theorem for "interval networks" in [6]. It is a partly combinatorial result about the consistency of systems of linear inequalities, suited in particular for application to the dual convex programs in [11].

Note. Since this paper was submitted, P. Camion has informed us that a more general form of Theorem 3 is proved in his unpublished thesis [23] in terms of modules over totally ordered integral domains. The thesis also contains results equivalent to Theorems 1 and 6, which we display below as corollaries of basic theorems in convex analysis, and there are ideas similar to those in Section 7 about using the simplex algorithm to determine which of the alternatives in Theorem 6 holds in a given case. Some results from Camion's thesis are stated without proof in an appendix to [22].

2. ELEMENTARY VECTORS AND SUPPORTS

It will be helpful to think of the vectors $X = (x_1, \ldots, x_N)$ in R^N as real-valued functions on a certain finite set $E = \{e_1, \ldots, e_N\}$, with $X(e_i) = x_i$. The *support* of X is then a certain subset of E, namely the set of e_i's such that $x_i \neq 0$. An *elementary vector* of K is defined to be a non-zero vector of K whose support is minimal, i.e. does not properly contain the support of any other non-zero vector of K. The system of subsets of E consisting of the supports of the elementary vectors of K (which we call the *elementary supports of* K) is, of course, the matroid associated with K.

It is important to keep in mind that two elementary vectors X and X' of K having the same support have to be scalar multiples of each other. Indeed, if $\lambda \in R$ is chosen so that one of the non-zero components of λX equals the corresponding component of X', then $X' - \lambda X$ is a vector of K whose support is properly smaller, so that $X' - \lambda X = 0$. Hence K has only *finitely many elementary vectors, up to scalar multiples*. The ratios between the components of an elementary vector do not depend on the arbitrary multiple. Thus a certain finite "ratio system" is uniquely and intrinsically defined by K. This "ratio system" determines K completely, because K is the subspace generated by its elementary vectors (see Section 3). An interesting "combinatorial" problem is to determine necessary and sufficient conditions on a "ratio system" in order that it arises in this fashion. This is closely related to the problem of characterizing the classes of real matrices combinatorially equivalent to each other in the sense of Tucker (see Section 6). Most of the results below concern necessary conditions on the patterns of signs of the ratios or matrix elements.

By a *signed set* in E, we shall mean a subset S which has been partitioned into two further subsets S^+ and S^- (possibly empty). We shall say that S contains an element e_i *positively* or *negatively*, according to whether $e_i \in S^+$ or $e_i \in S^-$. A signed subset can be represented in an obvious manner by an N-vector formed from the symbols $+$, $-$ and 0.

With each vector X of R^N, we associate a signed set S formed from the support of X, where S^+ consists of the elements e_i with $x_i > 0$, and S^- consists of the elements e_i with $x_i < 0$. We call this the *signed support* of X. A signed set which is the signed support of some vector in K is said to be a *signed support of* K. It is *elementary* if it actually comes from an elementary vector of K.

The system of elementary signed supports of K may be regarded as a sort of "signed matroid." Its properties include an extensive duality with the system of elementary signed supports of $K^\perp$, the orthogonal complement of K, as one would readily expect from ordinary matroid theory.

A special case to which we shall often appeal for motivation, and which therefore deserves a brief review, is the case where E is the set of arcs of a directed graph. Here we interpret the vectors in K as *circulations* in the graph, i.e. flows which are conservative at every vertex. Thus $X \in K$ if and only if X is orthogonal to every row of the (vertex vs. arc, signed) incidence matrix of the graph. For the general theory of such flows, we refer the reader to the exposition of Berge [2].

The *elementary* circulations are easy to determine. Given any elementary cycle S in the graph and a real number α, a "circulation of intensity α around S" is obtained by setting $x_i = \alpha$ if S contains e_i in the sense of its orientation, $x_i = -\alpha$ if S contains e_i in the opposite sense, and $x_i = 0$ if S does not contain e_i at all. On the other hand, a simple argument invoking the conservation condition at each vertex shows that every non-zero circulation contains some cycle in its support. It follows that the elementary vectors for this choice of K are precisely the circulations of non-zero intensity around elementary cycles. The elementary signed supports of K can be identified with the elementary cycles themselves.

In this example, $K^{\perp}$ is the subspace generated by the rows of the incidence matrix. Thus $Y \in K^{\perp}$ means that Y is a *tension* in the graph, i.e. that there exists some "potential" function on the vertices of the graph such that each y_i is obtained by subtracting the potential at the initial vertex of e_i from the potential at the final vertex of e_i. Given any elementary cocycle S in the graph and a real number α, one can construct a "tension of intensity α across S", much as above. The non-zero tensions of this form turn out to be precisely the elementary vectors of $K^{\perp}$, so that the elementary signed supports of $K^{\perp}$ are the elementary cocycles.

Much of the theory of linear inequalities, our other main source of motivation, concerns "linear systems of variables" rather than subspaces of R^N. But the two settings are really interchangeable. In the "linear variables" case, one deals with the pairs of vectors $U \in R^m$ and $V \in R^n$ satisfying $UA = V$, where A is a given $m \times n$ matrix. The set of such pairs $X = (U, V)$ forms, of course, a certain subspace K of R^N, with $N = m + n$. The orthogonal complement $K^{\perp}$ of this K consists of the pairs $Y = (U', V')$ such that $U' \in R^m$, $V' \in R^n$, and $V'A^T = -U'$, where A^T is the transpose of A. Tucker's theory of combinatorial equivalence tells us how to represent an arbitrary subspace K in this way by various matrices A. More will be said about this in Section 6.

Everything that follows would still be valid if R were replaced by *any* ordered field.

3. HARMONIOUS SUPERPOSITION

A known result about circulations in directed graphs is that every such circulation X can be represented (non-uniquely) as a superposition $X_1 + \ldots + X_r$, where each X_k is a circulation around an elementary cycle of the graph. Moreover, the cycles can be chosen so that the orientations of their arcs agree with the signs of the corresponding flow components in X, see [2, p. 145]. In particular, the support of X is then the union of the elementary cycles involved. This theorem can be generalized to arbitrary K, as we now show.

Let us say that two vectors X and X' in R^N are *dissonant*, if, for some i, the components x_i and x'_i are non-zero and opposite in sign. Thus X and X' are *in harmony* (i.e. fail to be dissonant) if and only if $x_i x'_i \geq 0$ for every i.

Theorem 1. *Let X be any non-zero vector in K. Then there exist elementary vectors $X_1, \ldots, X_r$ of K, such that $X = X_1 + \ldots + X_r$. These elementary vectors may be chosen such that each is in harmony with X and has its support contained in the support of X, but none has its support contained in the union of the supports of the others, and such that r does not exceed the dimension of K or the number of elements in the support of X.*

Proof. The conditions on r follow immediately from the conditions on the supports of $X_1, \ldots, X_r$, and they need not be mentioned further. It suffices to treat the theorem in the notationally simpler case where $X \geq 0$, i.e. $x_i \geq 0$ for all i. We must show that X can be expressed as the sum of non-negative elementary vectors of K, each of which has an element in its support not belonging to the support of any of the others. A preliminary step is to show that there exists at least one non-negative elementary vector whose support is contained in the support of X. Assume inductively that this fact has already been established for all non-zero non-negative vectors $X' \in K$ whose supports are properly smaller than that of X. Let X_0 be any elementary vector of K (not necessarily non-negative) whose support is contained in the support of X. Replacing X_0 by its negative if necessary, we can assume that X_0 has a positive component. Then there exists a largest positive scalar λ such that $\lambda X_0 \leq X$. If $\lambda X_0 = X$, X is itself a non-negative elementary vector. Otherwise, $X' = X - \lambda X_0$ is a non-negative vector of K whose support is contained in the support of X but does not contain the support of X_0. By induction, there exists a non-negative elementary vector of K whose support is contained in the support of X', and hence in the support of X. We can proceed now to prove the theorem itself in the same way. Assume inductively that the theorem has already been established for all non-zero non-negative vectors $X' \in K$ whose supports are properly smaller than the support of X. Repeat the argument above, but this time taking $X_0 \geq 0$, as has just been shown possible. The induction hypothesis yields a decomposition

$$X_2 + \ldots + X_r = X' = X - \lambda X_0 .$$

Setting $X_1 = \lambda X_0$, we get the desired decomposition of X.

Corollary. *The elementary vectors of K generate K algebraically.*

Theorem 1 has been depicted as an extension of a result about graphs, but it is actually equivalent to a fundamental theorem in convex analysis. The theorem in question says that each non-zero vector in a polyhedral convex cone containing no whole lines may be expressed as a sum of r extreme vectors of the cone, where r need not exceed the dimension of the face of the cone in which the given vector lies.

It is not hard to deduce Theorem 1 from this cone theorem. One argues that the set of non-negative vectors of K is a polyhedral convex cone K_+ containing no whole lines, whose extreme vectors are elementary. The faces of K_+ correspond to the "non-negative" signed supports of K. It is just as easy, on the other hand, to deduce the cone theorem from Theorem 1. This is even a convenient route for attaining various important facts about polyhedral convex cones, since the direct proof furnished above for Theorem 1 is so elementary. Recall that, by definition, a polyhedral convex cone C in R^m can be represented as the inverse image of the non-negative orthant of some R^N under some linear transformation T. If C contains no whole lines, T is one-to-one from R^m onto a certain subspace K (the range space of T), and T carries C onto K_+. Application of Theorem 1 to K_+ yields the facts about C.

The study of signed sets is greatly aided by Theorem 1. We can define, in the obvious parallel way, what we mean by two *signed sets* being *dissonant* or *in harmony*. If $S_1, \ldots, S_r$ are signed sets pairwise in harmony, a new signed set S, the *harmonious union* of $S_1, \ldots, S_r$, can be formed by taking

$$S^+ = S_1^+ \cup \ldots \cup S_r^+ \quad \text{and} \quad S^- = S_1^- \cup \ldots \cup S_r^-$$

(In the dissonant case, this S^+ and S^- would overlap, so that there would be no natural way of introducing signs in the union.) If vectors $X_1, \ldots, X_r$ are pairwise in harmony, so are their signed supports, and vice versa. The harmonious union of these signed supports is then the signed support of $X_1 + \ldots + X_r$. Theorem 1 immediately yields the following result, according to which the properties of the signed supports of K can entirely be deduced from those of the elementary signed supports.

Theorem 2. *Every signed support of K is a harmonious union of elementary signed supports of K. On the other hand, every such harmonious union is a signed support of K.*

4. FUNDAMENTAL EXISTENCE THEOREM

In applications of flow theory, the question often comes up as to whether there exists a circulation X whose components x_i lie within certain given ranges I_i depending on the arcs e_i. It may be required for some arcs, say, that $0 \leq x_i \leq k_i$, where k_i is the "capacity" of the arc, while for other arcs x_i is to assume a constant value specified in advance. Some existence theorems pertaining to closed intervals, for instance, are presented by Berge [2, p. 157–160]. These are all really special cases of a theorem of Minty [6] for arbitrary intervals (i.e. non-empty connected sets of real numbers, not necessarily closed or open or bounded, possibly degenerating to a single point).

We shall now prove that Minty's theorem is valid for arbitrary K, if reformulated in terms of elementary vectors.

Theorem 3. *Let $I_1, \ldots, I_N$ be arbitrary real intervals. Then one of the following alternatives holds, but not both:*

(a) *There exists a vector X of K such that $x_i \in I_i$ for $i = 1, \ldots, N$;*

(b) *There exists an elementary vector Y of $K^\perp$ such that $y_1 I_1 + \ldots + y_N I_N > 0$ (i.e. the interval obtained by letting $y_1 x_1 + \ldots + y_N x_N$ vary over all choices of $x_i \in I_i$ lies entirely to the right of 0).*

Proof. The conditions are mutually exclusive, because $y_1 x_1 + \ldots + y_N x_N > 0$ is impossible when $X \in K$ and $Y \in K^\perp$. Let Q be the set of all vectors $X \in R^N$ such that $x_i \in I_i$ for $i = 1, \ldots, N$. In the terminology of [10], Q is a partial polyhedral convex set. If condition (a) fails, Q does not meet K, and a certain separation theorem of the writer [10] may be applied.

This gives the existence of a vector $Y \in K^{\perp}$, such that $y_1 x_1 + \ldots + y_N x_N > 0$ for every $X \in Q$, i.e. $y_1 I_1 + \ldots + y_N I_N > 0$. We must demonstrate that this Y can actually be replaced by an *elementary* vector of $K^{\perp}$. Theorem 1 allows us to set $Y = Y_1 + \ldots + Y_r$, where the vectors $Y_j = (y_{j1}, \ldots, y_{jN})$ are elementary vectors of $K^{\perp}$ pairwise in harmony with each other. The distributive law $(\lambda_1 + \lambda_2)I = \lambda_1 I + \lambda_2 I$ holds for any interval I provided $\lambda_1 \lambda_2 \geq 0$. Therefore

$$y_1 I_1 + \ldots + y_N I_N = \sum_{j=1}^{r} (y_{j1} I_1 + \ldots + y_{jN} I_N)$$

by "harmony." The interval represented on the left lies wholely in the positive part of R, so the same must be true of one of the r intervals corresponding to $Y_1, \ldots, Y_r$ on the right. (If all r intervals contained a non-positive number, then so would their sum.) Thus

$$y_{j1} I_1 + \ldots + y_{jN} I_N > 0$$

for some elementary vector Y_j of $K^{\perp}$, which is what was to be proved.

Notice that (b) in Theorem 3 is a combinatorial condition, in that there are essentially only *finitely* many possibilities to test. Up to positive multiples, $K^{\perp}$ has only finitely many elementary vectors, and a positive multiple of Y makes no difference in (b). In the graph example, the elementary vectors of $K^{\perp}$ correspond to cocycles, and the multiple can always be chosen so that all the components y_i of Y are $+1$, -1 or 0. Then condition (a) holds if and only if, for every elementary cocycle of the graph,

$$0 \in y_1 I_1 + \ldots + y_N I_N = \textstyle\sum_1 I_i - \sum_2 I_i ,$$

where $\sum_1$ is the sum over the indices i such that the given cocycle contains the arc e_i in the direction of its orientation, and $\sum_2$ is the sum over the indices such that e_i is contained in the opposite direction. The "max-flow-min-cut" theorem is readily deduced from this, as has been explained by Minty.

Theorem 3 has been derived from a separation theorem of convex analysis which is stronger than the well-known lemma of Farkas. Actually, this separation theorem can be derived in turn from Theorem 3, using the fact that, by definition, every partial polyhedral convex set is the inverse image under some linear transformation of a set of "paralellopiped form"

$$\{x | x_i \in I_i \quad \text{for every} \quad i\} ,$$

where the I_i are intervals.

Existence alternatives for inequalities involving the variables in a "linear system" $UA = V$ can be obtained by applying Theorem 3 to the subspace K described at the end of the Section 2. Tucker's results in [13] can be established this way. Some special cases will be considered below.

As an immediate combinatorial application of Theorem 3, we shall show how the signed supports of K may be constructed directly from those of $K^\perp$. From matroid theory it is known, of course, how to construct the elementary supports if signs are disregarded. One takes the collection of non-empty subsets S of E such that no elementary support of K meets S in just a single element; the minimal sets among these are the elementary supports of $K^\perp$. The following theorem shows what modification works for the *signed* supports.

Theorem 4. *Let S be a signed set in E. In order that S be a signed support of $K^\perp$, it is necessary and sufficient that every elementary signed support of K not disjoint from S be dissonant with S.*

Proof. The necessity is on the surface. For if $X \in K$ and $Y \in K^\perp$ had signed supports in harmony and not disjoint, then $x_i y_i \geq 0$ for every i with strict inequality for at least one i, contradicting $x_1 y_1 + \ldots + x_N y_N = 0$. To prove the sufficiency, we apply Theorem 3, with the roles of K and $K^\perp$ reversed, to the case where $I_i = (0, +\infty)$ for $e_i \in S^+$, $I_i = (-\infty, 0)$ for $e_i \in S^-$, and $I_i = \{0\}$ for $e_i \in S$. If S is not a signed support of K, that means there is no $Y \in K^\perp$ such that $y_i \in I_i$ for every i. Then by Theorem 3 there exists an elementary vector $X \in K$, such that

$$x_1 I_1 + \ldots + x_N I_N > 0 .$$

This implies that $x_i \geq 0$ for $e_i \in S^+$ and $x_i \leq 0$ for $e_i \in S^-$, with strict inequality for at least one $e_i \notin S$. The signed support of X is then an elementary signed support of K in harmony with S, but not disjoint from S.

5. PAINTINGS

Certain combinatorial problems in graphs involve a specified partitioning of the set of arcs into several subsets. A happy way of describing the partitioning, which has been exploited by Minty, is to say that the arcs have been "painted" various colors. One can then speak of a "black and red cocycle", meaning a cocycle constructed exclusively of "black" arcs and "red" arcs, and so forth. (A black and red cocycle could be entirely black or entirely red.)

Here we shall present several results about the existence of signed supports matching a given "painting." The first is a complementarity theorem.

Theorem 5. *Let each of the elements e_i of E arbitrarily be painted white, green or red (where any of the colors can remain unused). Then there exist a green and white signed support S of K and a red and white signed support S' of $K^\perp$, such that S and S' have no element in common, but every white element is contained in S positively or in S' positively."*

Proof. From among the vectors $X \in K$ such that $x_i \geq 0$ for e_i white and $x_i = 0$ for e_i red, choose one whose support contains a maximal number of white elements. Call it X_0, and let S be its support. Take $I_i = (0, +\infty)$ for e_i white and not in the support of X_0, $I_i = (-\infty, +\infty)$ for e_i red, and $I_i = \{0\}$ for every other i. If there exists a vector $Y \in K^\perp$ such that $y_i \in I_i$ for every i, the support S' of Y, along with S, meets the requirements of the theorem. Suppose, therefore, that no such Y exists. We shall show that leads to a contradiction. By Theorem 3 (with K and $K^\perp$ reversed), there alternatively exists some $X \in K$, such that

$$x_1 I_1 + \ldots + x_N I_N > 0 .$$

The choice of intervals forces $x_i = 0$ for e_i red and $x_i \geq 0$ for e_i white and not in the support of X_0, with $x_i > 0$ for at least one of the latter elements. Then $X + \lambda X_0$, for λ positive and sufficiently large, has a green and white signed support containing no white element negatively and containing at least one more white element than was the case with X_0. This conflicts with the maximality in the selection of X_0.

Corollary. *There exist non-negative vectors $X \in K$ and $Y \in K^\perp$ which are complementary, i.e. such that $x_i y_i = 0$ and $x_i + y_i > 0$ for every i.*

Proof. Paint every element white.

This corollary is a well-known complementary slackness theorem of Tucker [13], which can be made the cornerstone of linear programming theory. Theorem 5 itself could be deduced without much trouble from Tucker's many results about dual linear systems of variables in [13], so only its formulation here, as a combinatorial theorem concerning dual systems of signed sets, is really new. The interesting thing about this formulation, however, is that it leads quickly to the following generalization of Minty's fundamental "colored arc lemma" [6] for directed graphs.

Theorem 6. *Let one of the elements e_i of E be painted black, and let each of the other elements arbitrarily be painted white, green or red. Then one of the following alternatives holds, but not both:*

(a) *There exists an elementary signed support of K containing the black element and otherwise only green and white elements, with the black and white elements contained positively;*

(b) *There exists an elementary signed support of $K^{\perp}$ containing the black element and otherwise only red and white elements, with the black and white elements contained positively.*

Proof. If both conditions could be satisfied simultaneously, one would have overlapping signed supports of K and $K^{\perp}$ in harmony, contrary to Theorem 4. Thus (a) and (b) are mutually exclusive. On the other hand, suppose the black element is repainted white and apply Theorem 5. The S obtained can be expressed as a harmonious union of green and white elementary signed supports of K by Theorem 2, and similarly for S' with "red" in place of green. The previously black element belongs to either S or S' and hence to one of the elementary signed supports in these decompositions. That signed support satisfies either (a) or (b).

Corollary. *Each element of E belongs either to some non-negative (i.e. $S^+ = S$, $S^- = \phi$) elementary signed support of K or to some non-negative elementary signed support of $K^{\perp}$, but not both.*

Proof. Paint the element in question black and every other element of E white.

In the directed graph case, the corollary reduces to the fact that every arc belongs either to some "unidirectional" elementary cycle or to some "unidirectional" elementary cocycle.

Minty has demonstrated in [8] that the simpler "unsigned" version of the property in Theorem 6, namely in which one omits the color white and all mention of signs, may be adopted as a fundamental axiom of matroid theory. He has not developed the signed version as an axiom, although he has shown it is valid for his "digraphoids". According to Theorem 6, the signed version is actually valid for a much broader class of systems than "digraphoids."

An important virtue of the "colored arc lemma" in Minty's convex programming theory for monotone networks [6] is that an efficient combinatorial algorithm actually constructs an elementary cycle or cocycle satisfying alternative (a) or (b). This prompts one to ask whether a constructive procedure exists for the more general case of Theorem 6, too. The proof we have given here is not constructive. We shall see below, however, that the construction can be effected by the simplex algorithm of linear programming.

6. MATRIX REPRESENTATIONS

The relationship between Tucker's combinatorial theory for linear systems of variables, "digraphoids," and the study of elementary vectors and signed supports will now be explained. The results described below are all known, in one way or another, but they need to be worked up together in a certain way as preparation for their use in the next section.

Suppose that, for a certain $m \times n$ matrix $A = (a_{ij})$, K is given by $UA = V$ as at the end of Section 2. The vectors X in K are then precisely the ones whose components satisfy

$$\sum_{i=1}^{m} x_i a_{ij} = x_{m+j} \quad \text{for} \quad j = 1, \ldots, n.$$

Here the values of $x_1, \ldots, x_m$ can be specified arbitrarity, and the values of the remaining components $x_{m+1}, \ldots, x_{m+n} = x_N$ are then explicitly given. At the same time, the vectors Y in $K^{\perp}$ are precisely the ones whose components satisfy

$$\sum_{j=1}^{n} a_{ij} y_{m+j} = - y_i \quad \text{for} \quad i = 1, \ldots, m.$$

These dual systems of equations may conveniently by summarized in a tableau.

$$
\begin{array}{c|ccc|l}
 & y_{m+1} & & y_{m+n} & \\
\hline
x_1 & a_{11} & & a_{1n} & = -y_1 \\
 & & & & \\
 & & & & \\
x_m & a_{m1} & & a_{mn} & = -y_m \\
\hline
 & = x_{m+1} & & = x_{m+n} &
\end{array}
$$

We shall call such a tableau a *Tucker representation* of the subspaces K and $K^{\perp}$. For notational simplicity, we have only pictured a representation in which the symbols $x_1, \ldots, x_N$ occur in undisturbed order along the margins of the tableau. In reality, of course, there will usually be numerous representations, involving different arrangements of the symbols. Every such representation entails the partitioning of E into two subsets D and D', such that the components x_i of a vector X in K for e_i in D are uniquely determined by the components for e_i in D', while the latter components take on all possible combinations of values as X ranges over K. With respect to $K^{\perp}$, D and D' have the opposite property.

Tableaus which represent the same complementary pair of subspaces are said to be *combinatorially equivalent* (along with their corresponding matrices A). How to pass arithmetically from any given tableau to any other combinatorially equivalent tableau has been thoroughly clarified by Tucker [14, 15, 16]. "Pivoting" and rearranging are all that is required. A simple pivot step corresponds to a classical elimination procedure for the dual systems of equations. Any non-zero entry in the tableau may be selected as "pivot"; one then passes to an adjacent representation, in which D and D' are modified by interchanging the e_i of the pivot row with the e_i of the pivot column. As far as getting an initial representation is concerned, that is a very easy matter, at least if K is defined as the subspace orthogonal to a known finite set of vectors in R^N, or as the subspace generated by such a set. (That is the situation in the graph example.)

Tucker's theory grew out of studies of the simplex algorithm for linear programs. But it is also relevant to some ideas Tutte has exploited for representing matroids, as we shall now relate.

Thinking of the vectors X in K as functions on E, we

may restrict them to a given subset D of E. The restrictions may be viewed as vectors in R^M, where M is the number of elements in D. The Tucker representation corresponds to the case where the restriction mapping (a linear transformation) is one-to-one from K onto R^M. The mapping clearly is "one-to-one" if and only if no non-zero vector of K has its support disjoint from D. It is "onto" if and only if no non-zero vector of $K^\perp$ has its support contained in D. Indeed, in these conditions it is enough to speak of elementary vectors. The case where both conditions hold is where D is minimal with respect to the property that it meets every elementary support of K, or equivalently where D is maximal with respect to the property that it contains no elementary support of $K^\perp$.

A set D with the latter properties is called a *dendroid* of K by Tutte. Notice that the complement of a dendroid of K is a dendroid of $K^\perp$. In the example of a connected directed graph, of course, the dendroids of K are the sets of arcs maximal with respect to the property that their deletion would not disconnect the graph; the dendroids of $K^\perp$ are the maximal trees of the graph. In general, according to the analysis above, *the various partionings of $x_1, \ldots, x_N$ and $y_1, \ldots, y_N$ into "row symbols" and "column symbols" in the Tucker representations of K and $K^\perp$ correspond to the possible ways of partitioning E into a dendroid D of K and a dendroid D' of $K^\perp$.*

Given a Tucker representation of K and $K^\perp$ in the notationally simple form above, the $m \times N$ matrix $[I_m, A]$ (where I_m is the $m \times m$ identity matrix) is what Tutte calls a *standard representative matrix* for K (and its matroid). The rows of this matrix are evidently *elementary* vectors of K forming a basis of K. Likewise, $[-A^T, I_n]$ is a standard representative matrix for $K^\perp$ (and the dual matroid), and its rows are elementary vectors of $K^\perp$ forming a basis of $K^\perp$.

Two such standard representative matrices are implicit similarly in a general Tucker representation. They are obtained by applying to the columns of $[I_m, A]$ and $[-A^T, I_n]$ the permutation which is required to restore the symbols x_i from the order in which they occur, down the left side and across the bottom of the tableau, to the order $x_1, \ldots, x_N$. *Every Tucker representation thus yields a basis of elementary vectors for K and one for $K^\perp$.* The bases so obtained will be called *elementary bases.* (A basis consisting of elementary vectors is not actually an elementary basis unless one can also "select an identity matrix from the components.")

In matroid terms, a dendroid D of K yields a certain unique family of elementary supports of K (namely, those of the vectors in the corresponding elementary basis), each having exactly one element e_i in common with D. From matroid theory, it is known that, for any elementary support S of K and any $e_i \in S$, one can find a dendroid D giving rise this way to S and having e_i as its only element in common with S. We can state that result equivalently as follows: each elementary vector of K having a component equal to 1 belongs to some elementary basis of K, and therefore occurs in some Tucker representation. *Tucker's "pivoting" formulas thus serve to compute all the elementary vectors of K and $K^\perp$, up to scalar multiples.*

In a directed graph, for example, an elementary vector of K having some component equal to 1 is a circulation of intensity 1 around some elementary cycle; hence it is actually a representative vector for some elementary cycle, and all its components equal $+1$, -1 or 0. The matrices in the Tucker representations thus must have all their components equal to $+1$, -1 or 0. If the graph is connected, each Tucker representation corresponds to a certain maximal tree D' of E. The elementary basis of K which can be read from the tableau gives the fundamental basis of elementary cycles associated with the tree D'. Pivoting in the tableau is then an arithmetic expression of the purely combinatorial operation of passage to an adjacent tree. That is why the general algorithms of linear programming can be supplemented by simpler combinatorial algorithms, when network problems are involved; see Dantzig's comments [3, Chapter 17].

More generally, a simplified combinatorial approach with strong graph-theoretic analogies is possible in the context of Minty's "digraphoids" and "unimodularity." A matrix A is said to have the unimodular property, if every square submatrix of A has determinant equal to $+1$, -1 or 0. Actually, by Tucker's theory, this is equivalent to the property that every matrix combinatorially equivalent to A (including A itself) have only $+1$'s, -1's and 0's as components. The latter property would make a better definition of unimodularity, in the author's opinion, since it is the property that one is directly concerned with in linear programming applications. (If an initial linear programming tableau in Tucker's format has integral "margins", and if its "non-marginal" matrix has the unimodular property, then the arithmetic of the simplex algorithm will be trivial, and the solutions calculated will be integral.) According to the above, any *circulation matrix* of a directed graph (i.e., the tableau matrix A of some Tucker representation of the space K of circulations in a directed graph) has the unimodular property. The derivation given here for this well-known and important result hinged merely on the fact that, for graphs, every elementary vector of K is a multiple of a *primitive* vector, i.e. a vector having every component equal to $+1$, -1 or 0. In general, let us call a subspace K with the latter property a *unimodular* subspace of R^N. We can say then that K is unimodular if and only if the matrices A in its Tucker representations have the unimodular property. *The study of matrices with the unimodular property is thus equivalent to the study of certain subspaces of* R^N *and their elementary vectors.* Such unimodular subspaces are what Tutte would call "regular chain groups over the real numbers." Minty has shown [8, Appendix *A*] that the systems of elementary signed supports of such subspaces and their orthogonal complements are precisely the objects of his "digraphoid" theory. Minty's results may therefore be regarded as a contribution to the theory of matrices with the unimodular property, in which everything is built up axiomatically in analogy with graphs.

The class of matrices with the unimodular property is, of course, closed under many operations besides those of Tucker's combinatorial equivalence (pivoting, and permutation of rows and columns), notably the operations of

(a) taking submatrices;
(d) multiplying various rows or columns through by -1;
(c) taking transposes;
(b) appending a new row or column having only one non-zero component, and that a $+1$ or -1.

A typical way of proving that a given matrix A has the unimodular property is to show that A may be constructed by a sequence of such operations from a matrix A', which in turn may be interpreted as a circulation matrix of some directed graph. Although A may itself no longer correspond directly to a directed graph, it does correspond to one of Minty's "digraphoids." Linear programming manipulations of A therefore have graphlike interpretations, which might be an important conceptual aid.

Part of our interest has been to show that many such interpretations can even be extended from unimodular subspaces to arbitrary subspaces, in terms of systems of elementary signed supports. Of course, where computational algorithms are concerned, the entirely combinatorial approach which is so efficient in graph theory must give way to a more general linear programming approach.

7. LINEAR PROGRAMMING

The results about signed supports in earlier sections of this paper place certain limitations on the patterns of signs which can occur in an equivalence class of Tucker representations. As a matter of fact, so do Tucker's results concerning linear programs. We shall apply these results now to the study of signed supports.

Tucker has shown that, starting with any tableau representing K and $K^{\perp}$, one may pass by a pivoting algorithm to a representation having one of the patterns of signs in Figure 1. In these tableaus, the top row and the leftmost column are to correspond to the same two e_i's as in the starting tableau.

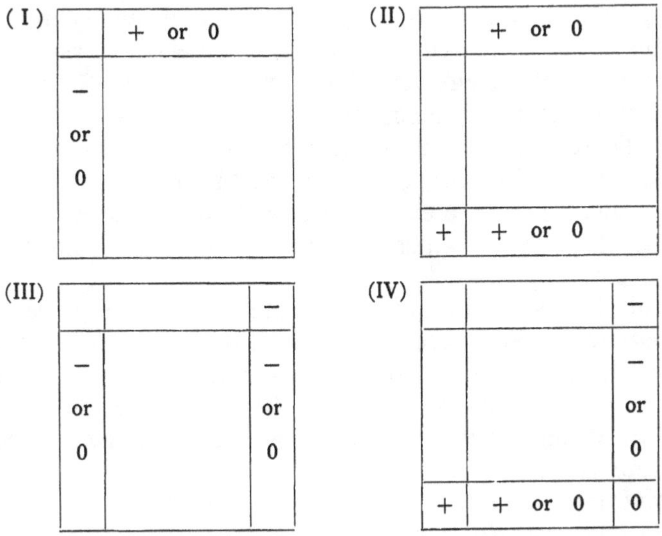

Figure 1

The four cases are mutually exclusive. In linear programming, they correspond to the cases where (I) the X problem and the Y problem have solutions, (II) the X problem is unbounded and the Y problem is inconsistent, (III) the X problem is inconsistent and the Y problem is unbounded, and (IV) the X and Y problems are both inconsistent.

The fact just described is a constructive version of the duality thorem for linear programs. But it may also be viewed, in the light of the observations of the last section, as essentially an assertion about elementary signed supports, and hence as a fundamental theorem about certain signed matroids. The bottom row in (II), for instance, corresponds to a vector in an elementary basis of K, whose support is a non-negative elementary signed support of K containing the e_i of the left-most column.

Taking (I) as alternative (a), and (II), (III) and (IV) together as (b), we can state the result as follows. Let one of the elements of E be distinguished as the "black" element and one as the "grey" element. (We have in mind the e_i of the top row in Tucker's terminal tableaus and the e_i of the left-most column, respectively.) Paint all the other elements white. Then one and only one of the following alternatives holds (and which one it is may be determined by an efficient algorithm):

(a) There exist an elementary signed support S of K containing the black element positively, and an elementary signed support S' of $K^{\perp}$ containing the grey element positively, such that no white element belongs negatively to S or to S', and no white element belongs both to S and to S'.

(b) There exists a non-negative elementary signed support of K containing the grey element but not the black element, or there exists a non-negative signed support of $K^{\perp}$ containing the black element but not the grey element, or both.

Actually, this is not quite completely contained in the result stated by Tucker, because there the black element and the grey element correspond to a row and a column initially. But that correspondence can always be arranged, unless (b) holds. For, if the grey element corresponds to a row in some Tucker representation, and that row is not entirely zeros, a simple pivoting step will calculate a new representation in which the grey element corresponds to a column. If, on the other hand, the row contains only zeros, the corresponding elementary basis of K has a vector whose support is the grey element along; this is a case of alternative (b). Similarly for the black element.

This somewhat mysterious, purely combinatorial result

about signed sets, let us emphasize, has the celebrated duality theorem for linear programs as a corollary. We must therefore regard it as one of the deepest theorems possible about the signed matroids arising from subspaces of R^N. Here is an even more elaborate result, which corresponds in linear programming to the case where some constraints are equations and some variables are unconstrained.

Theorem 7. *Let one of the elements of E be painted black and one grey. Let each of the remaining elements be painted white, green or red. Then one of the following alternatives holds, but not both:*

(a) *There exists an elementary signed support S of K containing the black element positively and no red elements, and an elementary signed support S' of $K^\perp$ containing the grey element positively and no green elements, such that no white element belongs negatively to S or to S', and no white element belongs both to S and to S'.*

(b) *There exists an elementary signed support of K containing the grey element and otherwise only green or white elements, with the grey and white elements contained positively; or there exists an elementary signed support of $K^\perp$ containing the black element and otherwise only red and white elements, with the black and white elements contained positively; or both.*

Proof. This theorem must be considered known as regards linear programming, although Tucker has not discussed equality constraints or free variables explicitly in terms of his terminal tableaus. Computationally, one can decide between (a) and (b) (and construct the elementary signed supports in question) using some extension of the simplex algorithm to this more general case, such as the extension described by the writer [9]. Details will not be given here. For the sake of proving Theorem 7, however, it seems appropriate to indicate how the general case may be reduced constructively to the one previously dealt with.

Starting from an arbitrary Tucker representation, we first arrange, by simple pivoting if necessary, that the black element corresponds to a row and the grey element to a column (henceforth the "black" row and the "grey" column, etc.) (If this is not possible, then alternative (b) holds, as already explained.) We continue with simple pivoting, choosing at each step as pivot a non-zero entry in a green row and white column, or in a green row and red column, or in a white row and red column. (The consequence is that the number of red elements in D plus the number of green elements in D' in the dendroid partition of E is increased at each step.) After finitely many steps, a Tucker representation of the sort in Figure 2 is obtained (upon rearrangement of the rows and columns). The 0's mark submatrices all of whose entries are 0. (In any given example, of course, one would expect a degenerate version of this tableau, without any green rows at all, say.)

	grey	white	red	green
black			?	
white			0	
green	?	0	0	
red				

Figure 2

At this point, we look to see whether the entries in the grey column and green rows are all zero. If not, one of the green rows furnishes an elementary vector whose support satisfies alternative (b) of Theorem 7. Similarly, if the black row has a non-zero entry in a red column, then (b) holds. Otherwise we proceed with Tucker's analysis of the black-gray-white subtableau, eventually transforming it to one of the four cases in Figure 1. (At each iteration, the whole tableau is transformed in accordance with what is happening in the subtableau. The transformations trivially preserve the indicated pattern of zeros.) The conclusion, in terms of elementary signed supports, can be read from the final tableau as before.

Theorem 7 reduces to our generalization of Minty's "colored arc lemma" (Theorem 6), if one simply omits everything having to do with there being a grey element. The simplex algorithm may then be employed in practically the same way to decide constructively between the alternatives. The terminal tableaus correspond to having (I) or (III) of

Figure 1 in the upper left of Figure 2, with leftmost columns deleted. In the purely black-and-white case, as Tucker has pointed out in [15], these alternative tableaus correspond to the alternatives in the classical lemma of Farkas.

Since Theorem 7 and its algorithm are so complicated (as, indeed, they have to be to cover so many cases), a more special illstration may be helpful. Let us demonstrate how the "unsigned" form of Minty's lemma (where nothing is painted white) may be decided for an arbitrary subspace K. Here we are given a painting of E, where one element is black, and all other elements e_i are red or green. We start with any Tucker representation of K. If the black element corresponds to a column of the tableau, we look for non-zero elements in that column. If one exists, pivoting on it will yield a representation of K in which the black element corresponds to a row. If none exists, then the set consisting of the black element alone is an elementary support of K, and alternative (b) holds. Assume now that the black element corresponds to a row. We pivot next on any non-zero entry in a green row and red column. This is kept up until there are no more such pivots, at which time the tableau has the form in Figure 3.

	red	green
black	?	
green	0	
red		

Figure 3

If now the black row has a non-zero entry in some red column, the e_i's corresponding to rows with non-zero entries in that column, along with the e_i of the column itself, form an elementary support of $K^\perp$ containing the black element and otherwise only red elements. If, on the other hand, the black row has only 0's in red columns, then an elementary support of K is given by the black element and the green elements corresponding to columns with non-zero entries in the black row. These are alternatives (a) and (b). (Note, incidentally, that this special case of the algorithm is valid for graphoids arising from subspaces of vector spaces over *arbitrary* fields.)

References

1. Berge, C. (1958). *Théorie des Graphes et ses Applications*, Dunod, Paris.
2. Berge, C. and Ghouila-Houri, A. (1962). *Programmes, Jeux et Reseaux dc Transport*, Dunod, Paris.
3. Dantzig, G. (1963). *Linear Programming and Extensions*, Princeton Univ. Press, Princeton.
4. Klee, V. L. (1968). "Maximal separation theorems for convex sets," *Trans. Amer. Math. Soc.*, to appear.
5. Kuhn, H. W. and Tucker, A. W. (ed.) (1956). *Linear Inequalities and Related Systems. Ann. of Math. Study* **38**, Princeton University Press, Princeton.
6. Minty, G. J. (1960). "Monotone Networks," *Proc. Roy. Soc. London Ser. A*, **257**, 194–212.
7. Minty, G. J. (1962). "On an Algorithm for Solving some Network-Programming Problems," *Operations Res.* **10**, 403–405.
8. Minty, G. J. (1966). "On the Axiomatic Foundations of the Theories of Directed Linear Graphs, Electrical Networks and Network-Programming, *J. of Math. and Mech.*, **15**, 485–520.
9. Rockafellar, R. T. (1964). "A Combinatorial Algorithm for Linear Programs in the General Mixed Form. *J. Soc. Indust. Appl. Math.*, **12**, 215–225.
10. Rockafellar, R. T. (1966). "Polyhedral Convex Sets with some Closed Faces Missing," unpublished. The separation theorem cited from this paper has recently been sharpened by Klee [4].
11. Rockafellar, R. T. (1966). "Convex Programming and Systems of Elementary Monotone Relations," *J. Math. Anal. and Appl.*, to appear.
12. Rockafellar, R. T. (1968). *Convex Analysis*, Princeton Univ. Press, Princeton.
13. Tucker, A. W. (1956). "Dual Systems of Homogeneous Linear Relations. [4], 3–18.
14. Tucker, A. W. (1960). "A Combinatorial Equivalence of matrices," *Combinatorial analysis*, R. Bellman and M. Hall (ed.), *Proceedings of Symposia in Applied Mathematics*, **10**, Amer, Math. Soc., 129–140.
15. Tucker, A. W. (1963). "Simplex Method and Theory," *Mathematical Optimization Techniques*, R. Bellman (ed.), U. of California Press, Berkeley, 213–231.
16. Tucker, A. W. (1963). "Combinatorial Theory Underlying Linear Programs," *Recent Advances in Mathematical Programming*, R. L. Graves and P. Wolfe (ed.), McGraw-Hill, New York, 1–16.
17. Tutte, W. T. (1956). "A Class of Abelian Groups," *Canad. J. Math.*, 8, 13–28.

18. Tutte, W. T. (1959). "Matroids and Graphs," *Trans. Amer. Math. Soc.*, **90**, 527–552.
19. Tutte, W. T. (1960). "An Algorithm for Determining Whether a Given Binary Matroid is Graphic," *Proc. Amer. Math. Soc.*, **11**, 905–917.
20. Tutte, W. T. (1965). "Lectures on Matroids," *J. Res. of Natl. Bur. Standards Ser. B.*, **69**, 1–47.
21. Whitney, H. (1935). "The Abstract Properties of Linear Dependence," *Amer. J. Math.*, **57**, 509–533.
22. Camion, P. (1965). "Application d'une généralisation du lemme de Minty à une problem d'infimum de fonction convexe," *Cahiers Centre d'Etudes Rech. Oper.*, **7**, 230–247.
23. Camion, P. (1963). "Matrices totalement unimodulaires et problèmes combinatoires," Thèse, Université de Bruxelles.

CHAPTER 8

A Combinatorial Problem in the Theory of Free Monoids[1]

A. LENTIN and M. P. SCHÜTZENBERGER,
Faculté des Sciences, Paris, France

1. INTRODUCTION

The combinatorial properties of free monoids play a role in several lemmas which are used in the theory of free groups (or free Lie algebras), formal languages, automata, variable length codes, and elsewhere.

The purpose of the present article is to determine a property of this type (Theorem 5 below).

2. SUMMARY OF PREVIOUS WORK

2.1. Freedom ; Primitiveness

We take a set $X = \{x, y, \ldots\}$, which generates the *free monoid* X^* whose elements are called *words*. The *length* of a word f is designated by $|f|$, the word of length zero by e. Every subset A of X^* generates a sub-monoid, written A^*, for which it forms a *system of generators*.

[1] This work has been supported by contract AF61(052)945 with the United States Air Force.

Example. Let $X = \{x, y\}$ and let

$$A_1 = \{xyx, x\},$$

$$A_2 = \{xy, yx, x\},$$

$$A_3 = \{x, y, xy\}.$$

It is seen that A_1^*, $A_2^* \subset X^*$ (strict inclusion),

whereas $$A_3^* = X^*.$$

Definition 1. *$B \subset X^*$ is called a base of A (or a code) if there exists a set X' and a surjection φ of X' on B which can be extended to a monomorphism of X'^* in X^*.*

More intuitively, this is equivalent to saying that B is a base iff every word of A^* has a unique factorization in terms of the words of B.

Example. A_1 is a base for A_1^*. A_2 is not a base for A_2^* since xyx is capable of two factorizations, namely, $(xy)x$ and $x(yx)$. A_3 is not a base for A_3^*; but $\{x, y\}$ is.

Definition 2. *The submonoid A^*, generated by the system of generators A, is called free if it has a base.*

Example. A_1^* is free and has the base A_1. A_3^* is free and has the base X. A_2^* is not free: in effect, if it had a base the latter would contain x, but not y, and would therefore necessarily contain xy and yx.

Remark. We have $A_2^* \subset X^*$, and **card** $(X) <$ **card** (A_2); note that this can be generalized. In what follows, whenever A^* is a free submonoid, A will always be its base.

Theorem 1. *A necessary and sufficient condition for A^* to be free and of base A is the following (condition L):*

$$(L): \quad \textit{For all} \quad h \in X^* \setminus A^*, \qquad hA^* \cap A^* \cap A^*h = \phi$$

Proof. We will prove the equivalent proposition

$$(A \text{ is not a base}) \Longleftrightarrow (\bar{L}).$$

We have

$$(\bar{L}): \quad \exists\, h \in X^* \setminus A^* \quad \text{such that} \quad hA^* \cap A^* \cap A^*h \neq \phi\,.$$

$$(A \text{ is not a base}) \Longrightarrow (\bar{L})\,.$$

Since A is not a base, A^* contains a set of words which are capable of at least two distinct factorizations, and in that set there is a subset consisting of minimal length words. Let

$$(1) \qquad m = a_{i_1} \ldots a_{i_n} = a_{j_1} \ldots a_{j_p}$$

be such a word. The minimality of m implies that $a_{i_1} \neq a_{j_1}$. We can then take $|a_{i_1}| < |a_{j_1}|$; whence

$$(2) \qquad a_{j_1} = a_{i_1}h\,, \qquad h \in X^*\,, \qquad |h| \neq 0\,.$$

Then

$$(3) \qquad a_{i_2} \ldots a_{i_n} = ha_{j_2} \ldots a_{j_p}\,.$$

But

$$(2) \Longrightarrow A^* \cap A^*h \neq \phi\,,$$

$$(3) \Longrightarrow hA^* \cap A^* \neq \phi\,,$$

whereas

$$\text{minimality and} \quad (3) \Longrightarrow h \in X^* \setminus A^*\,;$$

whence follows $(\bar{L})$.

$$(\bar{L}) \Longrightarrow (A \text{ is not a base})$$

$$(\bar{L}) \Longrightarrow \exists\, h \in X^* \setminus A^* \quad \text{and} \quad g, m, g' \in A^* \quad \text{such that} \quad hg = m = g'h\,.$$

By virtue of its belonging to $X^* \setminus A^*$, h is not empty, and so neither is m. By simplifying on the left in A (if necessary) we can arrange for m and g' not to have the same initial letter in A^*.

The double equality $g'hg = mg = g'm$ (or the one which is left after simplifying) proves that A is not a base.

Definition 3. *If A^* is a free submonoid of base A, we define the set $\pi(A^*)$ of A-primitive elements by the equivalence:*

$$f \in \pi(A^*) \Longleftrightarrow f \in A^* \quad \text{and} \quad f \neq g^p \quad \text{for any}$$
$$g \in A^* \quad \text{and any} \quad p \neq 0, 1 .$$

Instead of *X-primitive* we shall simply say *primitive.*

Remark. It is clear that A-imprimitiveness implies imprimitiveness, but the converse is not true. For instance:

$$A = \{xyx, y\}, \qquad f = xyxy, \qquad f \in \pi(A^*), \qquad f \notin \pi(X^*).$$

Every element of a submonoid A^* having the base A can be represented uniquely as the power of an A-primitive element.

Proof. Clearly, there exists such a representation. Suppose then that we have

$$f \in A^* ; \ f = g^p = h^q ; \ p, q \geqslant 1 ; \ g, h \in \pi(A^*) .$$

According to the hypothesis, f, g, and h each have a unique factorization. Proceeding by identification, one shows that $g = h$, whence $p = q$.

The following theorem, as well as its corollaries, are related to the concept of primitiveness (cf. [1]).

Theorem 2. *A necessary and sufficient condition for two words a, $b \in X^*$ to be two powers of the same word (which one can always suppose to be primitive) is that a power a^p of a and a power b^q of b contain a common left (right) factor of length*

$$|a| + |b| - (|a| \cap |b|) ,$$

where $|a| \cap |b|$ stands for the greatest common divisor.

Proof. Set $|a| = \alpha$, $|b| = \beta$. We first treat the case where $\alpha \cap \beta = 1$. Let $a = x_1 \ldots x_\alpha$ and $b = y_1 \ldots y_\beta$; we can take $\beta < \alpha$.

Sufficiency. We have the relations:

$$1 : x_1 = y_1 ,$$

$$2 : x_2 = y_2 ,$$

$$\alpha + \beta - 1 : x_\lambda = y_\mu .$$

Let us scan this set of relations in the following way:

i) If possible, add β to the number of the line one has just read;

ii) otherwise add $-\alpha + \beta$.

This scan is possible, for it is equivalent to uniting the vertices of a polygon in steps of β. Since β and α are relative primes, we exhaust the vertices. We have therefore $a = x_1^\alpha$, $b = x_1^\beta$.

Necessity. We have above a system of $\alpha + \beta - 1$ homogeneous equations in $(\alpha + \beta)$ unknowns. Let us add the relation:

$$\sum_i \lambda_i x_i + \sum_{j'} \mu_{j'} y_j = k$$

where $k \neq 0$ and the coefficients are not all zero. The system is then determined.

It is easily seen that one can form a determined system of the same rank by replacing the last relation with

$$x_\lambda = k_1 , \qquad y_\mu = k_2 .$$

The words a and b can be written with two types of letters, and since $\alpha \cap \beta = 1$, they are not powers of a same third word.

For $\alpha \cap \beta = \delta$, we take sections of length δ and apply the previous result.

Corollary 1. *A necessary and sufficient condition for $a, b \in X^*$ to be powers of the same word is that ab and ba contain a common left factor of length*

$$|a| \pm |b| - (|a| \cap |b|) .$$

Proof. (same notations). The theorem is trivially true for $\alpha = \beta$. Let us suppose $\beta < \alpha$. The hypothesis implies that ab is a left factor a^2, ba a word of the form $b^\lambda b_1$, hence a left factor of $b^{\lambda+1}$. We apply the theorem.

Corollary 2. *A necessary and sufficient conditon for $a, b \in X^*$ to be powers of the same word is that there exist in $\{a, b\}^*$ two distinct elements having no common factor in $\{a, b\}^*$ and having in X^* a common left factor of degree $|a| + |b| - (|a| \cap |b|)$.*

The proof is by case, in a way analagous to the preceding proof.

Corollary 3. *a, b is a base of $\{a, b\}^*$, and $\{a, b\}^*$ is free iff a and b are not powers of the same word.*

In later applications, we shall frequently use the following corollary.

Corollary 4. *For f, $g \in \pi(X^*)$, $h \in X^*$, $p, q > 1$, the hypothesis $f^p = g^q h$ implies that either $g = f$ and $h = f^{p-q}$, or else $(q-1) \cdot |g| < |f|$.*

To state the last corollary, we must give finally the definition of a fundamental concept.

Definition. *We shall call sesquipower on X^* a word f of the form:*

$$f = (uv)^k u\,, \quad k > 0\,, \qquad uv \in \pi(X^*)\,, \quad v \neq e\,.$$

A sesquipower such that $k \geqslant 2$ will be called a strong sesquipower.

Corollary 5. *For $k \geqslant 2$, a strong sesquipower $(uv)^k u$ has a unique representation as a strong sesquipower.*

Proof. Let $(uv)^k u = (wz)^j w$; then $(uv)^{k+1}$ and $(wz)^{j+1}$ are two powers of primitive words and have a common left factor of length $k|uv| + |u| = j|wz| + |w|$. If we subtract from the length of this common factor the sum of the lengths, we obtain:

$$k|uv| + |u| - |uv| - |wz| = (k-1)|uv| + |u| - \frac{k|uv| + |u|}{j + \theta}\,,$$
$$0 < \theta < 1\,.$$

This difference has the sign of:

$$[(k-1)(j+\theta) - k]|uv| + (j + \theta - 1)|u|\,.$$

The coefficient of $|uv|$ is:

$$k(j + \theta - 1) - (j + \theta)\,.$$

For $k = 2$ it becomes $j - 2 + \theta$, which is positive. Theorem 2 is now applicable.

2.1. Conjugacy

Definition. *If A^* is a free submonoid of base A, we define the relation of A-conjugacy by the equivalence*

$$f \; A\text{-conj. } g \Longleftrightarrow \exists\, h, h' \in A^* \quad \text{such that}$$
$$f = hh' \quad \text{and} \quad g = h'h\,.$$

Instead of *X-conjugacy* we shall simply speak of *conjugacy.*

Remark. It is clear that A-conjugacy implies conjugacy, but the converse is not true. For instance:

$A = \{xy, yx\}$; then xy and yx are conjugate, but not A-conjugate. It follows immediately from this definition that

1. f A-conjugate $g \Longrightarrow f, g \in A^*$.
2. i) A-conjugacy is *reflexive* (take $e = h'$).
   ii) A-conjugacy is *symmetric* (evident).
   iii) A-conjugacy is *transitive.*

Take

$$f = hh'\,, \qquad g = h'h\,; \qquad g = kk'\,, \qquad m = k'k\,.$$

Then we have in A^*

$$g = h'h = kk'\,.$$

Utilizing the uniqueness of the factorization in A^* (where A^* is free), we obtain, for example,

$$h' = k_1\,, \qquad h = k_2k'\,,$$

where $k = k_1k_2$; whence

$$f = k_2(k'k_1)\,, \qquad m = (k'k_1)k_2\,.$$

The relation of A-conjugacy is an equivalence.

3. A-conjugacy is compatible with the power mapping:

$$f \to f^p$$

In effect,

$$f = hh'\,, \qquad\qquad g = h'h\,,$$

$$f^p = h[(h'h)^{p-1}h'], \qquad g = [(h'h)^{p-1}h']h .$$

These different results can be synthesized in the following theorem :

Theorem 3. *For $f, g \in AA^*$, set*

$$C_A(f, g) = \{h \in A^* : fh = hg\} .$$

Then f A-conjugate $g \Longleftrightarrow C_A(f, g) \neq \phi$. Furthermore, for two different A-conjugate words there exists a unique positive integer p and a unique ordered pair $u, v \in A^*$ such that :

$$v \neq e ; \qquad uv, vu \in \pi(A^*) ; \qquad f = (uv)^p ; \qquad g = (vu)^p .$$

$$C_A(f, g) = u(vu)^* ; \qquad C_A(g, f) = v(uv)^*$$

Proof. *Necessity.*

$$f \text{ A-conjugate } g \Longrightarrow f = hh' , \qquad g = h'h ; \qquad h, h' \in A^*$$

$$fh = hg = hh'h ; \qquad h \in C_A(f, g) ;$$

$$C_A(f, g) \neq \phi .$$

Sufficiency. Let us suppose that $C_A(f, g)$ contains at least one word h ; then we have

$$fh = hg ,$$

$$fhg = hgg ,$$

$$ffh = hgg .$$

More generally, for all $m \geqslant 1$,

$$f^m h = hg^m .$$

However, there exists a unique integer n such that

$$n|f| \leqslant h < (n+1)|f| .$$

We have then :

$$f = f_1 f_2 , \qquad h = f^n f_1 ;$$

$$f^{n+1}f_1 = f^n f_1 g ;$$

$$f_2 f_1 = g .$$

It can be immediately verified that this solution, obtained from necessary conditions, verifies

$$fh = hg$$

Representation. We know that every word of A^* is a power of an A-primitive word, so that we have

$$f = f_1 f_2 = f_0^p ; \qquad p \geqslant 1 ;$$

$$f_0 = uv , \qquad f_1 = (uv)^i u , \qquad f_2 = v(uv)^j$$

$$i + j + 1 = p ,$$

and this uniquely. It follows that

$$f = f_1 f_2 = (uv)^p ; \qquad uv \in \pi(A^*) ,$$

$$g = f_2 f_1 = (vu)^p ; \qquad vu \in \pi(A^*) .$$

For $f \neq g$, we have $uv \neq vu$; hence $v \neq e$. The rest of the conclusion is evident.

Corollary 1. *For every $f \in AA^*$, the following properties are equivalent :*

(1) *f is A-primitive ;*
(2) *The class of A-conjugates of f containts an A-primitive word ;*
(3) *$C_A(f, f) = f^*$ and any relation*

$$f' f^p f'' = f^q \ \textit{implies that} \ f', f'' \in f^* .$$

(4) *If $f \in A^k$, the class of A-conjugates of A contains exactly k words.*

The proof presents no difficulties.

Finally, Theorem 2 yields the following theorem immediately by a " shift " :

Theorem 4. *A necessary and sufficient condition for the words f and g to be conjugate is that two powers f^p and g^q of these words contain a common factor of length $|f| + |g| - (|f| \cap |g|)$.*

2.3 Relation to other theories

To begin with, it is clear that, for $A = X$, the concepts of primitiveness and conjugacy originate, by restriction to the monoid X^*, in analogous concepts relative to the *free group* generated by X. They can be extended immediately to a base A with the help of the monoid X'^* and the monomorphism which were defined at the beginning. Furthermore, some concepts and results can be extended to other monoids. In order to better visualize these extensions, we give first a " geometrical " interpretation.

To each $f \in X^*$, let us associate the mapping $\hat{f}$ of the segment $[1, \ldots, |f|]$ in X which sends i onto the ith letter of f. Then with the product $h = fg$ (in the monoid) there is associated the mapping $\hat{h} = \widehat{fg} = \hat{f} \cdot \hat{g}$.

$$\hat{h}(i) = \begin{cases} \hat{f}(i)\,, & \text{for} \quad i \in [1, \ldots, |f|]\,; \\ \hat{h}(i - |f|)\,, & \text{for} \quad i \in [|f|, \ldots, |f| + |g|]\,. \end{cases}$$

In this construction, f and g are conjugate iff $\hat{g}$ can be deduced from $\hat{f}$ by a cyclic shift. In other words, there exists a fixed j such that:

$$\hat{f}(i) = \begin{cases} \hat{g}(i + j)\,, & \text{for} \quad i \in [1, \ldots, |g| - j]\,; \\ \hat{g}(i + j - |g|)\,, & \text{for} \quad i \in [|g| - j + 1, \ldots, |g|]\,. \end{cases}$$

In the same way, f is the pth power of g iff

$$|f| = p|g| \quad \text{and} \quad \hat{f}(i + k|g|) = \hat{g}(i) \quad \text{for}$$
$$i \in [1, \ldots, |g|] \quad \text{and} \quad k \in [0, 1, \ldots, p - 1]\,.$$

Thus, the *primitiveness* of a word is equivalent to the *aperiodicity* of the associated mapping onto its interval of definition.

Fine and Wilf have shown that most of these results can be extended to more general monoids consisting of continuous mappings in a topological set X of intervals of the real line, when these mappings are compounded by the product "." This is true in particular of Theorem 4: its extension shows that two periodic mappings are equal on the necessary and sufficient condition that they coincide on an interval whose length is equal to the sum of the lengths of their respective periods.

3. MAIN RESULTS

3.1. Statement

Theorem 5. *Let $A = \{a, b\}$ be a base such that each word of $a^*b \cap ab^*$ is primitive; then each A-primitive word of A^* is primitive.*

Actually, as we shall see, it suffices that each word of $a^*b \cap ab^*$ of length less than $3|ab|$ be primitive in order to guarantee the conclusion of the property. Also, at most one word of $a^*ab \cap abb^*$ can be imprimitive.

3.2. Terminology

We consider $A = \{a, b\} \subset X^*$, $a \neq b$. In view of the hypotheses, we have that a and b, elements of $a^*b \cup ab^*$ are primitive. For the sake of definiteness, we take $|b| \leqslant |a|$.

We introduce the following terminology :

$$\text{For} \quad d = d_1 d_2 \ldots d_k \in A^k \ (\text{i.e. } d_1, d_2, \ldots, d_k \in A),$$

we call an *A-factor of d* any product $d_i d_{i+1} \ldots d_j$ ($1 \leqslant i \leqslant j \leqslant k$) occurring in d. Furthermore, we say that $d' = d'_1 d'_2 \ldots d'_{k'} \in A^{k'}$ is a *principal segment of d* iff there exists f, $f' \in X^*$ such that $fd'f' = d$ with $|f| < |d_1|$; $|f'| < |d_k|$.

Further we say that c is *disjoint from d* iff

$$f, f' \neq e \quad \text{and for all} \quad j, j', \qquad fd'_1 \ldots d_{j'} \neq d_{1'} \ldots d_j \,.$$

Thus, if d' is a principal disjoint segment of d, any A-factor of d' (or of d, with the exception of $d_1, d_2, \ldots, d_k$, or $d_1 d_2 \ldots d_{k-1}$) is again a principal disjoint segment of a well defined A-factor of d (or of d').

3.3. Preliminary results

(1) *Let c, $d \in A^*$ be conjugate but not A-conjugate. Any A-factor of $c^n (n < 1)$ is a principal disjoint factor of an A-factor of d^n.*

Proof. We have $hc = dh$; hence for all positive integers n, $hc^n = d^n h$ with $h \in X^* \setminus A^*$. The hypothesis that an A-factor of c^n is not disjoint from d would imply that

$$c^n = c_1 c_2 \,, \qquad d^n = d_1 d_2 \,; \qquad c_1, c_2, d_1, d_2 \in A^* \,, \qquad hc_1 = d_1 \,;$$

$$c_2 = d_2h .$$

Thus we would have

$$hc_1c_2 = d_1c_2 = d_1d_2h \quad \text{with} \quad c_1c_2,\, d_1c_2,\, d_1d_2 \in A^* ;$$

hence $h\ A^* \cap A^*\ h \cap A^* \neq 0$ in contradiction to $h \notin A^*$ and, according to Theorem 1, the hypothesis that A^* is free.

(2) *For $p > 0$ and $q > 2$, $c = a^q$ cannot be a disjoint principal segment of $d = ab^pa$.*

Proof. Let $ab^pa = fa^qf'$ where $|f|, |f'| < |a|$. Since $q > 2$, at least one A-factor a of a^q is a principal disjoint segment of b^p, hence $p \geqslant 2$.

Now either b^p and a^q have a common segment of length $\geqslant |a| + |b|$ or they do not. In the first case, by Theorem 3, a and b are conjugate; then we have

$$a = uv , \qquad b = vu ; \qquad uv(vu)^puv = f(uv)^qf' ,$$

where the segments uv of $(vu)^p$ and of $(uv)^q$ must coincide because $p \geqslant 2$ and, by the hypothesis, $a,\ b \in \pi(X^*)$. Thus

$$uvvu = fuv ; \qquad (vu)^{p-2} = (vu)^{q-2} ; \qquad vuuv = uvf' .$$

The first (or the third) relation shows that $vu = uv$, i.e., $a = b$ in contradiction to the hypothesis $a \neq b$.

In the second case $|b^p| < |a| + |b|$. Because $|a| \geqslant |b|$ this implies $q = 3$ and we can write

$$a = fg = gh = h'g' = g'f' ,$$

so that

$$a^q = ghah'g' ; \qquad b^p = hah' \quad \text{with} \quad |h| + |h'| < b .$$

Thus at least one of h or h' (say h) has length $< |b|/2$; hence $1 < |a|/2$. By Theorem 3, the relation $a = fg = gh$ implies

$$f = u'v' ; \qquad h = v'u' , \qquad a = (u'v)^{r'}u' ,$$

where $r' \geqslant 2$ since $|h| < 1/2|a|$.

Thus a is a strong sesquipower and by Corollary 5, we can write in a unique manner

$$h = (vu)^s ; \qquad a = (uv)^r u \qquad (s > 0,\ r \geqslant r')$$
$$\text{with} \quad vu \in \pi(X^*) .$$

Then

$$b^p = (vu)^s (uv)^r u (uv)^{s'} g' ; \qquad (uv)^{s'} g' = h' ; \qquad |g'| < |uv| .$$

Again since $vu \in \pi(X^*)$, any segment of b^p equal to vu or to uv is in fact a $\{u, v\}$-factor. The inequality $|h| < |b|/2$ shows that $(vu)^s uv$ is a left factor of b. However, b^p has no other segment $vuuv$ except at its end, where it occurs in $uv(uv)^{s'} g'$. Now this last word is strictly shorter than b and the hypothesis $a \neq b$ implies that $vu \neq uv$. Thus there is a contradiction because $p \geqslant 2$ (as has been shown above).

(3) *For $p > 1$ the word $ab^p a$ cannot be a principal disjoint segment of $b^r a^2 b^s$. For $p = 1$, it is so only if $a^2 b$ is imprimitive.*

Proof. The hypothesis implies

$$b^r = b_1 b_2 ; \qquad b^s = b_3 b_4 ; \qquad ab^p a = b_2 a^2 b_3 \quad \text{with} \quad b_3 b_2 = b^p .$$

Thus we have $ab_3 b_2 a = b_2 aab_3$, showing that $ab^p a$, hence $a^2 b^p$, is imprimitive. For $p = 1$, the proposition is proved. For $p > 1$, the proposition will be proved by showing that $a^2 b^p = c^q$, $q > 1$, is incompatible with the hypothesis that a and b are not powers of the same word (Theorem 2).

According to Corollary 4 of Theorem 2, the conclusion is established for

$$|a| \geqslant |c| \quad \text{or} \quad (p-1)|b| \geqslant c .$$

Let us suppose that $|a| < |c|$ and $(p-1)|b| < |c|$. Then, in view of the equality

$$2|a| + p|b| = q|c| ,$$

these inequalities require that

$$2 + \frac{p}{p-1} > q ;$$

hence, $q = 2$ or 3. Let $q = 2$. For even p, the conclusion follows at once. For odd p, $|b|$ is necessarily even. We have $b = b_1b_2$ with $|b_1| = |b_2|$, which allows us to segment the equation and arrive at the conclusion. For $q = 3$, the calculation offers no difficulties in principle, but it is very long. For brevity we shall not give it here.

(4) *For* $p > 0$, ab^pa *cannot be a principal disjoint segment of* aad', *nor of* $d'aa$ $(d' \in A^*)$, *or of* $ab^{p'}a$ *or* $b^{p'}$.

Proof. Suppose $fab^paf' = aad'$. The hypothesis of disjointness implies that a is a principal disjoint segment of aa; hence by Corollary 1, $a \notin \pi(X^*)$, which is a contradiction. The same applies to $d'aa$.

In the two other cases, the same argument applies for b and bb.

(5) *Let* $c = ab^pa$ $(p > 0)$ *be a principal disjoint segment of* $d \in A^*$ *and suppose that* d *has no A-factor* $ab^{p'}a$ *with* $p' < p$. *Then either* d *has an A-factor of the form* b^rab^s $(r + s = p)$ *which is a principal disjoint segment of* c *or else* $p = 1$ *and* $d = ba^2b$.

Proof. Assume $d \neq ba^2b$. The case of $d \in b^*a^2ab^*$ is excluded by (2) and (3) above.

For $d = b^{r'}ab^{s'}$ the hypothesis of disjointness implies $r', s' > 1$; we must have

$$|b^{r'-1}ab^{s'-1}| < |ab^pa| < |b^{r'}ab^{s'}|;$$

hence $r' + s' > p + 1$, and $r' - 1 + s' - 1 \geqslant p$: The result is verified.

If $d \notin b^*aa^*b^*$, the case of $d \in b^*$ is excluded by (4) and d must have an A-factor of the form $ab^{p'}a$ where $p' \geqslant p$ by hypothesis. Again by (4), $d \neq ab^{p'}a$ so that either $d = ab^{p'}ad'$ or $d = d'ab^{p'}a$ with $d' \neq e$. The result is verified by taking b^pa or ab^p.

3.4. Conclusion of the proof

We consider $g, g' \in A^*$, conjugate but not A-conjugate, such that $rg^n = g'^nr$, $r \notin A^*$, for all n. Such a situation necessarily obtains when $g \in A^*$ is A-primitive without being primitive; $g = f^m$ $(f \in X^* \setminus A^*,\ m > 1)$ since $f^{m+1} = fg = gf$. According to (1), every A-factor of $g^n(g'^n)$ is a principal disjoint segment of some A-factor of $g'^n(g^n)$. Since a and b are primitive, we cannot have either $g, g' \in a^*$ or $g, g' \in b^*$. If $g \in a^*$ ($g \in b^*$), (5) shows that the only remaining possibility is $g' \in b^*$ ($g' \in a^*$). Thus we can assume now $g, g' \notin a^* \cup b^*$, and suppose that g^2 has an A-factor ab^pa with p positive such that g'^2 has no A-factor $ab^{p'}a$ with $p' < p$.

We shall show that under these conditions at least one word of $a^*ab \cup abb^*$ is imprimitive.

By (5) the principal disjoint segment d of g' that covers ab^pa has an A-segment ab^p or b^pa (unless $p = 1$ and $d = bab$ in which case we know already by (3) that a^2b is imprimitive). Then ab^p (or b^pa) is a proper principal segment of ab^pa; hence of ab^pab^p (b^pab^pa). Thus it is imprimitive.

3.5. Additions

From our proof it now follows that if the set $\{a, b, a^2b\} \cup abb^*$ consists only of primitive words, the only word pairs (if there are any) which are conjugate without being A-conjugates are of the form (a^n, b^n). We can establish the following more accurate result: if a and b are conjugate, $a^*b \cup ab^* \subset \pi(X^*)$; otherwise $a^*b \cup ab^*$ contains at most one imprimitive word.

For the first part of this result, one is led to examine

$$(uv)(vu)^\lambda = c^\mu ; \qquad \lambda \geqslant 1 ; \qquad \mu \geqslant 2 .$$

The case where $\lambda = 1$, $\mu = 2$ evidently contradicts the hypothesis of primitiveness. For $(\lambda - 1)|uv| > |c|$, the hypothesis of primitiveness is contradicted by Corollary 4 of Theorem 2. There remains the case $(\lambda - 1)|uv| < |c|$. From the equality $(\lambda + 1)|uv| = \mu c$, one obtains $2 > (\mu - 1)(\lambda - 1)$ and the conclusion follows easily.

We give an outline of the proof of the second part of the result.

(i) The following lemma is useful (we have already proven particular cases of it; cf. [2] and [3]):

The condition $a^m b^m = c^q$, $m, n, q \geqslant 2$ implies that a, b and c are imprimitive and powers of the same word.

By Theorem 2, we have only to consider the case where:

$$(m - 1)\alpha < \gamma - (\alpha \cap \gamma) , \qquad (n - 1)\beta < \gamma - (\beta \cap \gamma) ,$$

with $\alpha = |a|$, $\beta = |b|$, $\gamma = |c|$. From the equality

$$m\alpha + n\beta = q\gamma\,,$$

we obtain the condition

$$2 + \frac{1}{m-1} + \frac{1}{n-1} - \frac{m}{m-1}\frac{\alpha \cap \gamma}{\gamma} - \frac{n}{n-1}\frac{\beta \cap \gamma}{\gamma} > q\,,$$

which characterizes the cases to be studied. We treat them directly.

(ii) We have seen that for $|a| \geqslant |b|$, the only imprimitive word of a^*a^2b is a^2b. By Corollary 4, $a^2b = f^m$ implies that $|a| > |f|$; hence $m = 2$. Solving $a^2b = f^2$ gives $a = (uv)^{k+1}u$, $b = vuuv$, and the technique of (2) above applies.

(iii) The lemma in (i) is applicable to the case $ab^m = f^p$, $ab^{m'} = g^q$ ($p, q \geqslant 2$; $m' > m \geqslant 1$). We have only to consider $m' = m + 1$; then $f^pb = g^q$.

We have the equalities

$$|a| + m|b| = p|f|\,;\ |a| + (m+1)|b| = q|g|\,;$$

and the inequalities

$$(m-1)|b| < |f|\,; \qquad m|b| < |g|\,;$$

and by Corollary 4

$$(p-1)|f| < |g|\,.$$

This system of equalities and inequalities has only the following solutions:

$m = 1, \quad p = 2, \quad q = 3;$

or

$m = 2, \quad p = 3, \quad q = 2;$

$m = 1, \quad q = 2, \quad p$ arbitary;

or

m arbitary, $\quad p = q = 2.$

The first two solutions contradict the hypothesis $|a| \geqslant |b|$.

There remain the following cases to consider:

$$ab = f^p\,; \quad ab^2 = g^2 \text{ and } ab^m = f^2\,; \quad ab^{m+1} = g^2\,.$$

The first case can be treated in the same way as the case $a^2b = f^2$ above. In the second case we can assume $m > 1$. Set $f = cb$, $g = db$, giving

$$ab^{m-1} = cbc\,, \qquad ab^m = dbd\,.$$

The relation

$$cbcb = dbd$$

cuts b into two words of equal length which we can show to be equal, and this contradicts the hypothesis of primitiveness.

References

1. Fine, N. J. and Wiff, H. S. "Uniqueness Theorems for Periodic Functions," *Proc. American Math. Soc.*, **16** (1965), 109–114.
2. Lentin, A. "Sur l'Equation $a^M = b^N c^P d^Q$ dans un Monoide Libre," *C. R. Academie Sci.*, **260** (1965), 3242–3244.
3. Lyndon, R. C. and Schützenberger, M. P. "On the Equation $a^M = b^N c^P$ in a Free Group," *Michigan Math. J.*, **9** (1962), 289–298.

CHAPTER 9

Generalized Measure of Dispersion and Analysis of Distributional Pattern

TOSIO KITAGAWA, *Kyushu University, Japan*

1. INTRODUCTION

For measuring the dispersion of a distribution of numbers of individuals in a spatial area and analyzing their distributional pattern, a certain index of dispersion has been advocated by Simpson [5] and has since been used by many research workers. Its theoretical aspects have been discussed in depth by an animal ecologist, Morishita [1], [2], [3] and [4]. The purpose of the present paper is to explore certain mathematical principles which will enable us to obtain a clearer understanding of a generalized measure of dispersion and which, at the same time, may provide us with a useful set of tools for making penetrating observations on distributional patterns.

In Section 2 we shall introduce three basic notions upon which our formulation of ϕ-dispersions will be established: (1) a multi-stage decomposition system, $D^{(m)}(A)$, of the set A; (2) x-variates associated with $D^{(m)}(A)$; (3) d-depth ϕ-dispersion of a function f on the set $A_{i_1 i_2 \cdots i_h}$ with respect to $D^{(d)}(A_{i_1 i_2 \cdots i_h})$. In Section 3 another essential concept is introduced which is called a distributional pattern with generalized one-parameter exponential type distributions associated with the decomposition $D^{(m)}(A)$. There are also two important notions, *homogenity* and *inheritance*, in this connection. A few examples given in Section 3, will serve to show that important special cases discussed by Morishita [1] and Smith [5] can be explained in terms of these notions in our present formulation. The main purpose of this paper is to evalute the conditional expectations of various ϕ-dispersions.

For this purpose there is a crucial property of the concurrence function ϕ with respect to the assigned additive family of generalized exponential type distributions; we shall define this property in Section 4 as being reproductive. After these preparations we shall obtain the conditional expectation of ϕ-dispersion for such a reproductive ϕ-function. The main result in this section is enunciated in Proposition 6, of which Propositions 4 and 5 are special cases. They show that these conditional expectations of ϕ-dispersion can be obtained in concise forms.

In Section 6 we discuss partitional patterns of additive parameters in the decomposition system $D^{(m)}(A)$ with particular reference to two extreme cases, *equipartition* and *monopolistic concentration*. In Section 7 we discuss ϕ-dispersion as a function of the size variable associated with the decomposition system $D^{(m)}(A)$. An interpolation procedure for decomposing the size variable into finer and global structures is explained in terms of the notions of homogenity and inheritance. Section 8 is devoted to the relations between our ϕ-dispersion and the current index of dispersion introduced and discussed by Morishita [1]-[4]. We shall show that the latter is a particular example of the former, apart from a certain multiplier which is suited for the Poisson distribution. Two-stage and three-stage distributional patterns are illustrated to show the uses of ϕ-dispersions as functions of size variables, with particular reference to binomial, negative binomial and Poisson distributions.

2. ϕ-DISPERSION IN MULTISTAGE DISTRIBUTIONAL PATTERN

There are three fundamental notions upon which a notion of ϕ-dispersions regarding x-variates in a set A can be established.

1. *Multistage decomposition system* $D^{(m)}(A)$ *of the set* A.

A set A is decomposed into the sum of $q_1q_2\ldots q_k$ mutually disjoint subsets $A_{i_1i_2\ldots i_m}$ ($i_j = 1, 2, \ldots, q_j$; $j = 1, 2, \ldots, m$) in such a way that

1) $$A = \sum_{i_1=1}^{q_1} A_{i_1}$$

2) $$A_{i_1i_2\ldots i_h} = \sum_{i_{h+1}=1}^{q_{h+1}} A_{i_1i_2\ldots i_hi_{h+1}}, \quad (i_j = 1, 2, \ldots, q_j;\ j = 1, 2, \ldots, h)$$

for $h = 1, 2, 3, \ldots, m-1$.

The set $\{A_{i_1i_2\ldots i_m}\}$ is said to constitute the (m-stage) decomposition system $D^{(m)}(A)$ of the set A.

The family of q_1 subsets enunciated in (1) and those of $q_1q_2\ldots q_h$ subsets enunciated in (2) are said to constitute the first stage decomposition $D^{(m)}(A)$ and the $(h+1)$th stage decomposition $D^{(m)}_{h+1}(A)$, $(h = 1, 2, \ldots, m-1)$, respectively. Within the m-stage decomposition system $D^{(m)}(A)$ are imbedded the lower stage decomposition systems $D^{(n)}(A)$ for $n = 1, 2, \ldots, m-1$, and it is immediate to observe that, for each subset $A_{i_1i_2\ldots i_h}$, we can define the l-stage decomposition system $D^{(l)}\cdot(A_{i_1i_2\ldots i_h})$ and its νth stage decomposition $D^{(l)}_{\nu}(A_{i_1i\ldots i_h})$ for $\nu = 1, 2, \ldots, l$; $l = 1, 2, \ldots, m-h$. In view of these facts, we may and we shall consider the set A as the unique subset of the set A at the zero-th stage decomposition, $D^{(m)}_0(A)$, and consider it as an element of the family of the sets $\{A_{i_1i_2}\ldots_{i_h}\}$ corresponding to the case $h = 0$.

2. *x-variates associated with an m-stage decomposition system $D^{(m)}(A)$.* A set of $q_1q_2\ldots q_m$ non-negative integers $\{x_{i_1i_2\ldots i_m}\}$, $i_j = 1, 2, \ldots, q_j$; $j = 1, 2, \ldots, m$, is assumed to be assigned, and the following set of x-variates is defined in correspondence with the decomposition system $D^{(m)}(A)$:

1) $$x = \sum_{i_1=1}^{q_1} x_{i_1}$$

2) $$x_{i_1i_2\ldots i_h} = \sum_{i_{h+1}=1}^{q_{h+1}} x_{i_1i_2\ldots i_hi_{h+1}}, \quad (i_j = 1, 2, \ldots, q_j;\ j = 1, 2, \ldots, h)$$

for $h = 1, 2, \ldots, m-1$.

3. *ϕ-dispersions.* A function $f(x)$ is assumed to be defind in the set $\mathcal{D}_f$ which is a subset of the set of all non-negative integers. Among such functions let us choose a function ϕ, with reference to which we introduce the following definition.

Definition 1. *A d-depth ϕ-measure of dispersion, abbreviated as ϕ-dispersion, of a function f on the set $A_{i_1 i_2 \dots i_h}$ with respect to the decomposition system $D^{(d)}(A_{i_1 i_2 \dots i_h})$* is defined by

$$\phi_d(f)(A_{i_1 i_2 \dots i_h}) \equiv \frac{1}{\phi(x_{i_1 i_2 \dots i_h})} \sum_{i_{h+1}=1}^{q_{h+1}} \sum_{i_{h+2}=1}^{q_{h+1}} \cdots \sum_{i_{h+d}=1}^{q_{h+d}} \phi(x_{i_1 \dots i_h i_{h+1} \dots i_{h+d}}) f(x_{i_1 \dots i_{h+d}}), \tag{2.1}$$

for every $i_j = 1, 2, \dots, q_j$; $j = 1, 2, \dots, h$; $h = 0, 1, 2, \dots, m - 1$.

In particular, when $f(x)$ is identically 1, the identity, over the domain $\mathcal{D}_f$, we shall write

$$\phi_d[A_{i_1 i_2 \dots i_h}] = \phi_d(1)(A_{i_1 i_2 \dots i_h}). \tag{2.2}$$

As a direct consequence of Definition 1, we have

Proposition 1. *We have*

$$\phi_{d_1}(\phi_{d_2}(f))(A_{i_1 i_2 \dots i_h}) = \phi_{d_1 + d_2}(f)(A_{i_1 i_2 \dots i_h}). \tag{2.3}$$

Among the various ϕ-dispersions of a function f there are important cases which are of particular interest. These occur when at least one of the following three conditions is satisfied: (i) $A_{i_1 i_2 \dots i_h} = A$, that is, the case $h = 0$; (ii) the function $f(x)$ is identically 1; (iii) the depth d is equal to 1. For illustration a few examples are given.

Example 2.1. The d-depth dispersion of the identity:

$$\phi_d(1)(A) = \phi_d[A] = \frac{1}{\phi(x)} \sum_{i_1=1}^{q_1} \sum_{i_2=1}^{q_2} \cdots \sum_{i_d=1}^{q_d} \phi(x_{i_1 i_2 \dots i_d}) \tag{2.4}$$

and

$$\begin{aligned} \phi_d(1)(A_{i_1 i_2 \dots i_h}) &= \phi_d[A_{i_1 i_2 \dots i_h}] \\ &= \frac{1}{\phi(x_{i_1 i_2 \dots i_h})} \sum_{i_{h+1}=1}^{q_{h+1}} \cdots \sum_{i_{h+d}=1}^{q_{h+d}} \phi(x_{i_1 \dots i_h i_{h+1} \dots i_{h+d}}). \end{aligned} \tag{2.5}$$

Example 2.2. The 1-depth dispersion of the function f:

(2.6) $$\phi_1(f)(A) = \frac{1}{\phi(x)} \sum_{i_1=1}^{q_1} \phi(x_{i_1}) f(x_{i_1})$$

(2.7) $$\phi_1(f)(A_{i_1 i_2 \ldots i_h}) = \frac{1}{\phi(x_{i_1 \ldots i_h})} \sum_{i_{h+1}=1}^{q_{h+1}} \phi(x_{i_1 \ldots i_h i_{h+1}}) f(x_{i_1 \ldots i_h i_{h+1}}) .$$

Example 2.3. There are numerous possibilities in choosing the basic function ϕ in defining our ϕ-dispersion (2.1). The following examples are important:

(2.8) $$\phi^{[j]}(x) = x(x-1) \ldots (x-j+1) = x^{[j]} , \quad \mathscr{D}_{\phi^{[j]}} = \{x \,;\, x \geqq j\}$$

for $j \geqq 2$. Indeed the current index of dispersion discussed by Morishita [1]-[4] is, in our notation,

(2.9) $$I_\delta = q_1 \phi_1^{[2]}[A] = \frac{q_1}{\phi_1^{[2]}(x)} \sum_{i_1=1}^{q_1} \phi(x_{i_1}) = I_1^{[2]}[A] .$$

More generally, a d-depth $[j]$-index of dispersion of the function f on the set $A_{i_1 i_2 \ldots i_h}$ is defined by

(2.10) $$I_d^{[j]}(f)(A_{i_1 i_2 \ldots i_h}) = (q_{h+1} q_{h+2} \ldots q_{h+d})^{j-1} \phi_d^{[j]}(A_{i_1 i_2 \ldots i_h})$$

by virtue of (2.1). In particular, when the function f is an identity, (2.10) becomes

(2.11) $$I_d^{[j]}(1)(A_{i_1 i_2 \ldots i_h}) = I_d^{[j]}[A_{i_1 i_2 \ldots i_h}] = \frac{(q_{h+1} q_{h+2} \ldots q_{h+d})^{j-1}}{x^{[j]}_{i_1 i_2 \ldots i_h}} \sum_{i_{h+1}=1}^{q_{h+1}} \cdots \sum_{i_{h+d}=1}^{q_{h+d}} x^{[j]}_{i_1 i_2 \ldots i_h i_{h+1} \ldots i_{h+d}} .$$

Multistage distributional patterns have been discussed by some ecologists. In fact, Morishita [1] discussed intraclump distribution in the analysis of distributional patterns by I_δ. In this connection it seems necessary to introduce a multistage decomposition system $\{A_{i_1 i_2 \ldots i_m}\}$ and a set of x-variates $\{x_{i_1 i_2 \ldots i_m}\}$, as we have done, in order to develop a mathematical tool with which to explore a multistage distributional pattern.

If no confusion arises, we shall use the notations $x_{i_1 i_2 \dots i_h \dots}$ and $x \dots$ in the place of $x_{i_1 i_2 \dots i_h}$ and x respectively. At the same time, we shall simplify our notation so as to concentrate our discussion in the lower stage decompositions associated with our original decomposition $D^{(m)}(A)$. Thus we shall use x and q instead of $x \dots$ and q_1, as originally defined in the system $D^{(m)}(A)$, when we are particularly concerned with the first stage decomposition.

3. DISTRIBUTIONAL PATTERN WITH GENERALIZED ONE PARAMETER EXPONENTIAL TYPE DISTRIBUTIONS ASSOCIATED WITH THE m-STAGE DECOMPOSITION SYSTEM $D^{(m)}(A)$

We start with the following definition.

Definition 2. *A set of* $q_1 q_2 \dots q_m$ *stochastic variables* $\{X_{i_1 i_2 \dots i_m}\}$, $i_h = 1, 2, \dots, q_h$; $h = 1, 2, \dots, m$, *is defined to satisfy the following conditions.*

(1) *Each variable* $X_{i_1 i_2 \dots i_m}$ *is associated with the set* $A_{i_1 i_2 \dots i_m}$.

(2) *The* $q_1 q_2 \dots q_m$ *variables* $\{X_{i_1 i_2 \dots i_m}\}$ *are mutually independent.*

(3) *The probability density function of each* $X_{i_1 i_2 \dots i_m}$ *belongs to the family of exponential type distributions with probability element*

$$(3.1) \qquad \begin{aligned} & g(x\,;\, k_{i_1 i_2 \dots i_m} | A_{i_1 i_2 \dots i_m}) d\mu_{i_1 i_2 \dots i_m}(x) \\ & \equiv \exp\{\lambda x + k_{i_1 i_2 \dots i_m} b(\lambda)\} a_{i_1 i_2 \dots i_m}(x)\, d\mu_{k_{i_1 i_2 \dots i_m}}(x) \end{aligned}$$

which will be denoted briefly by $g(x\,;\, k_{i_1 i_2 \dots i_m} | A_{i_1 i_2 \dots i_m})$, *where* $\mu_{k_{i_1 i_2 \dots i_m}}(x)$ *is a step function having jumps solely at integral values of* x.

(4) *The set of variables* $\{X_{i_1 i_2 \dots i_h}\}$, $h = 1, 2, \dots, m-1$, *and* X *are defined in the following way:*

1) $$X = \sum_{i_1=1}^{q_1} X_{i_1}$$

2) $$X_{i_1 i_2 \dots i_h} = \sum_{i_{h+1}=1}^{q_{h+1}} X_{i_1 i_2 \dots i_h i_{h+1}},$$

$$i_j = 1, 2, \dots, q_j\,;\; j = 1, 2, \dots, h,$$

for $h = 1, 2, 3, \ldots, m-1$.

(5) *The set of additive constants* $\{k_{i_1 i_2 \ldots i_h}\}$, $i_j = 1, 2, \ldots, q_j$; $j = 1, 2, \ldots, h$; $h = 1, 2, \ldots, m-1$, *and* k *is defined in the following way:*

1) $$k = \sum_{i_1=1}^{q_1} k_{i_1}$$

2) $$k_{i_1 i_2 \ldots i_h} = \sum_{i_{h+1}=1}^{q_{h+1}} k_{i_1 i_2 \ldots i_h i_{h+1}}\,,$$

$$i_j = 1, 2, \ldots, q_j\,;\; j = 1, 2, \ldots, h\,,$$

for $h = 1, 2, 3, \ldots, m-1$.

The set of $q_1 q_2 \ldots q_m$ *stochastic variables* $\{X_{i_1 i_2 \ldots i_m}\}$ *is said to constitute an additive family of one parameter exponential type distributions* $\{g(x\,;\, k_{i_1 i_2 \ldots i_m} | A_{i_1 i_2 \ldots i_m})\}$ *associated with the m-stage decomposition system* $D^{(m)}(A)$ *of the set A, and is denoted by*

(3.2) $$\{X_{i_1 i_2 \ldots i_m}\} \in E_\lambda(k_{i_1 i_2 \ldots i_m}\,;\, D^{(m)}(A))\,.$$

Now it is immediate to observe

Proposition 2. *Under the assumption* (3.2), *the set of* $q_1 q_2 \ldots q_h$ *stochastic variables* $\{X_{i_1 i_2 \ldots i_h}\}$ *constitutes an additive family of one parameter exponential type distributions* $\{g(x\,;\, k_{i_1 i_2 \ldots i_h} | A_{i_1 i_2 \ldots i_h})\}$, *i.e.,*

(3.3) $$\{X_{i_1 i_2 \ldots i_h}\} \in E_\lambda(k_{i_1 i_2 \ldots i_h}\,;\, D^{(h)}(A))\,,$$

for $h = 1, 2, \ldots, m-1$. *Moreover the stochastic variable* X *is distributed as* $g(x\,;\, k|A)$.

We shall give here a few examples which are important in our practical applications.

Example 3.1. *Binomial distribution.* This is the case when each stochastic variable $X_{i_1 i_2 \ldots i_m}$ is distributed according to the distribution

(3.4) $$\Pr\{X_{i_1 i_2 \ldots i_h} = x\} = \binom{k_{i_1 i_2 \ldots i_m}}{x} p^x (1-p)^{k_{i_1 \ldots i_m} - x}$$

$$= g_B(x\,;\, k_{i_1 i_2 \ldots i_m} | A_{i_1 i_2 \ldots i_m})\,, \quad \text{say},$$

for $x = 0, 1, \ldots, k_{i_1 i_2 \ldots i_m}$. The transformation of the parameter p into λ by means of $\log\{p/(1-p)\} = \lambda$ reduces (3.4) to the exponential type (3.1) with

$$b(\lambda) = \log(e^{\lambda} + 1)^{-1} \tag{3.5}$$

$$a_{k_{i_1 i_2 \ldots i_m}}(x) = \binom{k_{i_1 i_2 \ldots i_m}}{x} \tag{3.6}$$

and $d\mu_{k_{i_1 i_2 \ldots i_m}}(x) = 1$, for $x = 0, 1, 2, \ldots, k_{i_1 i_2 \ldots i_m}$ and zero otherwise.

Example 3.2. *Negative binomial distribution.* This is the case when each $X_{i_1 i_2 \ldots i_m}$ is distributed according to the distribution

$$\begin{aligned} \Pr\{X_{i_1 i_2 \ldots i_m} = x\} &= \binom{k_{i_1 i_2 \ldots i_m} + x - 1}{x} p^x (1-p)^{k_{i_1 i_2 \ldots i_m}} \\ &= g_{NB}(x\,;\, k_{i_1 i_2 \ldots i_m} | A_{i_1 i_2 \ldots i_m}) \end{aligned} \tag{3.7}$$

for $x = 0, 1, 2, \ldots$. The transformation of the parameter p into λ by means of $\lambda = \log p$ reduces (3.7) to the exponential type (3.1) with

$$b(\lambda) = \log(1 - e^{\lambda}) \tag{3.8}$$

$$a_{k_{i_1 i_2 \ldots i_m}}(x) = \binom{k_{i_1 i_2 \ldots i_m} + x - 1}{x} \tag{3.9}$$

and $d\mu_{k_{i_1 i_2 \ldots i_m}}(x) = 1$, for $x = 0, 1, 2, \ldots,$ and zero otherwise.

Example 3.3. *Poisson distribution.* This is the case when each $X_{i_1 i_2 \ldots i_m}$ is distributed according to the distribution

$$\begin{aligned} \Pr\{X_{i_1 i_2 \ldots i_m} = x\} &= \exp\{- k_{i_1 i_2 \ldots i_m}\theta\}(k_{i_1 i_2 \ldots i_m}\theta)^x / x! \\ &= g_P(x\,;\, k_{i_1 i_2 \ldots i_m} | A_{i_1 i_2 \ldots i_m}) \end{aligned} \tag{3.10}$$

for $x = 0, 1, 2, \ldots$. The transformation of the parameter θ into λ by means of $\log\theta = \lambda$ reduces (3.10) to the exponential type distribution with

$$b(\lambda) = - e^{\lambda} \tag{3.11}$$

(3.12) $$a_{k_{i_1 i_2 \ldots i_m}}(x) = \frac{k^x_{i_1 i_2 \ldots i_m}}{x!}.$$

Remarks on possible modifications of notations for $\{X_{i_1 i_2 \ldots i_h}\}$ and $\{k_{i_1 i_2 \ldots i_h}\}$ should be added here. Indeed we shall use, as necessary, the notations $X_{i_1 i_2 \ldots i_h}, \ldots, k_{i_1 i_2 \ldots i_h}, \ldots, x_{\ldots\ldots}$ and $k_{\ldots\ldots}$ instead of the original notations $X_{i_1 i_2 \ldots i_h}, k_{i_1 i_2 \ldots i_h}, x$ and k used alone.

We proceed to introduce

Definition 3. *A set of $q_1 q_2 \ldots q_m$ stochastic variables $\{X_{i_1 i_2 \ldots i_m}\}$, $i_h = 1, 2, \ldots, q_h$; $h = 1, 2, \ldots, m$, is said to have an m-stage distributional pattern of generalized one parameter exponential type distributions with the set of additive parameters $\{k^{(\nu)}_{i_1 i_2 \ldots i_\nu}\}$, $\nu = 1, 2, \ldots, m$, when the following conditions are satisfied:*

(1) *The set of q_1 stochastic variables $\{X_{i_1}\}$, $i_1 = 1, 2, \ldots, q_1$, has a joint probability element*

(3.13) $$\Pr\left\{\prod_{i_1=1}^{q_1} X_{i_1} = x_{i_1}\right\} = \prod_{i_1=1}^{q_1} g^{(1)}(x_{i_1};\ k^{(1)}_{i_1} | A_{i_1}) d\mu^{(1)}(x_{i_1})$$

(2) *The conditional joint probability element of the q_h stochastic variables $\{X_{i_1 i_2 \ldots i_{h-1} i_h}\}$, $i_h = 1, 2, \ldots, q_h$, is given by*

(3.14)
$$\Pr\left\{\prod_{i_h=1}^{q_h} (X_{i_1 i_2 \ldots i_h} = x_{i_1 i_2 \ldots i_h}) | X_{i_1 i_2 \ldots i_{h-1}} = x_{i_1 i_2 \ldots i_{h-1}}\right\}$$
$$= \frac{\prod_{i_h=1}^{q_h} g^{(h)}_{i_1 i_2 \ldots i_{h-1}}(x_{i_1 i_2 \ldots i_h};\ k^{(h)}_{i_1 i_2 \ldots i_h} | A_{i_1 i_2 \ldots i_h}) d\mu^{(h)}_{i_1 \ldots i_{h-1}}(x_{i_1 i_2 \ldots i_h})}{g^{(h)}_{i_1 i_2 \ldots i_{h-1}}(x_{i_1 i_2 \ldots i_{h-1}};\ k^{(h)}_{i_1 i_2 \ldots i_{h-1}} | A_{i_1 i_2 \ldots i_h}) d\mu^{(h)}_{i_1 \ldots i_{h-1}}(x_{i_1 i_2 \ldots i_{h-1}})}$$
$$= dg^{(h)}_{i_1 i_2 \ldots i_{h-1}}((x_{i_1 i_2 \ldots i_h}) | x_{i_1 i_2 \ldots i_{h-1}};\ k^{(h)}_{i_1 i_2 \ldots i_h} | A_{i_1 i_2 \ldots i_h}),$$

say.

(3) *The conditional joint probability element of the $q_{h_1} q_{h_1+1} \ldots q_{h_2}$ stochastic variables $\{X_{i_1 i_2 \ldots i_h}\}$, $i_j = 1, 2, \ldots, q_j$; $j = h_1, h_1 + 1, \ldots, h_2$, under the condition that $X_{i_1 i_2 \ldots i_{h_1-1}} = x_{i_1 i_2 \ldots i_{h_1-1}}$, for $i_j = 1, 2, \ldots, q_j$, $j = 1, 2, \ldots, h_1 - 1$, is given by*

(3.15) $$\Pr\{E^{(2)}(h_1, h_2) | E^{(1)}(h_1 - 1)\}$$

$$= \prod_{i_{h_1}=1}^{q_{h_1}} \prod_{i_{h_1+1}=1}^{q_{h_1+1}} \cdots \prod_{i_{h_2}=1}^{q_{h_2}} dG(i_1, i_2, \ldots, i_{h_1-1} | i_{h_1}, \ldots, i_{h_2}) ,$$

where the left-hand side of (3.15) *denotes the probability element of the event* $E^{(2)}(h_1, h_2)$ *under the condition that the event* $E^{(1)}(h_1 - 1)$ *occurs, with*

$$\text{(3.16)} \quad E^{(2)}(h_1, h_2) : \prod_{i_{h_1}=1}^{q_{h_1}} \cdots \prod_{i_{h_2}=1}^{q_{h_2}} (X_{i_1 \ldots i_{h_1-1} i_{h_1} \ldots i_{h_2}} = x_{i_1 \ldots i_{h_1-1} i_{h_1} \ldots i_{h_2}})$$

$$\text{(3.17)} \quad E^{(1)}(h_1 - 1) : \prod_{i_1=1}^{q_1} \cdots \prod_{i_{h_1-1}=1}^{q_{h_1-1}} (X_{i_1 i_2 \ldots i_{h_1-1}} = x_{i_1 i_2 \ldots i_{h_1-1}}) .$$

In the right-hand side of (3.15) *we have put*

$$\text{(3.18)} \quad dG(i_{h_1}, i_{h_1+1}, \ldots, i_{h_2} | i_1, i_2, \ldots, i_{h_1-1})$$
$$= \prod_{j=h_1}^{h_2} dg^{(j)}_{i_1 i_2 \ldots i_{j-1}}((x_{i_1 \ldots i_{h_1-1} i_{h_1} \ldots i_j}) | x_{i_1 \ldots i_{j-1}} ; k^{(j)}_{i_1 i_2 \ldots i_j} | A_{i_1 i_2 \ldots i_j}) .$$

It is readily seen that Definition 2 is a specialized case of Definition 3. Between these two cases there are many types of multistage distributional patterns. For a systematic description of these situations we introduce two fundamental notions, homegeneity and inheritance.

Definition 4. *In the set of* $q_1 q_2 \ldots q_m$ *stochastic variables* $\{X_{i_1 i_2 \ldots i_m}\}$, $i_h = 1, 2, \ldots, q_h$; $h = 1, 2, \ldots, m$, *having a multistage distributional pattern of generalized exponential type distribution with the set of additive parameters* $\{k^{(\nu)}_{i_1 i_2 \ldots i_\nu}\}$ $(\nu = 1, 2, \ldots, m)$, *a set of* $q_{h+1} \ldots q_{h+d}$ *stochastic variables* $\{X_{i_1 i_2 \ldots i_h \ldots i_{h+d}}\}$, $i_j = 1, 2, \ldots, q_j$; $j = h + 1, \ldots, h + d$, *is said to be d-depth homogenous in the set* $A_{i_1 i_2 \ldots i_h}$ *when there exists a sequence of functions* $\{g^{(j)}_{i_1 i_2 \ldots i_h}\}$ *and* $\{d\mu^{(j)}_{i_1 i_2 \ldots i_h}\}$ *such that*

$$\text{(3.19)} \quad dg^{(j)}_{i_1 i_2 \ldots i_{j-1}}((x_{i_1 \ldots i_j}) | x_{i_1 i_2 \ldots i_{j-1}} ; k^{(j)}_{i_1 i_2 \ldots i_j} | A_{i_1 i_2 \ldots i_j})$$
$$= dg^{(j)}_{i_1 i_2 \ldots i_h}((x_{i_1 i_2 \ldots i_j}) | x_{i_1 i_2 \ldots i_{j-1}} ; k^{(j)}_{i_1 i_2 \ldots i_j} | A_{i_1 i_2 \ldots i_j}) ,$$

for $i_j = 1, 2, \ldots, q_j$; $j = h + 1, \ldots, h + d$.

Definition 5. *A set of* $q_{h+1} \ldots q_{h+d}$ *stochastic variables* $\{X_{i_1 i_2 \ldots i_{h-1} i_h \ldots i_{h+d}}\}$, $i_j = 1, 2, \ldots, q_j$; $j = h + 1, \ldots, h + d$, *is said to be* *d-depth inherited in the set* $A_{i_1 i_2 \dots i_h}$ *when the following two conditions are satisfied:*

1) $k^{(j)}_{i_1 i_2 \dots i_{j-1}\cdot} = k^{(j-1)}_{i_1 i_2 \dots i_{j-1}}, \qquad j = h+1, \dots, h+d,$

2)
$$\begin{aligned} & g^{(j)}_{i_1 i_2 \dots i_{j-1}}(x_{i_1 i_2 \dots i_{j-1}};\ k^{(j)}_{i_1 i_2 \dots i_{j-1}} | A_{i_1 i_2 \dots i_{j-1}}) \\ & \quad \times d\mu^{(j)}_{i_1 i_2 \dots i_{j-1}}(x_{i_1 i_2 \dots i_{j-1}}) \\ & = g^{(j-1)}_{i_1 i_2 \dots i_{j-2}}(x_{i_1 i_2 \dots i_{j-1}};\ k^{(j-1)}_{i_1 i_2 \dots i_{j-1}} | A_{i_1 i_2 \dots i_{j-1}}) \\ & \quad \times d\mu^{(j)}_{i_1 i_2 \dots i_{j-2}}(x_{i_1 i_2 \dots i_{j-1}}) \end{aligned}$$

for every $x_{i_1 i_2 \dots i_{j-1}}$ *and* $k^{(j-1)}_{i_1 i_2 \dots i_{j-1}}$ *in their domains of definition, for* $j = h+1, \dots, h+d$.

In virtue of Definitions 3, 4 and 5 it is immediate to see

Proposition 3. *The set of* $q_1 q_2 \dots q_m$ *stochastic variables* $\{X_{i_1 i_2 \dots i_m}\} \in E_\lambda(k_{i_1 i_2 \dots i_m};\ D^{(m)}(A))$ *is both m-depth homogenous and m-depth inherited in the set A.*

Proof. This can be seen from the fact that the assumption (3.2) implies

$$k^{(h)}_{i_1 i_2 \dots i_h} = k_{i_1 i_2 \dots i_h}, \tag{3.20}$$

and

$$\begin{aligned} & g^{(h)}_{i_1 i_2 \dots i_{h-1}}(x_{i_1 i_2 \dots i_h};\ k^{(h)}_{i_1 i_2 \dots i_h} | A_{i_1 i_2 \dots i_h}) d\mu^{(h)}_{i_1 \dots i_{h-1}}(x_{i_1 i_2 \dots i_{h-1}}) \\ & = g(x_{i_1 i_2 \dots i_h};\ k_{i_1 i_2 \dots i_h} | A_{i_1 i_2 \dots i_h}) d\mu(x_{i_1 i_2 \dots i_{h-1}}) \end{aligned} \tag{3.21}$$

for $i_j = 1, 2, \dots, q_j;\ j = 1, 2, \dots, h;\ h = 1, 2, \dots, m$.

Particular examples of two and three stage distributional patterns are explained by adopting various cominations of binomial, negative binomial and Poisson distributions in defining g function in Definition 3.

Example 3.4. *Two stage distributional patterns.* Nine sets of two stage distributional patterns can be obtained from all the possible permutations by adopting one of the three fundamental probability density functions as shown in the following Table. We may and we shall denote these two stage distributional patterns by $B \times B$, $B \times NB$, $B \times P$, $NB \times B$, $NB \times NB$, $NB \times P$, $P \times B$, $P \times NB$ and $P \times P$.

Table 1 Probability Element Function (pef) in Two Stage Distributional Pattern

pef	Illustrative Examples
$dg^{(1)}(x; k_{i_1}^{(1)}\|A_{i_1})$	B, NB, P
$dg_{i_1}^{(2)}((x_{i_1i_2})\|x_{i_1}; k_{i_1i_2}^{(2)}\|A_{i_1i_2})$	B, NB, P

A direct consequence of Definitions 3 and 4 yields the following observations.

(i) Homogeneity. A two stage distributional pattern is homogeneous if and only if there is a function g such that

$$(3.22)\quad dg^{(2)}((x_{i_1i_2})|x_{i_1}\,;\; k_{i_1i_2}|A_{i_1i_2}) = dg_{i_1}^{(2)}((x_{i_1i_2})|x_{i_1}\,;\; k_{i_1i_2}|A_{i_1i_2})$$

for $i_j = 1, 2, \ldots, q_j\,;\; j = 1, 2.$

(ii) Inheritance. A two stage distributional pattern is inherited if and only if the following two conditions are satisfied :

$$1)\qquad k_{i_1}^{(2)} = k_{i_1}^{(1)} \qquad (i_1 = 1, 2, \ldots, q_1)$$

$$2)\qquad g_{i_1}^{(2)}(x_{i_1}\,;\; k_{i_1i_2}^{(2)}|A_{i_1i_2}) = g^{(1)}(x_{i_1}\,;\; k_{i_1i_2}^{(2)}|A_{i_1i_2})$$

for $i_1 = 1, 2, \ldots, q_j\,;\; j = 1, 2.$

Example 3.5. *Three stage distributional patterns.* Twenty-seven sets of three stage distributions, such as $B \times B \times B$, $B \times B \times NB$, $NB \times P \times B$, etc., can be obtained from all the possible permutations of the fundamental probability density functions as shown in the following Table 2.

Table 2 Probability Element Function (pef) in Three Stage Distributional Pattern

pef	Illustrative Examples
$dg^{(1)}(x; k_{i_1}^{(1)}\|A_{i_1})$	B, NB, P
$dg_{i_1}^{(2)}((x_{i_1i_2})\|x_{i_1}; k_{i_1i_2}^{(2)}\|A_{i_1i_2})$	B, NB, P
$dg_{i_1i_2}^{(3)}((x_{i_1i_2i_3})\|x_{i_1i_2}; k_{i_1i_2i_3}^{(3)}\|A_{i_1i_2i_3})$	B, NB, P

To conclude Section 3, let us introduce some notations for the conditional expected values of stochastic variables. Let z

$= \{z_{i_1 i_2 \ldots i_h}\}$, $i_j = 1, 2, \ldots, q_j$; $j = 1, 2, \ldots, k$, be a set of stochastic variables and let $P(z)$ be the multidimensional probability distribution function of z. Let $H(z)$ be a function of z with its domain of definition $\mathscr{D}(H)$. For an assigned set of non-negative integers, $i_1, i_2, \ldots, i_h$, an assigned value y and an assigned function u, let us denote by $\mathscr{D}_{u(i_1 i_2 \ldots i_h)(y)}(H)$ the subset of $\mathscr{D}(H)$ where $u(z_{i_1 i_2 \ldots i_h})$ has the value y.

Let us introduce the conditional expected value of the stochastic variable $u(z)$ such that

$$E^*\{H\} = \frac{\int_{z \in \mathscr{D}(H)} H(z) dP(z)}{\int_{z \in \mathscr{D}(H)} dP(z)} \tag{3.23}$$

and

$$E^*_{u(i_1 i_2 \ldots i_h)(y)}\{H\} = \frac{\int_{z \in \mathscr{D}_{u(i_1 \ldots i_h)(y)}(H)} H(z) dP(z)}{\int_{z \in \mathscr{D}_{u(i_1 \ldots i_h)(y)}(H)} dP(z)}, \tag{3.24}$$

provided that the right-hand sides of (3.23) and (3.24) exist.

We use also the notion $E^*_G\{H\}$ to specify an underlying probability function G, such as the binomial (B), negative binomial (NB) and Poisson distributions. For instance, $E^*_B\{H\}$, $E^*_{NB}\{H\}$ and $E^*_P\{H\}$ will be used to express the conditional expectations of H.

4. REPRODUCTIVE FUNCTIONS IN AN ADDITIVE FAMILY OF ONE PARAMETER EXPONENTIAL TYPE DISTRIBUTIONS

Let us introduce

Definition 6. *A function $\phi(x)$ is said to be reproductive in an additive family of one parameter exponential type distributions $\{g(x; k)\}$ with additive parameter k, if there exist a function $\varphi(k)$, a transformation T of x, a transformation S of k and a function $c(\phi)$, which is independent of k and x,* *such that, for any assigned set of x and k belonging to their respective domains of definition for specifying $g(x\,;\,k)$, we have*

$$\phi(x)g(x\,;\,k)d\mu(x) = \varphi(k)c(\phi)g(Tx\,;\,Sk)d\mu(Tx) \tag{4.1}$$

with the following properties:

1) *for any set of $x_1, x_2, \ldots, x_{q-1}$ and x_q belonging to the domain of definition of $g(x\,;\,k)$ and for any positive integer i in $1 \leqq i \leqq q$, we have*

$$T\left(\sum_{j=1}^{q} x_j\right) = \sum_{j=1}^{i-1} x_j + Tx_i + \sum_{l=i+1}^{q} x_l\,. \tag{4.2}$$

2) *for any set of non-negative integers $k_1, k_2, \ldots, k_{q-1}$ and k_q and any positive integer i in $1 \leqq i \leqq q$, we have*

$$S\left(\sum_{j=1}^{q} k_j\right) = \sum_{j=1}^{i-1} k_j + Sk_i + \sum_{l=i+1}^{q} k_l\,. \tag{4.3}$$

We observe

Corollary 1. *The function $\phi^{[j]}(x) = x(x-1)\ldots(x-j+1)$ is reproductive in the additive family of one parameter exponential type distributions $\{g(x\,;\,k)\}$ with additive parameter k when $g(x\,;\,k)$ is one of the three fundamental distributions, namely, binomial, negative binomial and Poisson.*

(a) *Binomial distribution $g_B(x\,;\,k)$. We have*

$$\phi^{[j]}(x)g_B(x\,;\,k) = \varphi_B(k)c_B(\phi)g_B(T_Bx\,;\,S_Bk)\,, \tag{4.4}$$

where

$$\varphi_B(k) = k(k-1)\ldots(k-j+1) \tag{4.5}$$

$$C_B(\phi) = p^j \tag{4.6}$$

$$T_Bx = x - j\,, \qquad S_Bk = k - j\,. \tag{4.7}$$

(b) *Negative binomial distribution $g_{NB}(x\,;\,k)$. We have*

$$\phi^{[j]}(x)g_{NB}(x\,;\,k) = \varphi_{NB}(k)c_{NB}(\phi)g_{NB}(T_{NB}x\,;\,S_{NB}k)\,, \tag{4.8}$$

where

$$\varphi_{NB}(k) = k(k+1)\ldots(k+j-1) \tag{4.9}$$

(4.10) $$c_{NB}(\phi) = p^j(1-p)^{-j}$$

(4.11) $$T_{NB}x = x - j\,, \qquad S_{NB} = k + j\,.$$

(c) *Poisson distribution* $g_p(x\,;\,k)$. *We have*

(4.12) $$\phi^{[j]}(x)g_p(x\,;\,k) = \varphi_p(k)c_p(\phi)g_p(T_px\,;\,S_pk)\,,$$

where

(4.13) $$\varphi_p(k) = k^j$$

(4.14) $$c_p(\phi) = \lambda^j$$

(4.15) $$T_px = x - j\,, \qquad S_pk = k\,.$$

It is convenient to have

Definition 7. *A function* $\phi(x)$ *is said to be reproductive in a multistage distributional pattern of generalized one parameter exponential type distributions with the set of additive parameters* $\{k^{(\nu)}_{i_1i_2\cdots i_\nu}\}$ $(\nu = 1, 2, \ldots, m)$, *or simply reproductive in the system* $D^{(m)}(A)$, *when all the concurrence-probability density functions are reproductive, that is to say, when the following relations hold true.*

(4.16) $$\begin{aligned}\phi(x)g^{(h)}_{i_1i_2\cdots i_{h-1}}&(x_{i_1i_2\cdots i_h}\,;\,k^{(h)}_{i_1i_2\cdots i_h}|A_{i_1i_2\cdots i_h})d\mu^{(h)}_{i_1\cdots i_{h-1}}(x_{i_1i_2\cdots i_h})\\ &= \varphi^{(h)}_{i_1i_2\cdots i_{h-1}}(k^{(h)}_{i_1i_2\cdots i_h})c^{(h)}_{i_1i_2\cdots i_{h-1}}(\phi)\\ &\quad\times g^{(h)}_{i_1i_2\cdots i_{h-1}}(T^{(h)}_{i_1i_2\cdots i_{h-1}}x_{i_1i_2\cdots i_h}\,;\,S^{(h)}_{i_1i_2\cdots i_{h-1}}k^{(h)}_{i_1i_2\cdots i_h})\\ &\quad\times d\mu^{(h)}_{i_1i_2\cdots i_{h-1}}(T^{(h)}_{i_1i_2\cdots i_{h-1}}x_{i_1i_2\cdots i_h})\,,\end{aligned}$$

where the functions $c^{(h)}_{i_1i_2\cdots i_{h-1}}(\phi)$, *transformations* $T^{(h)}_{i_1i_2\cdots i_{h-1}}$ *and* $S^{(h)}_{i_1i_2\cdots i_{h-1}}$ *satisfy the conditions enunciated in Definition 5.*

5. THE CONDITIONAL EXPECTATION OF ϕ-DISPERSION FOR REPRODUCTIVE ϕ-FUNCTIONS

It is the purpose of this section to evaluate various conditional expectations of ϕ-dispersions, such as $E^*\{\phi_d[A]\}$ and $E^*_c\{\phi_d[A_{i_1i_2\cdots i_h}]\}$ for $d \geqq 1$ and for a certain set of conditions c such as specifyed underlying probability density functions.

In order to make clear that the essential aspects of our evaluations are valid for fairly general situations, let us consider two simple examples.

Proposition 4. *For a reproductive function ϕ in the decomposition system $D^{(1)}(A)$ we have*

$$E^*\{\phi_1[A]\} = \frac{\sum_{i=1}^{q} \varphi(k_i)}{\varphi\left(\sum_{i=1}^{q} k_i\right)} = \frac{\sum_{i=1}^{q} \varphi(k_i)}{\varphi(k.)} . \tag{5.1}$$

Proof. We have, by definition,

$$E^*\{\phi_1[A]\} = \iint_{\{x_i\} \in \mathscr{D}\phi_1[A]} \frac{\sum_{i=1}^{q} \phi(x_i)}{\phi(x.)} \prod_{i=1}^{q} g(x_i\,;\, k_i) d\mu(x_i) , \tag{5.2}$$

where the domain of integration $\mathscr{D}\phi_1[A]$ is the set of all non-negative integers for which $\phi_1[A]$ is defined and the symbol $\{x_i\} \in \mathscr{D}\phi_1[A]$ denotes the set of $\{x_i\}$, $i = 1, 2, \ldots, q$ such that $(x_1, x_2, \ldots, x_q)$ belongs to $\mathscr{D}\phi_1[A]$.

Since the function ϕ is reproductive in $D^{(1)}(A)$, we have, for $i = 1, 2, \ldots, q$,

$$\begin{aligned} &\phi(x_i) \prod_{h=1}^{q} g(x_h\,;\, k_h) d\mu(x_h) \\ &= \phi(x_i) g(x_i\,;\, k_i) d\mu(x_h) \prod_{h \neq 1} g(x_h\,;\, k_h) \\ &= g(k_i) c(\phi) g(Tx_i\,;\, Sk_i) \prod_{h \neq i} d\mu(Tx_i) g(x_h\,;\, k_h) d\mu(x_h) , \end{aligned} \tag{5.3}$$

in virtue of (4.1). In view of (4.2) and (4.3), it is readily observed that, when $(x_1, x_2, \ldots, x_q)$ runs through the set for which $\sum_{i=1}^{q} x_i$ is equal to an assigned value x^*, the point $(x_1, x_2, \ldots, x_{i-1}, Tx_i, x_{i+1}, \ldots, x_q)$ belongs to the set for which the sum is equal to $T\left(\sum_{i=1}^{q} x_i\right) = Tx^*$ for any non-negative integer i in $1 \leqq i \leqq q$. Also the converse is true. After summing up the probability elements (5.3) with respect to $d\mu(x_1) \ldots d\mu(x_q)$ over the set of $(x_1, x_2, \ldots, x_q)$ for which $x = \sum_{i=1}^{q} x_i$ is assigned a particular value, we have

$$\varphi(k_i) c(\phi) g\left(T\left(\sum_{i=1}^{q} x_i\right);\, S\left(\sum_{i=1}^{q} k_i\right)\right) d\mu\left(T\left(\sum_{i=1}^{q} x_i\right)\right) , \tag{5.4}$$

which, apart from $\varphi(k_i)$, is independent of i. Now again, the fact that $\phi(x)$ is reproductive implies since $c(\phi)\varphi(k) \neq 0$,

$$(5.5) \qquad g(Tx\,;\; Sx) = \frac{\phi(x)g(x\,;\; k)}{c(\phi)\varphi(k)}d\mu(x)\,,$$

which, in combination with (5.4), leads us to

$$(5.6) \qquad E^*\{\phi_1[A]\} = \frac{\sum_{i=1}^{q}\varphi(k_i)}{\varphi(k)} \iint_{x \in \mathscr{D}\phi_1[A]}\!\!\cdots\!\int g(x\,;\; k)d\mu(x) \times \frac{1}{\iint_{x \in \mathscr{D}\phi_1[A]}\!\!\cdots\!\int g(x\,;\; k)d\mu(x)}$$

$$= \frac{\sum_{i=1}^{q}\varphi(k_i)}{\varphi(k)}\,,$$

as we were to prove.

Example 5.1. (a) *Binomial distribution*

$$(5.7) \quad E_B^*\{\phi_1^{[j]}[A]\} = \frac{\sum_{i=1}^{q}\varphi_B(k_i)}{\varphi_B(k)} = \frac{\sum_{i=1}^{q}k_i(k_i-1)\ldots(k_i-j+1)}{k(k-1)\ldots(k-j+1)}\,.$$

(b) *Negative binomial distribution*

$$(5.8) \quad E_{NB}^*\{\phi_1^{[j]}[A]\} = \frac{\sum_{i=1}^{q}\varphi_{NB}(k_i)}{\varphi_{NB}(k)} = \frac{\sum_{i=1}^{q}k_i(k_i+1)\ldots(k_i+j-1)}{k(k+1)\ldots(k+j-1)}\,.$$

(c) *Poisson distribution*

$$(5.9) \qquad E_P^*\{\phi_1^{[j]}[A]\} = \frac{\sum_{i=1}^{q}\varphi_P(k_i)}{\varphi_P(k)} = \frac{\sum_{i=1}^{q}k_i^j}{k^j}\,.$$

Proposition 5. *For a reproductive function ϕ, in the decomposition system $D^{(2)}(A)$, we have*

$$(5.10) \qquad E^*\{\phi_2[A]\} = \sum_{i_1=1}^{q_1}\frac{\varphi^{(1)}(k_{i_1}^{(1)})}{\varphi^{(1)}(k^{(1)})}\sum_{i_2=1}^{q_2}\frac{\varphi_{i_1}^{(1)}(k_{i_1 i_2}^{(2)})}{\varphi_{i_1}^{(2)}(k_{i_1}^{(2)})}\,.$$

Proof. The proof is quite similar to that of Proposition 3. First we note that, for each pair (i_1, i_2), we have

$$(5.11)\quad \phi(x_{i_1 i_2}) \prod_{j=1}^{q_2} g_{i_1}^{(2)}(x_{i_1 j}\,;\, k_{i_1 j}^{(2)}) d\mu_{i_1}^{(2)}(x_{i_1 j}^{(2)})$$

$$= \phi(x_{i_1 i_2}) g_{i_1}^{(2)}(x_{i_1 i_2}\,;\, k_{i_1 i_2}^{(2)}) d\mu_{i_1 j}^{(2)}(x_{i_1 i_2}^{(2)}) \prod_{j \neq i_2} g_{i_1}^{(2)}(x_{i_1 j}\,;\, k_{i_1 j}^{(2)}) d\mu_{i_1}^{(2)}(x_{i_1 j}^{(2)})$$

$$= c^{(2)}(\phi) \varphi_{i_1}^{(2)}(k_{i_1 i_2}^{(2)}) g_{i_1}^{(2)}(T_{i_1}^{(2)} x_{i_1 i_2}\,;\, S_{i_1}^{(2)} k_{i_1 i_2}) d\mu_i^{(2)}(T_{i_1}^{(2)} x_{i_1 i_2}^{(2)})$$

$$\times \prod_{h \neq i_2} g_{i_1}^{(2)}(x_{i_1 h}\,;\, k_{i_1 h}^{(2)}) d\mu_{i_1}^{(2)}(x_{i_1 j}^{(2)}) ,$$

which yields us

$$(5.12)\quad \sum_{i_2=1}^{q_2} \phi(x_{i_1 i_2}) \left\{ \prod_{i_2=1}^{q_2} g_{i_1}^{(2)}(x_{i_1 i_2}\,;\, k_{i_1 i_2}^{(2)}) d\mu_{i_1}^{(2)}(x_{i_1 i_2}) \right\}$$

$$= c^{(2)}(\phi) \sum_{i_2=1}^{q_2} \varphi_{i_1}^{(2)}(k_{i_1 i_2}^{(2)}) g_{i_1}^{(2)}(T_{i_1}^{(2)} x_{i_1 i_2}\,;\, S_{i_1}^{(2)} k_{i_1 i_2}^{(2)}) d\mu_i^{(2)}(T_{i_1}^{(2)} x_{i_1 i_2}^{(2)})$$

$$\times \prod_{h \neq i_2} \{ g_{i_1}^{(2)}(x_{i_1 h}\,;\, k_{i_1 h}^{(2)}) d\mu_{i_1}^{(2)}(x_{i_1 i_2}) \} .$$

In view of (4.2) and (4.3), after summing up the probability elements in (5.12), we have

(5.13)

$$g_{i_1}^{(2)}(T_{i_1}^{(2)} x_{i_1 i_2}\,;\, S_{i_1}^{(2)} k_{i_1 i_2}^{(2)}) d\mu_{i_1}^{(2)}(T_{i_1}^{(2)} x_{i_1 i_2}^{(2)}) \prod_{h \neq i_2} \{ g_i^{(2)}(x_{i_1 h}\,;\, k_{i_1 h}^{(2)}) d\mu_{i_1}^{(2)}(x_{i_1 i_2}) \}$$

with respect to $d\mu_{i_1}^{(2)}(x_{i_1 1}) \ldots d\mu_{i_1}^{(2)}(x_{i_1 q_2})$, and we obtain

$$(5.14)\quad g_{i_1}^{(2)}(T_{i_1}^{(2)} x_{i_1}\,;\, S_{i_1}^{(2)} k_{i_1}^{(2)}) d\mu_{i_1}^{(2)}(T_{i_1}^{(2)} x_{i_1})$$

when $(x_{i_1 1}, \ldots, x_{i_1 q_2})$ runs through the set for which $\sum_{i_2=1}^{q_2} x_{i_1 i_2}$ is equal to an assigned value x_{i_1}. (5.14) leads us to the probability element

$$(5.15)\quad c^{(2)}(\phi) \left\{ \sum_{i_2=1}^{q_2} \varphi_{i_1}^{(2)}(k_{i_1 i_2}^{(2)}) \right\} g_{i_1}^{(2)}(T_{i_1}^{(2)} x_{i_1}\,;\, S_{i_1}^{(2)} k_{i_1}^{(2)}) d\mu_{i_1}^{(2)}(T_{i_1}^{(2)} x_{i_1}) .$$

Again in virtue of the fact that ϕ is reproductive, (5.15) becomes

$$(5.16)\quad \frac{\sum_{i_1=1}^{q_2} \varphi_{i_1}^{(2)}(k_{i_1 i_2}^{(2)})}{\varphi_{i_1}^{(2)}(k_{i_1}^{(2)})} \phi(x_{i_1}) g_{i_1}^{(2)}(x_{i_1}\,;\, k_{i_1}^{(2)}) d\mu_{i_1}^{(2)}(x_{i_1}) .$$

As a consequence we have

$$(5.17)\quad E^*\{\phi_2[A]\}$$

$$= \iint\cdots\int_{\{x_{i_1i_2}\}\in\mathcal{D}\phi_2[A]} \frac{\sum_{i_1=1}^{q_1}\sum_{i_2=1}^{q_2}\phi(x_{i_1i_2})}{\phi(x)}$$

$$\times \frac{\prod_{i_1=1}^{q_1}\prod_{i_2=1}^{q_2} g_{i_1}^{(2)}(x_{i_1i_2};\ k_{i_1i_2}^{(2)})d\mu_{i_1}^{(2)}(x_{i_1i_2})}{\prod_{i_1=1}^{q_1} g_{i_1}^{(2)}(x_{i_1};\ k_{i_1}^{(2)})d\mu_{i_1}^{(2)}(x_{i_1})}$$

$$\times \prod_{i_1=1}^{q_1} g^{(1)}(x_{i_1};\ k_{i_1}^{(1)})d\mu^{(1)}(x_{i_1})\frac{1}{\Pr\{\mathcal{D}\phi_2[A]\}}$$

$$= \iint\cdots\int_{\{x_{i_1i_2}\}\in\mathcal{D}\phi_2[A]} \frac{1}{\phi(x)}\sum_{i_1=1}^{q_1}\frac{\phi(x_{i_1})}{\varphi_{i_1}^{(2)}(k_{i_1}^{(2)})}\prod_{i_1=1}^{q_1} g^{(1)}(x_{i_1};\ k_{i_1}^{(1)})d\mu^{(1)}(x_{i_1})$$

$$\times \sum_{i_2=1}^{q_2}\varphi_{i_1}^{(2)}(k_{i_1i_2}^{(2)})/\Pr\{\{x_{i_1i_2}\}\in\mathcal{D}\phi_2[A]\}\,.$$

Here we can deal with a transformation

$$(5.18)\quad \phi(x_{i_1})\prod_{i_1=1}^{q_1} g^{(1)}(x_{i_1};\ k_{i_1}^{(1)}):$$
$$c^{(1)}(\phi)\varphi^{(1)}(k_{i_1}^{(1)})g^{(1)}(T^{(1)}x_{i_1};\ S^{(1)}k_{i_1}^{(1)})\prod_{h\neq i_1} g^{(1)}(x_h;\ k_h^{(1)})$$

as in the proof of Proposition 5.

Consequently, after summing up the probability elements in (5.18) with respect to $d\mu^{(1)}(x_1)\ldots d\mu^{(1)}(x_{q_1})$ under the condition that $(x_1, x_2, \ldots, x_{q_1})$ run through the set for which $\sum_{i_1=1}^{q_1} x_{i_1}$ is equal to an assigned value $x.$, we have that the sum of

$$(5.19)\quad c^{(1)}(\phi)\sum_{i_1=1}^{q_1}\varphi^{(1)}(k_{i_1}^{(1)})g^{(1)}(T^{(1)}x_{i_1};\ S^{(1)}k_{i_1}^{(1)})d\mu^{(1)}(Tx_{i_1})$$
$$\times\prod_{h\neq i_1} g^{(1)}(x_h;\ k_h^{(1)})d\mu^{(1)}(x_h)$$

is equal to

$$(5.20)\quad \frac{\sum_{i_1=1}^{q_1}\varphi^{(1)}(k_{i_1}^{(1)})}{\varphi^{(1)}(k^{(1)})}\phi(x)g^{(1)}(x;\ k^{(1)})d\mu^{(1)}(x)\,,$$

which leads us to the result to be proved.

Now we are in the position to enunciate a general result on the conditional expectation of ϕ-dispersions.

Proposition 6. *For a reproductive function ϕ in the decomposition system $D^{(d)}(A_{i_1 i_2 \cdots i_h})$ we have*

$$(5.21)\quad E^*\{\phi d[A_{i_1 i_2 \cdots i_h}]\} = \sum_{i_{h+1}=1}^{q_{h+1}} \frac{\varphi^{(h+1)}_{i_1 i_2 \cdots i_h}(k^{(h+1)}_{i_1 i_2 \cdots i_{h+1}})}{\varphi^{(h+1)}_{i_1 i_2 \cdots i_h}(k^{(h+1)}_{i_1 \cdots i_h})} \sum_{i_{h+2}=1}^{q_{h+2}} \frac{\varphi^{(h+2)}_{i_1 \cdots i_{h+1}}(k^{(h+2)}_{i_1 \cdots i_{h+2}})}{\varphi^{(h+2)}_{i_1 \cdots i_{h+1}}(k^{(h+2)}_{i_1 \cdots i_{h+1}})} \cdots \times \sum_{i_{h+d}=1}^{q_{h+d}} \frac{\varphi^{(h+d)}_{i_1 \cdots i_{h+d-1}}(k^{(h+d)}_{i_1 i_2 \cdots i_{h+d}})}{\varphi^{(h+d)}_{i_1 \cdots i_{h+d-1}}(k^{(h+d)}_{i_1 i_2 \cdots i_{h+d-1}})},$$

where the case $h=0$ can be interpreted as

$$(5.22)\quad E^*\{\phi_d[A]\} = \sum_{i_1=1}^{q_1} \frac{\varphi^{(1)}(k^{(1)}_{i_1})}{\varphi^{(1)}(k^{(1)})} \sum_{i_2=1}^{q_2} \frac{\varphi^{(2)}_{i_1}(k^{(2)}_{i_1 i_2})}{\varphi^{(2)}_{i_1}(k^{(2)}_{i_1})} \sum_{i_3=1}^{q_3} \frac{\varphi^{(3)}_{i_1 i_2}(k^{(3)}_{i_1 i_2 i_3})}{\varphi^{(3)}_{i_1 i_2}(k^{(3)}_{i_1 i_2})} \times \cdots \sum_{i_d=1}^{q_d} \frac{\varphi^{(d)}_{i_1 i_2 \cdots i_{d-1}}(k^{(d)}_{i_1 i_2 \cdots i_d})}{\varphi^{(d)}_{i_1 i_2 \cdots i_{d-1}}(k^{(d)}_{i_1 i_2 \cdots i_{d-1}})}.$$

Example 5.2. *Two stage homogeneous distributional pattern.* This is characterized by two systems of probability distribution elements $g^{(1)}d\mu^{(1)}$ in (3.13) and $g^{(2)}d\mu^{(2)}$ in (3.14), which, in this case, will be denoted by $Gd\mu^{(1)} \times Ld\mu^{(2)}$, where G(global) and L(Local) correspond to $g^{(1)}$ and $g^{(2)}$ respectively.

In general we have, for such a two stage homogeneous distributional pattern,

$$(5.23)\quad E^*_{G\times L}\{\phi_2[A]\} = \sum_{i_1=1}^{q_1} \frac{\varphi_G(k^{(1)}_{i_1})}{\varphi_G(k^{(1)})} \sum_{i_2=1}^{q_2} \frac{\varphi_L(k^{(2)}_{i_1 i_2})}{\varphi_L(k^{(2)}_{i_1})},$$

which reduces to

$$(5.24)\quad E^*_{G\times L}\{\phi_2[A]\} = \frac{1}{\varphi_G(k^{(2)})} \sum_{i_1=1}^{q_1} \sum_{i_2=1}^{q_2} \varphi_G(k^{(2)}_{i_1 i_2})$$

when the two-stage distributional pattern is inherited, as can be readily seen from the fact that $\varphi_G(k) = \varphi_L(k)$ for every k in their common domain of definition.

(a) *Inherited binomial distributional pattern* $(B \times B)$.

$$E^*_{B\times B}\{\phi_2^{[j]}[A]\} = \sum_{i_1=1}^{q_1} \frac{\varphi_B(k^{(1)}_{i_1})}{\varphi_B(k^{(1)})} \sum_{i_2=1}^{q_2} \frac{\varphi_B(k^{(2)}_{i_1 i_2})}{\varphi_B(k^{(2)}_{i_1})} \tag{5.25}$$

(b) *Inherited negative binomial distributional pattern* ($NB \times NB$).

$$E^*_{NB\times NB}\{\phi_2^{[j]}[A]\} = \sum_{i_1=1}^{q_1} \frac{\varphi_{NB}(k^{(1)}_{i_1})}{\varphi_{NB}(k^{(1)})} \sum_{i_2=1}^{q_2} \frac{\varphi_{NB}(k^{(2)}_{i_1 i_2})}{\varphi_{NB}(k^{(2)}_{i_1})} \tag{5.26}$$

(c) *Inherited Poisson distributional pattern* ($P \times P$).

$$E^*_{P\times P}\{\phi_2^{[j]}[A]\} = \sum_{i=1}^{q_1} \frac{\varphi_P(k^{(1)}_{i_1})}{\varphi_P(k^{(1)})} \sum_{i_2=1}^{q_2} \frac{\varphi_P(k^{(2)}_{i_1 i_2})}{\varphi_P(k^{(2)}_{i_1})} \tag{5.27}$$

Example 5.3. *Three stage homogeneous distributional pattern.* This is characterized by three systems of probability distribution elements $g^{(i)}d\mu^{(i)}$ ($i = 1, 2, 3$), which will be denoted by $Gd\mu^{(1)} \times Sd\mu^{(2)} \times Ld\mu^{(3)}$ in a symbolic way, abbreviated by $G \times S \times L$, where G(global), SG(or S)(subglobal) and L(local) corresponds to $g^{(1)}$, $g^{(2)}$ and $g^{(3)}$ respectively.

$$E^*_{G\times S\times L}\{\phi_3[A]\} = \sum_{i_1=1}^{q_1} \frac{\varphi_G(k^{(1)}_{i_1})}{\varphi_G(k^{(1)})} \sum_{i_2=1}^{q_2} \frac{\varphi_S(k^{(2)}_{i_1 i_2})}{\varphi_S(k^{(2)}_{i_1})} \sum_{i_3=1}^{q_3} \frac{\varphi_L(k^{(3)}_{i_1 i_2 i_3})}{\varphi_L(k^{(3)}_{i_1 i_2})}, \tag{5.28}$$

which becomes

$$E^*_{G\times S\times L}\{\phi_3[A]\} = \frac{1}{\varphi_L(k^{(3)})} \sum_{i_1=1}^{q_1} \sum_{i_2=1}^{q_2} \sum_{i_3=1}^{q_3} \varphi_L(k^{(3)}_{i_1 i_2 i_3}), \tag{5.29}$$

when the three-stage distributional pattern is 3-depth inherited, because $\varphi_G(k) = \varphi_S(k) = \varphi_L(k)$ for every k in the common domain of definition and $k^{(j)}_{i_1 i_2 \ldots i_{j-1}} = k^{(j-1)}_{i_1 i_2 \ldots i_j}$.

6. PARTITIONAL PATTERN OF ADDITIVE PARAMETERS IN THE DECOMPOSITION SYSTEM $D^{(m)}(A)$

Definition 8. *A set of additive parameters* $\{k^{(\nu)}_{i_1 i_2 \ldots i_\nu}\}$ ($\nu = h + 1, h + 2, \ldots, h + d$) *in the decomposition system* $D^{(m)}(A)$ *is said to be equipartitioned at the d-depth in the set* $A_{i_1 i_2 \ldots i_h}$ *when*

(6.1) $$k^{(h+\nu)}_{i_1 i_2 \ldots i_h i_{h+1} \ldots i_{h+\nu}} = \frac{k^{(h+\nu)}_{i_1 i_2 \ldots i_h}}{q_{h+1} q_{h+2} \cdots q_{h+\nu}} = \overline{k^{(h+\nu)}_{i_1 i_2 \ldots i_h}}$$

for $i_j = 1, 2, \ldots, q_j$; $j = h+1, h+2, \ldots, h+\nu$; $\nu = 1, 2, \ldots, d$.

Definition 9. *A set of additive parameters* $\{k^{(\nu)}_{i_1 i_2 \ldots i_\nu}\}$ $(\nu = h+1, \ldots, h+d)$ *in the decomposition system* $D^{(m)}(A)$ *is said to be monopolistically concentrated in the set* $A_{i_1 i_2 \ldots i_h}$ *when there exists a sequence of* d *integers* $(i^0_{h+1}, i^0_{h+2}, \ldots, i^0_{h+d})$ *such that*

(6.2) $$k^{(h+\nu)}_{i_1 i_2 \ldots i_h i^0_{h+1} \ldots i^0_{h+\nu}} = k^{(h+\nu)}_{i_1 i_2 \ldots i_h}, \qquad \nu = 1, 2, \ldots, d$$

and hence

(6.3) $$k^{(h+\nu)}_{i_1 i_2 \ldots i_h i_{h+1} \ldots i_{h+\nu}} = 0, \qquad \nu = 1, 2, \ldots, d$$

for every $(i_{h+1}, i_{h+2}, \ldots, i_{h+\nu}) \neq (i^0_{h_1+1}, i^0_{h_1+2}, \ldots, i^0_{h+\nu})$.

We observe

Proposition 7. *Let a set of additive parameters* $\{k^{(\nu)}_{i_1 i_2 \ldots i_\nu}\}$ $(\nu = h+1, h+2, \ldots, h+d)$ *in the decomposition system* $D^{(m)}(A)$ *be equipartitioned at the d-depth in the set* $A_{i_1 i_2 \ldots i_h}$. *Then we have*

(6.4) $$E^*\{\phi_d[A_{i_1 i_2 \ldots i_h}]\} = \sum_{i_{h+1}=1}^{q_{h+1}} \frac{q_{h+1} \varphi^{(h+1)}_{i_1 \ldots i_h} \overline{(k^{(h+1)}_{i_1 i_2 \ldots i_h})}}{\varphi^{(h+1)}_{i_1 i_2 \ldots i_h} (k^{(h+1)}_{i_1 i_2 \ldots i_h})}$$
$$\times \sum_{i_{h+2}=1}^{q_{h+2}} \frac{q_{h+2} \varphi^{(h+2)}_{i_1 \ldots i_h i_{h+1}} \overline{(k^{(h+2)}_{i_1 i_2 \ldots i_{h+1}})}}{\varphi^{(h+2)}_{i_1 \ldots i_{h+1}} (k^{(h+2)}_{i_1 i_2 \ldots i_{h+1}})} \cdots$$
$$\times \sum_{i_{h+d}=1}^{q_{h+d}} \frac{q_{h+d} \varphi^{(h+d)}_{i_1 \ldots i_{h+d-1}} \overline{(k^{(h+d)}_{i_1 i_2 \ldots h+d-1})}}{\varphi^{(h+d)}_{i_1 i_2 \ldots i_{h+d-1}} (k^{(h+d)}_{i_1 i_2 \ldots i_{h+d-1}})}.$$

In particular when the set of $q_{h_1} \ldots q_{h_2}$ *stochastic variables* $\{X_{i_1 \ldots i_h i_{h+1} \ldots i_{h+d}}\}$ $(i_j = 1, 2, \ldots, q_j$; $j = h+1, \ldots, h+d)$ *is both d-depth homogeneous and inherited to the h-th distributional pattern in the set* $A_{i_1 i_2 \ldots i_h}$, *then we have*

(6.5) $$E^*\{\phi_d[A_{i_1 i_2 \ldots i_h}]\} = \frac{q_{h+1} \cdots q_{h+d} \varphi^{(h+1)}_{i_1 i_2 \ldots i_h} \overline{(k^{(h+d)}_{i_1 i_2 \ldots i_h})}}{\varphi^{(h+1)}_{i_1 i_2 \ldots i_h} (k^{(h+d)}_{i_1 i_2 \ldots i_h})}.$$

Proposition 8. *Let a set of additive parameters* $\{k^{(\nu)}_{i_1 i_2 \ldots i_\nu}\}$ $(\nu = h+1, h+2, \ldots, h+d)$ *in the decomposition system* $D^{(m)}(A)$ *be* *monopolistically concentrated in* $A_{i_1i_2\ldots i_h}$ *at the d-depth. Then we have*

(6.6) $$E^*\{\phi_d[A_{i_1i_2\ldots i_h}]\} = 1\,.$$

7. ϕ-DISPERSION AS A FUNCTION OF SIZE VARIABLE

Let us consider a sequence of ϕ-dispersions $\{\phi_d[A]\}$ ($d = 1, 2, \ldots, m$) for an assigned decomposition system $D^{(m)}(A)$ of an assigned set A. We investigate how $\phi_d[A]$ will change as a function of the depth d. In discussing such a problem there are two fundamental tools which we use.

1) Size variable associated with the decomposition system $D^{(m)}(A)$. Let us assume every subset $A_{i_1i_2\ldots i_m}$ is measurable. Let its measure be denoted by $m(A_{i_1i_2\ldots i_m})$. Let us denote by a_h the average of all $\{m(A_{i_1i_2\ldots i_h})\}$, $i_j = 1, 2, \ldots, q_j$; $j = 1, 2, \ldots, h$; $h = 1, 2, \ldots, m$, i.e.,

(7.1) $$a_h = \frac{1}{q_1q_2\ldots q_h}\sum_{i_1=1}^{q_1}\ldots\sum_{i_h=1}^{q_h} m(A_{i_1i_2\ldots i_h})\,,$$

which is called the h-th average size associated with $D^{(h)}(A)$.

For the sake of convenience we define

(7.2) $$a_0 = A\,.$$

Under these preparations we introduce

Definition 10. *A sequence of functions* $\{\Phi(a_i|A)\}$, $i = 0, 1, 2, \ldots, m$, *defined by*

(7.3) $$\Phi(a_0|A) = 1$$

(7.4) $$\Phi(a_1|A) = \sum_{i_1=1}^{q_1}\frac{\varphi^{(1)}(k_{i_1}^{(1)})}{\varphi^{(1)}(k_{\cdot}^{(1)})}$$

(7.5) $$\Phi(a_h|A) = \sum_{i_1=1}^{q_1}\frac{\varphi^{(1)}(k_{i_1}^{(1)})}{\varphi^{(1)}(k_{i_1}^{(1)})}\sum_{i_2=1}^{q_2}\frac{\varphi_{i_1}^{(2)}(k_{i_1i_2}^{(2)})}{\varphi_{i_1}^{(2)}(k_{i_1}^{(2)})}\cdots\sum_{i_h=1}^{q_h}\frac{\varphi_{i_1\ldots i_{h-1}}^{(h)}(k_{i_1i_2\ldots i_h}^{(h)})}{\varphi_{i_1\ldots i_{h-1}}^{(h)}(k_{i_1i_2\ldots i_{h-1}}^{(h)})}$$

for $h = 2, 3, \ldots, m$, *is said to constitute a* ϕ*-dispersion system as a function of the average size* a_h ($h = 0, 1, 2, \ldots, m$).

Particularly when all $A_{i_1 i_2 \cdots i_h}$ have the same measure, i.e.,

$$(7.6) \qquad m(A_{i_1 i_2 \cdots i_h}) = \frac{m(A)}{q_1 q_2 \ldots q_h} = a_h$$

for $i_j = 1, 2, \ldots, q_j$; $j = 1, 2, \ldots, h$; $h = 1, 2, \ldots, m$ each a_h is the size of each subset in the h-th stage decomposition.

In view of these facts we shall call a_h a size variable which is monotone decreasing: $a_0 \geqq a_1 \geqq \ldots \geqq a_m$

2) Homogeneous and inherited multistage distributional patterns interpolated between the h-th stage decomposition $D_h^{(m)}(A)$ and the $(h+1)$-th stage one $D_{h+1}^{(m)}(A)$. To explain the principal aspects of our techniques, let us consider the simplest ϕ-dispersion

$$(7.7) \qquad \phi_1[A] = \Phi(a_1 | A) = \frac{1}{\varphi^{(1)}(k^{(1)})} \sum_{i_1=1}^{q_1} \varphi^{(1)}(k_{i_1}^{(1)}) .$$

Now let us consider the case when the non-negative integer q_1 is decomposed into the product of l_1 non-negative integers $\{q_{1\nu}\}$ $(\nu = 1, 2, \ldots, l_1)$ such that $q_1 = q_{11} q_{12} \ldots q_{1l_1}$. Then it is possible to establish a one-to-one correspondence defined by $i = \Psi(i_1, i_2, \ldots, i_l)$ where the variable (integer) i ranges from 1 to q_1 while $i_j = 1, 2, \ldots, q_{1j}$; $j = 1, 2, \ldots, l_1$. In this connection we define $A_{i_1 i_2 \cdots i_{l_1}}^{(1)} = A_{\Psi(i_1, i_2, \ldots, i_{l_1})}$ and $k_{i_1 i_2 \cdots i_{l_1}}^{(1)} = k_{\Psi(i_1, i_2, \ldots, i_{l_1})}^{(1)}$.

As a consequence, we may rewrite (7.7) as

$$(7.8) \qquad \phi_1[A] = \frac{1}{\varphi^{(1)}(k^{(1)})} \sum_{i_1=1}^{q_{11}} \sum_{i_2=1}^{q_{12}} \ldots \sum_{i_{l_1}=1}^{q_{1l_1}} \varphi^{(1)}(k_{i_1 i_2 \cdots i_{l_1}}^{(1)}) ,$$

which can be reformed as

$$(7.9) \quad \phi_1[A] = \sum_{i_1=1}^{q_{11}} \frac{\varphi^{(1)}(k_{i_1}^{(1)})}{\varphi^{(1)}(k^{(1)})} \sum_{i_2=1}^{q_{12}} \frac{\varphi^{(1)}(k_{i_1 i_2}^{(1)})}{\varphi^{(1)}(k_{i_1}^{(1)})} \cdots \sum_{i_{l_1}=1}^{q_{1l_1}} \frac{\varphi^{(1)}(k_{i_1 i_2 \cdots i_l}^{(1)})}{\varphi^{(1)}(k_{i_1 i_2 \cdots i_{l-1}}^{(1)})} .$$

The expression (7.9) of $\phi_1[A]$ shows that the 1-depth ϕ-dispersion can be interpretated as the l_1-depth ϕ-dispersion, which is homogeneous and inherited in the l_1-depth in the set A, in the sense of Definitions 4 and 5.

In view of this fact let us now introduce a sequence of area sizes $\{a_\nu^{(1)}\}$ $(\nu = 1, 2, \ldots, l_1)$ defined by

$$(7.10) \qquad a_\nu^{(1)} = \frac{1}{q_{11} q_{12} \ldots q_{1\nu}} \sum_{i_1=1}^{q_{11}} \ldots \sum_{i_\nu=1}^{q_1} m(A_{i_1 i_2 \cdots i_\nu}^{(1)})$$

and, with reference to this sequence, let us define a sequence of ϕ-dispersions

$$(7.11) \quad \Phi(a_\nu^{(1)}|A) = \sum_{i_1=1}^{q_{11}} \frac{\varphi^{(1)}(k_{i_1}^{(1)})}{\varphi^{(1)}(k^{(1)})} \sum_{i_2=1}^{q_{12}} \frac{\varphi^{(1)}(k_{i_1 i_2}^{(1)})}{\varphi^{(1)}(k_{i_1}^{(1)})} \cdots \sum_{i_\nu=1}^{q_{1\nu}} \frac{\varphi^{(1)}(k_{i_1 i_2 \ldots i_\nu}^{(1)})}{\varphi^{(1)}(k_{i_1 i_2 \ldots i_{\nu-1}}^{(1)})}$$

$$= \frac{1}{\varphi^{(1)}(k^{(1)})} \sum_{i_1=1}^{q_{11}} \cdots \sum_{i_\nu=1}^{q_{1\nu}} \varphi^{(1)} k(_{i_1 i_2 \ldots i_\nu}^{(1)})$$

for $\nu = 1, 2, \ldots, l_1$.

It is noted that $a_0 \geqq a_1^{(1)} \geqq a_2^{(1)} \geqq \ldots \geqq a_{l_1-1}^{(1)} \geqq a_{l_1}^{(1)} = a_1$ and that the sequence $\{\Phi(a_\nu^{(1)}|A)\}$ $(\nu = 1, 2, \ldots, l_1)$ is an interpolation between $\Phi(a_0|A)$ and $\Phi(a_1|A)$.

The interpolation technique just now explained between to the first stage decomposition $D_1^{(m)}(A)$ and the second stage $D_2^{(m)}(A)$ can be similarly extended between the h-th stage $D_h^{(m)}(A)$ and the $(h+1)$-th stage $D_{h+1}^{(m)}(A)$ for $h = 2, 3, \ldots, m-1$. In this way it is possible to introduce a sequence of average sizes $\{a_\nu^{(h)}\}$ $(\nu = 1, 2, \ldots, l_h;\ h = 1, 2, \ldots, m)$ such that

$$(7.12) \qquad a_{h-1} \geqq a_1^{(h)} \geqq a_2^{(h)} \geqq \ldots \geqq a_{l_\nu}^{(h)} = a_h$$

and to define a sequence of ϕ-dispersions

$$(7.13) \qquad \Phi(a_s^{(h)}|A)$$

for $s = 1, 2, \ldots, l_h;\ h = 1, 2, \ldots, m$, as a whole. In this manner we can obtain a graph of the ϕ-dispersion $\Phi(a|A)$ as a function of a, which is actually defined for $a = a_s^{(h)}$ just mentioned.

3) Equipartition of additive parameters in the homogeneous and inherited multistage distributional patterns in the decomposition system $D^{(m)}(A)$ with $\{A_{i_1 i_2 \ldots i_m}\}$ of equal measure $m(A_{i_1 i_2 \ldots i_m}) = m(A)/(q_1 q_2 \ldots q_m)$.

In this case we have

$$(7.14) \qquad k_{i_1 i_2 \ldots i_h}^{(h)} = k_{i_1 i_2 \ldots i_h}^{(m)} = K/q_1 q_2 \ldots q_h$$

$$= K a_h a_0^{-1}$$

and

$$(7.15) \qquad \Phi(a_h|A) = \frac{a_h^{-1}\varphi(K a_h a_0^{-1})}{a_0^{-1}\varphi(K)} = \frac{\varphi(b_h K)}{b_h \varphi(K)}, \quad \text{say,}$$

where

$$b_h = a_h a_0^{-1} , \tag{7.16}$$

which is called a relative area variable, or an area-index. A relative area variable is convenient because, in this scale, the relative area of the set A is equal to 1.

A deep insight into the structure of multistage distributional patterns can be obtained by use of the ϕ-dispersion function $\Phi(a|A)$ as a function of the size a in the context just explained.

It is noted that the behaviors of the ϕ-dispersion functions $\Phi(a|A)$ for each of nine two-stage and twenty-seven three-stage distributional patterns given in Examples 3.4 and 3.5 may be examined in particular numerical investigations. Indeed, numerical investigations based upon our theoretical formulations yield some of the possible explanations of the experimental results on natural and artificial populations given by Morishita [1].

8. INDEX FUNCTIONS OF DISPERSION AS FUNCTIONS OF THE SIZE VARIABLE

In connection with current uses of an index of dispersion, it is useful to introduce

Definition 11. 1) $I_d^{[j]}[A_{i_1 i_2 \ldots i_h}]$, *defined in* (2.11), *is said to be the d-depth index of dispersion of j-th order in the set* $A_{i_1, i_2, \ldots, i_h}$.

2) *A sequence of functions* $I^{[j]}(a_i|A)$, $i = 0, 1, 2, \ldots, m$, *defined by*

$$I^{[j]}(a_0|A) = (q_1 q_2 \ldots q_m)^{j-1} \tag{8.1}$$

$$I^{[j]}(a_\nu|A) = E^*\{I_\nu^{[j]}[A]\} , \tag{8.2}$$

for $\nu = 1, 2, \ldots, m$, *is said to be an index function of dispersion of j-th order as a function of a size variable which takes the values* $\{a_\nu\}$ $(\nu = 0, 1, 2, \ldots, m)$.

The current index of dispersion used and advocated by Morishita [1]–[4] and others is the 1-depth index of dispersion of second order in our terminology. Apart from a multiplier $(q_1 q_2 \ldots q_m)^{j-1}$, the essential aspects of $I_d^{[j]}[A]$ can be observed from our general considerations on $\phi_d[A]$ and those of $I^{[j]}(a_i|A)$, from those of functions $\Phi(a_i|A)$. We do not think the multiplier is essential in our general approaches. Its sole merits, so far as the results of the present paper are concerned, are that

$$E_p^*\{I_d^{[j]}[A]\} = 1 \tag{8.3}$$

for a homogeneous and inherited multistage distributional pattern based upon Poisson distribution. Multipliers are partly due to tradition and partly due to the fact that theoretical emphasis may be placed upon the Poisson distribution, as we may observe from the fact that

$$E^*\{I_d^{[j]}[A]\} = 1 \tag{8.4}$$

for a homogeneous and inherited multistage distributional pattern based upon the Poisson distribution.

References

1. Morishita, M. "Measuring of the Dispersion of Individuals and Analysis of the Distributional Pattern," *Mem. Fac. Sci., Kyushu Univ., Ser. E (Biol.)*, **2** (1957), 215–235.
2. Morishita, M. "I_δ-index, A Measure of Dispersion of Individuals," *Res. Popul. Ecol.*, **4** (1962), 1–7.
3. Morishita, M. "Application of I_δ-index to Sampling Techniques," *Res. Popul. Ecol.*, **6** (1964), 43–53.
4. Morishita, M. "A Revision of the Methods for Estimating Population Values of the Index of Dispersion in the I_δ-method," *Res. Popul. Ecol.*, **7** (1965), 126–128.
5. Smith, E. H. "Measurement of Diversity," *Nature*, **163** (1949), 688.

Part II

COMBINATORIAL PROBLEMS OF EXPERIMENTAL DESIGNS

CHAPTER 10

Symmetric Block Designs with $\lambda = 2$

MARSHALL HALL, JR.,[1] *California Institute of Technology*

1. INTRODUCTION

Symmetric block designs with parameters $v = k(k-1)/2 + 1$, k, and $\lambda = 2$ have been constructed [1, 2, 4, 5, 6, 7, 8] for $k = 3$, 4, 5, 6, and 9. The values $k = 7$ and 8 are impossible by the Chowla-Ryser criterion [3], and the value $k = 7$ had been excluded earlier by a direct search by Husain [6].

A natural way to look for an infinite class of symmetric block designs with $\lambda = 2$ is to assume existence of a large group of automorphisms. This is in analogy with the existence of the Desarguesian projective planes; symmetric block designs with $\lambda = 1$.

Husain [6] has introduced an "affine" representation of symmetric designs with $\lambda = 2$ by associating permutations on $1, 2, \ldots, k$ with those blocks which do not include a particular variety X. The permutation is unique except that any cycle may be replaced by its inverse.

In certain cases a symmetric design with $\lambda = 2$, $k = q + 1$ admits the group $LF(2, q)$ as a group of automorphisms fixing a particular variety X. This may be done by taking as the

[1] This research was supported in part by National Science Foundation Grant No. 3909.

Husain permutations a class of conjugate elements in the representation of $LF(2, q)$ on $q + 1$ letters and identifying inverses. In this paper it is shown that this construction does not yield an infinite class of designs, succeeds only for the values $q = 2$, 4, 5, 8. This result lends support to the conjecture[2] that there are only a finite number of symmetric block designs with $\lambda = 2$.

2. THE HUSAIN CHAINS

Let D be a symmetric block design with $\lambda = 2$. Then the basic relation

$$k(k-1) = \lambda(v-1) \tag{2.1}$$

yields

$$v = k(k-1)/2 + 1 . \tag{2.2}$$

Let us designate a particular element (variety) as X and consider the k blocks of D containing $X, B_1, \ldots, B_k$.

$$\begin{aligned} B_1 &: \ X, a_{11}, \ldots, a_{1k-1} , \\ B_2 &: \ X, a_{21}, \ldots, a_{3k-1} , \\ B_k &: \ X, a_{k1}, \ldots, a_{k,k-1} . \end{aligned} \tag{2.3}$$

Each element u of D different from X occurs exactly twice with X and so appears in exactly two of the blocks of (2.3). If u is in the two blocks B_i and B_j then we assign as "coordinates" to u the unordered pair (i, j) and we write

$$u = (i, j) = (j, i) , \qquad i \neq j , \ j = 1, \ldots, k , \quad u \in B_i , \ u \in B_j . \tag{2.4}$$

Since any two distinct blocks B_i, B_j have X and exactly one further element in common, the rule (2.4) established a one-to-one correspondence between the $v - 1 = k(k-1)/2$ elements of D different from X and the unordered pairs $(i, j) = (j, i)$ $i \neq j$, $i, j = 1, \ldots, k$. In this representation let the elements of a further block B_m be

$$B_m : \ (r_1, s_1), (r_2, s_2), \ldots, (r_k, s_k) . \tag{2.5}$$

[2] Recently the writer has constructed a design with $k=11$, and M. Aschbacker another with $k=13$.

Since B_m intersects each of $B_1, \ldots, B_k$ exactly twice, each of $1, 2, \ldots, k$ occurs exactly twice in the coordinates in (2.5). Following Husain [6] let us form cyclical chains from the coordinates (2.5) taking a cyclical chain

$$C = (a_1, a_2, a_3, \ldots, a_t) \tag{2.6}$$

if $(a_1, a_2), (a_2, a_3), (a_3, a_4), \ldots, (a_{t-1}, a_t)$ and (a_t, a_1) are among the unordered pairs of (2.5). Here the cycle C or its inverse represents the same set of elements. Here all numbers $1, 2, \ldots, k$ fall into disjoint cycles and we may represent B_m by a permutation P_m on $1, 2, \ldots, k$ in cycle form.

$$P_m = (a_1, \ldots, a_t)(b_1, \ldots, b_n) \ldots . \tag{2.7}$$

Here B_m does not correspond to a unique permutation P_m, since we may replace any cycle of P_m by its inverse and still represent the same block. In particular P_m and P_m^{-1} represent the same block.

An easy verification, not given here, shows: A set of $v - k = (k-1)(k-2)/2$ permutations of $1, \ldots, k$, P_m, $m = k + 1, \ldots, v$, together with the initial blocks B_i, X, $(i, 1), (i, 2), \ldots, (i, i-1), (i, i+1), \ldots, (i, k)$, $i = 1, \ldots, k$, $(i, j) = (j, i)$ yields a symmetric block design with parameters $v = k(k-1)/2 + 1$, k, $\lambda = 2$ if and only if:

(i) Writting P_m in cycle form, no cycle has less than three letters, for $m = k + 1, \ldots, v$.

(ii) If $m \neq n$ we cannot have $P_m = (a, b, c, \ldots) \ldots$ and P_n or $P_n^{-1} = (a, b, c, \ldots) \ldots$, for three letters a, b, c.

(iii) If $m \neq n$ then $P_m P_n$ and $P_m P_n^{-1}$ together fix exactly two letters.

Here condition (ii) says that the pair of elements (a, b), (b, c) which are together in B_b do not occur together in more than one further block.

Condition (iii) says that two blocks B_m B_n, $m \neq n$, $m, n = k + 1, \ldots, v$ have exactly two elements (x, y) and (t, w) in common.

An automorphism α of a symmetric design D with parameters $v = k(k-1)/2 + 1, k$ and $\lambda = 2$ which fixes an element X of D, will permute the blocks $B_1, \ldots, B_k$ containing X in some manner. If this permutation is $\pi(\alpha) = \begin{pmatrix} 1, 2, \ldots, k \\ a_1, a_2, \ldots, a_k \end{pmatrix}$ then α maps the element (i, j) onto the element (a_i, a_j). Thus α maps a Husain chain P onto $\pi(\alpha)^{-1} P \pi(\alpha)$. We conclude that an automorphism α of D fixing an element X is associated with a permutation $\pi(\alpha)$ which takes the Husain chains associated with X into themselves by conjugation. Conversely it is easy to see that a permutation which takes the Husain chains associated with X into themselves by conjugation determines an automorphism of D fixing X.

We list Husain chains and the corresponding automorphisms for designs with $k = 3, 4, 5, 6, 9$.

1. $k = 3$, $v = 4$. $(0, 1, 2)$. Automorphisms S_3, the symmetric group on three letters.
2. $k = 4$, $v = 7$. $(0, 1, 2, 3)$
   $(0, 2, 3, 1)$
   $(0, 3, 1, 2)$
   Automorphism group S_4.
3. $k = 5$, $v = 11$. $(0, 1, 2, 3, 4)$ $(0, 2, 1, 4, 3)$
   $(0, 1, 3, 4, 2)$ $(0, 2, 3, 1, 4)$
   $(0, 1, 4, 2, 3)$ $(0, 3, 1, 2, 4)$
   Automorphism group A_5, the alternating group on 5 letters. Here $A_5 \cong LF(2, 4) \cong LF(2, 5)$, the linear fractional groups over $GF(4)$, $GF(5)$.
4. $k = 6$, $v = 16$. $(6, 0, 1)(2, 4, 3)$ $(6, 0, 3)(1, 2, 4)$
   $(6, 1, 2)(3, 0, 4)$ $(6, 1, 4)(2, 3, 0)$
   $(6, 2, 3)(4, 1, 0)$ $(6, 2, 0)(3, 4, 1)$
   $(6, 3, 4)(0, 2, 1)$ $(6, 3, 1)(4, 0, 2)$
   $(6, 4, 0)(1, 3, 2)$ $(6, 4, 2)(0, 1, 3)$
   Here the automorphism group is S_6. This includes $LF(2, 5)$ as the subgroup generated by $(6, 0, 1)(2, 4, 3)$ and $(6)(0, 1, 2, 3, 4)$.
5. $k = 9$, $v = 37$
   Here one chain is $(8, 0, 1)(2, 6, 4)(3, 5, 7)$.
   The automorphism group is $LF(2, 8)$ generated by $(8, 0, 1)(2, 6, 4)(3, 5, 7)$ and $(0)(8)(1, 2, 4, 3, 6, 7, 5)$.

3. THE MAIN THEOREM

Theorem. *Consider the group $LF(2, q)$ of transformations over the finite field $GF(q)$, these being* $x \to \dfrac{ax + b}{cx + d}$, $a, b, c, d \in GF(q)$, $ad - bc = 1$, *and let $LF(2, q)$ be represented as a permutation group on $q + 1$ letters, the q elements of $GF(q)$ and a further symbol ∞. Then a conjugate class of these permutations (identifying a permutation and its inverse) will form the Husain* *chains for a symmetric block design with* $k = q + 1$, $v = (q+1)q/2 + 1$, $\lambda = 2$, *when* $q = 2, 4, 5, 8$ *and in no other cases. The chains will be isomorphic to those listed in section 2.*

Proof. The special linear group $S(2, q)$ is the group of matrices over $GF(q)$

$$\begin{bmatrix} a & b \\ c & d \end{bmatrix}, \qquad ad - bc = 1, \tag{3.1}$$

Here if q is odd $S(2, q)$ has a center Z consisting of the identity and the matrix

$$\begin{bmatrix} -1 & 0 \\ 0 & -1 \end{bmatrix} \tag{3.2}$$

and $LF(2, q)$ is $S(2, q)/Z$. This amounts to identifying a matrix and its negative. If $q = 2^r$ the center is the identity and $LF(2, q) = S(2, q)$. $LF(2, q)$ is of order $(q+1)q(q-1)/2$ if q is odd and of order $(q+1)q(q-1)$ if q is even.

In the representation of $LF(2, q)$ as a permutation group on ∞, and the q elements of $GF(q)$, the characteristic polynomial of (3.1) is $f(x)$ where

$$f(x) = x^2 - (a+d)x + 1\,. \tag{3.3}$$

If $f(x)$ is irreducible over $GF(q)$ then the corresponding permutation moves all letters. If $f(x)$ has two distinct roots in $GF(q)$ then the permutation fixes exactly two letters, while if $f(x)$ has two equal roots, either $+1$'s or -1's, then the permutation fixes one letter. If q is odd the elements fixing one letter fall into two conjugate classes, one conjugate to $x \to x + b$ where b is a square and another where b is not a square. In all other cases the conjugate class is determined by the trace $s = a + d$ where s and $-s$ determine the same class.

The symbol ∞ has the formal properties $1/0 = \infty$, $1/\infty = 0$, $\infty + a = \infty$ and $\dfrac{a\infty + b}{c\infty + d} = \dfrac{a}{c}$.

Let C be the class of permutations from which we form the Husain chains. Since G is doubly transitive C, contains a permutation of G taking ∞ into 0. This is of the form

$$x \to \frac{b}{cx + s}, \qquad bc = -1 \tag{3.4}$$

and all permutations for s fixed are conjugate[3]. Let us consider the particular permutation P in this class

(3.5) $$x \to \frac{1}{-x+s}$$

and the further conjugates of P taking ∞ into 0

(3.6) $$P_u: \ x \to \frac{u}{-u^{-1}x+s}, \qquad u \neq 0, \ u \neq 1 .$$

Here PP_u^{-1} fixes the letter ∞. Also putting

(3.7) $$\frac{1}{-x+s} = \frac{u}{-u^{-1}x+s}$$

we find

(3.8) $$(u - u^{-1})x = us - s .$$

Here $u \neq 0, 1$. If also $u \neq -1$ then $u - u^{-1} \neq 0$ and we have one further letter, $(us - s)(u - u^{-1})^{-1}$, fixed by PP_u^{-1}. Here

(3.9) $$P_u^{-1}: \ x \to \frac{sx - u}{u^{-1}x}$$

and PP_u^{-1} fixes letters x which satisfy

(3.10) $$\frac{1}{-x+s} = \frac{sx-u}{u^{-1}x}$$

or

(3.11) $$sx^2 + (u^{-1} - u - s^2)x + us = 0 .$$

If $s = 0$, P is of order two, contrary to the condition (i) for Husain chains since then all cycles of P would be of length two.

The rest of the proof will be divided into two cases; according to whether q is odd or even.

Case 1. *$q = p^r$ is odd.*

Here for $u = -1$, PP_u^{-1} fixes ∞ but no further letter since

[3] Except for $s = \pm 2$. These values correspond to the identity and two classes of elements fixing a letter. We are concerned only with permutations moving all letters.

(3.8) has no solution. Hence to satisfy condition (iii) for Husain chains PP_u fixes exactly one letter. This means that putting $u = -1$ in (3.11) there is exactly one solution. This is

$$sx^2 - s^2x - s = 0\,, \tag{3.12}$$

and as $s \neq 0$ we have

$$x^2 - sx - 1 = 0\,. \tag{3.13}$$

Since this may have only one solution, it must be a double root and the discriminant must vanish, whence

$$s^2 + 4 = 0 \tag{3.14}$$

and we write $s = 2i$ for the value of s and (3.13) becomes

$$(x - i)^2 = 0\,. \tag{3.15}$$

For $u \neq 0, 1, -1$, PP_u^{-1} fixes two letters and so by condition (iii) PP_u fixes no letters which means that (3.11) has no solution. This is the condition that the discriminant d at (3.11) should not be a square for $u \neq 0, 1, -1$. This is

$$d = (u^{-1} - u - s^2)^2 - 4us^2\,. \tag{3.16}$$

Putting $s^2 = -4$, we may express d in the form

$$d = u^{-2}(u + 1)^2(u^2 + 6u + 1)\,. \tag{3.17}$$

Hence our condition reduces to the requirement that $u^2 + 6u + 1$ is not a square for $u \neq 0, 1, -1$. For $q = 3$ there is no value with $s^2 = -4$. For $q = 5$, we have only $u = 2$ and -2 to consider and here $u^2 + 6u + 1$ is $17 = 2$ and $-7 = 3$ respectively and the condition holds. This yields the case $k = 6$, listed in section 2. It remains to consider cases with $q > 5$. Here for $u = -6$, $u^2 + 6u + 1 = 1$ is a square, whence -6 must be the same as one of 0, -1, or $+1$. With $q = p^r$ this requires $p = 3$, 5, or 7. If $p = 3$ or 7 since $s^2 = -4$, -1 is a square and so $(-1)^{(q-1)/2} = +1$ whence $q \equiv 1 \pmod 4$ and so r is even. Hence $GF(3^2)$ or $GF(7^2)$ is a subfield of $GF(q)$. For $u = 2$, $u^2 + 6u + 1 = 17$ and as $17 = 2$ in $GF(3^2)$ $17 = 3$ in $GF(7^2)$ this is a square in both cases. Hence $q = 3^r$, $r \geq 1$ is to be excluded and also $q = 7^r$, $r \geq 1$. Only $q = 5^r$, $r > 1$ remains to be considered. In $GF(q)$ there are $(q - 1)/2$ non zero squares, which we shall designate as class A and $(q - 1)/2$ non-squares which we shall designate as class B. Also $-1 = 2^2$ is a square. Let us write y for the number of solutions of

$$b + 1 = a\,, \qquad b \in B,\ a \in A \tag{3.18}$$

and z for the number of solutions of

$$b + 1 = b'\,, \qquad b, b' \in B\,. \tag{3.19}$$

Then as -1 is a square, $b + 1 \neq 0$

and so

$$y + z = (q - 1)/2\,. \tag{3.20}$$

In (3.19), let us multiply both sides by b^{-1} giving

$$1 + b^{-1} = b^{-1}b' \in A\,. \tag{3.21}$$

Hence every solution of (3.19) corresponds to a solution of (3.18). Hence

$$y = z = (q - 1)/4\,. \tag{3.22}$$

Our condition is that for $u \neq 0,\ 1,\ -1$.

$$u^2 + 6u + 1 = (u + 3)^2 - 8 = (u + 3)^2 + 2 \in B\,. \tag{3.23}$$

In particular for $u = -3$, 2 is a non-square, (from which we could prove r odd). If we multiply (3.19) by 2 we have

$$2b + 2 = 2b', \qquad 2b \in A,\ 2b' \in A\,. \tag{3.24}$$

This means, writing $2b = w^2$, $2b' = t^2$.

$$w^2 + 2 = t^2 \tag{3.25}$$

has $(q - 1)/4$ solutions w^2, $t^2 \neq 0$. But as $(q - 1)/4 > 3$ there must be some value of $u \neq 0, -1, 1$ with $(u + 3)^2 + 2$ a square, and our condition (3.23) is not satisfied. Thus $q = 5$ is the only odd q satisfying the theorem and the theorem is proved for Case 1.

Case 2. *$q = 2^r$ is even.*

In a field $GF(2^r)$ every element is a square and so an equation $ax^2 + c = 0$ not identically zero is solvable if and only if $a \neq 0$. Solvability of $ax^2 + bx + c = 0$ does not depend on the discriminant $b^2 - 4ac = b^2$ but on an additive subgroup R of the additive group A of $GF(2^r)$.

Lemma 3.1. *The set of elements in* $GF(2^r)$ *of the form* $w^2 + w$ *is an additive subgroup* R *of index 2 in the additive group* A *of* $GF(2^r)$. *An equation* $ax^2 + bx + c = 0$, $b \neq 0$ *over* $GF(2^r)$ *has a solution* x *in* $GF(2^r)$ *if and only if* $ac/b^2 \in R$.

Proof. As $(w_1^2 + w_1) + (w_2^2 + w_2) = (w_1 + w_2)^2 + (w_1 + w_2)$ the elements $w^2 + w$ from an additive subgroup R of A. From $x^2 + x = w^2 + w$ we have $(x + w)^2 + (x + w) = 0$ and $x + w = 0$ or $x + w = 1$ whence $x + w$ or $w = 1$. Hence there are exactly 2^{r-1} distinct elements $w^2 + w$ in $GF(2^r)$ and so $[A : R] = 2$. If $ax^2 + bx + c = 0$, $ab \neq 0$, has a solution x in $GF(2^r)$, then $(ax/b)^2 + (ax/b) = ac/b^2$ whence $ac/b^2 \in R$, and if $ac/b^2 \in R$ there is a w in $GF(2^r)$ with $w^2 + w = ac/b^2$ whence $ax/b = w$, and $x = bw/a$ satisfies $ax^2 + bx + c = 0$. This proves the lemma.

Since $-1 = 1$ in $GF(2^r)$ the condition on our permutations is that (3.11) have no solution x in $GF(2^r)$ for $u \neq 0, 1$. Let us multiply (3.11) by u. It becomes

$$(s + x)u^2 + (sx^2 + s^2x)u + x = 0 . \tag{3.26}$$

Here if $u = 0$, then $x = 0$, while if $x = 0$, $su^2 = 0$ and as we have observed that $s \neq 0$ this implies $u = 0$. The value $u = 1$ gives $sx^2 + s^2x + s = 0$ and as $s \neq 0$ this is $x^2 + sx + 1 = 0$. If this has a solution x, this yields a value fixed by P^2 and P has a 2 cycle, contrary to condition (ii) on Husain chains. Thus $1/s^2 \notin R$. Hence our condition is that (3.26) have no solution x for any value of $u \neq 0$. But as $x = 0$ and $u = 0$ correspond, this is equivalent to saying that for any value of $x \neq 0$ there is no solution for u regarding (3.26) as a quadratic equation for u. The coefficient $sx^2 + s^2x = sx(x + s)$ is zero only if $x = 0$ or $x = s$. If $x = s$, (3.26) reduces to $s = 0$, which we have already shown impossible since the P^2 is the identity. Hence by Lemma 3.1, with $x \neq 0$, $x \neq s$ we require that

$$\frac{(s + x)x}{(sx^2 + s^2x)^2} = \frac{1}{s^2x(s + x)} \notin R , \qquad x \neq 0, s . \tag{3.27}$$

Putting $x = sy$ the condition becomes

$$\frac{1}{s^4(y^2 + y)} \in R , \qquad y \neq 0, 1 . \tag{3.28}$$

Here $y^2 + y$ ranges over the $2^{r-1} - 1$ elements of $R^\# = R - 0$.

Lemma 3.2. *If r_1, r_2, r_3 are three elements of $R^\#$ for which $r_1^{-1} + r_2^{-1} = r_3^{-1}$ then the condition of* (3.28) *cannot hold.*

Proof. If $(s^4 r_1)^{-1} \notin R$ and $(s^4 r_2)^{-1} \notin R$, then $(s^4 r_1)^{-1} + (s^4 r_2)^{-1} \in R$ as $[A : R] = 2$ and this would lead to $(s^4 r_3)^{-1} \in R$ contrary to condition (3.28).

If $q = 2$ condition (3.28) is vacuous. If $q = 4$, $GF(4)$ consists of elements $0, 1, w, w + 1$ where $w^2 + w = 1$ and R consists of the elements 0, 1. Here (3.28) reduces to $s^{-4} \neq 1$ and we may take $s = w$ or $w + 1$. The solutions for $q = 2$, $q = 4$, correspond to the Husain chains for $k = 3$, $k = 5$ given in Section 2. Henceforth we suppose $q \geq 8$.

First suppose $1 \in R$. Then the $2^{r-1} - 1 \geq 3$ element of $R^\#$ contain an element $r_1 \neq 0, 1$. Here $r_2 = r_1 + 1 \in R^\#$. But then $r_1^{-1} + r_2^{-1} = (r_1(r_1 + 1))^{-1}$ and $r_1(r_1 + 1) = r_1^2 + r_1 = r_3 \in R^\#$ and this conflicts with Lemma 3.2. Thus $1 \in R$ leads to a conflict and we conclude that $1 \notin R$.

We observed above that $s^{-2} = 1/s^2 \in R$. Hence $s^{-4} = s^{-2} + (s^{-2} + s^{-4}) \in R$. If $1 = s^{-4}(y^2 + y)^{-1}$ for some y then $y^2 + y = s^{-4}$ conflicting with the fact that $s^{-4} \notin R$. Hence the 2^{r-1} elements of $GF(q)$ not in R consist of 1 and the $2^{r-1} - 1$ elements $s^{-4} r_i^{-1}$, $r_i \in R^\#$. As $s^{-4} \notin R$, it follows that $s^{-4} = s^{-4} r_i^{-1}$, or $s^{-4} = 1$. The first alternative yields $r_i = 1$, which we may exclude since we have already shown that $1 \notin R$. Hence $s^{-4} = 1$ and so $s = 1$.

Let us designate by S the set of elements not in R. We have shown that S consists of the element 1 and the inverses of elements of $R^\#$. Hence the inverse of any element of $R^\#$ is in S and the inverse of any element of S except 1 is in $R^\#$.

Lemma 3.3. *If $w \in R^\#$ then $w^3 + w + 1 = 0$.*

Proof. Proof by denial. Suppose $w \in R^\#$ but $w^3 + w + 1 \neq 0$. As $1 \in S$, $w \in R$, $w + 1 \in S$ and $w^{-1} \in S$. As $w + 1 \neq 1 (w + 1)^{-1} \in R$. As $w^2 + w \in R$ also $w^2 + w + (w + 1)^{-1} = (w^3 + w + 1)/(w + 1) \in R$. $(w^2 + w)^{-1} \in S$, $w + 1 \in S$ and so $(w^2 + w)^{-1} + (w + 1) = (w^3 + w + 1)/(w^2 + w) \in R$. Since we have assumed $w^3 + w + 1 \neq 0$ it follows that $(w + 1)/(w^3 + w + 1) \in S$, $(w^2 + w)/(w^3 + w + 1) \in S$ whence

$$\begin{aligned} &(w + 1)/(w^3 + w + 1) + (w^2 + w)/(w^3 + w + 1) \\ &= (w^2 + 1)/(w^3 + w + 1) \in R. \end{aligned}$$

Since $w^2+1\neq 0$, we have $(w^3+w+1)/(w^2+1)\in S$, and as $w+1\in S$, $(w^3+w+1)/(w^2+1)+w+1=w^2/(w^2+1)\in R$ and so $(w^2+1)/w^2\in S$, and as $1\in S$, $1+(w^2+1)/w^2=w^{-2}\in R$. Also $w^{-1}+w^{-2}\in R$. But then $w^{-2}+(w^{-1}+w^{-2})=w^{-1}\in R$ which conflicts with $w^{-1}\in S$. Thus the assumption that $w^3+w+1\neq 0$ has led to a conflict and our lemma is proved.

The proof of the theorem for Case 2 is now immediate. Since $GF(q)$ contains at most 3 elements satisfying $w^3+w+1=0$, then as R^* contains $2^{r-1}-1$ elements we have $2^{r-1}-1\leq 3$ and $r\leq 3$. As we have previously disposed of $q=2$ and $q=4$, only the case $q=8$ remains and here the class is determined by the value $s=1$, giving the Husain chains listed in Section 2. Thus all parts of the theorem are now proved.

It may be worth noting that there is an alternate characterization for the additive set R in $GF(2^r)$. For $x\in GF(2^r)$ its *trace*, $\mathrm{tr}(x)$ is given by

$$\text{(3.29)}\qquad \mathrm{tr}(x)=x+x^2+x^4+\ldots+x^{2^i}+\ldots+x^{2^{r-1}}.$$

Here trivially $\mathrm{tr}(x+y)=\mathrm{tr}(x)+\mathrm{tr}(y)$ and as $x^{2^r}=x$, $\mathrm{tr}(x^2)=\mathrm{tr}(x)$. Hence if $x=w^2+w$ then $\mathrm{tr}(x)=\mathrm{tr}(w^2)+\mathrm{tr}(w)=\mathrm{tr}(w)+\mathrm{tr}(w)=0$. Thus if $x\in R$, $\mathrm{tr}(x)=0$. Also for any x, $\mathrm{tr}(x)^2=\mathrm{tr}(x)$ whence $\mathrm{tr}(x)=0$ or $\mathrm{tr}(x)=1$. As $\mathrm{tr}(x)$ is a polynomial of degree 2^{r-1} there are at most 2^{r-1} elements in $GF(2^r)$ with $\mathrm{tr}(x)=0$. Hence the 2^{r-1} elements of R are precisely those elements for which $\mathrm{tr}(x)=0$, and $\mathrm{tr}(x)=1$ for $x\notin R$.

References

1. Atiqullah, M. "Some New Solutions of Symmetrical Balanced Incomplete Block Designs with $\lambda=2$ and $k=9$," *Bull. Calcutta Nath. Soc.*, **50** (1958), 23-28.
2. Bose, R. C. "On the Construction of Balanced Incomplete Block Designs," *Annals of Eugenics*, **9** (1938), 353-399.
3. Chowla, S. and Ryser, H. J. "Combinatorial Problems," *Canad. J. Math.*, **2** (1950), 93-99.
4. Fisher, R. A. "An Examination of the Different Possible Solutions of a Problem in Incomplete Blocks," *Annals of Eugenics*, **10** (1940), 52-75.
5. Husain, Q. M. "Impossibility of the Symmetrical Incomplete Block Design with $\lambda=2$, $k=7$," *Sankhya*, **7** (1946), 317-322.
6. Husain, Q. M. "Symmetrical Incomplete Block Designs with $\lambda=2$, $k=8$ or 9," *Bull. Calcutta Math. Soc.*, **37** (1945), 115-123.

7. Kerawala, S. M. "Symmetrical Incomplete Block Designs with $\lambda = 2$," *The Scientist* (*Karachi*), **1** (1953), 1-24.
8. Zaidi, N. H. "Symmetrical Balanced Incomplete Block Designs with $\lambda=2$ and $\lambda=9$," *Bull. Calcutta Math. Soc.*, **55** (1963), 163-167.

Discussion on Professor Hall's Paper

PROFESSOR J. N. SRIVASTAVA: It is a matter of great pleasure for me to thank Professor Hall for his very stimulating paper. One could be interested in this paper from various points of view, including the problem of finiteness of the number of symmetric block designs with $\lambda = 2$, the use of the linear fractional groups (LFG) here, and in the related area of coding theory, and the generalizations of the problems of symmetric BIBD's to modern areas of design construction.

It is certainly intriguing to see a connection between the blocks (expressed as permutations) of a symmetric BIBD with $\lambda = 2$, and the representation of these permutations by the elements of the LFG over $GF(q)$, where $q = k - 1$, and k denotes block size. However, as the main theorem in the paper states, the method works for $q = 2$, 4, 5 and 8. The case $k = 4$, for which a design actually exists, is thus left out. This fact, and some others, lead one to wonder whether values of k of the form 2^{2^u} $(u \geq 1)$ are somewhat special, and whether designs exist for an infinite number of values of u.

The method introduced by Professor Hall may, with the exceptions of special cases, be generalizable to values of $\lambda \geq 3$, by using the association scheme (see for example, Bose and Srivastava: "Mathematical Theory of Factorial Design", *Bull. Intern. Stat. Inst.*, **40**, 2e, p. 784) arising in factorial designs, which reduces to the triangular scheme as a very special case. We notice that the triangular scheme arises (in the case $\lambda = 2$) between the treatments contained in those blocks in which a fixed variety occurs.

Finally, it seems relevant to remark that there exist important generalizations (of which solutions are urgently needed) of the problems of symmetric BIBD's in modern areas of design construction. One such example is the construction of balanced factorial designs (called "partially balanced arrays") which are saturated, i.e., in which the number of treatments combinations used equals the number of parameters to be estimated.

CHAPTER 11

On Some Composition and Extension Methods in the Construction of Block Designs from Association Matrices

I. M. CHAKRAVARTI[1] and W. C. BLACKWELDER
University of North Carolina at Chapel Hill

1. INTRODUCTION

An example of *combinatorial extension* is the familiar problem of adjoining $n - r$ rows and $n - s$ columns to an r by s Latin rectangle in n symbols, so that the resulting configuration is a Latin square of order n. One may ask whether an r by s Latin rectangle in n symbols can always be extended to a Latin square of order n and if so, in how many ways. An extensive literature [11] has grown out of the study of this problem.

A second example is the problem of extension of a Latin square of order n to a Latin square of order $n + m$. Yet another example is the problem of extending a pair of mutually orthogonal Latin squares of order n to a pair of mutually orthogonal Latin squares of order $n + m$. Significant results obtained in this area are to be found in [17], [2], and [7].

[1] This research was supported by the U.S. Army Research Office-Durham Grant No. DA-ARO-D-31-124-G814 and the Air Force Office of Scientific Research Grant No. AF-AFOSR-760-65.

Again, for a given orthogonal array there is the problem of extending it to one having a larger number of constraints or to one of higher strength [13].

Similarly, the problem of augmenting the incidence matrix of a balanced incomplete block design into a larger matrix which retains the property of being an incidence matrix of a balanced incomplete block design has received considerable attention [3], [8], and [12].

By *combinatorial composition* is meant a rule for combining two or more given designs or configurations. For instance, given two Latin squares of order u and v respectively, it is possible to give a rule for combining these two Latin squares into one of order uv [9]. Similarly, one might ask for a rule for combining the incidence matrices of two balanced incomplete block designs, into the incidence matrix of another balanced incomplete design. Sometimes this is achieved by defining a direct product of matrices (or configurations) or a linear combination of matrices.

In this paper, we give a characterization of association matrices by permutation matrices. Composition and extension methods have been used to construct incidence matrices of BIB designs from association matrices.

2. DEFINITIONS AND SOME PROPERTIES OF INCIDENCE AND ASSOCIATION MATRICES

A balanced incomplete block (BIB) design is an arrangement of v objects into b sets of k elements each called blocks such that every object occurs exactly r times and every pair of objects occurs exactly λ times in a block. Then it is well-known (see for instance, [3]) that

$$bk = rv\,, \qquad \lambda(v-1) = r(k-1)\,, \tag{2.1}$$

$$b \geq v\,, \tag{2.2}$$

an inequality due to R. A. Fisher.

For any given BIB design, we define a $v \times b$ matrix $N = (n_{ij})$ where

$n_{ij} = 1$ if the ith object occurs in the jth block,

$n_{ij} = 0$ otherwise.

From the properties of the design it follows that

(2.3) $$NN' = (r - \lambda)I + \lambda J$$

where N' is the transpose of N, I is a $v \times v$ unit matrix and J is a $v \times v$ matrix all of whose entries are unity.

A BIB design is called a symmetrical BIB (SBIB) if $v = b$; hence $r = k$ and $\lambda(v - 1) = k(k - 1)$. Putting $k - \lambda = n$, for a symmetrical BIB design, one can then show that [10],

(2.4) $$NN' = nI + \lambda J$$

(2.5) $$NJ = JN = kJ$$

(2.6) $$N^{-1} = \frac{1}{n}[N' - (\lambda/k)J] ,$$

and

(2.7) $$N'N = NN' = nI + \lambda J .$$

From (2.7) it follows that in a symmetrical BIB design any two blocks have λ objects in common, a result which was first obtained by R. A. Fisher.

An m-class association scheme with v objects is defined by the following conditions [5].

(i) Any two distinct objects are either first, second, ..., or mth associates.

(ii) Each object has n_i ith associates, $i = 1, 2, \ldots, m$.

(iii) For any pair of objects which are ith associates, the number p^i_{jk} of objects which are jth associates of the first and the kth associates of the second is independent of the pair of ith associates with which we start.

The following identities which can be derived from this definition are well-known.

(2.8)
$$\begin{aligned}
&\sum_{i=1}^{m} n_i = v - 1 \\
&p^i_{jk} = p^i_{kj} , \\
&\sum_{k=1}^{m} p^i_{jk} = n_j \qquad j \neq i , \\
&\qquad\quad = n_i - 1 , \qquad j = i , \\
&n_i p^i_{jk} = n_j p^j_{ik} , \qquad i, j, k = 1, 2, \ldots, m .
\end{aligned}$$

We also define

$$(2.9)\qquad \begin{aligned} p^0_{ij} &= n_i \quad &&\text{if} \quad i = j\ , \\ &= 0 &&\text{otherwise}\ . \\ p^i_{0k} &= 1 &&\text{if} \quad i = k\ , \\ &= 0 &&\text{otherwise}\ . \end{aligned}$$

The association matrices $B_0, B_1, \ldots, B_m$ of an m-class association scheme are matrices of order v defined by

$$(2.10)\qquad B_0 = I\ , \qquad B_i = (b^i_{\alpha\beta})\ , \qquad i = 1, 2, \ldots, m$$

where

$$\begin{aligned} b^i_{\alpha\beta} &= 1 \quad &&\text{if objects } \alpha \text{ and } \beta \text{ are } i\text{th associates}\ , \\ &= 0 &&\text{otherwise}\ . \end{aligned}$$

Clearly B_i is a symmetric matrix with all row and column sums equal to n_i .

Lemma 2.1. *A set of necessary and sufficient conditions for matrices* $A_0 = I, A_1, \ldots, A_m$ *to be association matrices of an* m*-class association scheme with parameters* $v,\ n_i = p^0_{ii}\ ,\ p^i_{jk}\ ,$ $i, j, k = 0, 1, 2, \ldots, m,$ *is*

(i) *each* A_i *is a symmetric* $v \times v$ *matrix of* 0's *and* 1's,

(2.11) (ii) $\sum_{i=0}^{m} A_i = J$, *the* $v \times v$ *matrix of* 1's,

(iii) $A_j A_k = \sum_{i=0}^{m} p^i_{jk} A_i\ , \qquad j, k = 0, 1, 2, \ldots, m\ .$

Proof is given in [4] and [15].

3. CHARACTERIZATION OF ASSOCIATION MATRICES BY PERMUTATION MATRICES

A *permutation matrix* P is defined in [9] as a (0, 1)-matrix of size m by n ($m \leq n$) such that $PP' = I$, where P' is the transpose of P and I is the unit matrix of order m. A permutation matrix of order m has a single entry 1 in each row and column and all other entries 0.

Lemma 3.1. *Let A be a $(0, 1)$-matrix of order n such that each row and column sum of A is equal to the positive integer k. Then*

$$A = P_1 + P_2 + \dots + P_k, \tag{3.1}$$

where the P_i are permutation matrices of order n.

This is a well-known result and for proof see, for instance, [11].

Since an association matrix B_i is a $(0, 1)$-matrix of order v with each row and column sum equal to the positive integer n_i, it follows from Lemma 3.1, that

$$B_i = P_{i1} + P_{i2} + \dots + P_{in_i}, \tag{3.2}$$

where the P_{ij} are permutation matrices of order v.

Since B_i and B_j cannot have an entry 1 at the same position, it follows that they can not have a permutation matrix in common.

Again, B_i is a symmetric matrix, hence,

$$B_i = B_i' = \left(\sum_{u=1}^{n_i} P_{iu}\right)' = \sum_{u=1}^{n_i} P_{iu}^{-1} = \sum_{u=1}^{n_i} P_{iu}. \tag{3.3}$$

Also,

$$\sum_{i=0}^{m} B_i = J = I + \sum_{i=1}^{m} \sum_{u=1}^{n_i} P_{iu}. \tag{3.4}$$

Further

$$\begin{aligned} B_i B_j = B_j B_i &= \left(\sum_{u=1}^{n_i} P_{iu}\right)\left(\sum_{v=1}^{n_j} P_{jv}\right) \\ &= \sum_{k=0}^{m} p_{ij}^k B_k = \sum_{k=0}^{m} p_{ij}^k \sum_{w=1}^{n_k} P_{kw}. \end{aligned} \tag{3.5}$$

It follows that if the matrix equation

$$P_{iu} P_{jv}^{-1} = P_{kw} \tag{3.6}$$

for given i, j, k, and w ($w = 1, 2, \dots, n_k$), is satisfied for p_{ij}^k pairs (u, v), $i, j, k = 1, 2, \dots, m$, then (3.5) must hold.

Now we are in a position to state and prove

Theorem 3.1. *An m-class association scheme with parameters*

$v, n_1, n_2, \ldots, n_m, \{p^i_{jk}\}, i, j, k = 1, 2, \ldots, m$ exists if there are $v - 1$ permutation matrices $\boldsymbol{P}_1, \boldsymbol{P}_2, \ldots, \boldsymbol{P}_{v-1}$ each of order v, such that

(i) $\sum_{j=1}^{v-1} \boldsymbol{P}_j = \boldsymbol{J} - \boldsymbol{I}$.

(ii) *$\boldsymbol{P}_1, \boldsymbol{P}_2, \ldots, \boldsymbol{P}_{v-1}$ can be divided into m-classes, n_i of the matrices belonging to the ith class. Let us reindex the matrices belonging to the ith class as $\boldsymbol{P}_{iu}$, $u = 1, 2, \ldots, n_i$ such that $\sum_{u=1}^{n_i} \boldsymbol{P}_{iu}$ is a symmetric matrix.*

(iii) *$\boldsymbol{P}_{iu}\boldsymbol{P}_{jv}^{-1} = \boldsymbol{P}_{kw}$ for given i, j, k, and w, is satisfied for p^k_{ij} pairs (u, v), $w = 1, 2, \ldots, n_k$, $i, j, k = 1, 2, \ldots, m$.*

Proof. If there exist $(v - 1)$ permutation matrices $\boldsymbol{P}_1, \boldsymbol{P}_2, \ldots, \boldsymbol{P}_{v-1}$ with the stated properties, then if we define matrices

$$\boldsymbol{B}_i = \sum_{u=1}^{n_i} \boldsymbol{P}_{iu}, \qquad i = 1, 2, \ldots, m$$

it is easy to verify that $\boldsymbol{B}_i$, $i = 1, 2, \ldots, m$, satisfy the conditions of Lemma 2.1 and hence define an association scheme with the given parameters.

This theorem is a partial analogue of Mann's theorem for symmetrical BIB designs [10]. We quote Mann's theorem:

A symmetric BIB design with parameters v, k, λ exists if and only if there are k permutations $P_1, \ldots, P_k$ of order v such that

(i) *The permutations $P_1, \ldots, P_k$ applied to $1, 2, \ldots, v$ yield a $k \times v$ Latin rectangle.*

(ii) *For $s \neq t$, $s, t = 1, 2, \ldots, v$ there are among the permutations $P_iP_j^{-1}$ exactly λ which carry s into t.*

4. BIB DESIGNS FROM ASSOCIATION MATRICES BY COMPOSITION METHOD

Let us define

$$N = c_1\boldsymbol{B}_1 + \ldots + c_m\boldsymbol{B}_m, \tag{4.1}$$

where $\boldsymbol{B}_1, \boldsymbol{B}_2, \ldots, \boldsymbol{B}_m$ are the association matrices of an m-class association scheme with the parameters $v, n_1, n_2, \ldots, n_m$, $\{p^i_{jk}\}$ and $c_i = 0$ or 1 $i = 1, 2, \ldots, m$. Then N is a symmetric (0, 1)-matrix of order v with each row and column sum equal to $\sum_{i=1}^{m} c_i n_i = k$ (say). Hence

$$NJ = JN = kJ\,. \tag{4.2}$$

Now

$$B_i = \sum_{u=1}^{n_i} P_{iu}\,, \qquad i = 1, 2, \ldots, m\,,$$

where P_{iu}, $u = 1, 2, \ldots, n_i$, $i = 1, 2, \ldots, m$ are permutation matrices of order v satisfying the conditions of Theorem 3.1. Hence,

$$\begin{aligned} NN' &= \Big(\sum_i c_i \sum_u P_{iu}\Big)\Big(\sum_j c_j \sum_v P_{jv}\Big)' \\ &= \Big(\sum_{i=1}^{m} c_i n_i\Big) I + \sum_{l=1}^{m} \Big(\sum_i c_i p^l_{ii} + \sum_{i \neq j} c_i c_j p^l_{ij}\Big)\Big(\sum_{w=1}^{n_l} P_{lw}\Big) \\ &= kI + \sum_{l=1}^{m} \lambda_l B_l \end{aligned} \tag{4.3}$$

where

$$\lambda_l = \sum_i c_i p^l_{ii} + \sum_{i \neq j} c_i c_j p^l_{ij}\,. \tag{4.4}$$

Thus if $\lambda_1 = \lambda_2 = \ldots = \lambda_m = \lambda$ (say), then

$$NN' = kI + \lambda(J - I)\,,$$

and hence, N is the incidence matrix of a symmetrical BIB (v, k, λ)-design.

On the other hand, if N as defined in (4.1) is the incidence matrix of a symmetrical BIB (v, k, λ)-design, then from

$$\begin{aligned} NN' &= kI + \lambda(J - I) \\ &= kI + \sum_{l=1}^{m} \lambda_l B_l \end{aligned} \tag{4.5}$$

it follows that $\lambda_1 = \lambda_2 = \ldots = \lambda_m = \lambda$, since $B_1, B_2, \ldots, B_m$ are linearly independent.

From (4.3) and (4.4), we get

(4.6)
$$\sum_l n_l\lambda_l = \sum_i c_i \sum_l n_l p^l_{ii} + \sum_{i \neq j} c_i c_j \sum_l n_l p^l_{ij}$$
$$= (\sum_i c_i n_i)^2 - \sum_i c_i n_i = \sum_i c_i n_i (\sum c_i n_i - 1) .$$

So if $\lambda_1 = \lambda_2 = \ldots = \lambda_m = \lambda$, $(v - 1)$ must divide $\sum c_i n_i (\sum c_i n_i - 1)$.

Thus in order to be able to construct the incidence matrix of a symmetrical BIB design from a given set of association matrices of an m-class association scheme, a necessary condition is the existence of a set of c_i's ($c_i = 0$ or 1) $i = 1, 2, \ldots, m$ such that $v - 1$ divides $(\sum_i c_i n_i)(\sum_i c_i n_i - 1)$.

These results were first obtained in [1]. The following example taken from [1] will illustrate the technique.

Suppose we have a set of $v = ln$ elements for some integers $l, n \geq 2$. We arrange the ln elements in a rectangular array with l rows and n columns. Two elements are defined to be first associates if they are in the same row; they are second associates if they are in the same column, otherwise, they are third associates. This defines a 3-class association scheme called a rectangular scheme [16] and it has the following parameters.

$$v = ln , \quad n_1 = n - 1 , \quad n_2 = l - 1 , \quad n_3 = (l - 1)(n - 1) ,$$

$$p^1_{11} = n - 2 , \quad p^1_{12} = p^1_{13} = p^1_{22} = 0 , \quad p^1_{23} = l - 1 ,$$
$$p^1_{33} = (l - 1)(n - 2) ,$$

(4.7)
$$p^2_{11} = 0 , \quad p^2_{12} = 0, \quad p^2_{13} = n - 1 , \quad p^2_{22} = l - 2 ,$$
$$p^2_{23} = 0 , \quad p^2_{33} = (l - 2)(n - 1) ,$$

$$p^3_{11} = 0 , \quad p^3_{12} = 1 , \quad p^3_{13} = n - 2 , \quad p^3_{22} = 0 ,$$
$$p^3_{23} = l - 2 , \quad p^3_{33} = (l - 2)(n - 2) .$$

Consider,

$$N = B_1 + B_2 .$$

Then

$$\lambda_1 = p^1_{11} + p^1_{22} + 2p^1_{12} = n - 2$$

$$\lambda_2 = p^2_{11} + p^2_{22} + 2p^2_{22} = l - 2$$

$$\lambda_3 = p^3_{11} + p^3_{22} + 2p^3_{12} = 2 .$$

For $\lambda_1 = \lambda_2 = \lambda_3$, $n = l = 4$. Thus $\boldsymbol{N} = \boldsymbol{B}_1 + \boldsymbol{B}_2$ is the incidence matrix for a symmetrical BIB design with $v = 16$, $k = 6$, $\lambda = 2$.

5. BIB DESIGNS FROM ASSOCIATION MATRICES BY EXTENSION METHOD

Consider the matrix

$$\boldsymbol{A} = [\boldsymbol{B}_1 \vdots \boldsymbol{B}_2 \vdots \dots \vdots \boldsymbol{B}_t] \qquad t \leq m , \tag{5.1}$$

where $\boldsymbol{B}_1, \boldsymbol{B}_2, \dots, \boldsymbol{B}_t$ are t association matrices from an m-class association scheme. Then $\boldsymbol{A}$ is a v by tv (0, 1)-matrix with each row sum equal to $\sum_{i=1}^{t} n_i = r$ (say). In order that the column sums be all equal, we need $n_1 = n_2 = \dots = n_t = k$ (say).

Now,

$$\begin{aligned} \boldsymbol{A}\boldsymbol{A}' &= \boldsymbol{B}_1^2 + \boldsymbol{B}_2^2 + \dots + \boldsymbol{B}_t^2 , \\ &= (\sum_i p_{ii}^0)\boldsymbol{B}_0 + (\sum_i p_{ii}^1)\boldsymbol{B}_1 + \dots + (\sum_i p_{ii}^m)\boldsymbol{B}_m \\ &= r\boldsymbol{I} + \mu_1\boldsymbol{B}_1 + \dots + \mu_m\boldsymbol{B}_m \end{aligned} \tag{5.2}$$

where

$$\mu_j = \sum p_{ii}^j , \qquad j = 1, 2, \dots, m .$$

Hence if $\mu_1 = \mu_2 = \dots = \mu_m = \lambda$ (say) and $n_1 = n_2 = \dots = n_t$, then $\boldsymbol{A}$ as defined by (5.1) is the incidence matrix of a BIB design with the parameters v, $b = tv$, $r = \sum n_i$, k, λ.

On the other hand, if $\boldsymbol{A}$ as defined by (5.1) is the incidence matrix of a BIB (v, k, λ)-design, then $n_1 = n_2 = \dots = n_t = k$ and $\mu_1 = \mu_2 = \dots = \lambda$.

An example is given here to illustrate the method of extension in the construction of a BIB design from a three-class association scheme based on an *orthogonal array*.

An *orthogonal array* (N, m, s, t) is an m by N rectangular array in s symbols $(0, 1, 2, \dots, s-1$, say), such that in any t-rowed submatrix of the array each of the s^t possible column vectors appears exactly λ times, whence $\lambda s^t = N$.

Consider an orthogonal array $(n^2, m_1 + m_2, n, 2)$ where n, m_1, m_2 are positive integers such that $n \geq 2$ and $m_1 + m_2 \leq n$. Here $\lambda = 1$. Each column of the array is an $(m_1 + m_2)$-vector. Let the $v = n^2$ columns of the array be identified with v objects. Define two objects α and β to be first associates if the corresponding column vectors coincide in exactly one position in the first m_1 rows; second associates if the corresponding column vectors coincide in exactly one position in the remaining m_2 rows; third associates, otherwise. Then this defines a three class association scheme [14] with parameters,

$$(5.3) \qquad v = n^2, \quad n_1 = m_1(n-1), \quad n_2 = m_2(n-1), \quad n_3 = m_3(n-1)$$

where $m_3 = n + 1 - (m_1 + m_2)$, and

$$\begin{aligned}
p^1_{11} &= (n-2) + (m_1-1)(m_1-2), & p^2_{11} &= m_1(m_1-1) \\
& & p^3_{11} &= m_1(m_1-1) \\
p^1_{12} &= m_2(m_1-1) & p^2_{12} &= m_1(m_2-1) \\
& & p^3_{12} &= m_1m_2 \\
p^1_{13} &= m_3(m_1-1) & p^2_{13} &= m_1m_3 \\
& & p^3_{13} &= m_1(m_3-1) \\
p^1_{22} &= m_2(m_2-1) & p^2_{22} &= n-2+(m_2-1)(m_2-2) \\
& & p^3_{22} &= m_2(m_2-1) \\
p^1_{23} &= m_2m_3 & p^2_{23} &= m_3(m_2-1) \\
& & p^3_{23} &= m_2(m_3-1) \\
p^1_{33} &= m_3(m_3-1) & p^2_{33} &= m_3(m_3-1) \\
& & p^3_{33} &= n-2+(m_3-1)(m_3-2).
\end{aligned}$$

Define

$$(5.4) \qquad A = [B_1 \ \vdots \ B_2 \ \vdots \ B_3].$$

For A to be the incidence matrix of a BIB design, we need $n_1 = n_2 = n_3$ which is satisfied if $m_1 = m_2 = m_3 = (n+1)/3$. We also require

$$p^1_{11} + p^1_{22} + p^1_{33} = p^2_{11} + p^2_{22} + p^2_{33} = p^3_{11} + p^3_{22} + p^3_{33} = \lambda \text{ (say)}$$

which is also satisfied for $m_1 = m_2 = m_3 = (n+1)/3$, and $\lambda = (n^2-4)/3$. Thus if n is greater than 2 and $n \equiv 2 \pmod 3$ and if the orthogonal array $(n^2, 2(n+1)/3, n, 2)$ exists, then (5.4) is the incidence matrix for a BIB design with parameters $v = n^2$, $b = 3n^2$, $r = n^2-1$, $k = (n^2-1)/3$, $\lambda = (n^2-4)/3$. For $n = 5$ we get the BIB (25, 75, 24, 8, 7).

References

1. Blackwelder, W. C. "Construction of Balanced Incomplete Block Designs from Association Matrices," *Institute of Statistics*, Mimeo Series No. 481, University of North Carolina, 1966.
2. Barra, J. R. and Guérin, R. "Utilisation pratique de la méthode de Yamamoto pour la construction systématique de carrés gréco-latins," *Publ. Inst. Statist., Univ. Paris*, **12** (1963), 131-136.
3. Bose, R. C. "On the Construction of Balanced Incomplete Block Designs," *Annals of Eugenics*, **9** (1939), 353-399.
4. Bose, R. C. and Mesner, D. M. "On Linear Associative Algebras Corresponding to Association Schemes of Partially Balanced Designs," *Ann. Math. Statist.*, **30** (1959), 21-38.
5. Bose, R. C. and Shimamoto, T. "Classification and Analysis of Partially Balanced Incomplete Block Designs with Two Associate Classes," *J. Amer. Statist. Assoc.*, **47** (1952), 151-184.
6. Bose, R. C. and Shrikhande, S. S. "On the Composition of Balanced Incomplete Block Designs," *Canad. J. Math.*, **12** (1960), 177-188.
7. Guérin, R. "Sur une généralisation de la méthode de Yamamoto pour la construction de carrés latins orthogonaux," *C. R. Acad. Sci.*, **256** (1963), 2097-2100.
8. Hall, M., Jr. and Connor, W. S. "An Embedding Theorem for Balanced Incomplete Block Designs," *Canad. J. Math.*, **6** (1954), 35-41.
9. Mann, H. B. "The Construction of Orthogonal Latin Squares," *Ann. Math. Statist.*, **13** (1942), 418-423.
10. Mann, H. B. "Balanced Incomplete Block Designs and Abelian Difference Sets," *Illinois J. Math.*, **8** (1964), 252-261.
11. Ryser, H. J. *Combinatorial Mathematics*, Carus Mathematical Monograph No. 14, Math. Assoc. Amer., John Wiley and Sons, 1963.
12. Schützenberger, M. P. "An Extension Problem in the Theory of Incomplete Block Designs," *J. Roy. Statist. Soc., Ser. B*, **13** (1951), 120-125.
13. Shrikhande, S. S. and Bhagwandas, "A Note on Embedding of Orthogonal Arrays of Strength Two," *Proceedings of the Conference on Combinatorial Mathematics and Its Applications*, University of North Carolina Press, 1968.
14. Singh, N. K. and Shukla, G. C. "The Non-Existence of Some Partially Balanced Incomplete Block Designs with Three Associate Classes," *J. Indian Statist. Assoc.*, **1** (1963), 71-77.
15. Thompson, W. A., Jr. "A Note on PBIB Design Matrices," *Ann. Math. Statist.*, **29** (1958), 919-922.
16. Vartak, M. N. "The Non-Existence of Certain Partially Balanced Incomplete Block Designs," *Ann. Math. Statist.*, **30** (1959), 1051-1062.

17. Yamamoto, K. "Generation Principles of Latin Squares," *Bull. Inst. Internat. Statist.*, **38** (**1960**), 73–76.

Discussion on the Paper by Professor Chakravarti and Mr. Blackwelder

DR. S. IKEDA: On the characterization of association matrices by permutation matrices, the authors have given an interesting result (Theorem 3.1), which may be modified and stated as follows: *In order that an m-class association scheme with parameters, v, n_i, p^i_{jk}, $i, j, k = 0, 1, \ldots, m$, exists, it is necessary and sufficient that there exists a set of v permutation matrices, $H = \{P_0 = I_v, P_1, \ldots, P_{v-1}\}$, divided into $m+1$ subsets, $H_0 = \{P_0\}$, $H_1, \ldots, H_m$, such that $|H_i| = n_i$, $i = 0, 1, \ldots, m$, and*

(i) $\sum_{\alpha=1}^{v-1} P_\alpha = J_v$,

(ii) $\sum_{H_i} P_\alpha = \sum_{H_i} P_\alpha^{-1}$, $\qquad i = 0, 1, \ldots, m$,

(iii) $\sum_{H_j} P_\beta \cdot \sum_{H_k} P_\gamma^{-1} = \sum_{i=0}^{m-1} p^i_{jk} \sum_{H_i} P_\alpha$, $\qquad j, k = 0, 1, \ldots, m$.

If, in particular, H forms a group of order v, then the condition (iii) *is equivalent, under the other two conditions, to the following:*

(iii)′ *The matric equation*

$$XY^{-1} = P, \qquad X \in H_j, \qquad Y \in H_k,$$

has p^i_{jk} solutions in (X, Y) for any given P in H_i, $i, j, k = 0, 1, \ldots, m$.

Since every finite group is represented by a regular permutation group and H in the theorem is a regular permutation group if it is a group, the above result might be useful in constructing an association scheme from a finite group.

PROFESSOR M. MASUYAMA: The problem posed by Professor Ikeda can be solved at least partly by the method described in my paper (Chapter 14). Let S be a group of order v, elements being $s_1, s_2, \ldots, s_v$. It is easily verified that $I_e N_i$, e being the unity element of S, can be written as $\sum N_i^* N_i$, in which N_i^* is the conjugate block of N_i. The product is the counterpart of the Schnirelman's sum in an Abelian group. Consider a subgroup of the group of automorphisms of S of order v, elements being $\tau_1, \tau_2, \ldots, \tau_h$. If we construct a $v \times h$ matrix with the element $\tau_j s_i$ in the ith row and jth column and let the block containing every distinct element (as an element of S) in the ith row be F_i, then any sum of F_i's which satisfies the condition $\sum F_i \in S$ along with others stated in my paper can be taken as N_i. Then from $\sum N_i^* N_i = \sum \lambda_a D_a$, we obtain the solution D_a which satisfies Professor Ikeda's requirement.

CHAPTER 12

On the Nonexistence of Certain Block Designs

JUNJIRO OGAWA, *Nihon University, Japan*

INTRODUCTION AND SUMMARY

An arrangement of v varieties to b blocks of size k each is called a balanced incomplete block (BIB) design, if the following three requirements are fulfilled:

(1) each block contains k ($\leqq v$) different treatments,

(2) each treatment occurs in exactly r blocks, and

(3) any pair of treatments occurs in exactly λ blocks.

Among the five parameters v, b, r, k and λ describing a BIB design, there are two algebraic relations, i.e.,

$$vr = bk\,, \qquad \lambda(v-1) = r(k-1)\,.$$

Hence only three of them can be algebraically independent. It is also known that

$$v \leqq b \quad \text{and therefore} \quad r \geqq k\,,$$

which is due to R. A. Fisher [5].

The research reported has been made possible through the support and sponsorship of the U.S. Army Research and Development Group (Far East), Department of the Army under Grant No. DA-CRD-AFE-S92-544-67-G85.

A BIB design is said to be symmetrical if $v = b$ and is said to be asymmetric if $v < b$. It should be noted that $v \geqq k$ in general and the design is said to be complete if $v = k$. If $k < v$, the design is incomplete in the sense that the whole set of varieties does not occur in one block, and consequently the treatment-totals adjusted by the mean are not orthogonal to the block totals. In other words, the BIB design belongs to the class of the so-called nonorthogonal designs. However, due to the fact that all treatment-contrasts can be estimated linearly with the same variance, provided the design is connected, the arrangements of this sort are quite useful in the statistical design of experiments.

In the early part of the 1930's, R. A. Fisher and F. Yates [6] listed all plausible BIB designs with the number of replication r being not greater than 10. However, later on the following designs among them turned out to be nonexistent:

$$(\alpha) \quad v = 15, \quad b = 21, \quad r = 7, \quad k = 5, \quad \lambda = 2$$

$$(\alpha^*) \quad v = 22, \quad b = 22, \quad r = 7, \quad k = 7, \quad \lambda = 2$$

$$(\beta) \quad v = 21, \quad b = 28, \quad r = 8, \quad k = 6, \quad \lambda = 2$$

$$(\beta^*) \quad v = 29, \quad b = 29, \quad r = 8, \quad k = 8, \quad \lambda = 2$$

$$(\gamma) \quad v = 36, \quad b = 45, \quad r = 10, \quad k = 8, \quad \lambda = 2$$

$$(\gamma^*) \quad v = 46, \quad b = 46, \quad r = 10, \quad k = 10, \quad \lambda = 2 .$$

The existence of the following two designs is yet to be decided:

$$(\delta) \quad v = 46, \quad b = 69, \quad r = 9, \quad k = 6, \quad \lambda = 1$$

$$(\varepsilon) \quad v = 51, \quad b = 85, \quad r = 10, \quad k = 6, \quad \lambda = 1 .$$

Thus nonexistence proofs of certain BIB designs came into the picture. Nonexistence problems of similar nature arose with partially balanced incomplete block (PBIB) designs.

The purpose of this article is to give a comprehensive survey of the development of nonexistence proofs of blook designs up to the present stage. Although there are some minor new points, this paper is mainly of an expository nature.

A brief exposition of the incidence matrix of the BIB design is given in Section 1. Since the main mathematical tool of nonexistence proofs is the Hasse-Minkowski p-invariant of the rational congruence, this is discussed in Section 2 to the extent that it is necessary in the later discussions. In Section 3, the necessary condition for the existence of a symmetrical BIB design is given in terms of the Hasse-Minkowski p-invariant as the prototype of the nonexistence proofs which follow. In Section 4, the properties of the association algebra of an association are presented with some examples of association schemes. Section 5 is devoted to the derivation of the necessary condition for existence of a regular and symmetrical PBIB design. The argument presented in Section 5 is immediately carried over in Section 6 to the derivation of the necessary condition for the existence of a certain class of asymmetric PBIB designs. Finally in Section 7, the relationship algebra of a PBIB design is explained and some discussions of the nonexistence proofs in connection with the relationship algebra are presented.

1. THE INCIDENCE MATRIX OF A BIB DESIGN

The experimental units, the number of which being $n = vr = bk$, are serially numbered from 1 through n in any arbitrary but fixed manner. Let the incidence vector of the αth treatment be

$$\boldsymbol{\zeta}'_\alpha = (\zeta_{\alpha 1}, \zeta_{\alpha 2}, \ldots, \zeta_{\alpha n}),$$

where

$$\zeta_{\alpha f} = \begin{cases} 1, & \text{if the } \alpha\text{th treatment occurs in the } f\text{th unit, and} \\ 0, & \text{otherwise.} \end{cases}$$

The $n \times v$ matrix

$$\boldsymbol{\Phi} = \| \boldsymbol{\zeta}_1, \boldsymbol{\zeta}_2, \ldots, \boldsymbol{\zeta}_v \|$$

is called the incidence matrix of the treatments.

Let the incidence vector of the ath block be

$$\boldsymbol{\eta}'_a = (\eta_{a1}, \eta_{a2}, \ldots, \eta_{an}),$$

where

$$\eta_{af} = \begin{cases} 1, & \text{if the } f\text{th unit belongs to the } a\text{th block,} \\ 0, & \text{otherwise.} \end{cases}$$

The $n \times b$ matrix

$$\boldsymbol{\Psi} = \| \boldsymbol{\eta}_1, \boldsymbol{\eta}_2, \ldots, \boldsymbol{\eta}_b \|$$

is called the incidence matrix of the blocks.

It is readily seen that

$$\boldsymbol{\Phi}'\boldsymbol{\Phi} = rI_v \quad \text{and} \quad \boldsymbol{\Psi}'\boldsymbol{\Psi} = kI_b \,.$$

Since

$$n_{\alpha a} \equiv \boldsymbol{\zeta}'_\alpha \boldsymbol{\eta}_a = \sum_{f=1}^{n} \zeta_{\alpha f}\eta_{af} = \begin{cases} 1, & \text{if the } \alpha\text{th treatment occurs in the } a\text{th block, and} \\ 0, & \text{otherwise,} \end{cases}$$

the $v \times b$ matrix

$$N \equiv \| n_{\alpha a} \| = \boldsymbol{\Phi}'\boldsymbol{\Psi}$$

is the incidence matrix of the BIB design. It is known that

$$NN' = (r - \lambda)I_v + \lambda G_v \,,$$

where G_v is the $v \times v$ matrix whose elements are all unity. Therefore

$$\det(NN') \equiv | NN' | = rk(r - \lambda)^{v-1} \,.$$

Since $| NN' |$ is the gramian determinant of the v row vectors of the incidence matrix N, they are linearly independent unless $r = \lambda$ and consequently $v \leqq b$, which is the Fisher inequality. Whereas if $r = \lambda$, then $v = k$ and consequently $b = r = \lambda$. Thus the BIB design degenerates to a complete block design.

If the BIB design is symmetrical, one can show [5] that

$$N'N = (r - \lambda)I_v + \lambda G_v \,.$$

This means that the incidence matrix of a symmetrical BIB design is a normal matrix, i.e., $NN' = N'N$ and any two blocks of a symmetrical BIB design have exactly λ treatments in common. Thus the matrix $N'N$ represents the so-called block structure of the BIB design.

Given a symmetrical BIB design with parameters $v = b$, $r = k$ and λ, by deleting one block and discarding all treatments contained in the deleted block one obtains a BIB design with parameters

$$v^* = v - k, \qquad b^* = b - 1, \qquad r^* = r,$$
$$k^* = k - \lambda, \qquad \lambda^* = \lambda .$$

This procedure of generating a new BIB design is called the *cut-off method.* (α), (β) and (γ) mentioned in the previous section would have been obtained from the symmetrical designs (α^*), (β^*) and (γ^*) respectively by the cut-off method, if (α^*), (β^*) and (γ^*) exist.

Since for a symmetrical BIB design, the incidence matrix N is a square (0, 1)-matrix, one concludes that $rk(r - \lambda)^{v-1}$ should be a perfect square and consequently that $(r - \lambda)^{v-1}$ should be a perfect square, since $r = k$. Hence, if v is even, then $r - \lambda$ itself should be a perfect square. Thus one can conclude that the symmetrical BIB designs (α^*) and (γ^*) are impossible.

In order to prove the impossibility of (β^*), we need a theorem due to H. Hasse which will be presented in the next section.

2. THE HASSE-MINKOWSKI p-INVARIANT OF THE RATIONAL CONGRUENCE

After a brief exposition of the p-adic number, we give the definition of the Hasse-Minkowski p-invariant and then those of its properties which will be needed in the sequel. Reference should be made to the original paper by H. Hasse [9] and to the book by Burton W. Jones [13].

An infinite series of the form

$$\alpha = a_{-r}p^{-r} + a_{-r+1}p^{-r+1} + \dots$$
$$+ a_{-1}p^{-1} + a_0 + a_1p + a_2p^2 + \dots ,$$

where p is a prime number and a_i is an integer (positive, negative or zero) is called a *p-adic number*. We say that two p-adic numbers α and β are equal if there is an integer K such that for every positive integer $k \geqq K$, $\alpha_k \equiv \beta_k \mod p^k$, where α_k is obtained from the expansion of α by deleting all terms after the one involving p^k and β_k is defined similarly. If all the coefficients a_i in the expansion of α are restricted by $0 \leq a_i < p$, then the expansion is called the *canonical representation.* If the two p-adic numbers α and β are represented in the canonical form, then $\alpha = \beta$ is equivalent to the relations $a_k = b_k$ for all k.

A p-adic number γ whose representation involves no term of negative power of p, i.e.,

$$\gamma = a_0 + a_1 p + a_2 p^2 + \ldots$$

is called a *p-adic integer*. A p-adic integer with $a_0 \neq 0$ is called a *p-adic unit*. If γ is a p-adic unit, then γ^{-1} is a p-adic integer. Conversely if γ^{-1} is a p-adic integer, then γ is a p-adic unit.

The totality of p-adic numbers is a field which is denoted by R_p where R denotes the field of all rational numbers. It is readily noted that $R \subseteqq R_p$. The field of all real numbers is denoted by R_{p_∞}.

For α and β non-zero p-adic numbers, we define the symbol $(\alpha, \beta)_p$ by

$$(\alpha, \beta)_p = \begin{cases} +1, & \text{if } \alpha x^2 + \beta y^2 = 1 \text{ has a } p\text{-adic solution, and} \\ -1, & \text{otherwise.} \end{cases}$$

This symbol is called the *Hilbert symbol*. We list the necessary properties of the Hilbert symbol.

(1) If $\alpha\beta \neq 0$, then $(\alpha, \beta)_{p_\infty} = +1$.

(2) $(\alpha, \beta)_p = (\beta, \alpha)_p$.

(3) $(\alpha\rho^2, \beta\sigma^2)_p = (\alpha, \beta)_p$.

(4) $(\alpha, -\alpha)_p = +1$.

(5) If $\alpha = p^a\alpha_1$, $\beta = p^b\beta_1$, where α_1 and β_1 are the p-adic units, then

(a) $(\alpha, \beta)_p = \left(\frac{-1}{p}\right)^{ab}\left(\frac{\alpha_1}{p}\right)^{b}\left(\frac{\beta_1}{p}\right)^{a}$

for all odd prime p, and

(b) $(\alpha, \beta)_2 = \left(\frac{2}{\alpha_1}\right)^{b}\left(\frac{2}{\beta_1}\right)^{a}(-1)^{(\alpha_1-1)(\beta_1-1)/4}$,

where for the p-adic units α_1 and β_1 with principal terms a_0 and b_0, respectively,

$$\left(\frac{\alpha_1}{p}\right)=\left(\frac{a_0}{p}\right),\quad \left(\frac{\beta_1}{p}\right)=\left(\frac{b_0}{p}\right):\quad \left(\frac{2}{\alpha_1}\right)=\left(\frac{2}{a_0}\right),\quad \left(\frac{2}{\beta_1}\right)=\left(\frac{2}{b_0}\right)$$

$$\frac{(\alpha_1-1)(\beta_1-1)}{4}=\frac{(a_0-1)(b_0-1)}{4}$$

and $\left(\frac{a}{p}\right)$ stands for the Legendre symbol.

(5′) If p is relatively prime to $2\alpha\beta$ and p is finite, then

$$(\alpha, \beta)_p = +1 .$$

(6) $(\alpha, \beta)_p(\alpha, \gamma)_p = (\alpha, \beta\gamma)_p$.

(7) $(\alpha, \alpha)_p = (-1, \alpha)_p$.

(8) $(\alpha\rho, \beta\rho)_p = (\alpha, \beta)_p(\rho, -\alpha\beta)_p$.

(9) For non-zero rational integers a and b

$$\prod_p (a, b)_p = +1 ,$$

where p ranges over all primes including p_∞.

(10) $(a, b)_p = (a+b, -ab)_p$.

(10′) For any integer n, one has

$$(n+1, n)_p = (n+1, -1)_p .$$

Two rational and symmetric matrices $\boldsymbol{A}$ and $\boldsymbol{B}$ are said to be rationally congruent, denoted by symbol $\boldsymbol{A} \sim \boldsymbol{B}$, if there exists a rational and non-singular matrix $\boldsymbol{C}$ such that

$$\boldsymbol{C}'\boldsymbol{A}\boldsymbol{C} = \boldsymbol{B} .$$

The following theorem due to H. Hasse [9] is fundamental in the theory of rational congruence.

Theorem. *The necessary and sufficient conditions for two rational and symmetric matrices* $\boldsymbol{A}$ *and* $\boldsymbol{B}$ *to be rationally congruent are that*

$$|\boldsymbol{A}| \sim |\boldsymbol{B}|, \qquad \textit{Index of } \boldsymbol{A} = \textit{Index of } \boldsymbol{B} ,$$

and further that

$$C_p(A) = C_p(B)$$

for all primes p including p_∞, where $C_p(A)$ is the Hasse-Minkowski p-invariant defined by

$$C_p(A) = (-1, -1)_p \prod_{i=1}^{n} (D_i, -D_{i-1})_p ,$$

with $D_0 = 1, D_1, D_2, \ldots, D_n$ the principal minors of the determinant of A of degree $0, 1, 2, \ldots, n$, respectively.

The useful properties of $C_p(A)$ are listed below [30]:

(1) $C_p(\omega A) = (-1, \omega)^{n(n+1)/2}(\omega, |A|)_p^{n-1} C_p(A)$.

(2) $C_p(A + B) = (-1, -1)_p(|A|, |B|)_p C_p(A) C_p(B)$.

(3) $C_p(\underset{n}{A} \times \underset{m}{B}) = (-1, -1)_p^{m+n-1}(-1, |A|)_p^{m(m-1)/2}$
$\cdot (-1, |B|)_p^{n(n-1)/2}$.

$$(|A|, |B|)_p^{mn-1} C_p^m(A) C_p^n(B) .$$

3. NECESSARY CONDITIONS FOR EXISTENCE OF A SYMMETRICAL BIB DESIGN

Since the incidence matrix N of a symmetrical BIB design is a square matrix, one gets the congruence relation

$$M \equiv (r - \lambda)I_v + \lambda G_v \sim I_v .$$

The matrix M has two distinct characteristic roots $r + (v - 1)\lambda = r^2$ and $r - \lambda$ with multiplicities 1 and $v - 1$, respectively. The vector $j' = (1, \ldots, 1)$ is the characteristic vector corresponding to the root r^2. All the vectors of v dimensions which are orthogonal to j are the characteristic vectors corresponding to the other root $r - \lambda$. We can take rational vectors which are linearly independent and orthogonal to the vector j and let H be the $v \times v$ matrix which has those characteristic vectors as its columns.

Then

$$H = \begin{Vmatrix} 1 & v-1 & 0 & \dots & 0 \\ 1 & -1 & v-2 & \dots & 0 \\ 1 & -1 & -1 & \dots & 0 \\ \cdot & \cdot & \cdot & & \cdot \\ \cdot & \cdot & \cdot & & \cdot \\ \cdot & \cdot & \cdot & & \cdot \\ 1 & -1 & -1 & \dots & 1 \\ 1 & -1 & -1 & \dots & -1 \end{Vmatrix}.$$

and

$$MH = H \cdot \begin{Vmatrix} r^2 & & & & & \\ & r-\lambda & & & & \\ & & \cdot & & & \\ & & & \cdot & & \\ & & & & \cdot & \\ & & & & & r-\lambda \end{Vmatrix}.$$

Hence one obtains the relation

$$H'MH = H'H \cdot \begin{Vmatrix} r^2 & & & & & 0 \\ & r-\lambda & & & & \\ & & \cdot & & & \\ & & & \cdot & & \\ & & & & \cdot & \\ 0 & & & & & r-\lambda \end{Vmatrix}.$$

$$= \begin{Vmatrix} r^2 v & 0 \\ 0 & (r-\lambda)Q \end{Vmatrix},$$

where Q is the gramian matrix of a set of rational basis vectors of the proper subspace corresponding to the characteristic root $r-\lambda$ of the matrix M, i.e.,

$$Q = \begin{Vmatrix} v(v-1) & & & & & 0 \\ & (v-1)(v-2) & & & & \\ & & \cdot & & & \\ & & & \cdot & & \\ & & & & \cdot & \\ 0 & & & & & 2 \cdot 1 \end{Vmatrix}.$$

One can calculate the Hasse-Minkowski p-invariant of $\boldsymbol{M}$ as follows:

$$C_p(\boldsymbol{M}) = (-1, -1)_p(-1, r^2v)_p(r^2v, (r-\lambda)^{v-1}\,|\,\boldsymbol{Q}\,|)_p$$
$$\cdot\,(-1, r-\lambda)_p^{v(v-1)/2}(r-\lambda, |\,\boldsymbol{Q}\,|)_p^v C_p(\boldsymbol{Q})$$
$$= (-1, -1)_p(-1, r-\lambda)_p^{v(v-1)/2}(v, r-\lambda)_p\,.$$

Since this should be equal to $C_p(\boldsymbol{I}_v) = (-1, -1)_p$, one obtains a necessary condition for existence of a symmetrical BIB design:

$$S_p \equiv (-1, r-\lambda)_p^{v(v-1)/2}(v, r-\lambda)_p = 1$$

for all primes p [22].

For the design (β^*) $v = b = 29$, $r = k = 8$, $\lambda = 2$, one gets

$$S_p = (29, 6)_p = (29, 2)_p(29, 3)_p\,,$$

which gives us for $p = 3$

$$S_3 = (29, 3)_3 = \left(\frac{29}{3}\right) = \left(\frac{-1}{3}\right) = -1\,.$$

Thus (β^*) is seen to be impossible.

4. THE ASSOCIATION ALGEBRA

In order to carry over the method described in the preceding section to regular and symmetrical PBIB designs, the concept of the association and the accociation algebra is explained in this section.

A relation defined in a set of v elements is called an association if it satisfies the following three requirements:

(1) Any two elements of the set are either first, second, ..., or mth associates with each other,

(2) Each element has n_i ith associates, and,

(3) For any pair of elements α and β which are ith associates, the number of elements which are jth associates of α and at the same time kth associates of β is p^i_{jk} and this number is independent of the pair (α, β) with which we started.

Since each element may be considered as 0th associate of itself, we can introduce the following notations

$$n_0 = 1, \qquad p_{0i}^k = \delta_{ik}, \qquad p_{ij}^0 = n_i\delta_{ij},$$

where δ_{ij} stands for the Kronecker delta.

The following relations among these parameters are known:

$$\sum_{i=0}^m n_i = v, \qquad p_{ij}^k = p_{ji}^k,$$

$$\sum_{j=0}^m p_{jk}^i = n_k,$$

$$n_i p_{jk}^i = n_j p_{ik}^j = n_k p_{ij}^k.$$

The ith association matrix A_i is defined as follows:

$$A_i = \| \alpha_{\alpha i}^{\beta} \|_{\alpha,\beta=1,\dots,v},$$

where

$$\alpha_{\alpha i}^{\beta} = \begin{cases} 1, & \text{if the two elements } \alpha \text{ and } \beta \text{ are } i\text{th associates, and} \\ 0, & \text{otherwise.} \end{cases}$$

It can be seen that

$$\sum_{i=0}^m A_i = G_v,$$

and

$$A_i A_j = A_j A_i = \sum_{k=0}^m p_{ij}^k A_k.$$

This means that the linear closure $\mathfrak{A}$ of the set of $(m+1)$ linearly independent matrices $A_0, A_1, \dots, A_m$ over the field of all rational numbers is a linear associative and commutative algebra. $\mathfrak{A}$ is called the association algebra [2].

Let

$$P_i = \| p_{\alpha i}^{\beta} \|_{\alpha,\beta=0,1,\dots,m}.$$

the mapping $A_i \to P_i$, $i = 0, 1, \dots, m$ generate the regular representation of the algebra $\mathfrak{A}$. If all the characteristic roots of P_i, $i = 0, 1, \dots, m$ are rational numbers, then the regular representation of $\mathfrak{A}$ is decomposed into $m+1$ inequivalent and linear representations in the field of all rational numbers. Since

$\mathfrak{A}$ is completely reducible, $\mathfrak{A}$ can be expressed as a linear combination of those $m+1$ linear representations with rational coefficients in the sense of the equivalence of representations.

Using the relations,

$$\boldsymbol{A}_i \boldsymbol{G}_v = \boldsymbol{G}_v \boldsymbol{A}_i = n_i \boldsymbol{G}_v \,,$$

one can find a nonsingular rational matrix of the form

$$\boldsymbol{C} = \left\| \begin{matrix} 1 & 1 & \dots & 1 \\ c_{10} & c_{11} & \dots & c_{1m} \\ c_{20} & c_{21} & \dots & c_{2m} \\ \cdot & \cdot & & \cdot \\ \cdot & \cdot & & \cdot \\ \cdot & \cdot & & \cdot \\ c_{m0} & c_{m1} & \dots & c_{mm} \end{matrix} \right\| ,$$

such that

$$\boldsymbol{C}\boldsymbol{P}_i\boldsymbol{C}^{-1} = \left\| \begin{matrix} z_{0i} & & & & & & \\ & z_{1i} & & & & & \\ & & z_{2i} & & & & \\ & & & \cdot & & & \\ & & & & \cdot & & \\ & & & & & \cdot & \\ & & & & & & z_{mi} \end{matrix} \right\| ,$$

$$i = 0, 1, \dots, m \,,$$

where $z_{0i} = n_i$, $i = 0, 1, \dots, m$.

Thus one can argue that

$$\boldsymbol{A}_u^{\sharp} = \left(\sum_{i=0}^{m} c_{ui} z_{ui} \right)^{-1} \sum_{i=0}^{m} c_{ui} \boldsymbol{A}_i \,, \qquad u = 0, 1, \dots, m$$

are mutually orthogonal and idempotent matrices and $\boldsymbol{A}_u^{\sharp}$ generates the linear representation (u): $\boldsymbol{A}_i \to z_{ui}$. The multiplicities $\alpha_0, \alpha_1, \dots, \alpha_m$ of the $m+1$ inequivalent linear representations in $\mathfrak{A}$ itself are obtained by the following set of linear equations:

$$\alpha_0 = 1 \,,$$

$$\left.\begin{array}{l} \alpha_1 + \alpha_2 + \ldots + \alpha_m = v - 1\,, \\ \alpha_1 z_{11} + \alpha_2 z_{21} + \ldots + \alpha_m z_{m1} = -n_1\,, \\ \alpha_1 z_{12} + \alpha_2 z_{22} + \ldots + \alpha_m z_{m2} = -n_2\,, \\ \cdot \quad \cdot \quad \cdot \quad \cdot \quad \cdot \\ \alpha_1 z_{1m} + \alpha_2 z_{2m} + \ldots + \alpha_m z_{mm} = -n_m\,. \end{array}\right\}$$

Some examples of the association algebra

(1) *The association of the group-divisible (GD) type.* The number of treatments is $v = ln$ and they are divided into l groups of n treatments each. Any two treatments belonging to the same group are first associates and two treatments belonging to the different groups are second associates. In this case the number of associate classes is $m = 2$ and

$$n_0 = 1\,, \qquad n_1 = n - 1\,, \qquad n_2 = (l-1)n\,.$$

$$\boldsymbol{P}_0 = \left\| \begin{array}{ccc} 1 & 0 & 0 \\ 0 & 1 & 0 \\ 0 & 0 & 1 \end{array} \right\|, \qquad \boldsymbol{P}_1 = \left\| \begin{array}{ccc} 0 & 1 & 0 \\ n-1 & n-2 & 0 \\ 0 & 0 & n-1 \end{array} \right\|,$$

$$\boldsymbol{P}_2 = \left\| \begin{array}{ccc} 0 & 0 & 1 \\ 0 & 0 & n-1 \\ (l-1)n & (l-1)n & (l-2)n \end{array} \right\|.$$

Transforming the $\boldsymbol{P}_i$'s by the matrix

$$\boldsymbol{C} = \left\| \begin{array}{ccc} 1 & 1 & 1 \\ 1 & 1 & -1/(l-1) \\ 1 & -1/(l-1) & 1 \end{array} \right\|,$$

one gets the diagonal matrices

$$\left\| \begin{array}{ccc} 1 & 0 & 0 \\ 0 & 1 & 0 \\ 0 & 1 & 1 \end{array} \right\|, \quad \left\| \begin{array}{ccc} n-1 & 0 & 0 \\ 0 & n-1 & 0 \\ 0 & 0 & -1 \end{array} \right\|, \quad \left\| \begin{array}{ccc} (l-1)n & 0 & 0 \\ 0 & -n & 0 \\ 0 & 0 & 0 \end{array} \right\|.$$

Hence the three mutually orthogonal and idempotent matrices are given by

$$\left.\begin{aligned} A_0^{\#} &= \frac{1}{ln}[A_0 + A_1 + A_2] \\ A_1^{\#} &= \frac{1}{ln}[(l-1)A_0 + (l-1)A_1 - A_2] \\ A_2^{\#} &= \frac{1}{n}[(n-1)A_0 - A_1] \end{aligned}\right\}$$

with respective ranks

$$\alpha_0 = \operatorname{tr} A_0^{\#} = 1\,, \quad \alpha_1 = \operatorname{tr} A_1^{\#} = l - 1\,, \quad \alpha_2 = \operatorname{tr} A_2^{\#} = l(n-1)\,.$$

This can immediately be carried over to the association scheme of m-associate GD type [10], [28].

(2) *The association of triangular* (T_2) *type.*
The number of elements in $v = n(n-1)/2$ in this case. Fill the positions above the main diagonal of an $n \times n$ square with the elements taken in order leaving the positions in the main diagonal blank. Then fill the positions below the main diagonal in such a way that the $n \times n$ square is symmetrical with respect to the main diagonal. Any two elements occuring in the same row or column are first associates and any two elements occuring in the different rows or columns are second associates. This association is called the triangular or T_2 type. The number of the associate classes is two, and

$$n_0 = 1\,, \qquad n_1 = 2n - 4\,, \qquad n_2 = (n-2)(n-3)/2\,.$$

The generating matrices of the regular representation of the association algebra are given by

$$P_0 = \begin{Vmatrix} 1 & 0 & 0 \\ 0 & 1 & 0 \\ 0 & 0 & 1 \end{Vmatrix}, \qquad P_1 = \begin{Vmatrix} 0 & 1 & 0 \\ 2n-4 & n-2 & 4 \\ 0 & n-3 & 2n-8 \end{Vmatrix},$$

$$P_2 = \begin{Vmatrix} 0 & 0 & 1 \\ 0 & n-3 & 2n-8 \\ \dfrac{(n-2)(n-3)}{2} & \dfrac{(n-3)(n-4)}{2} & \dfrac{(n-4)(n-5)}{2} \end{Vmatrix}.$$

Transforming these matrices by the non-singular matrix

$$C = \begin{Vmatrix} 1 & 1 & 1 \\ 2n-4 & n-4 & -4 \\ -(n-2)(n-3) & n-3 & -2 \end{Vmatrix},$$

one obtains the diagonal matrices

$$\begin{Vmatrix} 1 & 0 & 0 \\ 0 & 0 & 0 \\ 0 & 1 & 1 \end{Vmatrix}, \quad \begin{Vmatrix} 2n-4 & 0 & 0 \\ 0 & n-4 & 0 \\ 0 & 0 & -2 \end{Vmatrix},$$

$$\begin{Vmatrix} \frac{(n-2)(n-3)}{2} & 0 & 0 \\ 0 & -(n-3) & 0 \\ 0 & 0 & 1 \end{Vmatrix}.$$

Thus the mutually orthogonal and idempotent matrices are given by

$$\left.\begin{aligned} A_0^{\sharp} &= \frac{2}{n(n-1)}[A_0 + A_1 + A_2] \\ A_1^{\sharp} &= \frac{1}{n(n-2)}[(2n-4)A_0 + (n-4)A_1 - 4A_2] \\ A_2^{\sharp} &= \frac{-1}{(n-1)(n-2)}[(n-2)(n-3)A_0 - (n-3)A_1 + 2A_2] \end{aligned}\right\}$$

with respective ranks

$$\alpha_0 = \operatorname{tr} A_0^{\sharp} = 1, \qquad \alpha_1 = \operatorname{tr} A_1^{\sharp} = n-1,$$
$$\alpha_2 = \operatorname{tr} A_2^{\sharp} = n(n-3)/2.$$

This can be generalized to the association scheme of T_m type [14], [18], [32].

(3) *The association of T_m type.* The number of treatments is $v = \binom{n}{m}$, where n and m are positive integers such that $2m \leqq n$. Let us consider the totality of all $\binom{n}{m}$ different m-tuples $(r_1, \ldots, r_m)$'s taken out of $(1, 2, \ldots, n)$. We may identify the v treatments with these m-tuples. Two treatments φ_i and φ_j, which correspond to the m-tuples $(r_1, \ldots, r_m)$ and $(r_1', \ldots, r_m')$, respectively, are said to be the uth associates if

and only if $(r_1, \ldots, r_m)$ and $(r'_1, \ldots, r'_m)$ have exactly $m - u$ integers in common, $u = 0, 1, \ldots, m$. The number of the uth associates of each treatment is

$$n_u = \binom{m}{m-u}\binom{n-m}{u}, \qquad u = 0, 1, \ldots, m .$$

For any two treatments φ_i and φ_j which are tth associates, the number of the treatments which are sth associates of φ_i and at the same time uth associates of φ_j is given by

$$p^t_{su} = \sum_{a=0}^{m-s} \binom{m-t}{a}\binom{t}{m-s-a}\binom{t}{m-u-a}\binom{n-m-t}{s+u-m+a}$$
$$u, s, t = 0, 1, \ldots, m .$$

The regular representation of the association algebra is generated by the mappings

$$\boldsymbol{A}_i \rightarrow \boldsymbol{P}_i = \| p^t_{si} \| , \qquad i = 0, 1, \ldots, m .$$

Transforming the $\boldsymbol{P}_i$'s by the non-singular matrix

$$\boldsymbol{C} = \left\| \frac{z_{st}}{n_t} \right\| , \qquad s, t = 0, 1, \ldots m ,$$

where

$$z_{st} = \sum_{a=0}^{t} (-1)^{t-a} \binom{m-a}{m-t}\binom{m-s}{a}\binom{n-m-s+a}{a} ,$$

one gets

$$\boldsymbol{C}\boldsymbol{P}_i\boldsymbol{C}^{-1} = \begin{Vmatrix} z_{0i} & & & & \\ & z_{1i} & & & \\ & & \cdot & & \\ & & & \cdot & \\ & & & & \cdot \\ & & & & & z_{mi} \end{Vmatrix} , \qquad i = 0, 1, \ldots, m .$$

Thus one obtains the $m + 1$ mutually orthogonal and idempotent matrices

$$\boldsymbol{A}^*_u = \frac{\alpha_u}{v}\left(\sum_{i=0}^{m} \frac{z_{ui}}{n_i} \boldsymbol{A}_i\right) , \qquad u = 0, 1, \ldots, m ,$$

with respective ranks

$$\alpha_u = \operatorname{tr} A_u^\# = \binom{n}{u} - \binom{u-1}{n} = \frac{n+1-2u}{n+1-u}\binom{n}{u},$$
$$u = 0, 1, \ldots, m .$$

5. NECESSARY CONDITIONS FOR EXISTENCE OF A REGULAR AND SYMMETRICAL PBIB DESIGN

An arrangement of v treatments, among which an association is defined, into b blocks of size k each is said to be a partially balanced incomplete block (PBIB) design if the following three conditions are satisfied:

(1) Each block contains k different treatments,

(2) Each treatment occurs in exactly r blocks, and

(3) Each pair of treatments which are ith associates occurs in exactly λ_i blocks.

It is readily noted that

$$vr = bk \quad \text{and} \quad \sum_{i=0}^{m} n_i\lambda_i = rk ,$$

where we have put $\lambda_0 = 1$.

A PBIB design is said to be *symmetrical* if $v = b$ and asymmetrical otherwise.

It is known that

$$NN' = rA_0 + \lambda_1 A_1 + \ldots + \lambda_m A_m$$

and this can be expressed as a linear combination of the $m + 1$ mutually orthogonal, idempotent matrices of the association algebra as follows:

$$(*) \qquad NN' = \rho_0 A_0^\# + \rho_1 A_1^\# + \ldots + \rho_m A_m^\# ,$$

where

$$\rho_0 = rk , \quad \rho_u = \sum_{i=0}^{m} z_{ui}\lambda_i , \qquad u = 1, 2, \ldots, m .$$

If all the ρ_u's are positive, then the PBIB design is said to be *regular*. There will be no loss of generality in assuming that α_u linearly independent column vectors of $A_u^\#$ are

$$a^{(u)*}_{\alpha_0+\cdots+\alpha_{u-1}+1},\ a^{(u)*}_{\alpha_0+\cdots+\alpha_{u-1}+2},\ \ldots,\ a^{(u)*}_{\alpha_0+\cdots+\alpha_{u-1}+\alpha_u},\qquad u = 0, 1, \ldots, m\,.$$

These α_u rational vectors generate the proper subspace $\mathcal{L}_u$ of the matrix NN' corresponding to the characteristic root ρ_u of the matrix. In other words, the expression (*) above is the spectral decomposion of matrix NN'. Let the gramian matrix of the set of α_u rational vectors given above be Q_u, and let the $v \times v$ matrix whose columns are

$$a^{(0)*}_1,\ a^{(1)*}_2,\ \ldots,\ a^{(1)*}_{1+\alpha_1},\ \ldots,\ a^{(m)*}_{\alpha_0+\cdots+\alpha_{m-1}+1},\ \ldots,\ a^{(m)*}_{\alpha_0+\cdots+\alpha_{m-1}+\alpha_m}$$

be S, then

$$S'S = \left\| \begin{matrix} 1/v & 0 & 0 & \cdots & 0 \\ 0 & Q_1 & 0 & \cdots & 0 \\ 0 & 0 & Q_2 & \cdots & 0 \\ \cdot & \cdot & \cdot & \cdot & \cdot \\ \cdot & \cdot & \cdot & \cdot & \cdot \\ \cdot & \cdot & \cdot & \cdot & \cdot \\ 0 & 0 & 0 & & Q_m \end{matrix} \right\|.$$

Hence it follows that

$$(**)\qquad v\,|Q_1| \cdot |Q_2| \cdots |Q_m| \sim 1\,,$$

and

$$(***)\qquad (-1, -1)_p^m \prod_{i<j} (|Q_i|, |Q_j|)_p \prod_{u=1}^{m} C_p(Q_u) = 1\,.$$

On the other hand, since

$$S'NN'S = \left\| \begin{matrix} rk/v & 0 & \cdots & 0 \\ 0 & \rho_1 Q_1 & \cdots & 0 \\ \cdot & \cdot & \cdot & \cdot \\ \cdot & \cdot & \cdot & \cdot \\ \cdot & \cdot & \cdot & \cdot \\ 0 & 0 & \cdots & \rho_m Q_m \end{matrix} \right\|,$$

one gets the determinant condition for a regular symmetrical PBIB design that

$$v \prod_{u=1}^{m} \rho_u^{\alpha_u} \prod_{u=1}^{m} | \boldsymbol{Q}_u | \sim 1 ,$$

and consequently by (**) that

$$\prod_{u=1}^{m} \rho_u^{\alpha_u} \sim 1 .$$

Calculating the Hasse-Minkowski p-invariant of NN', one obtains

$$C_p(NN') = (-1, -1)_p \prod_{i<j} (\rho_i^{\alpha_i}, \rho_j^{\alpha_j})_p \prod_{u=1}^{m} (\rho_u, | \boldsymbol{Q} |)_p \cdot \prod_{u=1}^{m} (-1, \rho_u)_p^{\alpha_u(\alpha_u+1)/2} ,$$

where we have used the relation (***). Finally one obtains the necessary condition for existence of a regular and symmetrical PBIB design

$$O_p \equiv \prod_{u=1}^{m} (-1, \rho_u)_p^{\alpha_u(\alpha_u+1)/2} \prod_{i<j} (\rho_i^{\alpha_i}, \rho_j^{\alpha_j})_p \prod_{u=1}^{m} (\rho_u, | \boldsymbol{Q}_u |)_p = 1$$

for all primes p.

For the two associate class PBIB designs, the necessary conditions for existence are given by

$$\rho_1^{\alpha_1} \rho_2^{\alpha_2} \sim 1 ,$$

and

$$O_p \equiv (-1, \rho_1)_p^{\alpha_1(\alpha_1+1)/2} (-1, \rho_2)_p^{\alpha_2(\alpha_2+1)/2} (\rho_1^{\alpha_1}, \rho_2^{\alpha_2})_p \cdot (\rho_1, | \boldsymbol{Q}_1 |)_p (\rho_2, | \boldsymbol{Q}_2 |)_p = 1$$

for all primes p.

Some examples are listed below:

I. *Symmetrical PBIB designs of group-divisible type* [1].

$$\rho_1 = r^2 - v\lambda_2 , \qquad \rho_2 = r - \lambda_1 ,$$

$$\alpha_1 = l - 1 , \qquad \alpha_2 = l(n-1) ,$$

$$(r^2 - v\lambda)^{l-1} (r - \lambda_1)^{l(n-1)} \sim 1 ,$$

$$O_p \equiv (-1, \rho_1)_p^{l(l-1)} (-1, \rho_2)_p^{\{l(n-1)(ln-l+1)\}/2} \cdot (\rho_1, n)_p^l (\rho_2, n)_p^l (\rho_1, \lambda_2)_p = 1 .$$

II. *Symmetrical PBIB designs of triangular* (T_2) *type* [15].

$$\rho_1 = r + (n-4)\lambda_1 - (n-3)\lambda_2\,, \qquad \rho_2 = r - 2\lambda_1 + \lambda_2$$

$$\alpha_1 = n-1\,, \qquad \alpha_2 = n(n-3)/2\,,$$

$$\rho_1^{\alpha_1}\rho_2^{\alpha_2} \sim 1\,,$$

and

$$\begin{aligned} O_p \equiv\ & (-1, \rho_1)_p^{\{(n-1)(n-2)\}/2}(-1, \rho_2)_p^{\{n(n-1)(n-2)(n-3)\}/8} \\ & \cdot (\rho_1, n)_p(\rho_1, n-2)_p^{n-1}(\rho_2, 2)_p(\rho_2, n-1)_p \\ & \cdot (\rho_2, n-2)_p^{n-1} = 1\,. \end{aligned}$$

III. *Symmetrical PBIB designs of* L_t *type* [24], [33].

$$\rho_1 = r + (n-t)\lambda_1 - (n-t+1)\lambda_2\,,$$
$$\rho_2 = r - t\lambda_1 + (t-1)\lambda_2\,,$$

$$\alpha_1 = t(n-1)\,, \qquad \alpha_2 = (n-1)(n-t+1)\,,$$

$$\rho_1^{t(n-1)}\rho_2^{(n-1)(n-t+1)} \sim 1\,,$$

$$\begin{aligned} O_p \equiv\ & (-1, \rho_1)_p^{\{t(n-1)(n-t+1)\}/2}(-1, \rho_2)_p^{\{(n-1)(n-t+1)(2n-t-2)\}/2} \\ & \cdot (n, \rho_1\rho_2)_p^{nt} = 1\,. \end{aligned}$$

IV. *Symmetrical cubic designs:* $v = s^3$, [29].

$$\rho_1 = r + (2s-3)\lambda_1 + (s-1)(s-3)\lambda_2 - (s-1)^2\lambda_3\,,$$
$$\rho_2 = r + (s-3)\lambda_1 - (2s-3) + (s-1)\lambda_3\,,$$
$$\rho_3 = r - 3\lambda_1 + 3\lambda_2 - \lambda_3\,,$$

$$\alpha_1 = 3(s-1)\,, \qquad \alpha_2 = 3(s-1)_2\,, \qquad \alpha_3 = (s-1)^3\,,$$

$$\rho_1^{\alpha_1}\rho_2^{\alpha_2}\rho_3^{\alpha_3} \sim 1\,,$$

$$\begin{aligned} R_p \equiv\ & \prod_{i=1}^{3} (-1, \rho_i)_p^{\alpha_i(\alpha_i+1)/2}(\rho_1^{\alpha_1}, \rho_2^{\alpha_2})_p(\rho_1^{\alpha_1}, \rho_3^{\alpha_3})_p \\ & \cdot (\rho_2^{\alpha_2}, \rho_3^{\alpha_3})_p(s, \rho_1)_p(s, \rho_2\rho_3)_p^{s-1} = 1\,. \end{aligned}$$

V. *Symmetrical right angular designs,* $v = 2sl$, [27].

$$\rho_1 = r - \lambda_1 + s(\lambda_1 - \lambda_2) + s(l-1)(\lambda_3 - \lambda_4)\,,$$
$$\rho_2 = r - \lambda_1\,,$$

$$\rho_3 = r - \lambda_1 + s(\lambda_1 + \lambda_2 - \lambda_3 - \lambda_4)\,,$$
$$\rho_4 = r - \lambda_1 + s(\lambda_1 - \lambda_2 - \lambda_3 + \lambda_4)\,,$$

$$\alpha_1 = 1\,, \qquad \alpha_2 = 2l(s-1)\,,$$
$$\alpha_3 = l-1\,, \qquad \alpha_4 = l-1\,,$$

$$\rho_1\rho_3^{l-1}\rho_4^{l-1} \sim 1\,,$$

$$T_p \equiv (-1, \rho_1)_p(-1, \rho_2)_p^{l(s-1)}(-1, \rho_3\rho_4)_p^{l(l-1)/2} \cdot (\rho_3\rho_4, l(2s)^{l-1})_p^{l-2}(\rho_3^{l-1}, \rho_4^{l-1}l(2s)^{l-1})_p \cdot (l(2s)^{l-1}, \rho_4^{l-1})_p = 1\,.$$

VI. *Symmetrical group-divisible m-associate class designs* [28].

$$\rho_u = (r - \lambda_{m-u+1}) + (\lambda_1 - \lambda_{m-u+1})n_1 + \dots + (\lambda_{m-u} - \lambda_{m-u+1})n_{m-u}\,,$$

$$\alpha_u = N_1N_2\dots N_{u-1}(N_u - 1)\,, \qquad u = 1, 2, \dots, m\,.$$

$$\prod_{u=1}^{m} \rho_u^{\alpha_u} \sim 1\,.$$

We present the condition in terms of the p-invariant only for the case when $m = 3$.

$$R_p \equiv (-1, \rho_1)_p^{N_1(N_1+1)/2}(-1, \rho_2)_p^{N_1\{N_2(N_2+1)+(N_1+1)(N_2-1)\}/2} \cdot (-1, \rho_3)_p^{\{N_1N_2N_3(N_1+N_2+N_3+1)-N_1N_2(N_1+N_2)\}/2} \cdot (\lambda_2 - \lambda_3, \rho_1\rho_2)_p^{N_1}(\lambda_1 - \lambda_2, \rho_2\rho_3)_p^{N_1N_2}(\lambda_2, \rho_1)_p \cdot (\rho_1, \rho_2)_p^{N_1N_2}(\rho_1, \rho_3)_p^{N_1N_2(N_3-1)}(\rho_2, \rho_3)^{N_1N_2} = 1$$

for all primes p.

VII. *Symmetrical PBIB designs of T_m type:* $v = \binom{n}{m}$, $2m \leq n$, [14], [18], [32].

$$\rho_u = \sum_{i=0}^{m} z_{ui}\lambda_i\,, \quad \alpha_u = \frac{n+1-2u}{n+1-u}\binom{n}{u}\,, \qquad u = 1, 2, \dots, m$$

where

$$z_{ui} = \sum_{a=0}^{i} (-1)^{i-a}\binom{m-a}{m-i}\binom{m-u}{a}\binom{n-m-u+a}{a}\,,$$

and

$$|Q_u| \sim \prod_{j=0}^{u-1} [(u-j)(n-u+1-j)]^{\alpha_j} \binom{n-2u}{m-u}^{\alpha_u} \prod_{u=1}^{m} \rho_u^{\alpha_u} \sim 1 .$$

$$O_p \equiv \prod_{u=1}^{m} [(-1, \rho_u)_p^{\alpha_u(\alpha_u+1)/2} (\rho_u, |Q|)_p] \prod_{1 \leq u < k \leq m} (\rho_u^{\alpha_u}, \rho_k^{\alpha_k})_p = 1 .$$

6. NECESSARY CONDITION FOR EXISTENCE OF CERTAIN ASYMMETRIC PBIB DESIGNS

We consider an asymmetric PBIB design with $b < v$. In this case the matrix NN' is necessarily singular and therefore at least one ρ_u should vanish.

For the sake of simplicity of the explanation, we deal with two-associate class PBIB designs first.

Let us suppose that $\rho_2 = 0$. Then one gets the relation

$$S'NN'S = \begin{Vmatrix} rk/v & 0 & 0 \\ 0 & \rho_1 Q_1 & 0 \\ 0 & 0 & 0 \end{Vmatrix} .$$

This means that

$$S_1'NN'S_1 = \begin{Vmatrix} rk/v & 0 \\ 0 & \rho_1 Q_1 \end{Vmatrix} ,$$

where

$$S_1 = \| a_1^{(0)*} a_2^{(1)*} \ldots a_{\alpha_1+1}^{(1)*} \|$$

is the $v \times (\alpha_1 + 1)$ submatrix of S.

If in particular $b = \alpha_1 + 1$ or $v - \alpha_2 = b$, then one gets the relation

$$\begin{Vmatrix} rk/v & 0 \\ 0 & \rho_1 Q_1 \end{Vmatrix} \sim I_b ,$$

where it follows that

$$b\rho_1^{\alpha_1} | Q_1 | \sim 1 ,$$

and

$$(-1, \rho_1)_p^{\alpha_1(\alpha_1+1)/2} (\rho_1, |Q_1|)_p^{\alpha_1 - 1} C_p(Q_1) = (-1, -1)_p ,$$

for all primes p [26].

In a similar manner, one gets the analogous conditions for the case where $\rho_1 = 0$ and $v - \alpha_1 = b$:

$$b\rho_2^{\alpha_2} \mid \boldsymbol{Q}_2 \mid \sim 1 \, ,$$

$$(-1, \rho_2)_p^{\alpha_2(\alpha_2+1)/2}(\rho_2, \mid \boldsymbol{Q}_2 \mid)_p^{\alpha_2-1} C_p(\boldsymbol{Q}_2) = (-1, -1)_p$$

for all primes p [26].

It is known that the dual of an affine resolvable BIB design is a group-divisible PBIB design belonging to the class mentioned above [23]. It is also known that the dual of a BIB design with parameters

$$b = \frac{(k+1)(k+2)}{2} \, , \qquad v = \frac{k(k+1)}{2} \, ,$$

$$r = k + 2 \, , \qquad k \, , \qquad \lambda = 2$$

is a triangular type PBIB design belonging to the class mentioned above [19].

Applications of the above results to group-divisible designs, triaugular designs and to L_t designs are discussed by S. S. Shrikhande, D. Raghavarao and S. K. Tharthare [26].

7. THE RELATIONSHIP ALGEBRA OF A PBIB DESIGN [17]

In this final section the concept of the relationship algebra of a PBIB design is utilized in deriving a necessary condition for the existence of symmetrical PBIB designs. For the sake of simplicity of the explanation, we will be concerned with a symmetrical PBIB design of group-divisible type only. However, the essence of this argument can immediately be carried over to symmetrical PBIB designs of any known type.

Firstly we shall consider a regular and symmetrical GD design. It is known that the eleven matrices of order $N = vr = bk$,

$$\boldsymbol{I}, \boldsymbol{G}, \boldsymbol{B}, \boldsymbol{T}_1, \boldsymbol{T}_1\boldsymbol{B}_1, \boldsymbol{B}\boldsymbol{T}_1, \boldsymbol{B}\boldsymbol{T}_1\boldsymbol{B}, \boldsymbol{T}_2, \boldsymbol{T}_2\boldsymbol{B}, \boldsymbol{B}\boldsymbol{T}_2, \boldsymbol{B}\boldsymbol{T}_2\boldsymbol{B} \, ,$$

generate a linear associative algebra $\mathcal{R}$, called the relationship algebra of the design, of rank 11 over the field of all rational numbers, where

$$\boldsymbol{G} = \boldsymbol{j}\boldsymbol{j}', \quad \boldsymbol{B} = \boldsymbol{\Psi}\boldsymbol{\Psi}', \quad \boldsymbol{T} = \boldsymbol{\Phi}\boldsymbol{\Phi}', \quad \boldsymbol{T}_1 = \boldsymbol{\Phi}\boldsymbol{A}_1\boldsymbol{\Phi}', \text{ and } \boldsymbol{T}_2 = \boldsymbol{\Phi}\boldsymbol{A}_2\boldsymbol{\Phi}' \, .$$

$\mathcal{R}$ is completely reducible and has three inequivalent, linear and rational representations and two inequivalent, irreducible and rational representations of the second degree. Since $1^2 + 1^2 + 1^2 + 2^2 + 2^2 = 11$, there are no other irreducible representations of $\mathcal{R}$. All irreducible representations of $\mathcal{R}$ are shown in the following table together with their respective multiplicities in $\mathcal{R}$.

Let the invariant linear subspace of the N-dimensional vector space over the field of all rational numbers corresponding to the above listed irreducible representations be $\mathcal{L}_G^{(1)}$, $\mathcal{L}_0^{(N-b-v+1)}$, $\mathcal{L}_1^{(b-v)}$, $\mathcal{L}_{21}^{(2\alpha_1)}$, and $\mathcal{L}_{22}^{(2\alpha_2)}$, respectively, where the superscripts indicate the dimensionality of the subspaces. Then it is clear that $\mathcal{L}_G^{(1)}$ is generated by the vector $\boldsymbol{j}$ and $\mathcal{L}_0^{(N-b-v+1)}$ is generated by the common characteristic vector of the matrices $\boldsymbol{B}$ and $\boldsymbol{T}$ corresponding to the zero root. $\mathcal{L}_0^{(N-b-v+1)}$ is called the *error space* of the design.

The invariant subspace $\mathcal{L}_{2u}^{(2\alpha_u)}$ has the following rational basis:

Table 1. Irreducible representation of $\mathcal{R}$ of a regular GD design

Representation	Mapping	Multiplicities
$\mathcal{R}_G^{(1)}$	$\boldsymbol{I}\to 1,\ \boldsymbol{G}\to N,\ \boldsymbol{B}\to k,\ \boldsymbol{T}_1\to rn_1,\ \boldsymbol{T}_2\to rn_2,\ \boldsymbol{T}\to r$	1
$\mathcal{R}_D^{(1)}$	$\boldsymbol{I}\to 1,\ \boldsymbol{G}\to 0,\ \boldsymbol{B}\to 0,\ \boldsymbol{T}_1\to 0,\ \boldsymbol{T}_2\to 0,\ \boldsymbol{T}\to 0$	$N-b-v+1$
$\mathcal{R}_1^{(1)}$	$\boldsymbol{I}\to 1,\ \boldsymbol{G}\to 0,\ \boldsymbol{B}\to k,\ \boldsymbol{T}_1\to 0,\ \boldsymbol{T}_2\to 0,\ \boldsymbol{T}\to 0$	$b-v$
$\mathcal{R}_1^{(2)}$	$\boldsymbol{I}\to\begin{Vmatrix}1&0\\0&1\end{Vmatrix},\ \boldsymbol{G}\to\begin{Vmatrix}0&0\\0&0\end{Vmatrix},\ \boldsymbol{B}\to\begin{Vmatrix}0&0\\1&k\end{Vmatrix},$ $\boldsymbol{T}_1\to\begin{Vmatrix}rz_{11}&\rho_1 z_{11}\\0&0\end{Vmatrix},\ \boldsymbol{T}_2\to\begin{Vmatrix}rz_{12}&\rho_1 z_{12}\\0&0\end{Vmatrix},$ $\boldsymbol{T}\to\begin{Vmatrix}r&\rho_1\\0&0\end{Vmatrix}$	$\alpha_1=l-1$
$\mathcal{R}_1^{(2)}$	$\boldsymbol{I}\to\begin{Vmatrix}1&0\\0&1\end{Vmatrix},\ \boldsymbol{G}\to\begin{Vmatrix}0&0\\0&0\end{Vmatrix},\ \boldsymbol{B}\to\begin{Vmatrix}0&0\\0&k\end{Vmatrix},$ $\boldsymbol{T}_1\to\begin{Vmatrix}rz_{21}&\rho_2 z_{21}\\0&0\end{Vmatrix},\ \boldsymbol{T}_2\to\begin{Vmatrix}rz_{22}&\rho_2 z_{22}\\0&0\end{Vmatrix},$ $\boldsymbol{T}\to\begin{Vmatrix}r&\rho_2\\0&0\end{Vmatrix}$	$\alpha_2=l(n-1)$

$$\begin{aligned} \boldsymbol{T}\boldsymbol{x}_\alpha^{(u)} &= r\boldsymbol{x}_\alpha^{(u)}, \quad \boldsymbol{B}\boldsymbol{x}_\alpha^{(u)} = \boldsymbol{y}_\alpha^{(u)}, \\ \boldsymbol{T}\boldsymbol{y}_\alpha^{(u)} &= \rho_u\boldsymbol{x}_\alpha^{(u)}, \quad \boldsymbol{B}\boldsymbol{y}_\alpha^{(u)} = k\boldsymbol{y}_\alpha^{(u)}. \end{aligned} \qquad \alpha = 1, \ldots, \alpha_u\,;\ u = 1, 2\,.$$

Since

$$\boldsymbol{\varphi}_\alpha^{(u)} = \frac{\rho_u}{r}\boldsymbol{x}_\alpha^{(u)} - \boldsymbol{y}_\alpha^{(u)}, \qquad \alpha = 1, \ldots, \alpha_u\,,$$

are characteristic vectors of $\boldsymbol{T}$ corresponding to the root zero, they must be orthogonal to $\boldsymbol{x}_\alpha^{(u)}$, hence

$$\boldsymbol{x}_\alpha^{(u)\prime}\boldsymbol{y}_\beta^{(u)} = \frac{\rho_u}{r}\boldsymbol{x}_\alpha^{(u)\prime}\boldsymbol{x}_\beta^{(u)}\,.$$

Similarly, since $\boldsymbol{\phi}_\alpha^{(u)} = k\boldsymbol{x}_\alpha^{(u)} - \boldsymbol{y}_\alpha^{(u)}$ must be orthogonal to $\boldsymbol{y}_\alpha^{(u)}$, one gets

$$\boldsymbol{y}_\alpha^{(u)\prime}\boldsymbol{y}_\beta^{(u)} = k\boldsymbol{x}_\alpha^{(u)\prime}\boldsymbol{y}_\beta^{(u)}\,.$$

For a regular and symmetrical GD design, the linear representation $\mathscr{R}_1^{(1)}$ does not appear.

Let us put

$$\boldsymbol{A}^{(u)} \equiv \| a_{\alpha\beta}^{(u)} \| \quad \text{where} \quad a_{\alpha\beta}^{(u)} = \boldsymbol{x}_\alpha^{(u)\prime}\boldsymbol{x}_\beta^{(u)}\,, \qquad \alpha, \beta = 1, \ldots, \alpha_u\,,$$

then

$$\| \boldsymbol{y}_\alpha^{(u)\prime}\boldsymbol{y}_\beta^{(u)} \| = \rho_u\boldsymbol{A}^{(u)}\,.$$

Since

$$\boldsymbol{A}_1^* \equiv ln\boldsymbol{A}_1^* = l\boldsymbol{G}_n \times \boldsymbol{I}_l - \boldsymbol{G}_{ln}\,,$$

$$\boldsymbol{A}_2^* = n\boldsymbol{A}_2^* = n\boldsymbol{I}_{ln} - \boldsymbol{G}_n \times \boldsymbol{I}_l\,,$$

there are $\alpha_1 + \alpha_2 = ln - 1$ linearly independant rational vectors of $v = ln$ dimensions givenn as follows: Each of the vectors consists of l n-dimensional subvectors. For $s = 1, 2, \ldots, l-1$, the vector $\boldsymbol{a}_s^{(1)}$ has -1 in all positions except those of the sth subvector, which has $l-1$ in all positions. For $s = 1, 2, \ldots, l-1$; $t = 1, 2, \ldots, n-1$, the vector $\boldsymbol{a}_{(s-1)(n-1)+t}^{(2)}$ has zero in all positions except those of the sth subvector, which has $n-1$ in the tth position and -1 in the remaining $n-1$ positions.

It can then be seen that

$$\| \boldsymbol{a}_\alpha^{(1)\prime} \boldsymbol{a}_\beta^{(1)} \| = nl^2 \boldsymbol{I}_{l-1} - ln\boldsymbol{G}_{l-1} ,$$

$$\| \boldsymbol{a}_\alpha^{(2)\prime} \boldsymbol{a}_\beta^{(2)} \| = (n^2 \boldsymbol{I}_{n-1} - n\boldsymbol{G}_{n-1}) \times \boldsymbol{I}_l .$$

One may take as

$$\boldsymbol{x}_\alpha^{(u)} = \boldsymbol{\Phi a}_\alpha^{(u)} \quad \text{and} \quad \boldsymbol{y}_\alpha^{(u)} = \boldsymbol{\Psi N' a}_\alpha^{(u)} , \qquad \alpha = 1, \ldots, \alpha_u : \ u = 1, 2 .$$

Hence one gets

$$| \boldsymbol{A}^{(1)} | \sim r^{l-1} l n^{l-1} \quad \text{and} \quad | \boldsymbol{A}^{(2)} | \sim r^{l(n-1)} n^l .$$

If the symmetrical GD design under consideration does exist, by suitable relabeling of the N experimental units, two matrices $\boldsymbol{T}$ and $\boldsymbol{B}$ can be brought into the same matrix $\boldsymbol{G}_r \times \boldsymbol{I}_v$; in order words, there exists a permutation σ of the N experimental units such that

$$\boldsymbol{S}_\sigma \boldsymbol{B} \boldsymbol{S}_\sigma' = \boldsymbol{T} ,$$

where $\boldsymbol{S}_\sigma$ is the permutation matrix of order N. Thus $\boldsymbol{S}_\sigma \boldsymbol{y}_\alpha^{(u)}$ are the characteristic vectors of $\boldsymbol{S}_\sigma \boldsymbol{B} \boldsymbol{S}_\sigma'$ corresponding to the characteristic root r, and

$$\| \boldsymbol{y}_\alpha^{(u)\prime} \boldsymbol{S}_\sigma \boldsymbol{S}_\sigma \boldsymbol{y}_\beta^{(u)} \| = \| \boldsymbol{y}_\alpha^{(u)\prime} \boldsymbol{y}_\beta^{(u)} \| \sim \rho_u \boldsymbol{A}^{(u)} .$$

Since $\boldsymbol{x}_\alpha^{(u)}$, $\alpha = 1, \ldots, \alpha_u$: $u = 1, 2$ and $\boldsymbol{S}_\sigma \boldsymbol{y}_\alpha^{(u)}$, $u = 1, 2$ generate the same proper subspace corresponding to the characteristic root r of a symmetric matrix, one obtains the relation

$$\left\| \begin{matrix} \boldsymbol{A}^{(1)} & 0 \\ 0 & \boldsymbol{A}^{(2)} \end{matrix} \right\| \sim \left\| \begin{matrix} \rho_1 \boldsymbol{A}^{(1)} & 0 \\ 0 & \rho_2 \boldsymbol{A}^{(2)} \end{matrix} \right\| .$$

Whence one then has

$$\rho_1^{\alpha_1} \rho_2^{\alpha_2} \sim 1$$

and

$$(-1, \rho_1)_p^{l(l-1)/2} (-1, \rho_2)_p^{\{l(n-1)(ln-l+1)\}/2} (\rho_1, n)_p^l (\rho_2, n)^l (\rho_1, \lambda_2)_p = 1$$

for all primes p. The latter is equivalent to the Bose-Connor condition

$$(-1, \rho_1)_p^{l(l-1)/2} (-1, \rho_2)_p^{\{l(n-1)(l+n-1)\}/2} (\rho_1, n)_p^l (\rho_2, n)^l (\rho_1, \lambda_2)_p = 1$$

for all primes p.

Now we examine what will hapen if we carry over this method to a symmetrical singular GD design. A singular and symmetrical GD design is characterized by

$$\rho_0 = r^2 , \quad \rho_1 = r^2 - v\lambda_2 > 0 \qquad \text{and} \qquad \rho_2 = r - \lambda_1 = 0 .$$

One can see

$$\boldsymbol{T}_2^{\sharp}\boldsymbol{B} = \boldsymbol{B}\boldsymbol{T}_2^{\sharp} = \boldsymbol{B}\boldsymbol{T}_2^{\sharp}\boldsymbol{B} = 0 .$$

Thus the relationship algebra $\mathfrak{R}$ of a singular and symmetrical GD design degenerates to an algebra of rank 8, having four inequivalent linear representations and an irreducible representation of the second degree. They are shown in Table 2 together with their multiplicities in $\mathfrak{R}$.

Table 2. Table of irreducible representations of a symmetrical GD design in the singular case

Representation	Mapping	Multiplicities
$\mathfrak{R}_G^{(1)}$	$\boldsymbol{I}\to 1,\ \boldsymbol{G}\to N,\ \boldsymbol{B}\to r,\ \boldsymbol{T}_1\to rn_1,\ \boldsymbol{T}_2\to rn_2,\ \boldsymbol{T}\to r$	1
$\mathfrak{R}_D^{(1)}$	$\boldsymbol{I}\to 1,\ \boldsymbol{G}\to 0,\ \boldsymbol{B}\to 0,\ \boldsymbol{T}_1\to 0,\ \boldsymbol{T}_2\to 0,\ \boldsymbol{T}\to 0$	$N-2v+1$
$\mathfrak{R}_1^{(2)}$	$\boldsymbol{I}\to\begin{Vmatrix}1 & 0\\ 0 & 1\end{Vmatrix},\ \boldsymbol{G}\to\begin{Vmatrix}0 & 0\\ 0 & 0\end{Vmatrix},\ \boldsymbol{B}\to\begin{Vmatrix}0 & 0\\ 1 & r\end{Vmatrix},$ $\boldsymbol{T}_1\to\begin{Vmatrix}rz_{11} & \rho_1 z_{11}\\ 0 & 0\end{Vmatrix},\ \boldsymbol{T}_2\to\begin{Vmatrix}rz_{12} & \rho_1 z_{12}\\ 0 & 0\end{Vmatrix},$ $\boldsymbol{T}\to\begin{Vmatrix}r & \rho_1\\ 0 & 0\end{Vmatrix}$	$\alpha_1=l-1$
$\mathfrak{R}_{23}^{(1)}$	$\boldsymbol{I}\to 1,\ \boldsymbol{G}\to 0,\ \boldsymbol{B}\to 0,\ \boldsymbol{T}_1\to rz_{21},\ \boldsymbol{T}_2\to rz_{22},\ \boldsymbol{T}\to r$	$\alpha_2=l(n-1)$
$\mathfrak{R}_{24}^{(1)}$	$\boldsymbol{I}\to 1,\ \boldsymbol{G}\to 0,\ \boldsymbol{B}\to r,\ \boldsymbol{T}_1\to 0,\ \boldsymbol{T}_2\to 0,\ \boldsymbol{T}\to 0$	

Following a similar argument developed for the regular case, one may take

$$\boldsymbol{x}_\alpha^{(1)} = \boldsymbol{\Phi}\boldsymbol{a}_\alpha^{(1)} , \quad \boldsymbol{y}_\alpha^{(1)} = \boldsymbol{\Psi}\boldsymbol{N}'\boldsymbol{a}_\alpha^{(1)} , \qquad \alpha = 1,\ldots,\alpha_1$$

and

$$\boldsymbol{x}_\alpha^{(2)} = \boldsymbol{\Phi}\boldsymbol{a}_\alpha^{(2)} , \quad \boldsymbol{y}_\alpha^{(2)} = \boldsymbol{\Psi}\boldsymbol{a}_\alpha^{(2)} , \qquad \alpha = 1,\ldots,\alpha_2 .$$

Hence one gets the congruence relation

$$\begin{Vmatrix} A^{(1)} & 0 \\ 0 & A^{(2)} \end{Vmatrix} \sim \begin{Vmatrix} \rho_1 A^{(1)} & r \| a_\alpha^{(1)\prime} N a_\beta^{(2)} \| \\ r \| a_\beta^{(2)\prime} N' a_\alpha^{(1)} \| & A^{(2)} \end{Vmatrix}.$$

Since the right-hand side of the above expression is congruent rationally to the matrix

$$\begin{Vmatrix} \rho_1 A^{(1)} & 0 \\ 0 & K \end{Vmatrix},$$

where

$$K = A^{(2)} - \frac{r^2}{\rho_1} \| a_\beta^{(2)\prime} N' a_\alpha^{(1)} \| \cdot A^{(1)-1} \cdot \| a_\alpha^{(1)\prime} N a_\beta^{(2)} \|,$$

$$A^{(1)-1} = \frac{1}{ln}\left[\frac{1}{l} I_{l-1} + G_{l-1}\right],$$

one has to write down the matrix K completely in order to get the necessary condition in terms of the Hasse-Minkowski p-invariant. Thus one is led to the conclusion that the knowledge of the block structure of the design is necessary in one way or another in order to get the necessary condition in the case of singular GD design. In fact, since

$$\| a_1^{(1)} \ldots a_{\alpha_1}^{(1)} \| \, A^{(1)-1} \begin{Vmatrix} a_1^{(1)\prime} \\ \cdot \\ \cdot \\ \cdot \\ a_{\alpha_1}^{(1)\prime} \end{Vmatrix} = \frac{1}{r^2} A_1^{*},$$

one has to write down the matrix $N' A_1^{*} N$.

References

1. Bose, R. C. and Connor, W. S. "Combinatorial Properties of Group Divisible Incomplete Block Designs," *Ann. Math. Statist.*, **23** (1952), 367–383.
2. Bose, R. C. and Mesner, D. M. "On Linear Associative Algebras Corresponding to Association Schemes of Partially Balanced Designs," *Ann. Math. Statist.*, **30** (1959), 21–38.
3. Bruck, R. H. and Ryser, H. J. "Non-Existence of Certain Finite Projective Planes," *Canad. J. Math.*, **1** (1949), 88–93.

4. Connor, W. S., Jr. "On the Structure of Balanced Incomplete Block Designs," *Ann. Math. Statist.*, **23** (1952), 57-71.
5. Fisher, R. A. "An Examination of the Different Possible Solutions of a Problem in Incomplete Blocks," *Annals of Eugenics*, **10** (1940), 52-75.
6. Fisher, R. A. and Yates, F. *Statistical Tables for Biological, Agricultural and Medical Research*, Hafner Publishing Co., New York, 1949.
7. Guerin, R. "Vue d'Ensemble sur les Plans en Blocs Incomplets Equilibrés et Partiellement Equilibrés," *Rev. Inst. Internat. Statist.*, **33** (1965), 24-58.
8. Hanani, Haim. "The Existence and Construction of Balanced Incomplete Block Designs," *Ann. Math. Statist.*, **32** (1961), 361-386.
9. Hasse, H. "Uber die Aquivalenz quadratischer Formen im Körper rationalen Zahlen," *J. Reine Angew. Math.*, **152** (1923), 205-224.
10. Hinkelmann, K. "Extended Group Divisible Partially Balanced Incomplete Block Designs," *Ann. Math. Statist.*, **35** (1964), 681-695.
11. Husain, Q. M. "Symmetrical Incomplete Block Designs with $\lambda = 2$, $k = 8$ or 9," *Bull. Calcutta Math. Soc.*, **37** (1945), 115-123.
12. Husain, Q. M. "Impossibility of the Symmetrical Incomplete Block Design with $\lambda = 2$, $k = 7$," *Sankhyā*, **7** (1964), 317-322.
13. Jones, B. W. *The Arithmetic Theory of Quadratic Forms*, Carus Math. Monograph, No. 10, John Wiley and Sons, New York, 1950.
14. Kusumoto, K. "A Necessary Condition for Existence of Regular and Symmetrical PBIB Designs of T_3 Type," *Ann. Inst. Statist. Math.*, **17** (1965), 149-165.
15. Ogawa, J. "A Necessary Condition for Existence of Regular and Symmetrical Experimental Designs of Triangular Type, with Partially Balanced Incomplete Blocks," *Ann. Math. Statist.*, **30** (1959), 1963-1071.
16. Ogawa, J. "On a Unified Method of Deriving Necessary Conditions for Existence of Symmetrical Partially Balanced Incomplete Block Designs of Certain Types," *Bull. Inst. Internat. Statist.*, **38** (1961), IV, 43-57.
17. Ogawa, J. and Ishii, G. "The Relationship Algebra and Analysis of Variance of a Partially Balanced Incomplete Block Design," *Ann. Math. Statist.*, **36** (1965), 1815-1828.
18. Ogasawara, M. "A Necessary Condition for Existence of Regular and Symmetrical PBIBD of T_m Type," Institute of Statistics, Mimeo Series No. 418, University of North Carolina, 1965.
19. Raghavarai, D. "A Generalization of Group Divisible Designs," *Ann. Math. Statist.*, **31** (1960), 756-771.

20. Raghavarao, D. and Chandrasekharao, K. "Cubic Designs," *Ann. Math. Statist.*, **35** (1964), 389-397.
21. Seiden, Esther. "On Necessary Conditions for the Existence of Some Symmetrical and Unsymmetrical Triangular PBIB Designs," *Ann. Math. Statist.*, **34** (1963), 348-351.
22. Shrikhande, S. S. "The Impossibility of Certain Symmetrical Balanced Incomplete Block Designs," *Ann. Math. Statist.*, **21** (1950), 106-111.
23. Shrikhande, S. S. "Impossibility of Some Affine Resolvable Balanced Incomplete Block Designs," *Sankhyā*, **11** (1951), 185-186.
24. Shrikhande, S. S. "The Non-Existence of Certain Affine Resolvable Balanced Incomplete Block Designs," *Canad. J. Math.*, **5** (1953), 413-420.
25. Shrikhande, S. S. "The Uniqueness of the L_2 Association Scheme, *Ann. Math. Statist.*, **30** (1959), 781-789.
26. Shrikhande, S. S. "Relations between Certain Incomplete Block Designs," *Contributions to Probability and Statistics, Essays in Honor of Harold Hotelling*, Stanford University Press, 1960, pp. 388-395.
27. Shrikhande, S. S., Raghavarao, D. and Tharthare, S. K. "Non-Existence of Some Unsymmetrical PBIB Designs," *Canad. J. Math.*, **15** (1963), 686-701.
28. Singh, N. K. and Shukla, G. C. "Non-Existence of Some PBIBD," *J. Indian Statist. Assoc.*, **1** (1961), 71-78.
29. Tharthare, S. K. "Right Angular Designs," *Ann. Math. Statist.*, **34** (1963), 1057-1067.
30. Vartak, M. N. "On the Hasse-Minkowski Invariant of the Kronecker Product of Matrices," *Canad. J. Math.*, **10** (1958), 66-72.
31. Vartak, M. N. "The Non-Existence of Certain PBIB Designs," *Ann. Math. Statist.*, **30** (1959), 1051-1062.
32. Yamamoto, K. "A Necessary Condition for Existence of Partially Balanced Incomplete Block Designs with an m-Subset Association Scheme," *Mem. Fac. Sci. Kyushu Univ., Ser. A*, **19** (1965), 76-98.
33. Yamamoto, K. "On an Orthogonal Basis of the Eigenspaces Associated with Partially Balanced Incomplete Block Designs of a Latin Square Type Association Scheme," *Mem. Fac. Sci. Kyushu Univ., Ser. A*, **19** (1965), 99-104.

Discussion on Professor Ogawa's Paper

DR. S. IKEDA: My feeling is that the method of proving the non-existence of regular and symmetric designs which has been given in this paper is extremely powerful. Since, by using the inflation and replication method, one can derive an m-class, singular, symmetric, PBIB design from any given $(m-1)$-class PBIB design, it is strongly desirable to extend the method to singular cases. In this connection, one direction has been suggested in the paper, which shows that some information on the block structure of the design under consideration would be required for extending the method.

Let us look at this point in more detail for a class of asymmetric BIB designs: From a BIB design with parameters

$$(1) \qquad v, k, b, r, \lambda = 1,$$

one can construct a singular and symmetric PBIB design of group divisible type with parameters

$$(2) \qquad v^* = b^* = sv, \quad r^* = k^* = sk, \quad \lambda_1^* = tr, \quad \lambda_2^* = t,$$

where $s = b/(v, b)$, $t = v/(v, b)$, and (v, b) stands for the G.C.D. of v and b. Let $\boldsymbol{A}_i$, $i = 0, 1, 2$, be the association matrices of the association scheme of group divisible type defined on the new v^* treatments, and let $\boldsymbol{A}_i^\sharp$, $i = 0, 1, 2$, be the orthogonal idempotent matrices in the association algebra generated by the $\boldsymbol{A}_i$'s.

It is well-known that the b blocks of the BIB design (1) have a 2-class association sheme called the SLB association. Let B_i, $i = 0, 1, 2$, be the association matrices of this association.

For the incidence matrix $\boldsymbol{N}$ of the PBIB design (2), it is seen that

$$(3) \qquad \boldsymbol{N}'\boldsymbol{A}_1^\sharp\boldsymbol{N} = sk\boldsymbol{G}_t \times \left(\boldsymbol{I}_b - \frac{k}{v}\boldsymbol{G}_b + s\boldsymbol{B}_1\right),$$

where $\boldsymbol{G}_d$ denotes the $d \times d$ matrix whose elements are all unity. From this equation, we can see that all the information on $\boldsymbol{N}'\boldsymbol{A}_1^\sharp\boldsymbol{N}$ is obtained from the block structure of the original design (1) through the SLB association scheme on the blocks.

Hence, so far as the non-existence problem of a special class (1) of BIB designs is concerned, it might be more hopeful to investigate the construction and non-existence problem of the SLB association scheme itself in general.

CHAPTER 13

Calculus of Blocks Applied to PBIB Designs

MOTOSABURO MASUYAMA, *The Catholic University of America and Meteorological College, Japan*

We have tried thus far to formulate the method of differences in modules, [6], groups [8] and semi-groups [9] for constructing PBIB designs. To include all existing triangular designs in our formulation we have introduced the fractional development of initial blocks [9, 11] which suggests the possibility of removing any algebraic structure among elements. This set-theoretic approach is stated in our previous papers [12, 13, 14].

In Section 1 of this paper we introduce two new concepts: the product of two set-theoretic blocks and the block which contains objects with negative frequencies. In the following sections this symbolic method is applied to the theory of PBIB designs. We will show in the last section that the matrix theory given by Thompson [19] and Bose and Mesner [2] is a representation theory of our calculus.

1. INTRODUCTION

Let S be a set of n distinct objects (called either varieties, treatments, or blocks), say

$$(1.1) \qquad s_1, s_2, \ldots, s_n .$$

A block A of size m is a set of m objects of S, say

$$(1.2) \qquad a_1, a_2, \ldots, a_m ,$$

which are not necessarily distinct. It is denoted by

$$(1.3) \qquad A = \{a_1, a_2, \ldots, a_m\} ,$$

in which the order of presenting objects is immaterial. m is called its size and denoted by $s(A)$.

A block which contains all the objects of S exactly once is denoted by $\boldsymbol{S}$.

If the block A of (1.3) contains an objects s_i of S exactly f_i times, A may be denoted by a formal sum

$$(1.4) \qquad f_1 s_1 + f_2 s_2 + \ldots + f_n s_n ,$$

or a vector

$$(1.5) \qquad (f_1, f_2, \ldots, f_n) .$$

If two blocks A and B contain an object s_i of S exactly f_{ai} and f_{bi} times respectively, their sum $A + B$ is defined by a block which contains s_i of S exactly $(f_{ai} + f_{bi})$ times. Hence we have

$$(1.6) \qquad s(A + B) = s(A) + s(B) .$$

The sum is commutative and associative.

If $f_{ai} \leqq f_{bi}$ for all i, we denote it by

$$(1.7) \qquad A \subset B \quad \text{or} \quad B \supset A .$$

If

$$(1.8) \qquad A \subset B \quad \text{and} \quad A \supset B$$

hold at a time, it is denoted by

$$(1.9) \qquad A = B ,$$

which means $f_{ai} = f_{bi}$ for all i.

If $f_{bi} \geqq f_{ai}$ for all i, the difference $B - A$ is defined as a block which contains s_i of S exactly $(f_{bi} - f_{ai})$ times. It is sometimes convenient to remove the restriction on frequencies in this definition. In such a case a block A is defined as a formal sum (1.4), $f_1, f_2, \ldots, f_n$ being elements of a ring of rational integers Z.[1]

The null block ϕ is defined by

$$A + \phi = \phi + A = A \tag{1.10}$$

for any block A including ϕ itself. We have

$$s(\phi) = 0\,. \tag{1.11}$$

If $f_{bi} = cf_{ai}$ for a non-zero integer c and for all i, B is denoted by cA and A is denoted by B/c.

$0A$ is defined to be ϕ for any block A.

A sum of blocks

$$B_{a_1} + B_{a_2} + \ldots + B_{a_m} \tag{1.12}$$

is denoted by

$$\sum_{j=1}^{m} B_{a_j}\,. \tag{1.13}$$

If we denote a block of these indices a_j by A, then (1.13) may be denoted by

$$\sum_{a \in A} B_a\,, \tag{1.14}$$

or

$$AB_a\,. \tag{1.15}$$

If the size of B_{a_j} is independent of its index a_j, then we have

$$s(AB_a) = mk = s(A)s(B_a)\,, \tag{1.16}$$

with $k = s(B_{a_j})$.

The intersection $A \cap B$ is a block which contains s_i of S exactly min (f_{ai}, f_{bi}) times, f_{ai} and f_{bi} being non-negative integers.

2. PBIB(m) ASSOCIATION SCHEME

Given a set S of v distinct objects, $1, 2, \ldots, v$ a relation

[1] We do not use this convention except when explicitly stated. If $f_{ai} \geqq 0$, we may denote it by $A \geqq \phi$.

satisfying the following conditions is called an *association scheme with m classes* [3, 4].

(a) For any object u of S there are $m+1$ blocks

$$D_{u0}, D_{u1}, \ldots, D_{um} \qquad (m > 0) \tag{2.1}$$

such that

$$D_{u0} = \{u\} \tag{2.2}$$

and

$$\sum_{a=0}^{m} D_{ua} = S . \tag{2.3}$$

(2.3) implies that these $m+1$ blocks are mutually disjoint. Any object w of D_{ua} is said to be an ath associate of u. It may be denoted by

$$u \underset{a}{\sim} w .$$

(b) If $w \in D_{ua}$, then $u \in D_{wa}$, i.e., the relation of association is symmetric. But it is, in general, not transitive.

This relation may be denoted by two vertices u and w joined by an edge of the ath color in a linear graph.

(c) $s(D_{ua}) = n_a$ is independent of the initial u.

From (2.2) we have $n_0 = 1$.

(d) If $w \in D_{ua}$, then

$$s(D_{ub} \cap D_{wc}) = p^a_{bc} \tag{2.4}$$

is independent of the initial pair (u, w).

$$p^a_{bc} = p^a_{cb} . \tag{2.5}$$

From (2.4) or its linear graph we have

$$s(D_{ub} \cap D_{uc}) = p^0_{bc} = s(D_{ub})\delta_{bc} = n_b\delta_{bc} \tag{2.6}$$

and

$$s(D_{ub} \cap D_{w0}) = p^a_{b0} = s(D_{ub} \cap \{w\}) = \delta_{ab} . \tag{2.7}$$

Again from (2.4) we obtain

(2.8) $$\sum_{x \in D_{ub}} D_{xc} = \sum_{a=0}^{m} p_{bc}^{a} D_{ua} .$$

We now prove this. Let us fix an object w of D_{ua}. Then there exist exactly p_{bc}^{a} objects x which are bth associates of u and cth associates of w at a time. We note that w is contained in D_{xc} when and only when x is a cth associate of w and w is contained in D_{xc} at most once by (2.3). Hence (2.8) follows.

We may write

(2.9) $$D_{ub} D_{xc} = \sum_{a=0}^{m} p_{bc}^{a} D_{ua}$$

in place of (2.8) by (1.15).

If any object w of S is fixed, there are exactly n_a objects u which are ath associates of w. Hence by the symmetry in the relation of association w appears exactly n_a blocks D_{ua}, or

(2.10) $$\sum_{u \in S} D_{ua} = S D_{ua} = n_a S .$$

This is a generalization of the absorption law in [6].

3. PBIB(m) DESIGNS

If an association scheme with m classes is given, then we get a PBIB design with r replications and b blocks based on this association scheme, if we can arrange the v objects into b blocks,

(3.1) $$N_1, N_2, \ldots, N_b$$

such that

(i)

(3.2) $$N_i \subset S$$

and

(3.3) $$s(N_i) = k$$

independently of i;

(ii)

(3.4) $$\sum_{i=1}^{b} N_i = rS .$$

(iii)

$$\sum_{i \in I_u} N_i = I_u N_i = \sum_{a=0}^{m} \lambda_a D_{ua} , \tag{3.5}$$

where I_u is a set of indices i such that N_i contains u of S. λ_a is independent of u.

This is our version of the definition given in [1, 3, 4, 17].

We note that the condition (ii) is redundant. Suppose an object u of S appears exactly in r_u blocks of (3.1). Then we must have from (3.5)

$$I_u N_i = r_u\{u\} + \sum_{a=1}^{m} \lambda_a D_{ua} . \tag{3.6}$$

Taking the size of (3.6) we obtain

$$r_u k = r_u + \sum_{a=1}^{m} \lambda_a n_a , \tag{3.7}$$

or

$$r_u(k-1) = \sum_{a=1}^{m} \lambda_a n_a . \tag{3.8}$$

Since λ_a, n_a and k are all independent of u, r_u must be independent of u. The case $m = 1$ is stated in [18].

4. RELATIONS AMONG THE PARAMETERS

Most relations among the parameters are obtained as size relations of defining equations.

From (2.3) we have

$$\sum_{a=0}^{m} n_a = v . \tag{4.1}$$

From (2.8) we have

$$n_b n_c = \sum_{a=0}^{m} p_{bc}^{a} n_a . \tag{4.2}$$

From (3.4) we have

$$bk = rv . \tag{4.3}$$

From (3.5) or (3.8) we have

(4.4)
$$r(k-1) = \sum_{a=1}^{m} \lambda_a n_a \,.$$

If u is an ath associate of w, then from (2.3) we obtain

(4.5)
$$\boldsymbol{D}_{wb} \cap \boldsymbol{S} = \boldsymbol{D}_{wb} = \sum_{c=0}^{m} \boldsymbol{D}_{wb} \cap \boldsymbol{D}_{uc} \,.$$

The corresponding size relation is

(4.6)
$$n_b = \sum_{c=0}^{m} p_{bc}^{a} \,.$$

A more elaborate formula [2] is obtained from (2.5) and (2.8), i.e.,

(4.7)
$$\sum_{b=0}^{m} p_{de}^{b} p_{bc}^{a} = \sum_{f=0}^{m} p_{dc}^{f} p_{fe}^{a} \,.$$

To prove this we need a lemma:

(4.8)
$$\boldsymbol{D}_{uc} \boldsymbol{D}_{wb} = \boldsymbol{D}_{ub} \boldsymbol{D}_{tc} \,,$$

which is easily obtained from (2.5) and (2.8). This means that the totality of objects x which are bth assaciates of w, w being a cth associate of u, coincides with those of objects y which are cth associates of t, t being a bth associate of u. This may be denoted by

(4.9)
$$u \underset{b}{\sim} t \underset{c}{\sim} y \equiv x \underset{b}{\sim} w \underset{c}{\sim} u \,.$$

Now

(4.10)
$$\begin{aligned} \sum_{b=0}^{m} \sum_{a=0}^{m} p_{de}^{b} p_{bc}^{a} \boldsymbol{D}_{uc} &= \sum_{b=0}^{m} p_{de}^{b} \boldsymbol{D}_{uc} \boldsymbol{D}_{wb} \\ &= \boldsymbol{D}_{uc} \sum_{c=0}^{m} p_{de}^{b} \boldsymbol{D}_{wb} \\ &= \boldsymbol{D}_{uc} \boldsymbol{D}_{we} \boldsymbol{D}_{td} \,. \end{aligned}$$

This is the totality of objects x such that

(4.11)
$$u \underset{c}{\sim} w \underset{e}{\sim} t \underset{d}{\sim} x \,.$$

By (4.9) we have

(4.12)
$$u \underset{e}{\sim} v \underset{c}{\sim} s \underset{d}{\sim} y \equiv x \,.$$

Hence we obtain

$$(4.13)\qquad (4.10) = \sum_{f=0}^{m} \sum_{d=0}^{m} p_{dc}^{f} p_{fe}^{a} D_{ua} .$$

Since $D_{u0}, D_{u1}, \ldots, D_{um}$ are disjoint, we must have (4.7). The same equality is obtained by counting directly the number of certain circuits which consist of four edges in the linear graph of the association scheme [10].

By setting $a = 0$ in (4.7), we have

$$(4.14)\qquad \sum_{b=0}^{m} p_{de}^{b} \delta_{bc} n_b = p_{de}^{c} n_c = \sum_{f=0}^{m} p_{dc}^{f} \delta_{fe} n_e = p_{dc}^{e} n_e .$$

If we apply this to (4.2), it becomes trivial.

5. PBIB DESIGNS ARISING FROM AN ASSOCIATION SCHEME

If a is fixed for which $n_a \geq 2$,

$$(5.1)\qquad D_{1a}, D_{2a}, \ldots, D_{va}$$

constitute a symmetric PBIB design [13].

To prove this statement we need only to check (i)–(iii) in Section 3. From 2.3

$$(\mathrm{i})\qquad D_{ua} \subset S$$

and (c) says that

$$s(D_{ua}) = n_a$$

is independent of u. From (2.10)

$$(\mathrm{ii})\qquad SD_{ua} = n_a S .$$

As a special case of (2.8) we have

$$(\mathrm{iii})\qquad D_{ua} D_{wa} = \sum_{b=0}^{m} p_{aa}^{b} D_{ub}$$

$$(5.2)\qquad = r_0\{u\} + \sum_{b=1}^{m} \lambda_{0b} D_{ub} ,$$

with $r_0 = n_a$ and $\lambda_{0b} = p_{aa}^{b}$.

The association scheme is either the same as the original one or its degenerate form. Let us consider the simplest case in which only the amalgamation of $\boldsymbol{D}_{u1}$ and $\boldsymbol{D}_{u2}$ occurs. It can happen only when

$$(5.3) \qquad \lambda_{01} = \lambda_{02} \quad \text{or} \quad p^1_{aa} = p^2_{aa} \,.$$

(i) From

$$(5.4) \qquad (\boldsymbol{D}_{u1} + \boldsymbol{D}_{u2})(\boldsymbol{D}_{w1} + \boldsymbol{D}_{w2}) = \sum_{b=0}^{m} (p^b_{11} + 2p^b_{12} + p^b_{22})\boldsymbol{D}_{ub}$$

we must have

$$(5.5) \qquad p^1_{11} + 2p^1_{12} + p^1_{22} = p^2_{11} + 2p^2_{12} + p^2_{22} \,.$$

(ii) From

$$(5.6) \qquad (\boldsymbol{D}_{u1} + \boldsymbol{D}_{u2})\boldsymbol{D}_{wc} = \sum_{b=0}^{m} (p^b_{1c} + p^b_{2c})\boldsymbol{D}_{ub} \,, \qquad c \neq 1, 2 \,,$$

we must have

$$(5.7) \qquad p^1_{1c} + p^1_{2c} = p^2_{1c} + p^2_{2c} \,,$$

for $c = 0, 3, 4, \ldots, m$.

(iii) From

$$(5.8) \qquad \boldsymbol{D}_{uc}\boldsymbol{D}_{wd} = \sum_{b=0}^{m} p^b_{cd}\boldsymbol{D}_{ub} \,, \qquad c, d \neq 1, 2 \,,$$

we must have

$$(5.9) \qquad p^1_{cd} = p^2_{cd}$$

for $c, d = 0, 3, 4, \ldots, m$.

It is clear that (5.3), (5.5), (5.7) and (5.9) are the sufficient conditions for the amalgamation of $\boldsymbol{D}_{u1}$ and $\boldsymbol{D}_{u2}$ only.

If $a \neq b$ $(m \geq 2)$,

$$(5.10) \qquad \boldsymbol{D}_{1a} + \boldsymbol{D}_{1b} \,,\; \boldsymbol{D}_{2a} + \boldsymbol{D}_{2b} \,,\; \ldots \,,\; \boldsymbol{D}_{va} + \boldsymbol{D}_{vb}$$

constitutes a PBIB design. The association scheme is either the same as the original one or its degenerate scheme. The method of proof is in the same vein. We may add more than two in the same manner.

6. AN APPLICATION TO PBIB DESIGNS OF PARTIAL GEOMETRIC TYPE

A partial geometry with parameters (ρ, κ, t) has been introduced by R. C. Bose [3, 4] as a generalization of a net. It consists of v distinct points and b distinct lines which satisfy the following incidence relations.

A.1. Any two distinct points are incident with at most one line.

A.2. Each point is incident with exactly ρ distinct lines, $\rho \geq 2$.

A.3. Each line is incident with exactly κ points, $\kappa \geq 2$.

A.4. If the point P is not incident with the line L, there are exactly t lines which are incident with P and also incident with some point incident with L, $t \geqslant 1$.

We may use the usual terminology "contain" or "be contained" instead of "be incident with".

As is pointed out by Mesner [16], a group divisible association scheme with m groups of n objects can be regarded as a trivial case of a partial geometry, the groups comprising the lines, i.e., $\rho = 1$, $\kappa = n$, $t = 0$. This suggests to us that the geometric method used in proving the Bose-Connor inequality for PBIB designs of GD scheme [15] can be applied to a PBIB design of partial geometry type with some modification.

Let L_a be any fixed line in a partial geometry with given parameters which is incident with a point a. Then we obtain

$$\text{(6.1)} \qquad \boldsymbol{L}_a\boldsymbol{D}_{u1} = t(\boldsymbol{S} - \boldsymbol{L}_a) + (\kappa - 1)\boldsymbol{L}_a\,,$$

which will be proved presently.

Consider a point z which is not incident with L_a. Then by **A.4** there exists exactly t lines which are incident with z and t distinct points of L_a. Hence each point of $(\boldsymbol{S} - \boldsymbol{L}_a)$ is contained in $\boldsymbol{L}_a\boldsymbol{D}_{u1}$ t times. Each point of L_a appears $(\kappa - 1)$ times in it by **A.3**. This completes the proof.

Similarly, we have

$$\text{(6.1}'\text{)} \qquad \boldsymbol{L}_a\boldsymbol{D}_{u2} = (\kappa - t)(\boldsymbol{S} - \boldsymbol{L}_a)$$

and

$$\text{(6.1}''\text{)} \qquad \boldsymbol{D}_a\{u\} = \boldsymbol{L}_a\,.$$

Now from (2.3) and (3.6) we get

$$\begin{aligned}(6.2)\qquad \boldsymbol{I}_u\boldsymbol{N}_i &= r\{u\} + \lambda_1\boldsymbol{D}_{u1} + \lambda_2\boldsymbol{D}_{u2}\\ &= (r-\lambda_2)\{u\} + (\lambda_1-\lambda_2)\boldsymbol{D}_{u1} + \lambda_2\boldsymbol{S}\,,\end{aligned}$$

in which we may require the convention defined by (1.4) with $f_i \in Z$, since $\lambda_1 - \lambda_2$ may be negative.

Then

$$\begin{aligned}(6.3)\qquad \boldsymbol{L}_a\boldsymbol{I}_u\boldsymbol{N}_i &= (r-\lambda_2)\boldsymbol{L}_a + (\lambda_1-\lambda_2)[t(\boldsymbol{S}-\boldsymbol{L}_a) + (\kappa-1)\boldsymbol{L}_a]\\ &\qquad + \lambda_2\kappa\boldsymbol{S}\\ &= [r+\lambda_1(\kappa-1)]\boldsymbol{L}_a + [(\lambda_1-\lambda_2)t + \lambda_2\kappa](\boldsymbol{S}-\boldsymbol{L}_a)\,.\end{aligned}$$

The result can be obtained by counting directly the frequencies of an object of L_a and an object of any other line L_b contained in the left hand side of (6.3), which are equal to $r + \lambda_1(\kappa - 1)$ and $\lambda_1 t + \lambda_2(\kappa - t)$ respectively.

Let us set

$$(6.4)\qquad \begin{aligned}A &= \kappa[r + \lambda_1(\kappa-1)]\,,\\ B &= \kappa[(\lambda_1-\lambda_2)t + \lambda_2\kappa]\,.\end{aligned}$$

Now let the number of blocks in (3.1) which contain x points of L_a and y points of L_b ($\neq L_a$) together be $f(x, y)$, for which

$$(6.5)\qquad x \geq 0\,, \qquad y \geq 0\,, \qquad x + y \leq k\,.$$

Then the total number of blocks in (6.3) which contain at least a point of L_a is

$$(6.6)\qquad \sum_x \sum_y x f(x, y)\,,$$

the summation being taken over all the possible combinations of x and y under the condition (6.5). Hence the total number of points in (6.3) which are incident with L_a is

$$(6.7)\qquad \sum_x \sum_y x^2 f(x, y)\,,$$

Similarly the total number of points in (6.3) which are incident with L_b is

$$(6.8)\qquad \sum_x \sum_y xy f(x, y)\,.$$

Hence we have

$$A = \sum_x \sum_y x^2 f(x, y) \quad \text{and} \quad B = \sum_x \sum_y xy f(x, y) . \tag{6.9}$$

If possible, let us suppose that $A < B$, then

$$\sum_x \sum_y x^2 f(x, y) < \sum_x \sum_y xy f(x, y) . \tag{6.10}$$

By interchanging the role of a and b, we would have

$$\sum_x \sum_y y^2 f(x, y) < \sum_y \sum_x yx f(x, y) , \tag{6.11}$$

and hence

$$\sum \sum x^2 f(x, y) \sum \sum y^2 f(x, y) < [\sum \sum xy f(x, y)]^2 , \tag{6.12}$$

which violates the Buniakovsky-Cauchy inequality. Thus we obtain

$$A - B \geqslant 0 , \tag{6.13}$$

or

$$\begin{aligned} r + \lambda_1(\kappa - 1) - [(\lambda_1 - \lambda_2)t + \lambda_2\kappa] \\ = r - \lambda_1 + (\lambda_1 - \lambda_2)(\kappa - t) \geqslant 0 . \end{aligned} \tag{6.14}$$

The equality holds when and only when for any two distinct lines L_a and L_b

$$f(x, y) = 0 , \quad \text{if} \quad x \neq y . \tag{6.15}$$

Since $f(x, y)$ denotes the number of blocks which contain x points of L_a and y points of L_b ($\neq L_a$) at a time, each block N_i must contain the same number of points, say c, from each line of the geometry. Let us find the explicit form of c.

From (4.1) and (4.4) we get

$$\begin{aligned} r(k - 1) &= \lambda_1 n_1 + \lambda_2 n_2 \\ &= \lambda_1(v - 1 - n_2) + \lambda_2 n_2 \\ &= \lambda_1 v - \lambda_1 + (\lambda_2 - \lambda_1)n_2 , \end{aligned} \tag{6.16}$$

or

$$rk - \lambda_1 v = r - \lambda_1 + (\lambda_2 - \lambda_1)n_2 ,$$

which, by (6.14), can be expressed as

$$rk - \lambda_1 v = r - \lambda_1 + (r - \lambda_1)n_2/(\kappa - t) = (r - \lambda_1)v/\kappa > 0 , \tag{6.17}$$

because of

$$n_2 = (\rho - 1)(\kappa - 1)(\kappa - t)/t \qquad \text{and}$$
$$(6.18) \qquad v = \kappa[(\rho - 1)(\kappa - 1) + t]/t$$

[3, 4].

From (6.17) we get

$$(6.19) \qquad (\kappa - 1)\lambda_1 = r(\kappa k - v)/v = r\kappa k/v - r .$$

Hence

$$(6.20) \qquad A = \kappa[r + \lambda_1(\kappa - 1)] = \kappa^2 rk/v = c^2 f(c, c)$$

with

$$(6.21) \qquad f(c, c) = b .$$

From (6.20)

$$(6.22) \qquad \kappa^2 rk/v = c^2 b = c^2 vr/k$$

or

$$(6.23) \qquad c = \kappa k/v ,$$

which must be an integer [16]. In this case (6.3) is reduced to

$$(6.24) \qquad L_a L_u N_i = crS .$$

7. THE REPRESENTATION THEORY OF CALCULUS OF BLOCKS

It is already given by Thompson [19] and Bose and Mesner [2], since their ath association matrix is defined by $B_a = (b_{ia}^j)$ in which its uth row vector corresponds to our block D_{ua} in the sense of (1.4) and (1.5). The ith column vector of their incidence matrix N is our block N_i and their NN' corresponds to our $I_u N_i$, because I_u corresponds to the uth row vector of N.

The author believes that using our set-theoretic approach makes it easier to grasp the nature of formulas and often simplifies their deduction, since it treats the objects directly.

Two augmentative methods of constructing PBIB designs by mapping with this symbolic approach are given in [13, 14].

Referenccs

1. Bose, R. C. and Nair, K. R. "Partially Balanced Incomplete Block Designs," *Sankhyā*, **4** (1939), 337-372.
2. Bose, R. C. and Mesner, D. M. "On Linear Associative Algebras Corresponding to Association Schemes of Partially Balanced Designs," *Ann. Math. Statist.*, **30** (1959), 21-38.
3. Bose, R. C. "Strongly Regular Graphs, Partial Geometries and Partially Balanced Designs," *Pacific J. Math.*, **13** (1963), 389-419.
4. Bose, R. C. "Combinatorial Properties of Partially Balanced Designs and Association Schemes," *Sankhyā, Ser. A*, **25** (1963), 109-136.
5. Masuyama, M. "Cyclic Difference Sets which Generate Orthogonal Arrays, I and II," *Rep. Statist. Appl. Res.*, **6** (1959), 47-53, and **8** (1961), 70-76.
6. Masuyama, M. "Calculus of Blocks and a Class of Partially Balanced Incomplete Block Designs," *Rep. Statist. Appl. Res.*, **8** (1961), 50-69.
7. Masuyama, M. "Le Calcul des Block et ses Applications aux Plans d'Expérience," Colloques Internationaux du C. N. R. S., No. 110, Le Plan d'Expériences, 51-60.
8. Masuyama, M. "Calculus of Non-Commutative Blocks and its Applications to Experimental Designs," *Rep. Statist. Appl. Res.*, **10** (1963), 237-240.
9. Masuyama, M. "Construction of PBIB Designs by Fractional Development," *Rep. Statist. Appl. Res.*, **11** (1964), 47-53.
10. Masuyama, M. "Linear Graphs of PBIB Designs," *Rep. Statist. Appl. Res.*, **11** (1964), 147-151.
11. Masuyama, M. "Cyclic Generation of Triangular PBIB Designs," *Rep. Statist. Appl. Res.*, **12** (1965), 73-81.
12. Masuyama, M. "Construction Cyclique de Blocs Incomplets Partiellement Equilibrés," *Revue Inst. Internat. Statist.*, to appear, 1967.
13. Masuyama, M. "Symbolic Calculus of Blocks Applied to PBIB Designs," *Rep. Statist., Appl. Res.*, **13** (1966), 1-11.
14. Masuyama, M. "Calculus of Blocks and its Applications to Experimental Designs," *Department of Statistics*, University of North Carolina, mimeographed notes, 1966.
15. Masuyama, M. "A Combinatorial Proof of the Bose-Connor Inequality for PBIB Designs of GD Scheme," to appear in *Rep. Statist. Appl. Res.*, 1967.
16. Mesner, D. M. "On the Block Structure of Certain PBIB Designs of Partial Geometry Type," *Institute of Statistics*, Mimeo Series No. 457, University of North Carolina, 1966.
17. Nair, K. R. and Rao, C. R. "A Note of Partially Balanced Incomplete Block Designs, *Science and Culture*, **7** (1942), 568-569.

18. Stanton, R. G. and Mullen, R. C. "Inductive Methods for Balanced Incomplete Block Designs," *Ann. Math. Statist.*, **37** (1966), 1348-1354.
19. Thompson, W. A., Jr. "A Note on PBIB Design Matrices," *Ann. Math. Statist.*, **29** (1958), 919-922.

Discussion on Professor Masuyama's Paper

PROFESSOR K. TAKEUCHI: I think Professor Masuyama's "block calculus" is useful and especially effective in the construction of new PBIB designs or other arrays. But I think it is doubtful whether by "treating the objects directly, the deduction becomes simpler," though it must be true that "it is easier to grasp the meaning for the objective world." The "block calculus" can be translated into the terms of vectors and matrices of ordinary linear algebra, and the "block calculus" version of the association algebra for PBIB designs is not necessarily easier to manipulate than the ordinary way of interpretation. My own feeling is that the "block calculus" is of more value in the construction of specific designs than in the mathematical deduction of properties of several types of designs.

Incidentally, I would like to ask whether PBIB association schemes have any practical meaning in actual situations. Of course, some classes, for example, GD schemes, have a practical meaning; but generally, PBIB schemes may not have a plausible counterpart in reality, and on the other hand, the relations among actual varieties and/or treatments may not form an algebra.

It is well established that BIB designs are optimum in some (or in almost any) sense when there are no specific relations among varieties, but when BIB designs do not exist for given situations, what does one do? Very limited classes of PBIB designs may be proved to be the next best in such cases, but it may be fruitless to seek from one class of PBIB designs to another which meets the conditions of the parameters. Instead, I want to recommend that we randomize among some class of "nearly balanced" designs (the meaning of which must be defined in some way), but this is quite another story in combinatorial mathematics. (I myself have analyzed the properties of such "random balanced" designs in a series of papers, and the problem of "nearly balanced" designs will be discussed in a forthcoming paper.)

PROFESSOR M. MASUYAMA replied as follows: I understand that Professor K. Takeuchi does not deny the use of blocks in constructing designs (as was shown in his own paper and in Menon's papers) when a certain algebraic structure is introduced among the objects. I should like to note that if we use our (but not my) notations, the same relations hold for a module, a group, and a semi-group. If we use incidence matrices we would lose the naturalness of its extension in presenting formulae, since the incidence matrix is its representation and not the object itself. I wonder how Professor Takeuchi could use a $(0-1)$ matrix in place of a point in the geometric problem mentioned in my paper? I completely agree with him in that representation theory is powerful in many problems. I can add at least two more examples which lead to two new classes of PBIB designs in my paper [13]. It would be a good exercise to see what happens if they are treated by the incidence matrix method.

As for the general theory of PBIB designs, I agree partly with Professor Takeuchi in the sense that the optimality of PBIB designs was proved only for certain subclasses, i.e., BIB designs and Shrikande-Hartley squares, including Youden squares by myself and others and group divisible PBIB designs with $\lambda_2 = \lambda_1 + 1$ for $p_{12}^1 = 0$ by Professor Takeuchi. However, history has shown us that not only BIB designs but also Latin squares were introduced before their optimality was proved. I hope that the optimality of PBIB designs in general or in other subclasses of them will be proved in the near future. The idea of "nearly balanced" designs is interesting and ought to be published.

CHAPTER 14

On the Application of Half-Norms to Cyclic Difference Sets

KOICHI YAMAMOTO, *Florida Atlantic University*

1. INTRODUCTION

We denote the rational number field by Q, and the cyclotomic number field of $\zeta_v = e^{2\pi i/v}$ by Q_v. The concept of a half-norm was first introduced and investigated systematically by Hasse [1], and plays an important role in our study of cyclic difference sets. The half-norm $\mathfrak{N}\alpha$ of α in Q_v is simply $\alpha\bar{\alpha}$, the norm to the maximal real subfield R_v of Q_v. If D is a (v, k, λ)—difference set, and $D(x) = \sum_{a \in D} x^a$ its generating polynomial, then $\mathfrak{N}D(\zeta) = n = k - \lambda$ for all vth roots of unity $\zeta \neq 1$ is necessary. Conversely the condition implies that D is a difference set provided that $(v-1)\lambda = k(k-1)$.

2. THE HALF-NORM

We treat the case when $v = p$ is a prime in the present paper. Let $\alpha = \sum_{j=1}^{p-1} a_j\zeta_p^j$, $a_j \in Q$. We define the *Lagrange resolvent*

$$x_t = x_t(\alpha) = -\sum_{j=1}^{p-1} \vartheta^t(j)a_j \tag{1}$$

for any t, where ϑ is a primitive $(p-1)$th power residue character of p. This is Q-linear in α, belongs to Q_d with $d = \frac{p-1}{(p-1, t)}$. If ρ_r denotes the automorphism $\zeta_{p-1} \to \zeta_{p-1}^r$ of Q_{p-1}, then we have

$$(2) \qquad x_{rt} = x_t^{\rho_r}.$$

Thus in particular

$$(3) \qquad x_{-t} = \bar{x}_t .$$

Similarly if σ_r denotes the automorphism $\zeta_p \to \zeta_p^r$ of Q_p, then

$$(4) \qquad x_t(\alpha^{\sigma_r}) = \vartheta^t(r) x_t(\alpha) .$$

By the orthogonality relation of the characters we have

$$(5) \qquad (p-1)\alpha = \sum_{t=0}^{p-2} x_t \tau(\vartheta^{-t}) ,$$

where

$$(6) \qquad \tau(\vartheta_1) = -\sum_{x=1}^{p-1} \vartheta_1(x) \zeta_p^x$$

is the Gaussian sum for the character ϑ_1. We note the norm relation

$$(7) \qquad \tau(\vartheta_1)\overline{\tau(\vartheta_1)} = p \quad \text{for} \quad \vartheta_1 \neq I ,$$

$\tau(I) = 1$, where I is the principal character. Now from (4) it follows that

$$(p-1)\bar{\alpha} = \sum_{t=0}^{p-2} (-1)^t x_t \tau(\vartheta^{-t}) ,$$

$$(8) \qquad (p-1)^2 \mathfrak{N}\alpha = \sum_{t=0}^{p-2} \sum_{s=0}^{p-2} (-1)^t x_s x_t \tau(\vartheta^{-s}) \tau(\vartheta^{-t})$$
$$= x_0^2 + p \sum_{s=1}^{p-2}{}' x_s x_{-s}$$
$$+ \sum_{s=1}^{p-2} \left(\sum_{t=0}^{p-2} (-1)^t \pi(\vartheta^{-t}, \vartheta^{-(s-t)}) x_t x_{s-t} \right) \tau(\vartheta^{-s}) ,$$

where

$$\pi(\vartheta_1, \vartheta_2) = - \sum_{x+y \equiv 1 \ (\mathrm{mod}\ p)} \vartheta_1(x) \vartheta_2(y)$$

is the Jacobi sum for the characters ϑ_1 and ϑ_2.

Because of a transformation formula $\pi(\vartheta^t, \vartheta^{s-t}) = (-1)^t\pi(\vartheta^t, \vartheta^s)$ we have that

$$(9) \qquad (p-1)^2\mathfrak{N}\alpha = x_0^2 + p\sum_{s=1}^{p-2}{}^s x_s x_{-s} + \sum_{\substack{s=2\\ s\ \text{even}}}^{p-3}\left(\sum_{t=0}^{p-2}\bar{\pi}(\vartheta^s, \vartheta^t)x_s x_{s-t}\right)\tau(\vartheta^{-s}).$$

Since $\pi(\vartheta_1, \vartheta_2)$ and x_t belong to Q_{p-1}, and since $1, \tau(\bar{\vartheta}), \ldots, \tau(\bar{\vartheta}^{p-2})$ are linearly independent over Q_{p-1}, the equation (9) shows that $\mathfrak{N}\alpha$ is rational only if

$$(10) \qquad \sum_{t=0}^{p-2}\bar{\pi}(\vartheta^s, \vartheta^t)x_t x_{s-t} = 0$$

for $s = 2, 4, \ldots, p-3$. Moreover then we have

$$(p-1)^2\mathfrak{N}\alpha = x_0^2 + p\sum_{s=1}^{p-2}\mathfrak{N}^* x_s$$

where $\mathfrak{N}^*$ is the half-norm of the field Q_{p-1}.

3. DIFFERENCE SETS WITH A GIVEN MULTIPLIER GROUP INDEX

Suppose that D is a nontrivial (p, k, λ)—difference set with the multiplier group E of index e. Then $p = ef + 1$, e is even, f is odd and $f > 1$. D is a union of several cosets of E or the complementary set D^* of D has that property. Assume that $D = \bigcup_{c \in S} cE$ is the disjoint union representation of D. If $\chi = \vartheta^f$, and $\{\chi(c)\} = B$, then D is characterized as the complete inverse image under χ of B. Now, the Lagrange resolvent x_t is $= 0$ unless $f \mid t$, and

$$x_{ft} = -fK_t,$$

where $K_t = \sum_{\beta \in B}\beta^t$ is the tth power sum of the basic set B. Indeed if ε_A denotes the characteristic function of a set A, then

$$a_j = \sum_{c \in S}\varepsilon_{cE}(j)$$

and

$$\varepsilon_{cE} = \frac{1}{e}\sum_{m=0}^{e-1}\bar{\chi}^m(c)\chi^m.$$

So we have

$$x_t = -\sum_{j=1}^{p-1} \vartheta^t(j) \sum_{c \in S} \frac{1}{e} \sum_{m=0}^{e-1} \bar{\chi}^m(c)\chi^m(j)$$

$$= -\frac{1}{e} \sum_{c \in S} \sum_{m=0}^{e-1} \chi^{-m}(c) \sum_{j=1}^{p-1} \vartheta^{t+mf}(j)\,,$$

which equals to 0 unless $f|t$, and

$$x_{ft} = -\frac{p-1}{e} \sum_{c \in S} \chi^t(c) = -f \sum_{\beta \in B} \beta^t = -fK_t\,.$$

Therefore the condition (10) reads now

$$\text{(11)} \qquad \sum_{t=0}^{e-1} \bar{\pi}(\chi^s, \chi^t) K_t K_{s-t} = 0$$

for $s = 2, 4, \ldots, e-2$.

4. THE MAIN THEOREM

Consider now the complementary set D^* of D. We have $K_t^* = -K_t$ for $e \nmid t$ and $K_0^* = e - K_0$. Therefore the condition (11) above for D^* is equivalent to

$$\sum_{t=0}^{e-1} \bar{\pi}(\chi^s, \chi^t) K_t K_{s-t} = 2eK_s\,.$$

Combining we have

Theorem. *The set D or the set $D \cup 0$ is a difference set if and only if*

$$\sum_{t=0}^{e-1} \bar{\pi}(\chi^s, \chi^t) K_t K_{s-t} = 2deK_s$$

for $s = 2, 4, \ldots, e-2$, where $d = 1$ or $d = 0$ according as 0 is contained or not.

Example. E or $E \cup 0$ is a difference set if and only if

$$\sum_{t=0}^{e-1} \bar{\pi}(\chi^t, \chi^s) = 2de$$

for $s = 2, 4, \ldots, e-2$. If we define the bilinear form $\pi(\alpha, \beta) = \sum_{x+y\equiv 1 \pmod p} \alpha(x)\beta(y)$, then the equation above is equivalent to

$$\pi(\varepsilon_E, \varepsilon_A) = \frac{f + 2d - 1}{e} - d\delta_{E,A}$$

with the Kronecker delta, where $\pi(\varepsilon_A, \varepsilon_B)$ is the cyclotomic number for cosets A, B of E.

5. SOME PREVIOUS RESULTS

It was shown in [2] that, if $\mathfrak{p}$ is the prime ideal divisor of p in Q_e such that $\chi(x) \equiv x^f \pmod{\mathfrak{p}}$ for all x, then the condition in the theorem is equivalent to

$$\sum_{t=0}^{s} \bar{\pi}(\chi^t, \chi^s) K_t K_{s-t} \equiv 2deK_s \pmod{\mathfrak{p}}$$

for $s = 2, 4, \ldots, e-2$. Moreover there are exactly $\varphi(e)/2$ multiplicatively independent Jacobi sums if $e \geqslant 4$, so that we can eliminate $\bar{\pi}(\chi^t, \chi^s)$ from the above congruences. The result is a system of congruences involving K_t and the torsion solution of the Γ-functional equation.

6. ILLUSTRATIONS

We illustrate this with $e = 2q$, q an odd prime and $B = 1$. Let $\xi = \chi(16)$. If $e = 2q$, q an odd prime and if either E or $E \cup 0$ is a difference set, then the system of congruences

$$2W_0W_2 + \xi W_1^2 \equiv 0\,,$$

$$2W_0W_4 + 2\xi^2 W_1W_3 + W_2^2 \equiv 0\,,$$

$$2W_0W_6 + 2\xi^3 W_1W_5 + 2W_2W_4 + \xi^3 W_3^2 \equiv 0\,,$$

$$\cdots\cdots\cdots\cdots\cdots\cdots$$

$$2W_0W_{2q-2} + 2\xi^{q-1}W_1W_{2q-3} + 2W_2W_{2q-4} + \cdots + 2\xi^{q-1}W_{q-2}W_q + W_{q-1}^2 \equiv 0\,,$$

$$W_2 \equiv W_1W_{q+1}\,,$$

$$W_4 \equiv W_2W_{q+2}\,,$$

$$\vdots$$

$$W_{2q-2} \equiv W_{q-1}W_{2q-1}\,,$$

$$W_1 W_{2q-1} \equiv W_2 W_{2q-2} \equiv \ldots \equiv W_{q-1} W_{q+1} \equiv 1 ,$$
$$W_1 \equiv 1$$

has a solution. Put $E_j = \dfrac{W_{2j}}{W_0}$. Then any W_j is expressed in terms of $E_1, E_2, \ldots, E_{(q-1)/2}$.

(a) *The case* $q = 3$. We have

$$\left.\begin{aligned} &2 + \xi E_1^3 \equiv 0 , \\ &2 + 2\xi^2 E_1^3 + E_1^3 \equiv 0 . \end{aligned}\right\} \pmod{\mathfrak{p}}$$

Elimination of E_1^3 results in

$$\Phi_1(\xi) = 2\xi^2 - \xi + 1 \equiv 0 \pmod{\mathfrak{p}}$$

However $\Phi_1(1) = 2$, and $N\Phi_1(\zeta_3) = 7$. This shows that except for the trivial case $p = 7$, neither the set of all 6th power residues, nor the set joined with 0, is a difference set.

(b) *The case* $q = 5$. We have

$$2 + \xi E_1 E_2^2 \equiv 0 ,$$
$$2 + 2\xi^2 \frac{E_1^2}{E_2} + \frac{E_1^2}{E_2} \equiv 0 ,$$
$$2 + 2\xi^3 E_1 E_2^2 + 2E_1 E_2^2 + \xi^3 \frac{E_1^2}{E_2} \equiv 0 ,$$
$$2 + 2\xi^4 E_1 E_2^2 + 2\frac{E_1^2}{E_2} + 2\xi^4 \frac{E_1^2}{E_2} + E_1 E_2^2 \equiv 0$$

From the first two we have

$$E_1 E_2^2 \equiv -2\xi^4 , \qquad \frac{E_1^2}{E_2} = -\frac{2}{2\xi^2 + 1} ,$$

and substitution in the last two results in

$$\left.\begin{aligned} &\Phi_1(\xi) = 6\xi^4 + \xi^3 + 4\xi - 1 \equiv 0 , \\ &\Phi_2(\xi) = 3\xi^4 + 2\xi^3 - 2\xi^2 + 2\xi + 5 \equiv 0 . \end{aligned}\right\} \pmod{\mathfrak{p}}$$

However

$$\Phi_1(1) = \Phi_2(1) = 10\,, \qquad N\Phi_1(\zeta_5) = 605\,, \qquad N\Phi_2(\zeta_5) = 1210\,.$$

The only prime divisor $p \equiv 1 \pmod{10}$ is 11. Thus except for the trivial case $p = 11$, neither the set of all 10th power residues, nor the set joined with 0, is not a difference set.

(c) *The case* $q = 7$. We have

$$2 + \xi E_1 E_3^2 \equiv 0\,,$$

$$2 + 2\xi^2 E_1 E_3^2 + \frac{E_1^2}{E_2} \equiv 0\,,$$

$$2 + 2\xi^3 \frac{E_1^2}{E_2} + 2\frac{E_1 E_2}{E_3} + \xi^3 E_2^2 E_3 \equiv 0\,,$$

$$2 + 2\xi^4 E_1 E_3^2 + 2E_1 E_3^2 + 2\xi^4 E_1 E_3^2 + E_2^2 E_3 \equiv 0\,,$$

$$2 + 2\xi^5 E_2^2 E_3 + 2\frac{E_1 E_2}{E_3} + 2\xi^5 E_2^2 E_3 + 2E_2^2 E_3 + \xi^5 \frac{E_1^2}{E_2} \equiv 0\,,$$

$$2 + 2\xi^6 \frac{E_1^2}{E_2} + 2\frac{E_1^2}{E_2} + 2\xi^6 E_2^2 E_3 + 2\frac{E_1 E_2}{E_3} + 2\xi^6 \frac{E_1^2}{E_2} + E_1 E_3^2 \equiv 0\,.$$

Solving the first four congruences we have

$$E_1 E_3^2 \equiv -2\xi^6\,, \qquad \frac{E_1^2}{E_2} \equiv -2(1 - 2\xi)\,,$$

$$E_2^2 E_3 \equiv -2(1 - 4\xi^3 - 2\xi^6)\,, \qquad \frac{E_1 E_2}{E_3} \equiv (1 + 2\xi^2 3\xi^3 + 4\xi^4 + \xi^6)\,,$$

and by substituting in the last two we have

$$\Phi_1(\xi) = 2 - 16\xi + 2\xi^2 - 11\xi^3 - 4\xi^4 + 5\xi^5 - 2\xi^6 \equiv 0 \pmod{\mathfrak{p}},$$

$$\Phi_2(\xi) \equiv 6 + 4\xi + 6\xi^2 + 3\xi^3 - 4\xi^4 + 4\xi^5 - 11\xi^6 \equiv 0 \pmod{\mathfrak{p}}.$$

Moreover $(E_1 E_3^2)\dfrac{E_1 E_2}{E_3} = (E_1^2 E_3)\dfrac{E_1}{E_2}$ yields

$$\Phi_3(\xi) = 10 - 6\xi + 3\xi^2 - 12\xi^3 + 16\xi^4 - 4\xi^5 - 5\xi^6 \equiv 0 \pmod{\mathfrak{p}}.$$

However

$$\Phi_1(1) = -24\,, \qquad \Phi_2(1) = 8\,, \qquad \Phi_3(1) = 2$$

and

$$N\Phi_1(\zeta_7) = 30706649\,, \qquad N\Phi_2(\zeta_7) = 32268853\,,$$
$$N\Phi_3(\zeta_7) = 9998311\,.$$

The last three norms are relatively prime. So there is no difference set consisting of all the 14th power residues, nor of all the 14th power residues and of 0.

References

1. Hasse, H. "Der 2^nte Potenzcharakter von 2 im Körper der 2^nten Einheitswurzeln," *Rend. Circ. Mat. Palermo*, 7 (1958), 185–244.
2. Yamamoto, K. "On Jacobi Sums and Difference Sets," to appear in *J. Combinatorial Theory*.

Discussion on Professor Yamamoto's Paper

PROFESSOR J. N. SRIVASTAVA: It is a great pleasure for me to thank Dr. Yamamoto for his stimulating paper, which in my view contains a significant development in the theory of difference sets, viz., the application to difference sets of the properties of half-norms introduced by Hasse in his 1958 paper (in German), published in the Italian journal *Rendiconti del Circolo Matematico di Palermo.* Dr. Yamamoto's approach may go a long way toward providing workable necessary and sufficient conditions for the existence of difference sets.

Necessary and sufficient conditions have been given by Emma Lehmer in order that the eth power residues may form a difference set. Thus let (i, j) denote a cyclotomic number, i.e., the number of solutions of the equation

$$1 + g^{e\mu+i} = g^{e\nu+j} \pmod p\,;$$

$$0 \le i, j \le n-1\,;\ 0 \le \nu, \mu \le f-1\,;$$

where p is a prime, $p = ef + 1$, p odd, f odd, e even. Then the class of eth power residues form a difference set if and only if $(i, 0) = (f-1)/e$, $i = 0, 1, \ldots, e-1$.

However if we wish to answer the question, as to what are the values of p for which eth power residues form a difference set the above result is not always very helpful. Professor Yamamoto's work treats this in a different way. For example, to test whether a Hall's multiplier group E (of index $e = 10$) forms a difference set, we only try to solve certain congruences (as in the example given in the text), at the end of which we find that the only permissible case is $p = 11$.

At the moment the only difficulty under the present approach is that the set of congruences to be solved will become large and cumbersome for large q. Methods have to be found therefore to eliminate the Jacobi sums by some theoretical approach which gives directly the polynomials $\Phi_i(\xi)$.

CHAPTER 15

On Embedding of Orthogonal Arrays of Strength Two

S. S. SHRIKHANDE AND BHAGWANDAS,
University of Bombay, India

1. INTRODUCTION

An *orthogonal array of strength two* is a matrix $A = A(n, q, \mu)$ of q rows and $n^2\mu$ columns in $n \geq 2$ symbols such that in any two rows of A each of the n^2 ordered pairs of these n symbols occurs exactly μ times where $\mu \geq 1$. It is then obvious that each row of A contains every symbol exactly $n\mu$ times. The number q is called the number of constraints of the array. It is known [3] that

$$q \leq \frac{n^2\mu - 1}{n - 1}.$$

For a nonnegative integer t and

(1.1) $$\mu = \mu^* = (n - 1)t + 1$$

the above inequality reduces to

(1.2) $$q \leq n^2t + n + 1 = q^*.$$

An array $A(n, q^*, \mu^*)$ may be called a maximal array and hence the nonnegative integer

$$d = q^* - q \tag{1.3}$$

may be called the deficiency of an array $A(n, q, \mu^*)$.

For $\mu = 1$, the orthogonal array $A(n, q, \mu)$ of strength two is equivalent to a set of $q - 2$ mutually orthogonal latin squares (m.o.l.s.) of order n [6, 9] and the inequality

$$q \leq \frac{n^2\mu - 1}{n - 1} = n + 1$$

implies that the maximum number of m.o.l.s. of order n is $n - 1$. Such a set ot $n - 1$ m.o.l.s. is called a complete set of m.o.l.s. of order n.

A set of $n - 1 - d$ m.o.l.s. of order n will be called a set of m.o.l.s. with deficiency d. It is well known that a set of m.o.l.s. with $d = 1$ can always be embedded in a complete set. For a given value of d, Bruck [5] has shown that this embedding is possible if

$$n > \frac{1}{2}(d^4 - 2d^3 + 2d^2 + d - 2) .$$

The particular case $d = 2$ was first given by Shrikhande [12].

The main purpose of this note is to generalize these results to the case $\mu > 1$, and to show that (i) for any value of n an array $A(n, q^* - 1, \mu^*)$ and (ii) for $n = 2$ or 3 an array $A(n, q^* - 2, \mu^*)$ can each be embedded in the corresponding maximal array.

2. PRELIMINARY RESULTS

A balanced incomplete block design (BIBD) with parameters v, b, r, k, λ is an arrangement of v symbols or treatments in b sets or blocks each containing k ($< v$) different treatments such that any pair of different treatments occurs in exactly λ blocks. It is then obvious that each treatment occurs in exactly r blocks and then

$$vr = bk ,$$

$$\lambda(v-1) = r(k-1) .$$

It is also known [8] that a necessary condition for the existence of a BIBD is that

$$b \geq v .$$

A BIBD is called a *resolvable* BIBD (RBIBD) if the b blocks can be partitioned into r sets of n blocks each such that each set is a complete replication of the v treatments. Obviously then

$$b = nr , \qquad v = nk .$$

Bose [2] has shown that for a RBIBD, the inequality $b \geq v$ can be strengthened to

$$b \geq v + r - 1 .$$

A RBIBD is called an *affine resolvable* BIBD (ARBIBD) if any two blocks of different replications have the same number μ of treatments in common. It is known [2] that the parameters of an ARBIBD can be expressed in terms of integers n and t, $n \geq 2$, $t \geq 0$ by

$$v = nk = n^2\mu^* , \qquad b = nr = nq^* , \qquad \lambda = nt + 1 , \qquad \mu = \mu^* ,$$

μ^* and q^* being given by (1.1) and (1.2).

An *affine resolvable design* (ARD) is an arrangement of v treatments in b blocks of k ($< v$) different treatments if the b blocks can be partitioned into r sets such that each set is a complete replication of the v treatments and any two blocks of different sets have exactly μ treatments in common. If n is the number of block in each set then obviously $v = nk$. Since the k treatments of any block of a set are distributed into n disjoint subsets corresponding to the n blocks of a different set we further have $k = n\mu$ and hence $v = n^2\mu$. We denote the parameters of such a design by

$$v = nk = n^2\mu , \qquad b = nr; \mu .$$

A *group divisible design* (GDD) is an arrangement of $v = mn$ treatments (partitioned into m groups of n each) in b blocks of $k < v$ different treatments satisfying the following conditions: (i) each treatment occurs in exactly r of the b

blocks (ii) any two treatments of the same group occurs together in λ_1 blocks, (iii) any two treatments of different groups occur together in λ_2 ($\neq \lambda_1$) blocks. Necessary conditions [4] for the existence of a GDD are $r - \lambda_1 \geq 0$, $rk - \lambda_2 v \geq 0$. A GDD is called a *semi-regular* GDD (SRGDD) if $rk - \lambda_2 v = 0$ and in this case it is known [4] that each block contains the same number of treatments from each of the m groups. The parameters of a GDD may be denoted by

$$v, b, r, k, \lambda_1, \lambda_2; m, n .$$

Lemma 1. *The existence of any one of the three following configurations implies the existence of the other two*

(i) ARD: $v = nk = n^2\mu$, $b = nr$; μ.

(ii) *an array* $A(n, r, \mu)$.

(iii) SRGDD *with parameters* $v_1, b_1, r_1, k_1, \lambda_1, \lambda_2; m, n$ *with* $v_1 = nk_1 = nr$, $b_1 = nr_1 = n^2\mu$, $\lambda_1 = 0$, $\lambda_2 = \mu$; $m = r$, $n = n$.

Proof. It is obvious that configurations (i) and (iii) are duals of each other and hence the existence of any one of them implies the existence of the other. Hence it is sufficient to prove that (i) and (ii) either both exist or both do not exist.

Suppose an ARD with parameters of (i) exists. Number the blocks of each replication from 1 to n and the treatments from 1 to v in any arbitrary manner. We construct a matrix A of r rows and v columns where ith replication of (i) corresponds to the ith row of the A in the following manner. Number the columns of A corresponding to the v treatments and in row i put the symbol j in position x if the jth block of the ith replication contains the treatment numbered x. It is then obvious that each row of A contains each of the n symbols exactly $n\mu$ times. Further since two blocks coming from different replications of (i) intersect in μ treatments it is obvious that in any two rows of A all the ordered n^2 pairs occur each μ times which implies that the matrix A is an array $A(n, r, \mu)$ of (ii).

Conversely, reversing the process it is easily seen that an orthogonal array (ii) leads to an ARD given by (i).

This completes the proof.

It has been shown by Bose [2] that a RBIBD with parameters v, b, r, k, λ is an ARBIBD if and only if $b = v + r - 1$. Now suppose that an ARD with parameters

(2.1) $$v = nk = n^2\mu^* , \qquad b = nr; \mu^*$$

exists where $r < q^*$ and it is possible by adding suitable additional blocks to embed this design in a RBIBD with parameters,

(2.2) $$v = nk = n^2\mu^* , \quad b = nr = nq^* , \quad \lambda = nt + 1 .$$

Since the equality $b = v + r - 1$ holds for (2.2) the design thus obtained is an ARBIBD. Hence from Lemma 1, we can embed the array $A(n, r, \mu^*)$ into the corresponding maximal array $A(n, q^*, \mu^*)$ where the actual embedding can be done as indicated in the latter half of the proof of Lemma 1. We thus have the following result.

Lemma 2. *An orthogonal array $A(n, r, \mu^*)$ with $r < q^*$ can be embedded in the maximal array $A^*(n, q^*, \mu^*)$ if and only if an ARD with parameters* (2.1) *can be embedded in a RBIBD with parameters* (2.2).

Lemma 3 [1]. *For a SRGDD with parameters*

$$v, b, r, k, \lambda_1, \lambda_2; m, n$$

the number of treatments common between any two blocks of the design satisfies the inequality

$$\max (0, 2k - v, k - r + \lambda_1) \leq x$$
$$\leq \min \left(k, r - \lambda_1 - k + 2\frac{r(k-1) + \lambda_1}{b}\right) .$$

Now consider an ARD with parameters (2.1). As noted in Lemma 1, its dual is an SRGDD and from the above lemma, we can find bounds for x, the number of treatments common to any two blocks of the SRGDD. This number x is obviously the number of times any two distinct treatments of the ARD (2.1) occur together in a block of (2.1). It is now easy to verify the following corollaries:

Corollary 1. *In an ARD with parameters*

(2.3) $$v = nk = n^2\mu^* , \quad b = nr = n(q^* - 1) , \quad \mu = \mu^*$$

with respect to any given treatment θ there are n_i treatments each of which occurs with θ, x_i times, $i = 1, 2$, where

$$
\begin{aligned}
n_1 &= (n-1)k\,, & x_1 &= nt+1\,, \\
n_2 &= k-1\,, & x_2 &= nt\,.
\end{aligned}
\tag{2.4}
$$

Proof. The values x_1 and x_2 follow from Lemma 3. If n_1 and n_2 are the corresponding multiplicities then obviously

$$n_1 + n_2 = v - 1$$

and

$$r(k-1) = n_1 x_1 + n_2 x_2\,.$$

These equations have the unique solution for n_1 and n_2 as given above.

Corollary 2. *If $n = 2, 3$ then in an ARD with parameters*

$$v = nk = n^2\mu^*\,,\quad b = nr = n(q^* - 2)\,,\quad \mu = \mu^*\,, \tag{2.5}$$

with respect to any treatment θ there are n_i treatments each of which occurs with θ, x_i times, $i = 1, 2, 3$ where

$$
\begin{aligned}
n_1 &= 2(n-1)\mu^*\,, & x_1 &= nt\,, \\
n_2 &= (n-1)^2\mu^*\,, & x_2 &= nt+1\,, \\
n_3 &= \mu^* - 1\,, & x_3 &= nt-1\,.
\end{aligned}
\tag{2.6}
$$

Proof. The values x_1, x_2, x_3 follow from the use of Lemma 3. As in the above corollary, we have

$$\sum_1^3 n_i = v - 1$$

$$\sum_1^3 n_i x_i = r(k-1)\,.$$

Now consider the r blocks containing a given treatment θ. Each of the $r(r-1)$ intersections taken two at a time contain $\mu^* - 1$ treatments besides the treatment θ. In the aggregate of $r(r-1)(\mu^* - 1)$ treatments thus obtained each treatment occuring x_i times with θ occurs $x_i(x_i - 1)$ times and hence we have an additional relation

$$\sum_1^3 x_i(x_i - 1)n_i = r(r-1)(\mu^* - 1)\,.$$

The unique solution of the three above equations gives the values n_i as indicated in the corollary.

3. EMBEDDING OF AN ARRAY OF DEFICIENCY ONE

Consider an ARD of deficiency 1 with parameters

$$v = nk = n^2\mu^* , \quad b = nr = n(q^* - 1) , \quad \mu = \mu^* . \tag{3.1}$$

Let $N = (n_{ij})$ denote the usual $(v \times b)$ incidence matrix of the design (3.1) where $n_{ij} = 1$ or 0 according as treatment i occurs or does not occur in block numbered j. Then N' is the incidence matrix of its dual which is a SRGDD witn parameters $v_1, b_1, r_1, k_1, \lambda_1, \lambda_2; m, n$ where

$$v_1 = nk_1 = n(q^* - 1) , \quad b_1 = nr_1 = n^2\mu^* , \quad \lambda_1 = 0 , \quad \lambda_2 = \mu^* ;$$
$$m = k_1 , \quad n = n .$$

From [7] it follows that $N'N$ has characteristic roots $n^2(nt + 1)\mu^*$, $n\mu^*$ and 0 with corresponding multiplicities 1, $n(n-1) \cdot (nt+1)$ and $q^* - 2$. Hence NN' has the same characteristic roots with multiplicities 1, $n(n-1)(nt+1)$ and $n-1$.

From Corollary 1, it is obvious that

$$NN' = rI + (nt+1)A + nt(J - I - A)$$

where I is the identity matrix of order v and J is the square matrix of order v with all elements 1 and A is a regular adjacency matrix [13] of valence $(n-1)k$. Further since NN', I and J commute with each other the characteristic roots of A are $(n-1)k$, 0 and $-\mu^*$ with respective multiplicities 1, $n(n-1)(nt+1)$ and $n-1$. It now follows [Lemma 7, 13] that N is the incidence matrix of a PBIBD with two associate classes with

$$\lambda_1 = nt + 1 , \qquad \lambda_2 = nt , \qquad n_1 = (n-1)k ,$$
$$p^1_{11} = (n-2)k , \qquad p^2_{11} = (n-1)k .$$

Since

$$p^2_{12} = n_1 - p^2_{11} = 0$$

this implies that ARD is a GDD with n groups of k treatments each, where two treatments belonging to the same group occur together nt times and two treatments belonging to different groups occur together $nt+1$ times.

To the blocks of the ARD we now add n blocks of the size k corresponding to the n groups. It is then obvious that we get a RBIBD with parameters (2.2). Hence from Lemma 2, we have the following theorem.

Theorem 1. *For any values of n and μ^* an array $A(n, q^*-1, \mu^*)$ can always be embedded in the corresponding maximal array $A(n, q^*, \mu^*)$.*

The particular case of the above theorem for $n=2$ is obtained in [14].

4. EMBEDDING OF AN ARRAY OF DEFICIENCY TWO WHEN $n = 2, 3$

Consider an ARD with $n=2$ or 3 and parameters

$$v = nk = n^2\mu^*\,, \quad b = nr = n(q^*-2)\,, \quad \mu = \mu^*\,, \tag{4.1}$$

and let $\boldsymbol{N}$ be its incidence matrix. Then $\boldsymbol{N}'$ is the incidence matrix of its dual which is a SRGDD according to Lemma 1. Without loss of generality assume that the blocks of the jth replication of (4.1) are numbered from $(j-1)n+1$ to jn and hence

$$\boldsymbol{N}'\boldsymbol{N} = \begin{pmatrix} \boldsymbol{B} & \boldsymbol{C} & \boldsymbol{C} & \dots & \boldsymbol{C} \\ \boldsymbol{C} & \boldsymbol{B} & \boldsymbol{C} & \dots & \boldsymbol{C} \\ \cdot & \cdot & \cdot & & \cdot \\ \cdot & \cdot & \cdot & & \cdot \\ \cdot & \cdot & \cdot & & \cdot \\ \boldsymbol{C} & \boldsymbol{C} & \boldsymbol{C} & & \boldsymbol{B} \end{pmatrix}, \tag{4.2}$$

where $\boldsymbol{B}$ and $\boldsymbol{C}$ are square matrices of order n and $\boldsymbol{B} = k\boldsymbol{I}$ and $\boldsymbol{C} = \lambda_2\boldsymbol{J}$ where $\lambda_2 = \mu^*$.

We will be interested in the structure of the matrix $(\boldsymbol{N}\boldsymbol{N}')\cdot(\boldsymbol{N}\boldsymbol{N}') = \boldsymbol{N}(\boldsymbol{N}'\boldsymbol{N})\boldsymbol{N}'$. From Corollary 2, we know that with respect to any row of $\boldsymbol{N}$ there are n_i other rows of $\boldsymbol{N}$ such that the given row and any one of the n_i rows have exactly x_i components both equal to 1 where the values of n_i and x_i, $i = 1, 2, 3$ are given by (2.6).

Let the scalar product of two different rows of N be x_3. Without loss of generality assume that they are the first two rows y_1 and y_2 of N. We now find the element in position (1, 2) of $(NN')(NN') = N(N'N)N'$, i.e. the value of $y_1(N'N)y_2'$. Again, without any loss of generality assume that y_1 has $q^* - 2$ components of order n each consisting of $(1\,0\,0\ldots0)$ and y_2 has first x_3 components each of the form $(1\,0\,0\ldots0)$ and the remaining $n\mu^*$ components of order n each of the form $(0\,0\ldots0\,1)$. Utilizing (4.2) it is easy to verify that

$$y_1(N'N)y_2' = \mu^*((q^* - 2)(q^* - 3) + n(nt - 1)) . \tag{4.3}$$

We now calculate the above element in position (1, 2) of $(NN')(NN')$ by another method. From (2.6) it is obvious that the first two rows of NN' are

$$u_1 = (r \quad x_3 \quad z_1)$$

$$u_2 = (x_3 \quad r \quad z_2)$$

and hence

$$u_1u_2' = y_1(N'N)y_2' \tag{4.4}$$

where z_1 and z_2 each have $n_3 - 1$ components equal to x_3, n_1 components equal to x_1 and n_2 components equal to x_2. Hence from (4.3) and (4.4)

$$z_1z_2' = y_1(N'N)y_2' - 2rx_3 .$$

It is easily verified that

$$z_1z_2' = z_1z_1' = z_2z_2'$$

and hence

$$z_1z_2' = ((z_1z_1')(z_2z_2'))^{1/2} .$$

From the Cauchy-Schwartz inequality this implies that the vectors z_1 and z_2 are proportional and hence actually equal.

If we call two treatments of (4.1) as i-associates if they occur together x_i times $i = 1, 2, 3$, then the above implies that the v treatments of (4.1) can be partitioned into n^2 sets S_i of μ^* treatments each such that (i) any two treatments of the same set S_i are 3-associates, and (ii) with respect to any treatment of S_i, the treatments of a set S_j, $j \neq i$ are either all

1-associates or all 2-associates of that treatment. If we pick arbitrarily a treatment θ_i from S_i, then we have n^2 treatments and with respect to any treatment θ_i there are exactly $2(n-1)$ treatments which are its 1-associates and the remaining $(n-1)^2$ treatments are its 2-associates.

We now consider the two cases $n = 2$ and $n = 3$ separately.

If $n = 2$, then there are 4 treatments θ_i, $i = 1, \ldots, 4$ and each θ_i has exactly two 1-associates and one 2-associate: it is easy to see that if θ_1 and θ_4 are 2-associates, then we can exhibit these treatments in the scheme

$$\begin{matrix} \theta_1 & \theta_2 \\ \theta_3 & \theta_4 \end{matrix}$$

such that any two treatments are 1-associates if and only if they come from the same row or same column. Hence the four sets S_i can be written in the scheme

$$\begin{matrix} S_1 & S_2 \\ S_3 & S_4 \end{matrix}$$

where two treatments from the same S_i occur together $2t-1$ times and two treatments belonging to S_i and S_j respectively $(i \neq j)$ occur together $2t$ or $2t+1$ times according as the corresponding S_i and S_j are in the same row or column or not. It is now obvious that by adding the two sets of blocks

$$[(S_1 \quad S_2)\,, \quad (S_3 \quad S_4)]$$

and

$$[(S_1 \quad S_3)\,, \quad (S_2 \quad S_4)]$$

to the blocks of (4.1) we get a RBIBD

$$v = 2k = 4(t+1)\,, \qquad b = 2r = 2(4t+3)\,, \qquad \lambda = 2t+1\,.$$

Now consider the case $n = 3$. In this case there are 9 sets S_i each of $2t+1$ treatments. Number the treatments of (4.1) so that the treatments of the set S_i are numbered from $(i-1) \cdot (2t+1)+1$ to $i(2t+1)$. Let $\boldsymbol{x}_1$ and $\boldsymbol{x}_2$ be the rows of N corresponding to the first treatment of S_1 and S_2 respectively and assume that the corresponding treatments of (4.1) occur together $3t$ times. Then utilizing (4.2)

$$(4.5)\qquad x_1(N'N)x_2' = (2t+1)(81t^2+36t+2)\,.$$

Obviously the above expression equals the element in position $(1, 2t+2)$ in the matrix $(NN')(NN')$ and hence

$$(4.6)\qquad x_1(N'N)x_2' = w_1 w_2'$$

where w_1 and w_2 are the rows numbered 1 and $2t+2$ in NN'. The first $2(2t+1)$ positions in w_1 and w_2 respectively are then

$$\begin{matrix} 9t+2 & 3t-1\ldots 3t-1 & 3t & 3t & \ldots 3t \\ 3t & 3t & \ldots 3t & 9t+2 & 3t-1\ldots 3t-1 \end{matrix}$$

and the remaining positions in w_1 and w_2 each consist of 3 components of order $2t+1$ each containing all entries $3t$ and 4 components of order $2t+1$ each containing all entries $3t+1$. Now assume that in these 7 components of order $2t+1$ the vector w_2 has α components with entries $3t$ under the 3 components of w_1 with these entries where $\alpha = 0$, 1, 2, or 3. It is then easy to verify that

$$(4.7)\qquad w_1 w_2' = (2t+1)(81t^2+36t+1+\alpha)\,.$$

Hence from (4.5), (4.6) and (4.7) we get $\alpha = 1$. If the elements θ_i are arbitrarily chosen from S_i as before $i = 1, 2, \ldots, 9$ then this means that each θ_i has four 1-associates and four 2-associates and if θ_i and θ_j are 1-associates then the number p^1_{11} of common 1-associates of both θ_i and θ_j is 1.

Now assume that θ_1 has 1-associates θ_2, θ_3, θ_4 and θ_7 whereas the remaining treatments θ_5, θ_6, θ_8, θ_9 are its 2-associates. Let the common 1-associates of θ_1 and θ_2 be θ_3. Then θ_4 and θ_7 are 2-associates of both θ_2 and θ_3. Hence θ_7 is the common 1-associate of θ_1 and θ_4. Now take any 2-associate of θ_1, say θ_5. Treatments θ_2 and θ_3 are 1-associates and they both cannot be 1-associate of θ_5 for otherwise θ_2 and θ_3 will have θ_1 and θ_5 as its two common 1-associates. Now suppose that θ_5 is 2-associate of both θ_2 and θ_3. Then θ_2 and θ_3 each have θ_4 and θ_5 and θ_7 as its 2-associates and hence θ_2 has a set of two treatments from θ_6, θ_8, θ_9 as its 1-associates and similarly θ_3 has a set of two treatments from θ_6, θ_8, θ_9 as its 1-associates. Since these two sets of two treatments will necessarily have one treatment in common we get a contradiction since θ_1 is the only common 1-associate of θ_2 and θ_3. Thus θ_5 has exactly one 1-associate from θ_2 and θ_3 and for a similar reason exactly one

1-associate from θ_4 and θ_7. Thus for any two treatments which are 2-associates the number p_{11}^2 of treatments which are common 1-associates of both is 2. It now follows from [11] that these nine treatments have the L_2 association scheme

$$\begin{matrix} \theta_1 & \theta_2 & \theta_3 \\ \theta_4 & \theta_5 & \theta_6 \\ \theta_7 & \theta_8 & \theta_9 . \end{matrix}$$

It is now obvious that if we add as in the case $n = 2$ the two sets of three blocks corresponding to the rows and columns of

$$\begin{matrix} S_1 & S_2 & S_3 \\ S_4 & S_5 & S_6 \\ S_7 & S_8 & S_9 \end{matrix}$$

to the blocks of (4.1) we get a RBIBD with

$$v = 3k = 9(2t + 1) , \qquad b = 3r = 3(9t + 4) , \qquad \lambda = 3t + 1 .$$

We can then, utilizing Lemma 2, state the following theorem.

Theorem 2. *If $n = 2$ or 3 an array $A(n, q^* - 2, \mu^*)$ can always be embedded in the corresponding maximal array $A(n, q^*, \mu^*)$.*

5. BOUNDS FOR q IN $A(n, q, \mu^*)$

It has been shown in [10] that a maximal array $A(n, q^*, \mu^*)$ does not exist for certain values of n and t. Utilizing this result and making use of the above two theorems, we can state the following results

Theorem 3. *For an array $A(n, q, \mu^*)$,*

$$q < q^* - 1$$

if

(i) *n and t are odd and either*
 (a) *$n\mu^*$ is not a perfect square, or*

(b) $n\mu^*$ *is a perfect square and* $nt \equiv 1 \pmod 4$ *and the square free part of* n *contains a prime* $\equiv 3 \pmod 4$

or

(ii) n *is odd and* t *is even and either*

(a) μ^* *is not a perfect square, or*

(b) μ^* *is a perfect square and* $n + t \equiv 1 \pmod 4$ *and the square free part of* n *contains a prime* $\equiv 3 \pmod 4$,

or

(iii) *if* $n \equiv 2 \pmod 4$ *and the square free part of* n *contains a prime* $\equiv 3 \pmod 4$.

Theorem 4. *For an array* $A(3, q, 2t + 1)$,

$$q < 9t + 2$$

if

(i) t *is odd and either*

(a) $3(2t + 1)$ *is not a perfect square, or*

(b) $3(2t + 1)$ *is a perfect square and* $t \equiv 3 \pmod 4$

or

(ii) t *is even and either*

(a) $2t + 1$ *is not a perfect square, or*

(b) $2t + 1$ *is a perfect square and* $t \equiv 2 \pmod 4$

6. SOME PARTIAL RESULTS

Consider an ARD with parameters

$$(6.1) \qquad v = nk = n^2\mu^* , \quad b = nr = n(q^* - 2) , \quad \mu = \mu^*$$

where $n > 4$. Utilizing Lemma 3, it can be shown that any two treatments of the design occur together x times where $nt - 1 \leq x \leq nt + 2$. Now if no pair of treatments occur together $nt + 2$ times then it can be proved as in Corollary 2 that with respect to any treatment θ there are n_i treatments which occur with θ, x_i times where x_i and n_i are given by (2.6). As in Section 4, it can be proved that the v treatments can be divided into n^2 sets S_i of μ^* treatments such that any two treatments of the same set S_i are 3-associate and with respect to any treatment of S_i, the treatments of a set S_j, $j \neq 1$ are either all 1-associates or all 2-associates. Defining α in a manner analogous to that for the case $n = 3$ in Section 4, we can prove in a similar manner that $\alpha = n - 2$. Similarly it can be shown that if $i \neq j$ and any treatment of S_i is a 2-associate of any treatment of S_j then there are exactly two sets S_l such that each treatment of S_l is a 1-associate of every treatment of S_i and S_j. Hence taking any arbitrary treatment θ_i from each S_i we have a set of n^2 treatments with $n_1 = 2(n-1)$ and $p^1_{11} = n - 2$, $p^2_{11} = 2$. Since $n > 4$, this implies [11] that the n^2 treatments have an L_2 association scheme and hence the v treatments can be exhibited in the scheme

$$\begin{array}{l} S_{11} \quad S_{12} \ldots S_{1n} \\ S_{21} \quad S_{22} \ldots S_{2n} \\ \cdots\cdots\cdots\cdots \\ S_{n1} \quad S_{n2} \ldots S_{nn} \end{array}$$

where (i) any two treatments from the same S_{ij} occur together $nt - 1$ times, (ii) any treatment from S_{ij} and any treatment from $S_{ij'}$, $j \neq j'$, or any treatment from S_{ij} and any treatment from $S_{i'j}$, $i' \neq i$, occur together nt times, and (iii) any treatment from S_{ij} and any treatment from $S_{i'j'}$, $i \neq i'$, $j \neq j'$, occur together $nt + 1$ times. It is then obvious that by adding two replications corresponding to the rows and columns of the above scheme we get ARBIBD.

$$v = nk = n^2\mu^* , \qquad b = nr = nq^* , \qquad \lambda = nt + 1 .$$

We note that the number of coincidences between any two columns of an array, i.e., the number of rows in which the symbols occuring in the two columns have the same value is also equal to the number of times the corresponding treatments occur together in the ARD obtained from the array with the help of Lemma 1. Hence for $n > 4$, an array $A(n, q^* - 2, \mu^*)$ of deficiency two can be embedded in the maximal array with q^* rows if no two columns of the array have $nt + 2$ coincidences.

Now consider an ARD with parameters

$$(6.2) \qquad v = 2k = 4(t+1) , \quad b = 2r = 2(4t) , \quad \mu = t + 1$$

corresponding to an array $A(2, 4t, t + 1)$. There from Lemma 3, it can be shown that any two treatments of (6.2) occur together x times where $2t - 2 \leq x \leq 2t + 1$. Assuming that no two treatments in (6.2) occur together $2t + 1$ times, it can be shown as in Corollary 2, that with respect to any treatment θ

of (6.2) any other treatment occurs with it either $2t-2$ or $2t$ times and the number of treatments which occur with θ $2t$ times is $3(t+1)$. We can then write

$$NN' = 4tI + 2tA + (2t-2)(J - I - A)$$

where A is a regular adjacency matrix of valency $3(t+1)$. Since the characteristic roots of $N'N$ are $8t(t+1)$, $2(t+1)$, 0 with the first root of multiplicity one, NN' also has these characteristic roots with $8t(t+1)$ of multiplicity one. It can now be seen that A has a characteristic root $3(t+1)$ of multiplicity one whereas the other two roots are 0 and $-(t+1)$. It now follows from [Lemma 7, 13] that N is the incidence matrix of a GDD with 4 groups of $t+1$ each such that any two treatments of the same group occur together $2t-2$ times while any two treatments coming from different groups occur together $2t$ times. If S_1, S_2, S_3, S_4 are the 4 groups, then by adjoining three replications corresponding to the sets of blocks $[(S_1S_2)\cdot(S_3S_4)]$, $[(S_1S_3)(S_2S_4)]$, $[(S_1S_4)(S_2S_3)]$ to the blocks of (6.2) we get an ARBIBD with

$$v = 2k = 4(t+1)\,, \qquad b = 2r = 2(4t+3)\,, \qquad \lambda = 2t+1\,.$$

Hence an array of $A(2, 4t, 4t+1)$ of deficiency 3 can be embedded in the maximal array of $(4t+3)$ rows if no two columns of the array have $2t+1$ coincidences.

References

1. Agarwal, Hiralal. "On the Bounds of the Number of Common Treatments between Blocks of Semiregular Group Divisible Designs," *J. Amer. Statist. Assoc.*, **59** (1964), 867-871.
2. Bose, R. C. "A Note on the Resolvability of Balanced Incomplete Block Designs," *Sankhyā*, **6** (1962), 105-110.
3. Bose, R. C. and Bush, K. A. "Orthogonal Arrays of Strength Two and Three," *Ann. Math. Statist.*, **23** (1952), 508-524.
4. Bose, R. C. and Connor, W. S. "Combinatorial Properties of Group Divisible Incomplete Block Designs," *Ann. Math. Statist.*, **23** (1952), 367-383.
5. Bruck, R. H. "Finite Nets II, Uniqueness and Embedding," *Pacific J. Math.*, **13** (1963), 421-457.
6. Bush, K. A. "Orthogonal Arrays of Index Unity," *Ann. Math. Statist.*, **23** (1952), 426-434.
7. Connor, W. S. and Clatworthy, W. H. "Some Theorems for

Partially Balanced Designs," *Ann. Math. Statist.*, **25** (1956), 100-112.
8. Fisher, R. A. "An Examination of Different Possible Solutions of a Problem in Incomplete Blocks," *Annals of Eugenics*, **10** (1940), 52-75.
9. Rao, C. R. "On a Class of Arrangements," *Proc. Edinburgh Math. Soc., Ser. 2*, **8** (1947), 119-125.
10. Shrikhande, S. S. "The Non-Existence of Certain Affine Resolvable Balanced Incomplete Block Designs," *Canad. J. Math.*, **5** (1953), 413-420.
11. Shrikhande, S. S. "The Uniqueness of the L_2 Association Scheme," *Ann. Math. Statist.*, **30** (1959), 781-798.
12. Shrikhande, S. S. "A Note on Mutually Orthogonal Latin Squares," *Sankhyā, Ser. A*, **23** (1961), 115-116.
13. Shrikhande, S. S. and Bhagwandas. "Duals of Incomplete Block Designs," *J. Indian Statist. Assoc.*, **3** (1965), 30-37.
14. Shrikhande, S. S. and Bhagwandas. "A Note Embedding for Hadamard Matrices," to appear in *Contributions to Statistics and Probability. Essays in Memory of Samarendra Nath Roy*, University of North Carolina Press, 1967.

Discussion on the Paper of Professor Shrikhande and Mr. Bhagwandas

PROFESSOR K. TAKEUCHI: I would like to ask three questions and make one remark. First, to what extent is the principle which Professor Shrikhande uses in his paper applicable in establishing a method of augmenting a design belonging to some class, in order to obtain a larger one belonging to the same class by considering the dual of the first, enlarging the dual, and then taking the dual again? Is there any scope for general applicability?

Secondly, it is generally easier to obtain a complete array directly if possible than to construct one by the method of this paper from an array of deficiency one or two. Is there any special case when Professor Shrikhande's theorem can be used in a constructive way?

Finally, the main theorem is chiefly applied to establish the nonexistence of an array with deficiency one or two from the nonexistence of a complete array. But upon examining the proofs of nonexistence of complete arrays, one notes that it is usually not necessary to be able to establish the complete array from an array of deficiency one or two in order to establish the nonexistence of the latter. Actually it may be sufficient to obtain idempotents and multiplicities. Thus the non-existence can be established in cases when the theorem of this paper is inapplicable. Isn't there any line of approach adhering to this viewpoint?

The main theorem relies heavily on the theorem that a GDRD is an ARD if and only if $b = v + r - 1$, so that the situation is very delicate, and it is usually very difficult to obtain a method of embedding an array with deficiency d into one with deficiency $d - 1$.

PROFESSOR SHRIKHANDE replied as follows:

In reply to the first question I would like to point out two applications of the method developed in this paper. This first application of this method is in embedding a BIBD with parameters

$$v = \binom{n-1}{2}, \quad b = \binom{n}{2}, \quad r = n, \quad k = n - 2, \quad \lambda = 2$$

in the corresponding symmetric BIBD with parameters

$$v = b = \binom{n-1}{2} + 1, \quad r = k = n, \quad \lambda = 2$$

which was given by the author ["Relation between Certain Incomplete Block Designs", *Contributions to Probability and Statistics*, Stanford U. Press (1960), pp. 388–395]. Yet another application due to the author is the embedding of an ARBIBD with parameters

$$v = nk = n^2((n-1)t + 1), \quad b = nr = n(n^2t + n + 1), \quad \lambda = nt + 1$$

in the symmetric BIBD with parameters

$$v = b = n^2(nt + 1) + n + 1, \quad r = k = n^2t + n + 1, \quad \lambda = nt + 1,$$

which can be done if a symmetric BIBD with parameters

$$v = b = n^2t + n + 1, \quad r = k = nt + 1, \quad \lambda = t$$

exists. ["On the Nonexistence of Affine Resolvable Balanced Incomplete Block Designs", *Sankhyā*, **11** (1951), 185–186]. Other applications will depend upon the existence of tactical configurations with similar "nice" properties.

About the second question I am afraid I see no way at present to use the methods developed here in a constructive manner. Restricting ourselves to the case of mutually orthogonal latin squares this question amounts to the following: Given positive integers n and d,

$$n > (d-1)(d^3 - d^2 + d + 2)/2$$

such that there is no known method of constructing an affine plane of order n, is there any method of constructing a set of $n - 1 - d$ mutually orthogonal latin squares of order n? This is indeed a difficult question and the answer lies in discovering a suitable system which will deliver the goods.

About the final question, I may add that improved results will depend upon (i) more powerful results on the nonexistence of complete arrays, (ii) better results on embedding an array of deficiency d into the corresponding complete array. They can also be improved if one can embed an array of deficiency d into an array of deficiency $d - 1$, which as Professor Takeuchi himself notes is very difficult, and if one can prove the nonexistence of an array of deficiency $d - 1$ by a method not depending upon this paper.

Part III

ERROR-CORRECTING CODES AND OTHER PROBLEMS OF INFORMATION THEORY

CHAPTER 16

File Organization of Records with Multiple-Valued Attributes for Multi-Attribute Queries

R. C. BOSE[1], *University of North Carolina at Chapel Hill,*
C. T. ABRAHAM and S. P. GHOSH, *IBM Watson Research Center*

1. INTRODUCTION

1. A large volume of data may be stored in a number of different ways for different purposes. In certain situations each item of the data may be represented by an n-vector, each component of which is a number providing information with respect to one of a set of n attributes $A_1, A_2, \dots, A_n$. This number may be called the value of the corresponding attribute. In this paper we shall only consider the case when each attribute can take one of q values, where q is a prime power. The q different values which an item can have with respect to any attribute may be identified with the q different elements of a Galois field $GF(q)$. Each item will also have an identifying number i, different for different items. We may thus speak of the ith item or record. If a_{ij} is the value of the ith item with respect to the jth attribute A_j, then we

[1] This work was supported in part by Contract AF 30(602)-4088.

shall call

$$f(i) = (a_{i1}, a_{i2}, \ldots, a_{ij}, \ldots, a_{in}),$$

the attribute vector of the ith item. The identifying number i together with attribute vector $f(i)$ constitute the record of the ith item. The set of all records constitutes the file I. We shall denote by N, the number of records in the file I. The attribute vectors are then elements of the vector space of n-vectors over the Galois field $GF(q)$.

2. A retrieval request or a query Q is a request to retrieve from the file, the subset of all records for which a certain subset of attributes possess certain specific values. Thus we may want to retrieve all records for which the attributes $A_{j_1}, A_{j_2}, \ldots, A_{j_g}$ possess the values $v_{j_1}, v_{j_2}, \ldots, v_{j_g}$. We shall then say that the query relates to the attributes $A_{j_1}, A_{j_2}, \ldots, A_{j_g}$. We shall denote this query by

$$Q\begin{pmatrix} A_{j_1} & A_{j_2} & \ldots & A_{j_g} \\ v_{j_1} & v_{j_2} & \ldots & v_{j_g} \end{pmatrix}.$$

The records for which the attributes $A_{j_1}, A_{j_2}, \ldots, A_{j_g}$ have the values $v_{j_1}, v_{j_2}, \ldots, v_{j_g}$ shall be said to satisfy the query, or pertain to the query.

3. A file organization scheme consists of arranging the records according to a scheme which will reduce the time needed for searching records, for a given class of queries. The problem of file organization is fairly simple when queries relate to only one attribute. A summary of this work has been given by Buchholz (1963). Prywes et. al. (1961) attempted the problem of minimizing search time for multiple attribute queries by grouping attributes into composite attributes and forming a tree structure, and Davis and Lin (1965) suggested the formation of partition classes by considering possible values of logical fields. Abraham, Ghosh and Ray-Chaudhuri (1965) were the first to use finite geometry to construct combinatorial filing schemes. Their methods were developed for binary-valued attributes and consisted of forming groups of records in such a manner that the group containing records pertaining to a given query could be determined algebraically, thus expediting the search. Ray-Chaudhuri (1966) discussed some further combinatorial properties of file organization schemes for binary-valued attributes. Ghosh and Abraham (1966) developed the theory for file organization schemes for multiple-valued attributes, where the attributes have equal number of possible values. However, these schemes were limited to queries involving only two values from two different attributes. In this paper we shall develop filling schemes, when queries relate to some g attributes where $g \leq t \leq n$, t being a fixed positive integer. The case when the number of values which two different attributes can take is not nceessarily equal will be considered in a subsequent paper.

4. In most computerized filing systems the records are stored in some comparatively slow permanent memory. The address of the permanent memory for a record is called the *accession number* of the record, and is usually much smaller in size than the complete record.

A set of addresses of the comparatively faster memory is reserved for storing the accession numbers. Let this set be M. File organization schemes have two features. Firstly there is a rule, the *storage rule* which defines the subset $\sigma(i)$ of the elements of M, where the accession number of the ith record is to be stored. Again there is a *retrieval rule* for finding out the elements of M where the accession numbers of records pertaining to any given query Q relating to any $g \leq t$ attributes are stored. The storage and retrieval rules must be designed to facilitate the retrieval of records pertaining to any query Q of the type indicated.

2. MATHEMATICAL REPRESENTATION OF ATTRIBUTES

1. In this section we shall develop a mathematical representation of the attributes, which we shall use as a basis of our file organization scheme. Let the attributes be $A_1, A_2, \ldots, A_n$. As explained in the introduction we shall consider the situation when each query relates to t or fewer attributes. Let

$$\text{(2.1.1)} \qquad \boldsymbol{H} = \begin{pmatrix} h_{11} & h_{12} & \ldots & h_{1r} \\ h_{21} & h_{22} & \ldots & h_{2r} \\ \ldots & \ldots & \ldots & \ldots \\ h_{n1} & h_{n2} & \ldots & h_{nr} \end{pmatrix}$$

be an $n \times r$ matrix of rank $r < n$, with elements from $GF(q)$, such that any t rows are independent. In this case the matrix is said to possess the property (P_t). Such matrices were first considered by one of the authors [Bose (1947)] in connection with the problem of confounding in symmetrical factorial designs, and are now quite familiar from their use in the construction of error correcting codes [Bose and Ray-Chaudhuri (1960 *a*) and (1960 *b*), and Peterson (1960 and 1961)], even though the problem of finding the most efficient matrix $\boldsymbol{H}$ with the property (P_t), i.e., a matrix with the minimum value of r when q and n are given, is far from solved. One method of obtaining $\boldsymbol{H}$ will be described in the next paragraph of this section. For the present we shall assume that $\boldsymbol{H}$ is known and proceed to give a representation for the attributes.

The jth attribute A_j $(j = 1, 2, \ldots, n)$ will be represented by the linear form

$$L_j = h_{j1}x_1 + h_{j2}x_2 + \ldots + h_{jr}x_r \tag{2.1.2}$$

where $(h_{j1}, h_{j2}, \ldots, h_{jr})$ is the jth row of the matrix $\boldsymbol{H}$ given by (2.1.1). To any query

$$\boldsymbol{Q}\begin{pmatrix} A_{j_1} & A_{j_2} & \ldots & A_{j_g} \\ v_{j_1} & v_{j_2} & \ldots & v_{j_g} \end{pmatrix}, \tag{2.1.3}$$

$g \leq t$, we shall associate the set of equations

$$L_{j_1} = v_{j_1}, \quad L_{j_2} = v_{j_2}, \quad \ldots, \quad L_{j_g} = v_{j_g} \tag{2.1.4}$$

which can be written out in full as

$$\begin{aligned} h_{j_1 1}x_1 + h_{j_1 2}x_2 + \ldots + h_{j_1 r}x_r &= v_{j_1} \\ h_{j_2 1}x_1 + h_{n_2 2}x_2 + \ldots + h_{j_2 r}x_r &= v_{j_2} \\ \ldots\ldots\ldots\ldots\ldots \\ h_{j_g 1}x_1 + h_{j_g 2}x_2 + \ldots + h_{j_g r}x_r &= v_{j_g}\,. \end{aligned} \tag{2.1.5}$$

Let Ω_0 be the set of r-vectors over $GF(q)$ for which not more than t of the coordinates are non-null. The number of vectors in Ω_0 is

$$b_0 = 1 + (q-1)\binom{r}{1} + (q-1)^2\binom{r}{2} + \ldots + (q-1)^t\binom{r}{t} \tag{2.1.6}$$

since there are $(q-1)^s\binom{r}{s}$ distinct r-vector over $GF(q)$ for which exactly s of the coordinates are non-null. We shall now show that there is at least one vector in Ω_0 which satisfies (2.1.5). Since $\boldsymbol{H}$ has the property (P_t), the matrix $\boldsymbol{G}$ of the coefficients on the left side of (2.1.5) is of rank g. Hence there is at least one $g \times g$ submatrix of $\boldsymbol{G}$ which is non-singular. Let this submatrix consist of the columns $\boldsymbol{k}_1, \boldsymbol{k}_2, \ldots, \boldsymbol{k}_g$ of $\boldsymbol{G}$. In the equations (2.1.5) we can put all the variates x_k for which $k \neq k_1, k_2, \ldots,$ or k_g equal to zero, and solve the resulting equations for $x_{k_1}, x_{k_2}, \ldots, x_{k_g}$, obtaining

$$x_{k_1} = u_{k_1}, \quad x_{k_2} = u_{k_2}, \quad \ldots, \quad x_{k_g} = u_{k_g}.$$

We then get a solution

$$\boldsymbol{u}' = (u_1, u_2, \ldots, u_n),$$

of (2.1.5) for which all the coordinates other than the k_1th, k_2th, ..., k_gth are equal to zero. Hence $\boldsymbol{u}'$ belongs to Ω_0.

We have now shown that there is at least one r-vector in Ω_0 for which the equations corresponding to the query $\boldsymbol{Q}$ given by (2.1.3) are satisfied.

There will in general be many vectors of Ω_0, which satisfy (2.1.5). We shall now give a procedure for selecting in a unique manner one of these vectors. The set of solutions of the equations (2.1.5) is not changed by elementary row operations of the following types:

(I) Multiplication of any equation by a non-zero element of the field.

(II) Addition of any two equations.

By using operations of type I and II, (i) any equation may be replaced by the sum of itself and a multiple of any other equation (ii) any two equations may be interchanged. Also in view of the fact that the rank of $\boldsymbol{G}$ is g, the left hand side in any equation will not vanish during these operations. By using these operations the equations can be brought to a unique echelon form satisfying the following conditions:

(i) The first non-zero coefficient on the left hand side of each equation is unity.

(ii) If the first non-zero coefficient in the ith equation is the coefficient of x_{k_i}, then $k_1 < k_2 < \ldots < k_g$.

(iii) The coefficient of x_{k_i} is zero in every equation except the ith, $i \leq g$.

Let the new equations be

$$
(2.1.7)\qquad
\begin{aligned}
b_{11}x_1 + b_{12}x_2 + \ldots + b_{1r}x_r &= u_{k_1}\\
b_{21}x_1 + b_{22}x_2 + \ldots + b_{2r}x_r &= u_{k_2}\\
\ldots\ldots\ldots\ldots\ldots &\\
b_{g1}x_1 + b_{g2}x_2 + \ldots + b_{gr}x_r &= u_{k_g}
\end{aligned}
$$

where

$$
(2.1.8)\qquad b_{ij} = 0 \text{ if } j < k_i,\quad b_{ik_i} = 1,\quad b_{sk_i} = 0 \text{ if } s \neq k_i;\quad i \leq g,\quad j \leq r.
$$

If we put $x_s = 0$ if $s \neq k_1$ or $k_2 \ldots$ or k_g we get the canonical solution

$$
(2.1.9)\qquad \boldsymbol{u}' = (u_1, u_2, \ldots, u_r),
$$

where $u_s = 0$ for $s \neq k_1$ or $k_2 \ldots$ or k_g.

For any r-vector $\boldsymbol{u}'$, the linear form L_j given by (2.1.2) may be said to attain the value

$$
(2.1.10)\qquad L_j(\boldsymbol{u}') = h_{j1}u_1 + h_{j2}u_2 + \ldots + h_{jr}u_r
$$

at $\boldsymbol{u}'$. Hence we state:

Theorem 2.1. *To any query*

$$
\boldsymbol{Q}\begin{pmatrix} A_{j_1} & A_{j_2} & \ldots & A_{j_g} \\ v_{j_1} & v_{j_2} & \ldots & v_{j_g} \end{pmatrix},
$$

we can make correspond a unique r-vector $\boldsymbol{u}'$ of Ω_0, viz. the canonical solution of the equations associated to the query, such that the linear forms $L_{j_1}, L_{j_2}, \ldots, L_{j_g}$ attain the values $v_{j_1}, v_{j_2}, \ldots, v_{j_g}$ at $\boldsymbol{u}'$.

The total number of queries is

$$
(2.1.11)\qquad q\binom{n}{1} + q^2\binom{n}{2} + \ldots + q^t\binom{n}{t}.
$$

Since $n > r$, this number is larger than the number b_0 of vectors in Ω_0, given by (2.1.6). Hence the correspondence described above is not (1, 1). In general the same vector of Ω_0 will correspond to many different queries, though it is possible that there are vectors in Ω_0 not corresponding to any query.

Let Ω be the subset of Ω_0, such that to any vector of Ω there corresponds at least one query. If b is the number of vectors in Ω, then

$$(2.1.12) \qquad b \leq b_0 = 1 + (q-1)\binom{r}{1} + (q-1)^2\binom{r}{2} + \ldots + (q-1)^t\binom{r}{t}.$$

2. We shall now briefly describe a method of obtaining $n \times r$ matrices with elements $GF(q)$, which are of rank r and have the property (P_t) such that no t rows are dependent. Let $t < n \leq q^m - 1$, where t, n, and m are integers. Extend the field $GF(q)$ to $GF(q^m)$ [For further details on Galois fields see Carmichael (1937)]. Let V_m be the vector space of m-vectors with elements from $GF(q)$. We can institute a correspondence between the element $(a_0, a_1, \ldots, a_{m-1})$ of V_m, and the element $a_0 + a_1\alpha + \ldots + a_{m-1}\alpha^{m-1}$ or $GF(q^m)$, where α is a given primitive element of $GF(q^m)$. This is a (1, 1) correspondence in which the null vector of V_m corresponds to the null element of $GF(q^m)$ and the sum of any two vectors of V_m corresponds to the sum of the corresponding elements of $GF(q^m)$. We can therefore identify an element α of $GF(q^m)$ with the corresponding vector of V_m. This in effect defines a multiplication of the vectors of V_m and converts it into a field. In particular we can speak of the powers of vectors of V_m, (which are now identified with the elements of $GF(q^m)$). Let $\alpha_1, \alpha_2, \ldots, \alpha_n$ be distinct non-zero elements of $GF(q^m)$. In particular we can choose $\alpha_i = \alpha^{i-1}$. Let

$$\boldsymbol{H}_0 = \begin{pmatrix} \alpha_1 & \alpha_1^2 & \ldots & \alpha_1^t \\ \alpha_2 & \alpha_2^2 & \ldots & \alpha_2^t \\ \ldots & \ldots & \ldots & \ldots \\ \ldots & \ldots & \ldots & \ldots \\ \alpha_n & \alpha_n^2 & \ldots & \alpha_n^t \end{pmatrix}.$$

Then no t rows of $\boldsymbol{H}_0$ are dependent over $GF(q^m)$. This follows by noting that the determinant of the submatrix of $\boldsymbol{H}_0$ formed by the i_1th, i_2th, $\ldots$, i_tth rows is

$$\prod (\alpha_{i_u} - \alpha_{i_v}), \qquad u, v = 1, 2, \ldots, t; \quad u \neq v$$

and is therefore non-zero.

Since $x \to x^q$ is an automorphism of $GF(q^m)$ and $c^q = c$, if c is an element of $GF(q)$, it follows that if we delete from $\boldsymbol{H}_0$ the columns headed by $\alpha_1^q, \alpha_1^{2q}, \ldots, \alpha_1^{sq}$ where $s = [t/q]$, then any t rows of the resulting matrix $\boldsymbol{H}_1$ will still be independent over $GF(q)$, i.e., no linear combination of t rows of $\boldsymbol{H}_1$, with coefficients from $GF(q)$, will vanish. If we now regard the elements of $\boldsymbol{H}_1$ as row vectors of V_m we have an $n \times m(t - [t/q])$ matrix with elements from $GF(q)$, which has the property (P_t), that no t rows are dependent. Here $[x]$ denotes the largest integer not exceeding x. If there are any columns in $\boldsymbol{H}_1$, which are dependent on others we can drop them, and obtain a matrix $\boldsymbol{H}$ still having the property (P_t). Let r be the number of columns in $\boldsymbol{H}$. Since they are independent $\boldsymbol{H}$ has rank r. Hence we have the following theorem:

Theorem 2.2. *If $t < n \leq q^m - 1$, then we can find an $n \times r$ matrix $\boldsymbol{H}$ of rank r with elements from $GF(q)$ and having the property (P_t), where $r \leq m(t - [t/q])$.*

3. The results of the previous two paragraphs are illustrated by the following example.

Example 1. Let $q = 3$, $t = 4$, $n = 8$. If we choose $m = 2$, the condition $n \leq q^m - 1$ is satisfied. Let us extend $GF(3)$ to $GF(3^2)$. The field has four primitive elements of which two are roots of $x^2 + x + 2 = 0$. Let α be one of these roots; then the non-null elements of $GF(3^2)$ can be expressed as powers of α, and using the relation $\alpha^2 + \alpha + 2 = 0$ can be expressed in the form $a_0 + a_1\alpha$ where a_0 and a_1 belong to $GF(3)$. Then $a_0 + a_1\alpha$ can be made to correspond to the vector (a_0, a_1) of the vector space V_2. This can be exhibited in the following table.

Table (2.3.1)

$$\begin{aligned}
\alpha^0 &= 1 &&= (1,0) = \alpha_1\\
\alpha &= \quad\alpha &&= (0,1) = \alpha_2\\
\alpha^2 &= 1 + 2\alpha &&= (1,2) = \alpha_3\\
\alpha^3 &= 2 + 2\alpha &&= (2,2) = \alpha_4\\
\alpha^4 &= 2 &&= (2,0) = \alpha_5\\
\alpha^5 &= \quad 2\alpha &&= (0,2) = \alpha_6\\
\alpha^6 &= 2 + \alpha &&= (2,1) = \alpha_7\\
\alpha^7 &= 1 + \alpha &&= (1,1) = \alpha_8
\end{aligned}$$

We can drop the third column of $\boldsymbol{H}_0$ and remembering that $\alpha^8 = 1$, we can write

$$H_1 = \begin{pmatrix} 1 & 1 & 1 \\ \alpha & \alpha^2 & \alpha^4 \\ \alpha^2 & \alpha^4 & 1 \\ \alpha^3 & \alpha^6 & \alpha^4 \\ \alpha^4 & 1 & 1 \\ \alpha^5 & \alpha^2 & \alpha^4 \\ \alpha^6 & \alpha^4 & 1 \\ \alpha^7 & \alpha^6 & \alpha^4 \end{pmatrix} = \begin{pmatrix} 1 & 0 & 1 & 0 & 1 & 0 \\ 0 & 1 & 1 & 2 & 2 & 0 \\ 1 & 2 & 2 & 0 & 1 & 0 \\ 2 & 2 & 2 & 1 & 2 & 0 \\ 2 & 0 & 1 & 0 & 1 & 0 \\ 0 & 2 & 1 & 2 & 2 & 0 \\ 2 & 1 & 2 & 0 & 1 & 0 \\ 1 & 1 & 2 & 1 & 2 & 0 \end{pmatrix}.$$

Finally we can drop the last column of H_1, when it is written as a matrix over $GF(3)$. We thus get

(2.3.2)
$$H = \begin{pmatrix} 1 & 0 & 1 & 0 & 1 \\ 0 & 1 & 1 & 2 & 2 \\ 1 & 2 & 2 & 0 & 1 \\ 2 & 2 & 2 & 1 & 2 \\ 2 & 0 & 1 & 0 & 1 \\ 0 & 2 & 1 & 2 & 2 \\ 2 & 1 & 2 & 0 & 1 \\ 1 & 1 & 2 & 1 & 2 \end{pmatrix}$$

where H is an 8×5 matrix over $GF(3)$, which has the property (P_4) that no four rows are dependent. Note that $r = 5 < 2(4 - [4/3])$.

We can now represent the $n = 8$ attributes by the linear forms shown in following table:

Table (2.3.3)

Name of the attribute	*Linear form*
L_1	$x_1 + x_3 + x_5$.
L_2	$x_2 + x_3 + 2x_4 + 2x_5$.
L_3	$x_1 + 2x_2 + 2x_3 + x_5$.
L_4	$2x_1 + 2x_2 + 2x_3 + x_4 + 2x_5$.
L_5	$2x_1 + x_3 + x_5$.
L_6	$2x_2 + x_3 + 2x_4 + 2x_5$.
L_7	$2x_1 + x_2 + 2x_3 + x_5$.
L_8	$x_1 + x_2 + 2x_3 + x_4 + 2x_5$.

It is easily checked that any four of these linear forms are independent over $GF(3)$.

We will now illustrate the procedure by which we can associate with every query a unique point of Ω.

(i) Consider the query

$$Q\begin{pmatrix} A_1 & A_2 & A_3 & A_4 \\ v_1 & v_2 & v_3 & v_4 \end{pmatrix}. \tag{2.3.4}$$

The associated equations are

$$\begin{aligned} x_1 \quad\quad + x_3 \quad\quad + x_5 &= v_1 \\ x_2 + x_3 + 2x_4 + 2x_5 &= v_2 \\ x_1 + 2x_2 + 2x_3 \quad\quad + x_5 &= v_3 \\ 2x_1 + 2x_2 + 2x_3 + x_4 + 2x_5 &= v_4 \,. \end{aligned}$$

Reducing these equations to the echelon form we get

$$\begin{aligned} x_1 \quad + 2x_5 &= 2v_2 + 2v_4 \\ x_2 \quad + 2x_5 &= 2v_1 + v_2 + 2v_3 + v_4 \\ x_3 \quad + 2x_5 &= v_1 + v_2 + v_4 \\ x_4 + 2x_5 &= v_2 + 2v_3 + 2v_4 \,. \end{aligned}$$

Putting $x_5 = 0$, we get for the vector of Ω corresponding to the query

$$\begin{aligned} u' = (&2v_2 + 2v_4, 2v_1 + v_2 + 2v_3 + v_4, \\ &v_1 + v_2 + v_4, v_2 + 2v_3 + 2v_4, 0) \,. \end{aligned} \tag{2.3.5}$$

In any particular case v_1, v_2, v_3, v_4 will have some definite values. For example the vector of Ω corresponding to the query

$$Q\begin{pmatrix} A_1 & A_2 & A_3 & A_4 \\ 1 & 0 & 2 & 1 \end{pmatrix}, \tag{2.3.6}$$

is $(2, 1, 2, 0, 0)$.

(ii) As another example consider the query

(2.3.7)
$$Q\begin{pmatrix} A_5 & A_6 & A_7 \\ v_5 & v_6 & v_7 \end{pmatrix}.$$

The associated equations are

$$\begin{aligned} 2x_1 \quad\quad + x_3 \quad\quad + x_5 &= v_5 \\ 2x_2 + x_3 + 2x_4 + 2x_5 &= v_6 \\ 2x_1 + x_2 + 2x_3 \quad\quad + x_5 &= v_7 . \end{aligned}$$

Reducing these equations to the echelon form we get

$$\begin{aligned} x_1 \quad\quad + x_4 \quad\quad &= 2v_6 + 2v_7 \\ x_2 \quad + 2x_4 + 2x_5 &= v_5 + v_6 + 2v_7 \\ x_3 + x_4 + x_5 &= v_5 + 2v_6 + 2v_7 . \end{aligned}$$

Putting $x_4 = 0$, $x_5 = 0$, we get for the vector Ω corresponding to the query

(2.3.8) $\quad u' = (2v_6 + 2v_7, v_5 + v_6 + 2v_7, v_5 + 2v_6 + 2v_7, 0, 0)$.

In particular suppose $v_5 = 0$, $v_6 = 1$, $v_7 = 0$. Then the vector of Ω corresponding to the query

(2.3.9)
$$Q\begin{pmatrix} A_5 & A_6 & A_7 \\ 0 & 1 & 0 \end{pmatrix},$$

is (2, 1, 2, 0, 0).

Thus the queries (2.3.6) and (2.3.9) correspond to the same vector of Ω, illustrating the fact that the correspondence between queries and vectors is not (1, 1).

3. GENERALIZED MULTIPLE-VALUED FILING SCHEME

1. Let u' be an element of Ω. Thus u' is an r-vector over $GF(q)$, having not more than t non-zero coordinates, and corresponding to at least one query relating to t or fewer attributes. To u' let there correspond an n-vector

$$B(u') = (b_1, b_2, \ldots, b_n) ,$$

over $GF(q)$ such that $b_i = L_i(\boldsymbol{u}')$. Thus the ith coordinate of $\boldsymbol{u}'$ is the value attained at $\boldsymbol{u}'$ by the linear form L_i, which represents the attribute A_i. $\boldsymbol{B}(\boldsymbol{u}')$ will be called the *block* corresponding to $\boldsymbol{u}'$.

To each element of Ω there corresponds a unique block but not every n-vector over $GF(q)$ is a block, since the number of n-vectors is in general much larger than the number of blocks.

We shall now show that the blocks corresponding to different r-vectors of Ω must be different. If possible let $\boldsymbol{u}_1'$, $\boldsymbol{u}_2'$ be different elements of Ω such that $\boldsymbol{B}(\boldsymbol{u}_1') = \boldsymbol{B}(\boldsymbol{u}_2')$. Then $L_i(\boldsymbol{u}_1') = L_i(\boldsymbol{u}_2')$ for $i = 1, 2, \ldots, n$. Hence $L_i(\boldsymbol{u}_1' - \boldsymbol{u}_2') = 0$. Now the rank of $\boldsymbol{H}$ is r. Hence there exists at least one set of r rows of $\boldsymbol{H}$ which are independent, say the j_1th, j_2th, ..., j_rth rows. Hence

$$(3.1.1) \qquad L_{j_1}(\boldsymbol{u}_1' - \boldsymbol{u}_2') = 0\,, \quad L_{j_2}(\boldsymbol{u}_1' - \boldsymbol{u}_2') = 0\,, \quad \ldots, \quad L_{j_r}(\boldsymbol{u}_1' - \boldsymbol{u}_2') = 0\,.$$

This shows that $\boldsymbol{u}_1' - \boldsymbol{u}_2'$ is a solution of the set of r independent linear homogeneous equations (in r variables)

$$(3.1.2) \qquad L_{j_1} = 0\,, \quad L_{j_2} = 0\,, \quad \ldots, \quad L_{j_r} = 0\,.$$

Hence $\boldsymbol{u}_1' - \boldsymbol{u}_2' = 0$ or $\boldsymbol{u}_1' = \boldsymbol{u}_2'$ which is a contradiction. Thus the correspondence between the vectors of Ω and the blocks is a (1, 1) correspondence. The number of blocks is $b \le b_0$, where b_0 is given by (2.1.6).

We can now make a correspondence between queries and blocks. If to a query $\boldsymbol{Q}$ there corresponds the vector $\boldsymbol{u}'$ of Ω, then we say that $\boldsymbol{B}(\boldsymbol{u}')$ is the block corresponding to $\boldsymbol{Q}$. To each query $\boldsymbol{Q}$ relating to t or fewer attributes there corresponds a unique block, but to each block there correspond in general many queries.

Theorem 3.1. *If the block* $(b_1, b_2, \ldots, b_n)$ *corresponds to the query*

$$(3.1.3) \qquad \boldsymbol{Q}\begin{pmatrix} A_{j_1} & A_{j_2} & \ldots & A_{j_g} \\ v_{j_1} & v_{j_2} & \ldots & v_{j_g} \end{pmatrix}$$

then

$$(3.1.4) \qquad b_{j_1} = v_{j_1}\,, \quad b_{j_2} = v_{j_2}\,, \quad \ldots, \quad b_{j_g} = v_{j_g}\,.$$

Let $u' = (u_1, u_2, \ldots, u_n)$ be the vector of Ω corresponding to the query Q given by (3.1.3). Then u' is a solution of the equations

$$L_{j_1} = v_{j_1}, \quad L_{j_2} = v_{j_2}, \quad \ldots, \quad L_{j_g} = v_{j_g}.$$

Hence $L_{j_s}(u') = v_{j_s}$, $s = 1, 2, \ldots, g$. Again from the definition of the block $B(u')$, $b_i = L_i(u')$ for any i. The required result follows by taking $i = j_1, j_2, \ldots, j_g$.

Example 2. Let q, t, n and L_i, $i = 1, 2, \ldots, 8$ be as in Example 1. We have seen that the vector of Ω corresponding to the query

$$Q\begin{pmatrix} A_1 & A_2 & A_3 & A_4 \\ 1 & 0 & 2 & 1 \end{pmatrix} \quad \text{or} \quad Q\begin{pmatrix} A_5 & A_6 & A_7 \\ 0 & 1 & 0 \end{pmatrix}$$

is $u' = (2, 1, 2, 0, 0)$. Calculating $L_i(u')$ for $i = 1, 2, \ldots, 8$, we find that

$$B(2, 1, 2, 0, 0) = (1, 0, 2, 1, 0, 1, 0, 1).$$

Notice that the Theorem 3.1 is satisfied.

2. Let M be the set of addresses of the memory which is reserved for storing the accession numbers of records. M can be divided into b subsets corresponding to the b blocks, the subset of M corresponding to the block $B(u')$ being denoted by $M(u')$. These subsets of M will be called *buckets*. The vector u' of Ω will be called the *label* of the bucket $M(u')$.

Let $A = \{A_1, A_2, \ldots, A_n\}$ be the set of n attributes. $M(u')$ is divided into $2^n - 1$ disjoint subsets, called *sub-buckets*, one sub-bucket corresponding to each non-empty subset A. The sub-bucket of $M(u')$ which corresponds to the subset A' of A may be denoted by $M(u', A')$.

Let $B(u') = (b_1, b_2, \ldots, b_n)$ be the block corresponding to the vector u' of Ω and let $f(i) = (a_{i1}, a_{i2}, \ldots, a_{in})$ be the attribute vector of the ith record. Let A' be a subset of A such that $A_k \in A'$ if and only if $a_{ik} = b_k$. We shall then use the notation $B(u') \otimes f(i) = A'$. It is clear from the definition of a block that $A_k \in A'$ if and only if the value of A_k for the ith record is the same as the value attained by the linear form L_k (representing A_k) at u'.

Our storage rule can now be stated as follows: The accession number of the ith record is not stored in the bucket $M(\boldsymbol{u}')$ if $\boldsymbol{B}(\boldsymbol{u}') \otimes \boldsymbol{f}(i)$ is empty. If $\boldsymbol{B}(\boldsymbol{u}') \otimes \boldsymbol{f}(i) = A'$ is non-empty then the accession number of the ith record is stored in the sub-bucket $M(\boldsymbol{u}', A')$ of the bucket $M(\boldsymbol{u}')$. It is clear that the accession number of any record is not stored more than once in any bucket $M(\boldsymbol{u}')$ according to this storage rule.

3. Let $\sigma(i)$ denote the set of elements of M where the accession number of the ith record is stored. The average of $|\sigma(i)|$ is defined to be the redundancy factor R corresponding to any storage rule. The value of R will in general depend on the distribution of the attribute vectors. We shall now calculate the value of R under the hypothesis of uniform distribution, and corresponding to the storage rule described in the previous paragraph.

There are q^n distinct n-vectors over $GF(q)$. Each is a possible attribute vector. Suppose the number of records having any given attribute vector is c, and is independent of the vector. Then the total number of records is $N = cq^n$. If $\boldsymbol{f}(i)$ is the attribute vector of ith record, then the accession number of this record does not appear in the bucket $M(\boldsymbol{u}')$ if and only if $\boldsymbol{B}(\boldsymbol{u}') \otimes \boldsymbol{f}(i)$ is empty. $\boldsymbol{B}(\boldsymbol{u}')$ being given there are $(q-1)^n$ ways of choosing $(a_{i1}, a_{i2}, \ldots, a_{in})$ such that $a_{i1} \neq b_1$, $a_{i2} \neq b_2, \ldots, a_{in} \neq b_n$. Hence there are $c(q-1)^n$ records whose accession numbers do not occur in the bucket $M(\boldsymbol{u}')$. The number of records whose accession numbers occur in the bucket is therefore $cq^n - c(q-1)^n$. Since there are b buckets the total of the accession numbers stored in M is $bc\{q^n - (q-1)^n\}$. Hence

$$
\begin{aligned}
R &= bc\{q^n - (q-1)^n\}/N \\
&= b\left\{1 - \left(1 - \frac{1}{q}\right)^n\right\} \\
&\leq \left\{1 - \left(1 - \frac{1}{q}\right)^n\right\}\left\{1 + (q-1)\binom{r}{1} + (q-1)^2\binom{r}{2}\right. \\
&\qquad \left. + \ldots + (q-1)^t\binom{r}{t}\right\}
\end{aligned}
$$

from (2.1.12).

Example 3. This is a continuation of Examples 1 and 2, and we shall assume that the values of q, t and n are as before and that the linear forms L_i are those given in Table (2.3.3).

Consider the block $\boldsymbol{B}(\boldsymbol{u}') = (1, 0, 2, 1, 0, 1, 0, 1)$ corresponding to the vector $\boldsymbol{u}' = (2, 1, 2, 0, 0)$ of Ω. Let the attribute vector of the ith record be $\boldsymbol{f}(i) = (0, 1, 2, 1, 1, 0, 2, 1)$. Now $\boldsymbol{B}(\boldsymbol{u}') \otimes \boldsymbol{f}(i) = \{A_3, A_4, A_8\}$. Since this is non-empty the accession number of the ith record should be stored in the sub-bucket $M(\boldsymbol{u}', A')$ of $M(\boldsymbol{u}')$ where $A' = \{A_3, A_4, A_8\}$. On the other hand the accession number of the ith record with the attribute vector $\boldsymbol{f}(i') = (0, 1, 1, 2, 2, 0, 2, 0)$ is not stored in the bucket $M(\boldsymbol{u}')$ since $\boldsymbol{B}(\boldsymbol{u}') \otimes \boldsymbol{f}(i')$ is empty.

An upper bound for the redundancy factor R for our storage rule, under the hypothesis of uniform distribution is given by $R \leq 203$.

4. We shall now give the *retrieval rule* for retrieving all records satisfying the query

$$(3.4.1) \qquad \boldsymbol{Q}\begin{pmatrix} A_{j_1}, & A_{j_2}, \dots, & A_{j_g} \\ v_{j_1}, & v_{j_2}, \dots, & v_{j_g} \end{pmatrix}, \qquad 0 < g \leq t\,.$$

We shall denote by $Q(A)$, the subset $\{A_{j_1}, A_{j_2}, \dots, A_{j_g}\}$ of the attributes to which the query relates. Let $\boldsymbol{u}'$ be the uniquely determined vector of Ω corresponding to the query and let $\boldsymbol{B}(\boldsymbol{u}') = (b_1, b_2, \dots, b_n)$ be the block corresponding to $\boldsymbol{u}'$. Then the bucket $M(\boldsymbol{u}')$ corresponding to $\boldsymbol{B}(\boldsymbol{u}')$ is defined to be the bucket corresponding to the query. From Theorem (3.1)

$$b_{j_1} = v_{j_1}, \quad b_{j_2} = v_{j_2}, \quad \dots, \quad b_{j_g} = v_{j_g}\,.$$

If the ith record satisfies the query, and its attribute vector is $\boldsymbol{f}(i) = (a_{i1}, a_{i2}, \dots, a_{in})$, then

$$a_{ij_1} = v_{j_1}, \quad a_{ij_2} = v_{j_2}, \quad \dots, \quad a_{ij_g} = v_{j_g}\,.$$

Hence

$$(3.4.2) \qquad \boldsymbol{B}(\boldsymbol{u}') \otimes \boldsymbol{f}(i) \supseteq Q(A)\,.$$

Since $Q(A)$ is non-empty, the accession number of the ith record has been stored in the sub-bucket $M(\boldsymbol{u}', A')$ of the bucket $M(\boldsymbol{u}')$, where $A' = \boldsymbol{B}(\boldsymbol{u}') \otimes \boldsymbol{f}(i)$.

Conversely, let A' be a subset of A such that $A' \supseteq Q(A)$. It is clear that every record whose accession number has been stored in the sub-bucket $M(\boldsymbol{u}', A')$ satisfies the query. Thus to find the accession numbers of all records satisfying the query we have to take all the accession numbers stored in

$$M(Q) = \bigcup_{A \supseteq A' \supseteq Q(A)} M(\boldsymbol{u}', A') .$$

It is readily seen that the number of sub-buckets whose union is $M(Q)$ is 2^{n-g}.

Hence the retireval rule may be stated as follows: Let $\boldsymbol{Q}$ be any query relating to the attributes of the subset $Q(A)$ of A. Determine the bucket $M(\boldsymbol{u}')$ corresponding to $\boldsymbol{Q}$. Then the accession numbers of all records satisfying $\boldsymbol{Q}$ are found in $M(Q)$ given by (3.4.3). Conversely any record whose accession number is in $M(Q)$ satisfies the query Q. Once the accession numbers have been found we can retrieve the complete records.

Retrieval can be further facilitated by grouping the buckets into *super-buckets*. Two buckets $M(\boldsymbol{u}_1')$ and $M(\boldsymbol{u}_2')$ will be said to belong to the same super-bucket if the bucket labels $\boldsymbol{u}_1'$ and $\boldsymbol{u}_2'$ are such that corresponding coordinates are both zero or both non-zero. Each super-bucket can be labelled by an r-vector whose coordinates are 0 or 1, according as the jth coordinates of all the buckets contained in the super-bucket are zero or non-zero.

If $\boldsymbol{u}'$ is given then the sub-bucket $M(\boldsymbol{u}', A')$ is completely specified by the subset A' of A. Hence a sub-bucket can be labelled by an n-vector for which the jth coordinate is unity if the attribute $A_j \in A'$, and is zero otherwise. The address of a sub-bucket then consists of its label and the labels of the bucket and the super-bucket containing it. The labels of the sub-buckets entering in $M(Q)$ are easy to compute.

The filing scheme described by the storage and retrieval rules given above may be called a *generalized multiple-valued filing scheme.*

4. The time needed for retrieving the records satisfying a given query will be made up of a number of components:

1. T_1 = Time needed for coding physical attributes into linear forms and the values of the attributes into the elements of $GF(q)$.

2. T_2 = Time needed for reducing the query equations into the echelon form and determining the canonical solution $\boldsymbol{u}'$, which will determine both the super-bucket and the bucket labels.

3. T_3 = Time needed for matching the computed super-bucket label with the stored super-bucket labels.

4. T_4 = Time needed for matching the computed bucket label with the stored bucket labels in the specified super-bucket.

5. T_5 = Time needed for computing the labels of the sub-buckets entering in $M(Q)$.

6. T_6 = Time needed for matching the computed sub-bucket labels with the stored sub-bucket labels in the specified bucket.

7. T_7 = Time needed for physically retrieving the records.

Thus the total time is

$$T = T_1 + T_2 + T_3 + T_4 + T_5 + T_6 + T_7 .$$

T_1, T_2, T_5 and T_7 will depend on the parameters of the computer system used. A special device suggested by Abraham, Ghosh and Ray-Chaudhuri (1966) can perform row echelon reductions much faster than general purpose computers. Its use can materially reduce T_2.

The super-bucket labels can be ordered among themselves. The bucket labels can be ordered within a super-bucket, and the sub-bucket labels can be ordered within a bucket. Thus T_3, T_4 and T_6 can be greatly reduced by performing a binary search [Demuth (1956)].

Example 4. Continuing Examples 1, 2, 3 let us now recount the steps necessary for retrieving the records satisfying the query

$$Q\begin{pmatrix} A_1 & A_2 & A_3 & A_4 \\ 1 & 0 & 2 & 1 \end{pmatrix}.$$

Writing down the associated query equations, and reducing them to the echelon form we find as in Example 1, that the corresponding vector of Ω is

$$u' = (2, 1, 2, 0, 0) .$$

Thus the label of the super-bucket corresponding to the query is (1, 1, 1, 0, 0) and the label of the bucket is (2, 1, 2, 0, 0). By a search among the super-bucket labels the relevant super-bucket label is located, and by search within the super-bucket the relevant bucket label is located. The sub-buckets of $M(\boldsymbol{u}')$ which contain accession numbers of the records satisfying the query correspond to subsets containing the subset $Q(A) = \{A_1, A_2, A_3, A_4\}$. There are 16 such subsets, each corresponding to a sub-bucket with its appropriate label. For example the label of the sub-bucket $M(\boldsymbol{u}', A')$ where $A' = \{A_1, A_2, A_3, A_4, A_6, A_8\}$ is $(1, 1, 1, 1, 0, 1, 0, 1)$. These sixteen sub-buckets are then located by a search within the bucket, and finally the records whose accession numbers are contained in these sub-buckets are pulled out.

Acknowledgement

The authors would like to thank A. J. Hoffman of Thomas J. Watson Research Center, for valuable discussions during the preparation of this manuscript.

References

1. Abraham, C. T., Ghosh, S. P. and Ray-Chaudhuri, D. K. "File Organization Schemes Based on Finite Geometries," IBM Research Report No. 1459, 1965.
2. Abraham, C. T., Ghosh, S. P. and Ray-Chaudhuri, D. K. "An Information Storage and Retrieval System," Patent disclosure, 1966.
3. Bose, R. C. "Mathematical Theory of Symmetrical Factorial Designs," *Sankhyā*, **8** (1947), 107–166.
4. Bose, R. C. and Ray-Chaudhuri, D. K. "On a Class of Error Correcting Binary Group Codes," *Info. and Control*, **3** (1960), 68–79.
5. Bose, R. C. and Ray-Chaudhuri, D. K. "Further Results on Error Correcting Binary Group Codes," *Info. and Control*, **3** (1960), 279–290.
6. Buchholz, W. "File Organization and Addressing," *IBM System J.*, **2** (1963), 86–111.
7. Carmichael, R. D. *Introduction to the Theory of Groups of Finite Order*, Ginn and Co., Boston, 1937.
8. Davis, D. R. and Lin, A. D. "Secondary Key Retrieval Using an IBM 7090-1301 System," *Comm. A. C. M.*, 8 (1965), 243–246.
9. Demuth, H. B. "A Report on Electronic Data Sorting," Stanford Research Institute, 1956.
10. Ghosh, S. P. and Abraham, C. T. "Application of Finite Geometry in File Organization of Records with Multiple Valued Attributes," IBM Research Report No. 1561, 1966.

11. Gray, H. J., Landuer, W. I., Lefkowitz, D., Litwin, S. and Prywers, N. S. "The Multiple List System," University of Pennsylvania Report, 1961.
12. Peterson, W. W. *Error-Correcting Codes*, MIT Press, Cambridge, 1961.
13. Peterson, W. W. "Encoding and Error Correcting Procedures for Bose-Chaudhuri Codes," *IRE Trans.*, **IT-6** (1960), 459–470.
14. Ray-Chaudhuri, D. K. "Combinatorial Information Retrieval Systems for Files," IBM Research Report No. 1554, 1966.

Discussion on the Paper of Professor Bose, Dr. Abraham and Dr. Ghosh

MR. G. G. KOCH: This paper has an important place in the development of filing systems for efficient information retrieval. The filing scheme which the authors have constructed is of importance because it represents one of the first systems which allows rapid retrieval of records satisfying general multi-attribute queries involving multiple-valued attributes.

Until recently, the most well-known type of filing scheme was the inverted filing system (IFS). With the IFS, a bucket corresponded to each level of each attribute. An individual's record was stored in each of the buckets corresponding to the levels of attributes which he possessed. Such systems allow efficient retrieval of queries involving one attribute—indeed, one simply retrieves all records in the bucket corresponding to the level of that particular attribute specified in the query. However, to retrieve a query involving two attributes, one must first extract all records in each of the two corresponding buckets, and then find the records common to the two groups by matching the accession numbers (i.e., the individual's identification numbers). This matching process can take a large amount of computer time. It also increases as the size of the file increases since there are more records to be examined in the matching. For queries involving more than two attributes, the retrieval problem with the IFS becomes more and more serious.

Because of the previously mentioned disadvantages associated with the IFS, a need arose for the development of filing systems permitting more efficient information retrieval. In recent years, research directed at the application of combinatorial mathematics to the design of filing schemes was started at the IBM Thomas J. Watson Research Center. Abraham, Ghosh, and Ray-Chaudhuri (1965) used the theory of finite geometries to form systems which could deal with certain types of queries involving pairs of binary attributes. With these, attributes corresponded to points and buckets to lines. To retrieve a query involving a pair of attributes, one determines the bucket corresponding to the line through the points. Abraham and Ghosh (1966) used deleted finite geometries to construct similar filing schemes capable of efficient retrieval of two-fold queries involving multiple-valued attributes. Following a different approach, Ray-Chaudhuri (1966) proposed a mathematical model useful in developing systems which could handle multi-fold queries for binary attributes. The construction problem with which he was concerned involved finding a minimal set of blocks which covered all t-plets of attributes. He gave some solutions to this problem using the theory of coverings in finite geometries and some of the properties of BCH codes.

The present paper provides a filing system which can deal with the general situation of multi-fold queries involving multiple-valued attributes. The scheme that the authors propose also represents a different point of view in that they associate linear functions (or flat spaces) with attributes and points (so to speak, like the $B(\boldsymbol{u})$) with blocks (buckets). Also, these blocks represent a covering of t-plets, to some extent, in the sense of the work of Ray-Chaudhuri. Most important, however, the system they have developed is one that is readily adapted and mechanized by a computer. It is also reasonable to believe that its retrieval time is quite efficient. In particular, the retrieval time is independent of the size of the file. Finally, one may find an alternative method of determining the bucket which satisfies a query to be of interest. For example, one such procedure is as follows:

1. Let $\boldsymbol{H}$ be the matrix of coefficients specific to the relevant attributes and let $\boldsymbol{v}$ be the relevant levels.
2. To determine the bucket, solve the equations $\boldsymbol{Hx} = \boldsymbol{v}$ with $\boldsymbol{x} = \boldsymbol{H}'(\boldsymbol{HH}')^{-1}\boldsymbol{v}$; i.e., the projection of the solution space on $\boldsymbol{H}$.

The vectors $\boldsymbol{x}$ may be easier to compute than the canonical solution of the authors. On the other hand, the use of this type of solution may require many more buckets and hence produce a scheme which is not as efficient for retrieval.

Even though the authors' paper has provided a solution

to some aspects of the general problem of file designs, a great deal of further research is still needed. Systems which are efficient for cases in which different attributes assume different numbers of levels are of interest. Also, compromise designs which are suitable for one type of query involving some sets of attributes and other types of queries involving other sets of attributes need to be developed for the cases where they are applicable. Finally, before the different systems currently in existence can be effectively compared with one another, the concepts of retrieval time and redundancy need to be more precisely defined. When this is achieved, then one will be able to specify more completely the type of properties which are desirable for filing systems.

CHAPTER 17

Negacyclic Codes for the Lee Metric

ELWYN R. BERLEKAMP[1], *Bell Telephone Laboratories*

1. INTRODUCTION

One of the central problems in coding theory is the design of error correcting codes of given block length, N, over an alphabet of given size, q, and given information rate, $R = k/N$. Ideally, one would like to find the code with the lowest probability of error for a given memoryless channel. For certain types of highly symmetric channels, this problem can be greatly simplyfied by the introduction of an appropriate metric to measure the "distance" between different codewords. The most commonly used such metric is the *Hamming metric*, according to which the distance between two sequences is the number of positions in which they differ. For example, the Hamming distance between the two quintary sequences in Figure 1 is 3:

Figure 1. Two quintary codewords:

Sequence 1:	2, 1, 0, 3, 0, 4, 2, 4, 3, 1
Sequence 2:	2, 1, 1, 3, 0, 2, 1, 4, 3, 1

It is convenient to introduce an arithmetic structure on

[1] This research was partially supported by National Science Foundation Grant No. GP-5790.

the alphabet of q symbols. In general, one may use the ring of integers mod q. If q is a prime power, one may instead take the structure to be the Galois Field, $GF(q)$. In either case, it is convenient to associate each sequence of N letters with a polynomial of degree less than N over the underlying ring (or field). For example, the two sequences in Figure 1 correspond to the polynomials

$$2 + x + \quad + 3x^3 + 4x^5 + 2x^6 + 4x^7 + 3x^8 + 1x^9$$

and

$$2 + x + x^2 + 3x^3 + 2x^5 + 1x^6 + 4x^7 + 3x^8 + 1x^9$$

whose difference is: $\quad 4x^2 + 2x^5 + x^6\,.$

(Here we take the difference mod 5). After defining the Hamming weight of each digit, A, by

$$w_H(A) = \begin{cases} 0 & \text{if} \quad A = 0\,, \\ 1 & \text{if} \quad A \neq 0\,, \end{cases}$$

one may then define the Hamming weight of a codeword as the sum of the Hamming weights of the coefficients of the associated polynomial. The Hamming distance between two codewords is then defined as the Hamming weight of the difference. If the channel has the property that all errors are equally likely, then the probability of receiving one sequence when another sequence is sent is a monotonic decreasing function of the Hamming distance from the transmitted sequence to the received sequence. For this reason, the problem of finding a code with a low probability of error is approximated by the problem of finding a code which will correct all patterns of t or fewer errors. This latter problem has proved considerably more tractable than the former. Following the pioneering work of Shannon (1948), Hamming (1950), and Slepian (1956 and 1960); Hocquenghem (1959) and (independently) Bose and Chaudhuri (1960) presented a remarkable class of binary group codes, which have become known as BCH codes. Shortly thereafter, Peterson (1961) presented a decoding algorithm for these codes, and showed them to be a subset of the "cyclic codes" which had been first studied by Prange (1958). Gorenstein and Zierler (1962) then succeeded in generalizing the codes to the nonbinary case. Additional properties of these codes were discovered by

Mattson and Solomon (1962), and a refined decoding algorithm to include erasures was presented by Forney (1965). In 1966 Berlekamp discovered another decoding algorithm which greatly reduced the complexity of the decoder. Although research still continues on many unanswered questions about the weight distributions of the codewords, the weight distributions of the coset leaders, the permutation groups which leave the codes invariant, and algorithms to decode more than t errors, the BCH codes have already proved themselves to be of considerable value in some situations. The applications of the BCH codes will no doubt increase as the knowledge of the latest decoding procedures spreads. Research also continues on the more general class of cyclic codes, of which the BCH codes are a proper subset.

All of this work has been based upon Hamming's notion of distance. For some channels, this notion is a reasonable representation of reality. For example, if the letters of the input alphabet correspond to the different members of a set of orthogonal or simplex signals which are transmitted through additive white Gaussian noise, then all pairs of different letter-letter error transitions are equally likely. For other channels, however, the Hamming metric is a poor choice. For example, the letters of the input alphabet may correspond to different phases of a sinusoidal signal of fixed amplitude and frequency, to which white Gaussian noise is added. Such a channel has an inherent ordering of the letters of the input alphabet, and transitions between adjacent letters are much more likely than transitions between distant input letters. For such a channel, a much more useful notion is the *Lee metric*. The letters of the alphabet are taken to be the residue classes of integers mod q. The Lee weight of each letter of the alphabet is then defined by

$$w_L(A) = |A| \,, \quad \text{where} \quad 0 \le |A| \le [q/2] \quad \text{and}$$
$$\text{either} \begin{cases} |A| \equiv A \bmod q & \text{or} \\ |A| \equiv -A \bmod q \,. \end{cases}$$

One then defines the Lee weight of a codeword as the sum of the Lee weights of the coefficients of the associated polynomial, and the Lee distance between two codewords as the Lee weight of the difference between the codewords. For example, mod 5 one has $|0| = 0$, $|1| = 1$, $|2| = 2$, $|3| = 2$, $|4| = 1$. The Lee distance between the two codewords in Figure 1 is seen to be

$|4| + |2| + |1| = 1 + 2 + 1 = 4$. The reader should have no trouble in verifying that the Lee metric is nonnegative, symmetric, and that it satisfies the triangle inequality. If $q = 2$ or $q = 3$, the Hamming metric and the Lee metric are identical; for larger q, these metrics differ.

The Lee metric also proves useful in coding for the amplitude modulated channel, in which the q input letters represent different amplitude levels of the same basic signal to which noise[2] is added. In many such examples, q may be rather large, say $q = 31$ or $q = 127$. In such cases, the Lee distance between two symbols near the middle of the alphabet is indeed a monotonic decreasing function of the probability of receiving one of them when the other is sent. However, the close Lee distance between the highest and lowest amplitudes belies the very small probability of confusing them. Nevertheless, codes which correct all patterns of not more than t errors in the Lee metric still prove quite useful for such channels, because every sufficiently probable channel error pattern corresponds to an error pattern with low Lee weight. (The converse of this statement, as we have noted, is flagrantly false.) Although the same can be said for the Hamming metric, the approximation is much cruder.

To the author's knowledge, no significant work has been done in coding with the Lee metric except Lee's original paper in 1958. After defining the metric, most of that paper was concerned with assertions on the nonexistence of perfect codes. Rather than pursue that question (which remains unsolved in both the Hamming and Lee metrics), we devote the remainder of this paper to the construction of good algebraic codes and decoding algorithms.

2. ERROR LOCATION NUMBERS AND THE ERROR POLYNOMIAL

For the remainder of this paper, we assume that the channel input alphabet consists of the elements of an odd prime

[2] We deliberately avoid restrictive statements concerning the noise distribution. The noise distribution most suitable for the Lee metric turns out to be the exponential distribution, according to which the noise assumes an average value between y and $y + dy$ with probability given by $1/2 \exp - |y| dy$. Many noise distributions, including the Guassian, may be approximated by this distribution, and Lee metric codes can be used on such channels. Although the approximations may seem crude, the Lee metric codes may often be as good as any others which are known!

field, $GF(p)$. The reader may at first think that this assumption is overly restrictive, since the Lee metric is defined for input alphabets of arbitrary size. Recall, however, that the Hamming metric is also defined for input alphabets of arbitrary size, yet no one has yet constructed any promising algebraic codes over the alphabet of six letters, nor over any alphabets whose size is not a prime-power. Unlike the Hamming metric, the Lee metric is defined in a modular way which forces the code-constructer to work in the ring of integers mod q, and not in some other structure, such as $GF(q)$. If the size of the input alphabet is not prime, this modular ring is not a field and our constructions fail.

Let us begin by considering the case of single errors. If the block length is N, then there are $2N$ possible single error patterns, since we may have a single Lee error of ± 1 in any

Figure 2. $GF(25)$, in terms of α, a root of $X^2 + X + 2$

	Conjugate	Minimum Function	Order
$\alpha^0 = \alpha^{-24} = 0\alpha + 1$	α^0	$X - 1$	1
$\alpha^1 = \alpha^{-23} = 1\alpha + 0$	α^5	$X^2 + X + 2$	24
$\alpha^2 = \alpha^{-22} = -1\alpha - 2$	α^{10}	$X^2 + 2x - 1$	12
$\alpha^3 = \alpha^{-21} = -1\alpha + 2$	α^{15}	$X^2 - 2$	8
$\alpha^4 = \alpha^{-20} = -2\alpha + 2$	α^{20}	$X^2 - X + 1$	6
$\alpha^5 = \alpha^{-19} = -1\alpha - 1$	α^1	$X^2 + X + 2$	24
$\alpha^6 = \alpha^{-18} = 0\alpha + 2$	α^6	$X - 2$	4
$\alpha^7 = \alpha^{-17} = 2\alpha + 0$	α^{11}	$X^2 + 2x - 2$	24
$\alpha^8 = \alpha^{-16} = -2\alpha + 1$	α^{16}	$X^2 + X + 1$	3
$\alpha^9 = \alpha^{-15} = -2\alpha - 1$	α^{21}	$X^2 + 2$	8
$\alpha^{10} = \alpha^{-14} = 1\alpha - 1$	α^2	$X^2 - 2x - 1$	12
$\alpha^{11} = \alpha^{-13} = -2\alpha - 2$	α^7	$X^2 + 2x + 2$	24
$\alpha^{12} = \alpha^{-12} = 0\alpha - 1$	α^{12}	$X + 1$	2
$\alpha^{13} = \alpha^{-11} = -1\alpha + 0$	α^{17}	$X^2 - X - 1$	24
$\alpha^{14} = \alpha^{-10} = 1\alpha + 2$	α^{22}	$X^2 + 2x - 1$	12
$\alpha^{15} = \alpha^{-9} = 1\alpha - 2$	α^3	$X^2 - 2$	8
$\alpha^{16} = \alpha^{-8} = 2\alpha - 2$	α^8	$X^2 + X + 1$	3
$\alpha^{17} = \alpha^{-7} = 1\alpha + 1$	α^{13}	$X^2 - X + 2$	24
$\alpha^{18} = \alpha^{-6} = 0\alpha - 2$	α^{18}	$X + 2$	4
$\alpha^{19} = \alpha^{-5} = -2\alpha + 0$	α^{23}	$X^2 - 2x - 2$	24
$\alpha^{20} = \alpha^{-4} = 2\alpha - 1$	α^4	$X^2 - X + 1$	6
$\alpha^{21} = \alpha^{-3} = 2\alpha + 1$	α^9	$X^2 + 2$	8
$\alpha^{22} = \alpha^{-2} = -1\alpha + 1$	α^{14}	$X^2 + 2x - 1$	12
$\alpha^{23} = \alpha^{-1} = 2\alpha + 2$	α^{19}	$X^2 - 2x - 2$	24

of the N positions. Including the all-zero pattern of no errors, we see that there are $2N + 1$ different error patterns of Lee weight ≤ 1. If we wish to correct all of these error patterns with a linear code containing r parity check digits, then each of these $2N + 1$ error patterns must have a different pattern of parity check failures and $2N + 1 \leq p^r$, or $N \leq (p^r - 1)/2$.

Following the essential idea of the constructions of Bose-Chaudhuri and Hocquenghem, we next label each digit of the code with a nonzero element in some extension field of $GF(p)$. The previous inequality suggests that we label each digit of the code with *two different* error location numbers from $GF(p^r)$. Upon considering the definition of the Lee metric, and our desire to correct an error of ± 1 in each of the N positions, we assign the two location numbers $\pm \alpha^{j-1}$ to the jth digit of the code, where α is a primitive element of $GF(p^r)$. Since $\alpha^N = -1$, no two different positions have a common location number.

As one of the simplest nontrivial examples, we take $p = 5$, $r = 2$, $N = (5^2 - 1)/2 = 12$. The 24 non-zero elements of $GF(25)$ may be represented as the successive powers of α, where α is a root of the primitive quadratic $X^2 + X + 2$. Alternatively, every element of $GF(25)$ may be represented in the form $A_1\alpha + A_0$, where A_1 and $A_0 \in GF(5)$. The relations between these representations are given in Figure 2. We label the twelve digits of the code as follows:

Code digit position:	1	2	3	4	5	6	7	8	9	10	11	12
Positive location number:	α^0	α^1	α^2	α^3	α^4	α^5	α^6	α^7	α^8	α^9	α^{10}	α^{11}
	0	1	−1	−1	−2	−1	0	2	−2	−2	1	−2
	1	0	−2	2	2	−1	2	0	1	−1	−1	−2
Negative location number:	α^{12}	α^{13}	α^{14}	α^{15}	α^{16}	α^{17}	α^{18}	α^{19}	α^{20}	α^{21}	α^{22}	α^{23}
	0	−1	1	1	2	1	0	−2	2	2	−1	2
	−1	0	2	−2	−2	1	−2	0	−1	1	1	2

We now claim that any error pattern of Lee weight t may be specified by giving the location of the t different errors, or by giving the polynomial, $\sigma(z)$, whose reciprocal roots are these errors. As examples, we consider these error patterns:

$e(X) = X^4$	$X_1 = \alpha^4$	$\sigma(z) = 1 - \alpha^4 z$
$e(X) = -X^8$	$X_1 = -\alpha^8 = \alpha^{20}$	$\sigma(z) = 1 - \alpha^{20} z$
$e(X) = X^3 + X^7$	$X_1 = \alpha^3,\ X_2 = \alpha^7$	$\sigma(z) = (1 - \alpha^3 z)(1 - \alpha^7 z)$
$e(X) = X^5 - X^9$	$X_1 = \alpha^5,\ X_2 = -\alpha^9 = \alpha^{21}$	$\sigma(z) = (1 - \alpha^5 z)(1 - \alpha^{21} z)$
$e(X) = 2X^6$	$X_1 = \alpha^6,\ X_2 = \alpha^6$	$\sigma(z) = (1 - \alpha^6 z)^2$
$e(X) = -X^7 - 2X^{10}$	$X_1 = -\alpha^7 = \alpha^{19}$, $X_2 = X_3 = -\alpha^{10} = \alpha^{22}$	$\sigma(z) = (1 - \alpha^{19} z)(1 - \alpha^{22} z)^2$

The crucial property of the error location numbers X_1, X_2, ... is that

$$e(\alpha^j) = \sum_i X_i^j = S_j \qquad \text{for all odd } j.$$

The reader who is familiar with the Gorenstein-Zierler Hamming-metric extension of the BCH codes must be warned that *no error values* are used in our present formulation. A multiple error in some position is evidenced by a multiple root of the error polynomial. It is clear that

Each distinct error pattern of Lee weight t corresponds to a distinct error polynomial, $\sigma(z)$, whose degree is t.

With appropriate restrictions, the converse is also true:

An error polynomial, $\sigma(z)$ of degree t, corresponds to an error pattern of Lee weight t iff all reciprocal roots of $\sigma(z)$ are $2N$th *roots of unity, and no root has multiplicity greater than* $(p-1)/2$, *and no two reciprocal roots of $\sigma(z)$ sum to zero.*

3. DOUBLE ERROR CORRECTING CODES

Let us now construct a code of block length 12 over $GF(5^2)$ which corrects double errors in the Lee metric. Following the BCH argument in a heuristic manner, we take the first two rows of the parity check matrix as the positive digit location numbers, and the second two rows as the cubes of the first two rows:

$$\boldsymbol{H} = \begin{pmatrix} 0 & 1 & -1 & -1 & -2 & -1 & 0 & 2 & -2 & -2 & 1 & -2 \\ 1 & 0 & -2 & 2 & 2 & -1 & 2 & 0 & 1 & -1 & -1 & -2 \\ 0 & -1 & 0 & -2 & 0 & 1 & 0 & 2 & 0 & -1 & 0 & -2 \\ 1 & 2 & 2 & -1 & -1 & -2 & -2 & 1 & 1 & 2 & 2 & -1 \end{pmatrix}.$$

The codewords are chosen to satisfy all four of these parity check equations. A codeword is transmitted and noise is added. From the first two parity check equations (the top two rows of the above matrix), the decoder may deduce the sum of the error locations, $S_1 = \sum X_i$; from the bottom two rows of the above equations, the decoder may deduce the sum of the cubes of the error equations, $S_3 = \sum X_i^3$. If there are no more than two errors, we have

$$S_1 = X_1 + X_2 \neq 0 \quad \text{unless} \quad X_1 = X_2 = 0$$

$$S_3 = X_1^3 + X_2^3$$

$$S_3/S_1 = X_1^2 - X_1X_2 + X_2^2$$

$$(S_3/S_1) - S_1^2 = -3X_1X_2$$

$$X_1X_2 = (S_1^3 - S_3)/3S_1$$

$$\sigma(z) = 1 - S_1z + [(S_1^3 - S_3)/3S_1]\,z^2 .$$

Thus, this code is capable of correcting double errors. To decode, we compute S_1 and S_3 from the parity check equations, and then perform the necessary arithmetic calculations in $GF(5^2)$ to find the error polynomial. If α^j is a reciprocal root of this polynomial, and $0 \leq j < N$, then there is an error of $+1$ in the $(j+1)$st digit of the received word. If α^j is a reciprocal root of the error polynomial, and $N \leq j < 2N$, then there is an error of -1 in the $(j+1-N)$th digit of the received word. A double error in any position of the code manifests itself in a double reciprocal root of the error polynomial. For these double error correcting codes, the quadratic error polynomial

$$1 - S_1z + [(S_1^3 - S_3)/3S_1]z^2$$

has repeated roots iff

$$S_1^2 = 4(S_1^3 - S_3)/3S_1 \quad \text{or} \quad 4S_3 = S_1^3 .$$

This condition also follows from the equations $S_1 = 2X_1$ and $S_3 = 2X_1^3$. Similarly, it may be seen that there is one error only if $(S_1^3 - S_3)/3S_1 = 0$ and zero errors only if $S_1 = 0$.

The reader may wonder why we selected the bottom two rows of the parity check matrix to be the cubes of the first two rows, rather than the squares. If instead one selects the squares, then one does *not* get the desired equations:

$$X_1 + X_2 = S_1 ,$$

$$X_1^2 + X_2^2 = S_2 ,$$

but instead one gets the formidable-looking equations:

$$X_1 + X_2 = S_1 ,$$

$$X_1|X_1| + X_2|X_2| = S_2 .$$

Here $|X_i| = \pm X_i$, accordingly as $\log_\alpha X_i$ is between 0 and $N-1$, or between N and $2N-1$. The difficulty arises because $(+X_i)^2 = (-X_i)^2 \neq -(X_i)^2$. A similar problem arises if one includes any even power of the error location numbers as rows of the parity check matrix. Although this may not necessarily represent a bad choice, it results in the formidable-looking equations which we prefer to avoid.

Our actual choice of the $\boldsymbol{H}$ matrix has another nice mathematical property which we shall now investigate. Since the successive columns of the first two rows of this matrix represent successive powers of α, the codeword polynomial, $c(x)$, of degree < 12, satisfies these parity check equations iff $c(\alpha) = 0$. This can happen iff the code polynomial, $c(x)$ is a multiple of the minimum function of α, which is $x^2 + x + 2$. Similarly, the code polynomial satisfies the last two parity checks iff $c(\alpha^3) = 0$, which can happen iff $c(x)$ is a multiple of the minimum function of α^3, which is $x^2 - 2$. Evidently, then, the polynomial $c(x)$ of degree < 12 represents a codeword iff $c(x)$ is a multiple of the product, $(x^2 + x + 2)(x^2 - 2)$. Since α is a primitive element of $GF(25)$, $\alpha^{24} = 1$ but $\alpha^k \neq 1$ for any k less than 24. Since $(\alpha^{12})^2 = 1$, and the only two square roots of 1 are ± 1, it is evident that $\alpha^{12} = -1$. A similar argument reveals that $(\alpha^j)^{12} = -1$ for every odd j, including $j = 3$. Thus, both the minimum function of α (namely $x^2 + x + 2$) and the minimum function of α^3 (namely $x^2 - 2$) must be divisors of $x^{12} + 1$. Therefore, if $c(x)$ is a codeword, represented by

$$c(x) = c_0 + c_1x + c_2x^2 + \ldots + c_{10}x^{10} + c_{11}x^{11} \,,$$

then $c(x)$ is multiple of $(x^2 + x + 2)(x^2 - 2)$, and so is $xc(x) - c_{11}(x^{12} + 1)$ which is $-c_{11} + c_0x + c_1x^2 + \ldots + c_9x^{10} + c_{10}x^{11}$. For this reason, we call this code a *negacyclic code.*

4. NEGACYCLIC CODES

In general, we define

A negacyclic code of block length N over $GF(p)$, (p an odd prime and N a nonmultiple of p) is the set of polynomials in an ideal of polynomials modulo $x^N + 1$ over $GF(p)$.

The monic polynomial of lowest degree in this ideal is called the *generator polynomial*, $g(x)$, and the quotient $(x^N + 1)/g(x)$ is called the *check polynomial*, $h(x)$.

Since $x^N+1=(x^{2N}-1)/(x^N-1)$, the roots of x^N+1 are the roots of $x^{2N}-1$ which are not roots of x^N-1. If α is a primitive root of $x^{2N}-1$, then the even powers of α are roots of x^N-1 and the odd powers of α are roots of x^N+1. We repeat, the roots of x^N+1 are the odd powers of a primitive $2N$th root of unity. Therefore, both the generator polynomial and the check polynomial of any negacyclic code of block length N may be conveniently described in terms of their roots, which are odd powers of a single primitive $2N$th root of unity. For the double-error correcting code of block length 12 over $GF(5)$, which we discussed in the previous section, the roots of the generator polynomial are α, α^3, and their quintary conjugates, α^5 and $(\alpha^3)^5=\alpha^{15}$. The other eight odd powers of α, namely α^7, α^9, α^{11}, α^{13}, α^{17}, α^{19}, α^{21}, and α^{23} are roots of the check polynomial.

This method of describing negacyclic codes is important because it immediately reveals the equations which must be solved in order to find the error polynomial. If α^j is a root of the generator polynomial, then the decoder may immediately compute the sum of the jth powers of the error locations by evaluating the received polynomial, $r(x)=c(x)+e(x)$ at $x=\alpha^j$, obtaining $r(\alpha^j)=0+e(\alpha^j)=\sum_i X_i^j=S_j$. Given these S_j, the decoder must then attempt to solve for the error polynomial, $\sigma(z)$, whose reciprocal roots give the error locations. The relation between $\sigma(z)$ and the S's is given by Newton's Identities, which may be readily derived as follows: Let $\sigma(z)=\prod_i(1-X_iz)$; X_i not necessarily distinct. Then $\sigma'(z)=-\sum_i X_i\prod_{j\neq i}(1-X_jz)$ and

$$\frac{-z\sigma'(z)}{\sigma(z)}=\sum_i(X_iz)/(1-X_iz)=\sum_i\sum_{k=1}^{\infty}(X_iz)^k$$

$$=\sum_{k=1}^{\infty}(\sum_i X_i^k)z^k=\sum_{k=1}^{\infty}S_kz^k=S(z)\,.$$

We thus have Newton's Identities in generating function notation:

$$S\sigma+z\sigma'=0\,.$$

In negacyclic codes, all of the coefficients of the even powers of z in the generating function for S are initially unknown. For this reason, it is helpful to eliminate these terms from the equations by separating Newton's Identities into their even and odd parts:

Letting

$$\hat{\sigma} = 1 + \sigma_2 z^2 + \sigma_4 z^4 + \dots ,$$

$$\tilde{\sigma} = \sigma_1 z + \sigma_3 z^3 + \dots ,$$

$$\hat{S} = S_2 z^2 + S_4 z^4 + \dots ,$$

$$\tilde{S} = S_1 z + S_3 z^3 + \dots ,$$

Newton's Identities become

$$(\hat{S} + \tilde{S})(\tilde{\sigma} + \hat{\sigma}) + z(\tilde{\sigma} + \hat{\sigma})' = 0 ,$$

which may be broken up into the two equations

$$\hat{S}\hat{\sigma} + \tilde{S}\tilde{\sigma} + z\hat{\sigma}' = 0 ,$$

and

$$\hat{S}\tilde{\sigma} + \tilde{S}\hat{\sigma} + z\tilde{\sigma}' = 0 .$$

Subtracting $\hat{\sigma}$ times the latter equation from $\tilde{\sigma}$ times the former gives

$$\tilde{S}(\tilde{\sigma}^2 - \hat{\sigma}^2) + z(\tilde{\sigma}\hat{\sigma}' - \hat{\sigma}\tilde{\sigma}') = 0 .$$

Under sufficiently restrictive circumstances, these equations may be solved. The major result is the following theorem:

Theorem 1. *If the roots of the generator polynomial of a negacyclic code over $GF(p)$ include $\alpha, \alpha^3, \alpha^5, \dots, \alpha^{2t-1}$, where $2t - 1 < p$, then that negacyclic code is capable of correcting all error patterns of Lee weight $\leq t$.*

Remarks. Before proving this theorem, we give a short table of parameters of some of the codes which this theorem promises. The codes which have relatively large t and a relatively small number of check digits, r, usually come in block lengths of the form $N = (p^m - 1)/2$. Nevertheless, there are also moderately good codes of this type having other block lengths; these codes are marked in the table by (*).

Error Correction:		1,	2,	3,	4,	5, ...
p	N	r				
5	2	1,	2			
5	6*	2,	3			

5	12	2,	4						
5	62	3,	6						
5	312	4,	8						
7	3	1,	2,	3					
7	24	2,	4,	6					
7	171	3,	6,	9					
11	5	1,	2,	3,	4,	5			
11	15*	2,	3,	5,	7,	8			
11	60	2,	4,	6,	8,	10			
11	665	3,	6,	9,	12,	15			
17	8	1,	2,	3,	4,	5,	6,	7,	8
17	24*	2,	3,	5,	7,	8,	10,	12,	13
17	72*	2,	4,	6,	8,	9,	11,	13,	15
17	144	2,	4,	6,	8,	10,	12,	14,	16
127	63	1,	2,	3,	4,	...	62,	63	
127	8064	2,	4,	6,	8,	...	124,	126	

Proof. Our proof is constructive; it consists of an efficient decoding procedure.

We begin with the equation

$$\tilde{S}(\tilde{\sigma}^2 - \hat{\sigma}^2) = z(\hat{\sigma}\tilde{\sigma}' - \tilde{\sigma}\hat{\sigma}') .$$

Since $\hat{\sigma}(0) = 1$, we may divide through by $\hat{\sigma}^2$ to obtain

$$\tilde{S}[(\tilde{\sigma}/\hat{\sigma})^2 - 1] = z[(\hat{\sigma}\tilde{\sigma}' - \tilde{\sigma}\hat{\sigma}')/\hat{\sigma}^2] = z(\tilde{\sigma}/\hat{\sigma})' .$$

Introducing the generating function $R = \tilde{\sigma}/\hat{\sigma}$, we have $\tilde{S}(R^2 - 1) = zR'$ or

$$R = \int \frac{1}{z}\tilde{S}(-1 + R^2) .$$

Although this equation may look formidable, its solution is, in fact, trivial, because each coefficient of R is specified in terms of certain coefficients of $\tilde{S}$ and previously computed coefficients of R:

$$(R_1 z + R_3 z^3 + R_5 z^5 + \ldots)$$
$$= \int \frac{(S_1 z + S_3 z^3 + S_5 z^5 + \ldots)}{z}(-1 + (R_1 z + R_3 z^3 + \ldots)^2)$$

$$R_1 = -S_1$$

$$R_3 = \tfrac{1}{3}(-S_3 + R_1^2 S_1)$$

$$R_5 = \tfrac{1}{5}(-S_5 + R_1^2 S_3 + 2R_1 R_3 S_1)$$

$$\vdots$$

Difficulty is encountered if we attempt to compute the coefficient R_p, since this would require dividing by the integer p, which is zero in $GF(p)$. However, under the hypotheses of the theorem, $2t - 1 < p$, and no such difficulty arises if we compute only $R_1, R_3, R_5, \ldots, R_{2t-1}$. Since R is an odd function of z, we may define the generating function T by the equation $T(z^2) = (1 + zR(z))^{-1} - 1$. It is evident that $T(0) = 0$, and that $(1 + T(z^2)) = (1 + zR(z))^{-1}$. Knowing the coefficients of $R_1, R_3, \ldots, R_{2t-1}$, this equation enables us to compute recursively the coefficients $T_1, T_2, \ldots, T_t$. Since $R = \tilde{\sigma}/\hat{\sigma}$ and $1 + zR = \hat{\sigma}^{-1}(\hat{\sigma} + z\tilde{\sigma})$, we define the polynomials

$$\omega(z^2) = \hat{\sigma}(z)\,; \qquad \xi(z^2) = \hat{\sigma}(z) + z\tilde{\sigma}(z)\,.$$

It is evident that

$$(1 + zR(z)) = \xi(z^2)/\omega(z^2)$$

and that

$$(1 + T(z^2)) = \omega(z^2)/\xi(z^2)\,,$$

so

$$(1 + T(z))\xi(z) \equiv \omega(z) \bmod z^{t+1}\,.$$

We further claim that if there are no more than t errors, then ξ and ω satisfy the additional conditions $\xi(0) = \omega(0) = 1$, $\deg \xi \leq (1 + t)/2$, $\deg \omega \leq t/2$, and ξ and ω are relatively prime. The last observation follows from the fact that no two reciprocal roots of σ may sum to zero. Therefore, σ may have no even factors of positive degree. This implies that $\tilde{\sigma}$ and $\hat{\sigma}$ are relatively prime, as are ξ and ω.

In view of these conditions, the equation

$$(1 + T)\xi \equiv \omega \bmod z^{t+1}$$

may be solved for ξ and ω by the iterative algorithm for decoding nonbinary BCH codes given by Berlekamp (1968). The solution is evidently given by

$$\xi = \xi^{(t)}, \qquad \omega = \omega^{(t)}$$

where $\xi^{(t)}$ and $\omega^{(t)}$ are generated by the algorithm. We may summarize the decoding procedure as follows:

1) Compute $R \bmod z^{2t}$ from the equation

$$R = \int (S/z)(R^2 - 1)dz\,;$$

2) Compute $(1 + T) \bmod z^{t+1}$ from the equation

$$(1 + T(z^2)) = (1 + zR(z))^{-1}\,;$$

3) Use the iterative algorithm for decoding nonbinary BCH codes to find $\xi^{(0)}, \omega^{(0)}, \tau^{(0)}, \gamma^{(0)}, \xi^{(1)}, \omega^{(1)}, \tau^{(1)}, \gamma^{(1)}, \ldots, \xi^{(t)}, \omega^{(t)}$, which solve the equations, $(1 + T)\xi^{(t)} \equiv \omega^{(t)} \bmod z^{t+1}$;

4) Set $\hat{\sigma}(z) = \omega^{(t)}(z^2)$; $\tilde{\sigma}(z) = (1/z)(\xi^{(t)}(z^2) - \omega^{(t)}(z^2))$; $\sigma = \hat{\sigma} + \tilde{\sigma}$;

5) Using a multiple Chien (1964) search, evaluate the polynomials $\hat{\sigma}(\alpha^{-j})$ and $\tilde{\sigma}(\alpha^{-j})$ for $j = 0, 1, \ldots, N-1$, where α is the primitive $2N$th root of unity whose successive powers give the code's successive location numbers. If $\hat{\sigma}(\alpha^{-j}) = -\tilde{\sigma}(\alpha^{-j})$, there is a positive error in the code's $(j-1)$st position; $\hat{\sigma}(\alpha^{-1}) = +\sigma(\alpha^{-j})$, there is a negative error in the code's $(j1)$st position. In either case, the multiplicity of the error may be determined by evaluating the derivatives of $\hat{\sigma}$ and $\tilde{\sigma}$.

If $\dfrac{d^{(i)}\hat{\sigma}(\alpha^{j})}{dz^{(i)}} = \pm \dfrac{d\tilde{\sigma}^{(2)}(\alpha^{j})}{dz^{(i)}}$ for $i < k$, but $\dfrac{d\hat{\sigma}^{(k)}(d^{j})}{dz^{(k)}} = \pm \dfrac{d\tilde{\sigma}^{(k)}(\alpha^{j})}{dz^{(k)}}$, then the error has multiplicity k. In these equations, the negative signs are used to determine the multiplicity of a positive error, and vice versa.

If high speed is required and computing registers are plentiful, then one may perform multiple Chien searchs on all of the polynomials $\hat{\sigma}, \tilde{\sigma}, \hat{\sigma}', \tilde{\sigma}', \hat{\sigma}'', \tilde{\sigma}'', \ldots, \dfrac{d^{(t-1)}\tilde{\sigma}}{dz^{(t-1)}}$.

On the other hand, if computing registers are scarce but time is available, one should perform Chien searches only on the polynomials $\hat{\sigma}$ and $\tilde{\sigma}$. By judicious programming it is possible to calculate the derivatives only when they are needed, without interupting the Chien searches on $\hat{\sigma}(\alpha^{-j})$ and $\tilde{\sigma}(\alpha^{-j})$.

Although Theorem 1 gives us a powerful class of negacyclic codes, together with a practical decoding algorithm, there is a strong desire to remove the restriction that $t \leq (p-1)/2$. One may wish to correct more than $(p-1)/2$ errors. The obvious conjecture to do this is the code whose roots include further successive odd powers of α. If $\alpha, \alpha^3, \ldots \alpha^{p-2}, \alpha^p$ prove insufficient, then possibly $\alpha, \alpha^3, \alpha^5, \ldots \alpha^p, \alpha^{p+2}$ might work? Unfortunately, this attempt does not work, and the situation is even worse than this. We cannot correct that $(p+1)/2$st error even if we use almost twice as much redundancy as we used to correct $(p-1)/2$ errors! The theorem is as follows:

Theorem 2. *The negacyclic code of block length $N = (p^m - 1)/2$ over $GF(p)$, $m > 1$, whose generator polynomial is the product of the distinct minimum functions of $\alpha, \alpha^3, \alpha^5, \ldots \alpha^{2p-5}, \alpha^{2p-3}$ has codewords of Lee weight p.*

Proof. We shall exhibit a set of p locations such that $S_j = \sum_{i=1}^{p} X_i^j = 0$ for all odd $j \leq 2p-3$. The word containing $(\pm)$ ones in these p locations must then be a codeword of Lee weight p.

For $i = 1, 2, \ldots, p$, let $X_i = \xi + (i-1)$ where $\xi \notin GF(p)$, but $\xi \in GF(p^m)$. Then the X_i represent distinct locations, for $X_i \neq X_j$ unless $i = j$, and if $X_i = -X_j$ then $\xi + (i-1) = -(\xi + (j-1))$ and $\xi = -(i+j-2)/2 \in GF(p)$.

The "error" polynomial is

$$\sigma(z) = \prod_{i=1}^{p} (1 - X_i z) = \prod_{i=1}^{p} (1 - \xi z - z(i-1)).$$

Recalling that

$$\prod_{i=1}^{p} (y - (i-1)) = y^p - y\,,$$

for every y, and that

$$\prod_{i=1}^{p} (y - z(i-1)) = y^p - z^{p-1} y\,,$$

we deduce that

$$\begin{aligned} \sigma(z) &= (1 - \xi z)^p - z^{p-1}(1 - \xi z) \\ &= 1 - z^{p-1} - (\xi^p - \xi) z^p \\ &= 1 - z^{p-1} - \phi z^p \end{aligned}$$

where we have defined $\phi = \xi^p - \xi$. Since $\sigma'(z) = z^{p-2}$, Newton's Identities become

$$S(z) = \frac{-z\sigma'(z)}{\sigma(z)} = \frac{-z^{p-1}}{1-(z^{p-1}+\psi z^p)}$$

$$= -(z^{p-1} + z^{2p-2} + \psi z^{2p-1} + z^{3p-3}$$

$$+ 2\psi z^{3p-2} + \psi^2 z^{3p-1} + \dots) .$$

Since $p-1$ is even, $S_j = 0$ for all odd $j \leq 2p-3$.

Remark. Theorem 2 may be generalized along the lines suggested by the Kasami-Peterson proof that Reed-Muller codes are cyclic. If the reciprocal roots of $\sigma(z)$ form an affine subspace of dimension k over $GF(p)$ (sometimes called a k-flat), then $S_n = 0$ whenever the sum of the digits of the p-ary representation of n is less than $k(p-1)$. Therefore the negacyclic code whose generator is the product of the distinct minimum functions of $\alpha, \alpha^3, \alpha^5, \dots, \alpha^{2p^k-3}$ has codewords of Lee weight p^k.

The argument of the preceding theorem can be continued to show that certain other S's must also vanish. This shows that certain other negacyclic codes, whose generators have these additional roots, still have codewords of weight p. We first continue the argument to derive an explicit formula for S_n: We have

$$S(z) = \frac{-z^{p-1}}{1-(z^{p-1}+\psi z^p)} = -z^{p-1}\sum_{K=0}^{\infty}(z^{p-1})^K(1+\psi z)^K$$

$$= -z^{p-1}\sum_{K=0}^{\infty}(z^{p-1})^K\sum_{I=0}^{\infty}\binom{K}{I}\psi^I z^I .$$

Letting $I = i(p-1) + j$,

$$S(z) = -z^{p-1}\sum_{K=0}^{\infty}\sum_{i=0}^{\infty}\sum_{j=0}^{p-2}\binom{K}{i(p-1)+j}\psi^{i(p-1)+j}z^{(p-1)(i+K)+j} .$$

Letting $k = i + K + 1$ gives

$$S(z) = -\sum_{k=1}^{\infty}\sum_{j=0}^{p-2}\psi^j\sum_{i=0}^{\infty}\binom{k-i-1}{i(p-1)+j}\psi^{(p-1)i}z^{(p-1)k+j} .$$

The binomial coefficient vanishes unless $i(p-1)+j \leq k-i-1$, or $i \leq (k-j-1)/p$, and hence

$$S_{k(p-1)+j} = -\psi^j\sum_{i=0}^{[(k-j-1)/p]}\binom{k-i-1}{i(p-1)+j}(\psi^{(p-1)})^i .$$

From this argument it is obvious that codewords of weight p occur even if the roots of the generator include $\alpha, \alpha^3, \ldots \alpha^{2p-3}, \alpha^{2p+1}, \alpha^{2p+3}, \ldots \alpha^{3p-4}, \alpha^{3p}, \alpha^{3p+2}, \ldots \alpha^{4p-5}, \alpha^{4p+1}, \ldots$. In some cases, it might be possible to choose $\psi^{(p-1)}$ in such a way that certain other (nonconsecutive) S's also vanish.

Certain other classes of weak negacyclic codes may be discredited by examining the obvious low weight factors of $x^N + 1$. For example, over $GF(5)$, one has $x^{12} + 1 = (x^6 + 2)(x^6 - 2)$, and $x^{12} + 1 = (x^4 + 1)(x^8 - x^4 + 1)$. From this one sees that any negacyclic code whose generator divides $(x^4 + 1)$ has distance ≤ 2; any code whose generator divides $x^6 + 2$ or $x^6 - 2$ or $x^8 - x^4 + 1$ has distance ≤ 3.

Nevertheless, there do exist negacyclic codes having very large minimum distances—if one is willing to transmit information at sufficiently low rates. We shall prove the following theorem:

Theorem 3. *Let $N = (p^m - 1)/2$ and let α be a primitive $2N$th root of unity over $GF(p)$. The negacyclic code whose check polynomial is the minimum function of α is equidistant; every non-zero codeword has Lee weight given by*

$$\omega_L = (p^2 - 1)p^{m-1}/8 .$$

Proof. This code has $p^m - 1 = 2N$ non-zero codewords, each of which is of the form $c(x) = M(x)g(x)$, where $\deg M < \deg h$. If some non-zero codeword had only k distinct negacyclic shifts, then

$$(x^k - 1)c(x) \equiv 0 \mod x^N + 1$$

or

$$\begin{aligned}(x^k - 1)c(x) &= (x^k - 1)M(x)g(x) \\ &= (x^k - 1)M(x)(x^N + 1)/h(x) \equiv 0 \mod (x^N + 1) ,\end{aligned}$$

which implies that h divides $(x^k - 1)M$. Since h is irreducible and $\deg M < \deg h$, h must divide $x^k - 1$. Since h is primitive, $k = 2N$ or some multiple thereof. This proves that all $2N$ negacyclic shifts of any non-zero codeword are distinct. Since there are only $2N$ non-zero codewords in the entire code, it follows that every non-zero codeword is a negacyclic shift of every other non-zero codeword, and that every non-zero codeword has the same Lee weight.

We may compute the average Lee weight of all of the codewords in any nontrivial linear code over $GF(p)$ as follows: We list each codeword as a row of a matrix, containing N columns and as many rows as there are codewords. Since the code is linear, every nontrivial column (excluding trivial columns containing all zeroes) must contain equal numbers of each of the p symbols. The average weight of each column is therefore

$$\frac{2}{p}\sum_{k=1}^{(p-1)/2} k = (p^2-1)/4p\,,$$

and the average weight of the entire code is $N(p^2-1)/4p$. Applying this result to the present case gives $(p^m-1)(p^2-1)/8p$ for the average weight of all of the p^m codewords, and $p^{m-1}(p^2-1)/8$ for the average weight of the p^m-1 non-zero codewords.

This theorem shows that certain of the codes guaranteed by our main theorem are actually much better than claimed. For example, the code with $p=127$, $N=63$, having 1 message digit and 62 check digits actually has Lee distance $32 \cdot 63$, so that it is capable of correcting almost 16 times as many errors as promised by the theorem!

These low-rate equidistant negacyclic codes are quite analogous to the low rate maximum length shift register cyclic codes. It may be possible to use these low rate negacyclic codes as the raw material from which shorter, higher rate, non-negacyclic codes may be manufactured via algebraic puncturing *à la* Solomon-Stiffler (1966).

References

Berlekamp, E. R. (1968). *Algebraic Coding Theory*, McGraw-Hill.

Bose, R. C. and Ray-Chaudhuri, D. K. (1960). "On a Class of Error-Correcting Binary Group Codes," *Info. and Control*, **3**, 68–79.

Bose, R. C. and Ray-Chaudhuri, D. K. (1960). "Further Results on Error-Correcting Binary Codes," *Info. and Control*, **3**, 279–290.

Chien, R. T. (1964). "Cyclic Decoding Procedures for Bose-Chaudhuri-Hocquenghem Codes, *IEEE Trans.*, **IT-10**, 357–362.

Forney, G. D. (1965). "On Decoding BCH Codes," *IEEE Trans.* **IT-11**, 549–557.

Gorenstein, D. and Zierler, N. (1961). "A Class of Error-Correcting Codes in p^m Symbols," *J. Soc. Indust. Appl. Math.*, **9**, 207–214.

Hamming, R. W. (1950). "Error Detecting and Error Correcting Codes," *Bell System Tech. J.*, **29**, 47-160.

Hocquenghem, A. (1959). "Codes Correcteurs d'Erreurs," *Chiffres*, **2**, 147-156.

Lee, C. Y. (1958). "Some Properties of Nonbinary Error-Correcting Codes," *IRE Trans.*, **IT-4**, 77-82.

Mattson, H. F. and Solomon, G. (1961). "A New Treatment of Bose-Chaudhuri Codes," *J. Soc. Indust. Appl. Math.*, **9**, 654-669.

Peterson, W. W. (1961). *Error-Correcting Codes*, MIT Press, Cambridge.

Prange, E. (1957). "Cyclic Error-Correcting Codes in Two Symbols," Scientific Report AFCRC-TN-57-103, Air Force Cambridge Research Center, Cambridge, Mass.

Solomon, G. and Stiffler, J. J. (1965). "Algebraically Punctured Cyclic Codes," *Info. and Control*, **8**, 170-179.

Shannon, C. E. and Weaver, W. (1949). *A Mathematical Theory of Communication*, University of Illinois Press, Urbana, Illinois.

Slepian, D. (1956). "A Class of Binary Signaling Alphabets," *Bell System Tech. J.*, **35**, 203-234.

Slepian, D. (1960). "Some Further Theory of Group Codes," *Bell System Tech. J.*, **39**, 1219-1252.

CHAPTER 18

Codes and Ideals in Group Algebras

MRS. JESSIE MacWILLIAMS, *Bell Telephone Laboratories*

Several years ago, Professor A. M. Gleason suggested that it would be worth while to look at the ideals in the group algebra of a noncyclic group from the point of view of coding theory. One may reasonably hope to find among them some new codes which retain most of the useful properties of cyclic codes, and have some new ones of their own. We propose to examine this suggestion.

1. Let F be the binary field $(0, 1)$ and let $G = \{g_1, g_2, \ldots, g_n\}$ be a finite group of order n, with the group operation written as multiplication. The group algebra FG of G over F is a vector space of dimension n over F, with basis $g_1, g_2, \ldots, g_n$. An element of FG is a vector $\sum_{i=1}^{n} a_i g_i$, $a_i \in F$. Addition of vectors is defined as usual; there is also a multiplication defined by means of the group multiplication:

$$\left(\sum_{i=1}^{n} a_i g_i\right)\left(\sum_{j=1}^{n} b_j g_j\right) = \sum_{i=1}^{n} \sum_{j=1}^{n} a_i b_j g_i g_j = \sum_{k=1}^{n} c_k g_k ,$$

where

$$c_k = \sum_{g_i g_j = g_k} a_i b_j .$$

A left ideal $\mathcal{A}$ of FG is (1) a subspace of the vector space FG, and (2) invariant under left multiplication by the group elements; $g\mathcal{A} = \mathcal{A}$ for all $g \in G$.

The left ideals are codes which are preserved by the permutation of the basis elements $g_i \rightarrow gg_i$ induced by any g in G.

The properties of $\mathcal{A}$ which are of interest in coding theory are, roughly speaking, the combinatorial properties of the set of vectors $\boldsymbol{a} = (a_1, a_2, \ldots, a_n)$ where $\sum a_i g_i \in \mathcal{A}$. In particular we are interested in the weight of $\boldsymbol{a}$, $w(\boldsymbol{a})$ which is the number of non-zero a_i, and in $d(\mathcal{A})$ which is the minimum of $w(\boldsymbol{a})$ for $\sum a_i g_i$ a non-zero element of $\mathcal{A}$.

If G is the cyclic group $\langle x \rangle$, $x^n = 1$, then FG is the polynomial ring $F[x]/(x^n - 1)$; all ideals are two sided, and the ideals in FG are cyclic codes. Many of the useful properties of cyclic codes depend on the fact that they are ideals in a group algebra, and not on the cyclic structure of G. Some examples are listed below.

(1) We start with a group of permutations (namely G) which preserve the code; this group is transitive on the coordinate positions, so that if a code contains a vector of weight one it contains the whole space.

(2) If $|G|$ is odd (prime to the characteristic of F) FG is a principal ideal ring [3]. However, it is probably not wise to cling to this restriction, since there is no reason to suppose (except for cyclic G) that this case contains the most interesting codes.

(3) The cosets of the code are residue classes of an ideal, that is, they can be multiplied as well as added. In the case of cyclic codes this is the basis of several methods for recovering synchronization.

(4) The dual code (parity check code) of $\mathcal{A}$ which is the foundation of all decoding algorithms, is also a left ideal of FG, and is closely related to $\mathcal{A}$.

Statement (4) is not quite obvious; a proof is given at the end of this section.

In view of all these advantages, we attempt to look at the combinatorial properties of left ideals in a group algebra FG, and immediately discover the crucial advantage we have given up with the cyclic structure of G—that is, the ability to find the left ideals in the form that we want to investigate.

Section 2 of this paper describes an effort to solve this problem for a class of groups which are a small but nontrivial step away from the cyclic groups. The effort is by no means completely successful.

To conclude this section we give a fuller explanation of statement (4) above.

The dual code $D(\mathcal{A})$ of $\mathcal{A}$ is the set of all vectors $\sum d_i g_i \in FG$, such that $\sum a_i d_i = 0$ for all $\sum a_i g_i \in \mathcal{A}$. The theoretical and practical properties of $\mathcal{A}$ are closely connected with those of $D(\mathcal{A})$. It is thus fortunate that $D(\mathcal{A})$ is also a left ideal of FG. This is obvious, since $\sum a_i d_i$ is not changed when the same permutation is applied to the a_i and the d_i.

The right annihilator of $\mathcal{A}$ is the right ideal

$$M(\mathcal{A}) = \{\sum m_i g_i \in FG; (\sum a_i g_i)(\sum m_j g_j) = 0, \quad \text{all } \sum a_i g_i \text{ of } \mathcal{A}\}.$$

The following theorem establishes the connection between $M(\mathcal{A})$ and $D(\mathcal{A})$.

Theorem 1. $\sum d_i g_i \in D(\mathcal{A})$ *if and only if* $\sum d_i g_i^{-1} \in M(\mathcal{A})$.

In words, $D(\mathcal{A})$ *is obtained from* $M(\mathcal{A})$ *by the transformation* $g_i \to g_i^{-1}$.

Proof. If $\sum d_i g_i^{-1} \in M(\mathcal{A})$, then $(\sum a_i g_i)(\sum d_i g_i^{-1}) = 0$ for all $\sum a_i g_i$ in $\mathcal{A}$. In particular, from the coefficient of the unit of G, $\sum a_i d_i = 0$, and $\sum d_i g_i \in D(\mathcal{A})$.

Suppose that $\sum d_i g_i \in D(\mathcal{A})$, and $\sum a_i g_i \in \mathcal{A}$. Let $g \sum a_i g_i = \sum a_i^{(g)} g_i$, where $a_i^{(g)}$ is the coefficient of $g^{-1} g_i$ in $\sum a_i g_i$. From the definition of $D(\mathcal{A})$ $\sum a_i^{(g)} d_i = 0$ for all $g \in G$. Now

$$(\sum a_i g_i)(\sum d_j g_j^{-1}) = \sum c_k g_k ,$$

with

$$c_k = \sum_{g_i g_j^{-1} = g_k} a_i d_j = \sum_{j=1}^{n} a_j^{(g_k^{-1})} d_j = 0 .$$

Hence

$$(\sum a_i g_i)(\sum d_j g_j^{-1}) = 0 \qquad \text{for all } \sum a_i g_i \in \mathcal{A} ,$$

and

$$\sum_{j=1}^{n} d_j g_j^{-1} \in M(\mathcal{A}) .$$

2. The particular class of groups considered are the groups on two generators,

$$G = \langle x, y \rangle\,, \quad x^n = 1\,, \quad y^2 = 1\,, \quad xy = yx^{n-1}\,.$$

FG will be denoted by R; it is a vector space of dimension $2n$ over F. X will denote the group algebra of the cyclic group $\langle x \rangle$ over F.

The basis elements of R will be written

$$1, x, \ldots, x^{n-1}, y, yx, \ldots, yx^{n-1}\,.$$

Multiplication by x produces a cyclic permutation in the sets $\{x^i\}$ and $\{yx^i\}$; these cyclic permutations go in opposite directions. This is illustrated by the group multiplication table for the case $n = 3$ which is given below.

	1	x	x^2	y	yx	yx^2
1	1	x	x^2	y	yx	yx^2
x	x	x^2	1	yx^2	y	yx
x^2	x^2	1	x	yx	yx^2	y
y	y	yx	yx^2	1	x	x^2
yx	yx	yx^2	y	x^2	1	x
yx^2	yx^2	y	yx	x	x^2	1

An element of R will be written $f(x) + yg(x)$, where $f(x)$, $g(x)$ are polynomials mod $x^n - 1$. If $f(x) + yg(x)$ belongs to a left ideal C of R, so does $x(f(x) + yg(x)) = xf(x) + yx^{n-1}g(x)$, and so does $y(f(x) + yg(x)) = g(x) + yf(x)$. Thus the sets of polynomials

$$\{f(x); f(x) + yg(x) \in C\} \quad \text{and} \quad \{g(x); f(x) + yg(x) \in C\}$$

belong to the same ideal $\mathcal{A}$ of X.

If C contains an element $f(x) + y \cdot 0$ it contains the whole ideal $X \cdot f(x)$; clearly it then contains all elements $a_1(x) + ya_2(x)$ where $a_1(x), a_2(x) \in X \cdot f(x)$. This is essentially the same as the direct sum of $X \cdot f(x)$ with itself, and is not interesting as a code.

Before going further, we give an example.

Let $n = 3$, and let C be a left ideal of R which contains the vector $c = (1 + x) + y(1 + x + x^2)$. Then C contains the following vectors:

	1	x	x^2	y	yx	yx^2
$1c$	1	1	0	1	1	1
xc	0	1	1	1	1	1
x^2c	1	0	1	1	1	1
$(1 + x + x^2)c$	0	0	0	1	1	1
yc	1	1	1	1	1	0
$(1 + y)c$	0	0	1	0	0	1
$(1 + x^2)(1 + y)c$	0	1	1	1	0	1

A little calculation shows that the smallest left ideal C which contains all of these is the direct sum of two left ideals shown below:

1	x	x^2	y	yx	yx^2
1	1	1	0	0	0
0	0	0	1	1	1
1	1	1	1	1	1

and

1	x	x^2	y	yx	yx^2
0	1	1	1	0	0
1	0	1	0	1	1
1	1	0	1	1	0

We concentrate on looking for the ideals C of R such that every non-zero element of C is of the form $a(x) + yb(x)$ where $a(x)$, $b(x)$ are both non-zero.

Now $a(x)$, $b(x)$ belong to the same ideal $\mathcal{A}$ of X; let $f(x)$ be the factor of $x^n + 1$ which generates $\mathcal{A}$. C must contain an element of the form

$$f(x) + yp(x)f(x) .$$

Let

$$g(x) = (x^n + 1)/f(x) .$$

C contains the element

$$g(x^{-1})f(x) + yg(x)p(x)f(x) = g(x^{-1})f(x)\,.$$

By hypothesis

$$g(x^{-1})f(x) = 0$$

or

$$g(x^{-1}) \equiv 0 \bmod g(x)\,.$$

(This means that if α is a zero of $g(x)$, so is α^{-1}.) By a similar argument, $p(x)$ must be prime to $g(x)$; this however, is not the end of the story.

Consider the element of C

$$\begin{aligned}&(yp(x) + 1)(f(x) + yp(x)f(x))\\ &= yp(x)f(x) + p(x^{-1})p(x)f(x) + f(x) + yp(x)f(x)\\ &= (p(x^{-1})p(x) + 1)f(x)\,.\end{aligned}$$

By hypothesis, $(p(x^{-1})p(x) + 1)f(x) = 0$, or

$$p(x^{-1})p(x) + 1 \equiv 0 \bmod g(x)\,.$$

Theorem 2. *Let C be the principal left ideal of R generated by $f(x) + yp(x)f(x)$. Suppose*

(1) *$f(x)$ is a factor of $x^n + 1$ such that $f(x^{-1}) \equiv 0 \bmod f(x)$.*

(2) *$1 + p(x)p(x^{-1}) \equiv 0 \bmod g(x)$, where $g(x) = (x^n + 1)/f(x)$.*

These conditions are necessary and sufficient to ensure that if $a(x) + yb(x)$ is a non-zero element of C, then both $a(x)$, $b(x)$ are non-zero.

Proof. We have already shown that the conditions are necessary.

Suppose there is an element $c(x) + yd(x) \in R$ such that

$$(c(x) + d(x^{-1})p(x))f(x) = a(x)\,,$$

$$(d(x) + c(x^{-1})p(x))f(x) = 0\,.$$

The second equation implies

$$d(x) + c(x^{-1})p(x) \equiv 0 \bmod g(x)\,.$$

By (1),

$$d(x^{-1}) + c(x)p(x^{-1}) \equiv 0 \bmod g(x) .$$

Multiply this equation by $p(x)$ and use (1). We obtain

$$d(x^{-1})p(x) + c(x) \equiv 0 \bmod g(x) .$$

This implies $a(x) = 0$, so that the conditions are also sufficient.

The problem of constructing $p(x)$ to satisfy the conditions of Theorem 2 is of considerable complexity, and is not completely solved, except in the case $f(x) = 1$. This case, however, is of especial interest from the coding theory point of view, as is shown by the following theorem:

Theorem 3. *If $f(x) = 1$ in Theorem 2, the code C is of dimension $n/2$ and is self-dual.*

Proof. C is the left ideal generated by $1 + yp(x)$, where $p(x) \cdot p(x^{-1}) = 1$. Clearly the dimension of C is $n/2$.

Since

$$(1 + yp(x))(1 + yp(x)) = 1 + p(x)p(x^{-1}) = 0 ,$$

the right annhilator of C is the right ideal generated by

$$1 + yp(x) = 1 + \sum_{i=0}^{n-1} a_i y x^i .$$

We apply Theorem 1 to obtain the dual of C.

$$(yx^i)^{-1} = x^{-i}y = yx^i ,$$

so that the dual of C is the left ideal generated by $1 + yp(x)$, that is, C itself.

We give in the appendix a description of how to construct polynomials $p(x)$ for various values of n, also some examples of codes obtained in this way.

We conclude this section by pointing out again that the discussion herein is far from complete. We have by no means found all the ideals of our group algebra, and perhaps (indeed probably) have missed some which are interesting as codes, this in spite of the fact that the group was chosen to give the easiest case. It is clear that a great deal more work is needed.

APPENDIX

We describe here, without proofs, methods for constructing the $p(x)$ of Theorem 3 [4].

(1) *n odd.*

The polynomial ring $X = F[x]/(x^n + 1)$ is then a semisimple algebra [3].

Let $x^n + 1 = (x + 1)\prod_{i=1}^{r} f_i(x)$ be the decomposition of $x^n + 1$ into distinct irreducible factors, f_i of degree k_i. Let $\mathcal{A}_i$ be the minimal ideal of X generated by $(x^n + 1)/f_i(x)$, $(f_0 = x + 1)$. $\mathcal{A}_i$ is isomorphic to $GF(2^{k_i})$.

If α is a zero of $f_i(x)$, α^{-1} is a zero of $f_i^*(x) = x^{k_i}f(\alpha^{-1})$.

It may happen that $f_i(x) = f_i^*(x)$; its degree k_i, $i \geq 1$, is then necessarily even, say $k_i = 2c_i$. We suppose $f_1, \ldots, f_t$ have this property, and arrange the others in pairs $g_j(x) = f_j(x)f_j^*(x)$. Let $2d_j = 2k_j$ be the degree of $g_j(x)$, $j = 1, \ldots, n$. We have then

$$(x^n + 1) = (x + 1)\prod_{i=1}^{t} f_i(x)\prod_{j=1}^{u} g_j(x)$$

and, with the obvious notation, X is the direct sum

$$X = \mathcal{A}_0 \oplus \sum_{i=1}^{t} \mathcal{A}_i \oplus \sum_{j=1}^{u} (\mathcal{A}_j + \mathcal{A}_j^*)$$

Theorem 4. $p(x) \in X$ is such that $p(x)p(x^{-1}) = 1$ if and only if

$$p(x) = a_0(x) + \sum_{i=1}^{t} a_i(x) + \sum_{j=1}^{u} (b_j(x) + b_j'(x)),$$

where

$$a_i(x) \in \mathcal{A}_i, \qquad b_j(x) \in \mathcal{A}_j, \qquad b_j'(x) \in \mathcal{A}_j^*,$$

and have the following properties

(i) $a_0(x) = \sum_{i=0}^{n-1} x^i$

(ii) $a_i(x)^{2^{c_i}+1}$ is the unit of $\mathcal{A}_i$.

(iii) $b_j(x)$ is any non-zero element of $\mathcal{A}_j$, and $b_j'(x^{-1})$ (which is in $\mathcal{A}_j$) is the unique inverse of $b_j(x)$ in $\mathcal{A}_j$.

Corollary 5. *The number of polynomials $p(x)$ with the desired property is*

$$\prod_{i=1}^{t} (2^{c_i} + 1) \prod_{j=1}^{u} (2^{d_j} - 1) .$$

We illustrate the construction for $n = 11$.

$$x^{11} + 1 = (x + 1)\left(\sum_{i=1}^{10} x^i\right) = f_0 f_1 .$$

Now $c_1 = 5$ so that the number of $p(x)$ is $2^5 + 1 = 33$, and

$$p(x) = \sum_{i=0}^{10} x^i + a_1(x) ,$$

where $a_1(x) \in \mathcal{A}_1$ and $a_1(x)^{33}$ is the unit of $\mathcal{A}_1$.

Let $\beta \neq 1$ be an 11th root of unity; the map $x \to \beta$ induces an isomorphism between $\mathcal{A}_1$ and $F[\beta] = GF(2^{11})$. We look for elements of order 33 in $F[\beta]$ and transfer them back to $\mathcal{A}_1$ by the inverse mapping, which is [2]

$$\sigma \to a(x) \quad \text{where} \quad a_i = \sum_{j=0}^{10} (\sigma\beta^{-i})^{2^j} .$$

The elements of $F[\beta]$ are of course polynomials

$$\sum_{i=0}^{9} \varepsilon_i \beta^i , \qquad \varepsilon_i \in F, \quad \text{where } \beta^{10} = \sum_{i=0}^{9} \beta^i .$$

β has order 11, and it is readily checked that $\beta + \beta^3 + \beta^4 + \beta^5 + \beta^9$ has order 3, so that the powers of

$$\beta(\beta + \beta^3 + \beta^4 + \beta^5 + \beta^9) = 1 + \beta + \beta^3 + \beta^7 + \beta^8 + \beta^9$$

are the required elements.

Part of the mapping of $F[\beta]$ onto $\mathcal{A}_1$ is shown below

	1	x	x^2	x^3	x^4	x^5	x^6	x^7	x^8	x^9	x^{10}
1	0	1	1	1	1	1	1	1	1	1	1
β	1	1	1	1	1	1	1	1	1	1	0
β^2	1	1	1	1	1	1	1	1	1	0	1
β^3	1	1	1	1	1	1	1	1	0	1	1
–	–	–	–	–	–	–	–	–	–	–	–
$1+\beta+\beta^3+\beta^7+\beta^8+\beta^9$	1	0	1	1	1	0	0	0	1	0	1

Thus $a(x) = 1 + x^2 + x^3 + x^4 + x^8 + x^{10}$ has order 33, and we may take

$$p_1(x) = x + x^5 + x^6 + x^7 + x^9 .$$

In fact the other possibilities for $p(x)$ are

$$p_2(x) = x^2 + x^4 + x^5 + x^6 + x^{10} ,$$

$$p_3(x) = 1 ,$$

and cyclic permutations of these, giving 33 in all.

The code of R generated by $1 + yp_1(x)$ is a (22, 11) code with minimum distance 6.

(2) $n = 2m$.

Let Y be the polynomial ring $F[y]/(y^m + 1)$. Let $Q(y) \in Y$ such that $Q(y)Q(y^{-1}) = 1 \bmod (y^m + 1)$. Let $Q(x)$ be the same polynomial regarded as an element of $X = F[x]/(x^{2m} + 1)$.

If $1 + Q(x)Q(x^{-1}) = r(x) + r(x^{-1})$ where $r(x) \equiv 0 \bmod x^m + 1$ in X, set

$$p(x) = Q(x) + r(x)/Q(x^{-1}) .$$

Then $p(x)p(x^{-1}) = 1 \bmod (x^{2m} + 1)$, and every such $p(x)$ can be obtained in this way.

If $m \not\equiv 0 \bmod 4$, then for every choice of $Q(y)$ we can find a suitable polynomial $r(x)$. If $m \equiv 0 \bmod 4$, $r(x)$ exists for only half the choices of $Q(y)$. In this case we write

$$1 + Q(x)Q(x^{-1}) = \sum_{i=1}^{4t-1} u_i x^i ,$$

and $r(x)$ exists if and only if

$$u_0 = u_t = u_{2t} = u_{3t} = 0 .$$

We illustrate for $n = 4$, $n = 8$.

For $n = 4$ the choice of $Q(y)$ is 1 or y. In both cases, $1 + Q(x)Q(x^{-1}) = 0$. We can take $r(x)$ to be

$$0 , \quad 1 + x^2 , \quad x + x^3 , \quad 1 + x + x^2 + x^3 .$$

We obtain, for $Q(x) = 1$,

$$p(x) = 1 , \quad x^2 , \quad 1 + x + x^3 , \quad x + x^2 + x^3 ,$$

and for $Q(x) = x$,

$$p(x) = x\,, \qquad x^3\,, \qquad 1 + x + x^3\,, \qquad 1 + x^2 + x^3\,.$$

The code generated by $1 + y(1 + x + x^2)$ is an $(8, 4)$ code with minimum distance 4. It is equivalent to the code obtained from the Hamming $(7, 3)$ code by adding an additional parity check.

Change notation so that $n = 4$ is the y ring; the choices for $Q(y)$ are 1, y, y^2 or $1 + y + y^2$, $y + y^2 + y^3$, $y^2 + y^3 + y^4$. If $Q(y)$ is 1, y or y^2, then $1 + Q(x)Q(x^{-1}) = 0$ and we may take $r(x)$ to be any of the rows labelled C in Table 1. The corresponding expressions for $p(x)$ are

$$1\,, \qquad 1 + x^2 + x^6\,, \qquad 1 + x + x^3 + x^5 + x^7\,,$$
$$1 + x + x^2 + x^3 + x^4 + x^5 + x^6$$

and cyclic permutation of these.

For the other choices of $Q(y)$ we obtain

$$1 + Q(x)Q(x^{-1}) = 1 + x^2 + x^6\,,$$

so there is no corresponding $p(x)$.

Table 1. The ideal generated by $x^4 + 1$ in $F[x]/(x^8 + 1)$.

1	x	x^2	x^3	x^4	x^5	x^6	x^7	
0	0	0	0	0	0	0	0	C
1	0	0	0	1	0	0	0	C
0	1	0	0	0	1	0	0	
0	0	1	0	0	0	1	0	C
0	0	0	1	0	0	0	1	
1	1	0	0	1	1	0	0	
1	0	1	0	1	0	1	0	C
1	0	0	1	1	0	0	1	
0	1	1	0	0	1	1	0	
0	1	0	1	0	1	0	1	C
0	0	1	1	0	0	1	1	
1	1	1	0	1	1	1	0	
1	1	0	1	1	1	0	1	C
1	0	1	1	1	0	1	1	
0	1	1	1	0	1	1	1	C
1	1	1	1	1	1	1	1	C

References

1. Berlekamp, E. R. "Distribution of Cyclic Matrices in a Finite Field," *Duke Math. J.*, **33** (1966), 45–48.

2. Mattson, H. F. and Solomon, G. A. "A New Treatment of Bose-Chaudhuri Codes," *J. Soc. Indust. Appl. Math.*, **9** (1961), 654–669.
3. Curtis, C. W. and Reiner, I. *Representation Theory of Finite Groups and Associative Algebras*, John Wiley and Sons, 1962.
4. MacWilliams, Mrs. Jessie. "Orthogonal Circulant Matrices over Finite Fields, and How to Find Them." (To be published.)

CHAPTER 19

Some New Results on Finite Fields and Their Application to the Theory of BCH Codes

W. W. PETERSON[1], *University of Hawaii*

1. INTRODUCTION

Let q be a power of a prime and consider $GF(q^m)$ as a vector space over $GF(q)$. A set of three very interesting theorems of Pele [3] give two one-to-one correspondences between subspaces of $GF(q^m)$ and polynomials over $GF(q^m)$ whose non-zero terms have exponents which are powers of q, and which split completely in $GF(q^m)$. These polynomials can be used to construct minimum weight code words in a large class of BCH codes to establish that the minimum weight in these codes meets the bound derived by Hocquenghem and Bose and Chaudhuri.

2. PELE'S THEOREMS

Let L denote the linear transformation which takes any element α into α^q.

[1] This work was supported in part by but does not necessarily constitute the opinion of the Air Force Cambridge Research Laboratories, Office of Aerospace Research, under contract AF-1d(628)-4379.

Let $f(L) = a_0 + a_1L + a_2L^2 + \ldots + a_2L^r$ denote a polynomial in L, and $\bar{f}(x) = f(L)x = a_0x + a_1x^q + a_2x^{q^2} + \ldots + a_rx^{q^r}$.

Theorem 1. *For any vector subspace V of dimension r there exists a unique monic polynomial in L, $f(L)$, of degree r such that if $\alpha \in V$, then $f(L)\alpha = 0$.*

Proof. Let $\alpha_1, \alpha_2, \ldots, \alpha_r$ be a basis for V. Let $f_1(L) = L - \alpha_1{}^{q-1}$ and $f_i(L) = (L - \beta_i)f_{i-1}(L)$ where $\beta_i = (f_{i-1}(L)\alpha_i)^{q-1}$ for $i > 1$.

Then it is easily verified that $f_1(L)\alpha_1 = 0$, and if $f_{i-1}(L)\alpha_j = 0$ for all $j < i$, then $f_i(L)\alpha_j = 0$ for $j \leqq i$. Thus by induction $f_r(L)\alpha_j = 0$ for $j \leqq r$. But since $f_r(L)$ is a linear transformation, $f_r(L)$ annihilates every linear combination of the basis vectors, and hence every vector in V. Consider the polynomial $f_r(L)x = \bar{f}(x)$. Since for each element α of V, $f(L)\alpha = 0$, all the q^r elements of V are roots of $\bar{f}(x)$. Since $\bar{f}(x)$ has degree q^r, it splits completely, and has all its roots the elements of V. Thus its coefficients are unique functions of V and $f(L)$ is uniquely determined by V.

Consider the set of polynomials in x over $GF(q^m)$ with the following rule for multiplication denoted by $*$, for x and any field elements and α:

$$x * \alpha = \alpha^q * x\,, \qquad x^i * x^j = x^{i+j}\,, \qquad \alpha * \beta = \alpha\beta\,.$$

It is easily verified that with this multiplication the set forms a non-commutative ring. With this definition, if $f(x) * g(x) = h(x)$, then $f(L)g(L) = h(L)$. It is, moreover, easily verified that one can do division on the left or the right, and hence there exist Euclidean algorithms:

$$\begin{aligned} f(x) &= q_L(x) * d(x) + r_L(x) \\ &= d(x) * q_R(x) + r_R(x) \end{aligned}$$

where r_L and r_R have degree less than the degree of $d(x)$. Finally, since for any field element α, $\alpha^{q^m} = \alpha$,

$$x^m * \alpha = \alpha * x^m\,,$$

x^m is in the center of this ring. And since for any element b of $GF(q)$ $b^q = b$, and thus $x * b = b * x$, b is in the center also. It follows that any polynomial in x^m with coefficients from $GF(q)$ is in the center of this ring.

Theorem 2. *Let V be any r-dimensional subspace of $GF(q^m)$ and let $f(x)$ be the polynomial of degree r for which $f(L)V = 0$. Then there exists a polynomial $g(x)$ such that $g(x) * f(x) = x^m - 1 = f(x) * g(x)$.*

Proof. Consider the ring of polynomials over $GF(q^m)$ with multiplication $*$, and consider the subset I of polynomials $h(x)$ such that $h(L)\alpha = 0$ for all $\alpha \in V$. Then I is a left ideal which contains $f(x)$ and also $x^m - 1$. Because there is a Euclidean algorithm, it is a principle left ideal. Furthermore $\bar{h}(x)$ has all the q^r elements of V as roots, and therefore has degree at least q^r. It follows that $h(x)$ has degree at least r, and hence $f(x)$ is a monic polynomial of minimum degree in I and is its generator. Therefore $x^m - 1$ is a left multiple of $f(x)$, i.e. there exists a unique $g(x)$ such that $g(x) * f(x) = x^m - 1$. Since $x^m - 1$ is in the center of the ring, $g(x) * f(x) * g(x) = (x^m - 1) * g(x) = g(x) * \cdot (x^m - 1)$, and dividing both sides by $g(x)$ on the left gives $f(x) * g(x) = x^m - 1$.

Theorem 3. *For any r dimensional subspace of $GF(q^m)$ there exists a unique polynomial $g(x)$ of degree $m - r$ such that $\bar{g}(x)$ splits completely in $GF(q^m)$ and such that $V = \{\bar{g}(\alpha) : \alpha \in GF(q^m)\}$.*

Proof. Since the dimension of the null space of $f(L)$ is r, the image space $U = \{f(\alpha) : \alpha \in GF(q^m)\}$ has dimension $m - r$. Since $g(L)f(L) = 0$, $G(L)\alpha = 0$ if $\alpha \in U$. Thus the null space of $g(L)$ contains U and has dimension at least $m - r$. But since $\bar{g}(x)$ has degree q^{m-r}, it must split completely and have the elements of U as all its roots. Since the null space of $g(L)$ is exactly U which has dimension $m - r$, the image space $W = \{g(\alpha) : \alpha \in GF(q^m)\}$ must have dimension exactly r. Since $f(L)g(L) = 0$, W is contained in the null space of $f(L)$, i.e. W is contained in V. But W and V have the same dimension, and so they are equal.

3. APPLICATION TO BCH CODES [2]

In the following, a BCH code whose generator polynomial has $\alpha, \alpha^2, \ldots, \alpha^{d-1}$ but not α^d as roots, is referred to as a "distance -d" BCH code, and if α is primitive, the code is called primitive.

Theorem 4. *Let x_i, $1 \leqq i \leqq w$ be the location numbers* [4] *of the non-zero components of a code word which has as components w ones and the rest zeros in a primitive distance-d BCH code of length $q^m - 1$, and let $f(x) = \prod_{i=1}^{w} (x - x_i) = \sum_{i=0}^{w} a_i x^{w-i}$. Then*

(A) *$f(x)$ is of degree w and splits completely in $GF(q^m)$,*
(B) *$a_i = 0$ if i is less than d and is relatively prime to q,*
(C) *$a_w \neq 0$.*

Conversely, if $f(x)$ is a polynomial which satisfies conditions (A), (B), *and* (C), *its w roots are the location numbers of a code word of* 1's *and* 0's *of weight w in a distance-d primitive BCH code of length $q^m - 1$.*

Proof. For the first part, (A) and (C) are obvious. If the x_i are location numbers of components of a code word which are ones, then

$$\sum_i x_i^j = S_j = 0 \qquad \text{if } 1 \leqq j < d\,,$$

and it follows from Newton's identities that $\sigma_j = 0$ if j is relatively prime to q, and hence property (B) holds also.

Now assume that (A), (B), and (C) hold. Call the roots of $f(x)$ $x_1, x_2, \ldots, x_w$. By (C), none is zero. By (B) $\sigma_j = 0$ if j is relatively prime to q. It can be shown then from Newton's identities that $S_j = 0$ for $1 \leqq j < d$, and hence a vectors with 1's in positions corresponding to $x_1, x_2, \ldots, x_w$ and zeros in the other positions is a code word.

Theorem 5. *A distance-$(q^i - 1)$ primitive BCH code of length $q^m - 1$ has minimum distance exactly $q^i - 1$. In fact, the non-zero elements of any i-dimensional subspace of $GF(q^m)$ are location numbers of the ones in a minimum weight code word consisting only of ones and zeros.*

Proof. The $f(x)$ given by Theorem 1 for an i-dimensional subspace V has x as a factor. However $\bar{f}(x)/x$ satisfies conditions (A), (B), and (C) of Theorem 4, and Theorem 5 follows.

Theorem 5 proves that the BCH bound gives the true minimum weight for an infinite subset of codes.

Thorem 6. *Suppose a primitive distance-d_0 BCH code of length $q^m - 1$ over $GF(q)$ has minimum distance exactly d_0, and suppose $d_0 + 1$ is divisible by p, the characteristic of the field. Then* *a distance* $((d_0+1)q^{m-h}-1)$ *primitive BCH code with* $h \geqq d_0$ *has minimum distance exactly* $((d_0+1)q^{m-h}-1)$. (Note that the condition that d_0+1 be divisible by p is satisfied by all binary BCH codes.)

Proof. By hypothesis and Theorem 4 there exists a polynomial $a(x) = \sum a_i x^{d_0-i}$ which splits completely in $GF(q^m)$, and has $a_w \neq 0$, $a_i = 0$ for each i relatively prime to q in the range from 1 to $d_0 - 1$.

Let V be any h-dimensional subspace containing all the roots of $a(x)$. Since $h \geqq d_0$, such subspaces exist. For this V from the polynomial $\bar{g}(x)$ given by Theorem 3. Then $\bar{g}(x)$ has the following form:

$$\bar{g}(x) = x^{q^{m-h}} + b_1 x^{q^{m-h-1}} + \ldots + b_{m-n} x\,.$$

Furthermore $\bar{g}(x)$ is a linear mapping. The set of elements mapped into zero is an $(m-h)$-dimensional subspace U, and the elements mapped into each particular element of V is a distinct coset of U.

Now consider $b(x) = xa(x)$. It has a linear term, and all other non-zero terms have exponents divisible by p. Next consider $b(\bar{g}(x))$. Because of the form of $\bar{g}(x)$, this also has the property that there is a linear term, and the exponents of all other non-zero terms are divisible by p. Thus $b(\bar{g}(x))/x$ satisfies conditions (B) and (C) of Theorem 4.

Now $a(\bar{g}(x))$ has degree d_0q^h. It has as roots all elements of each of the d_0 cosets which maps into a root of $a(x)$. Since each coset has q^h elements, $a(\bar{g}(x))$ has d_0q^h roots in $GF(q^m)$, i.e. it splits completely. Also $\bar{g}(x)$ splits completely. Therefore

$$b(\bar{g}(x))/x = \bar{g}(x)a(\bar{g}(x))/x$$

also splits completely in $GF(q^m)$, and so satisfies condition (A) in Theorem 4. Since $b(\bar{g}(x))/x$ has degree $(d_0+1)^{m-h}-1$, the theorem follows from Theorem 4.

As an example of the application of this theorem, consider the binary primitive BCH codes with $d_0 = 5$. All such codes are known to meet the bound [1]. If h is taken to be $m-1$, and $m \geqq 6$, the theorem implies that all primitive distance-11 codes of length at least 63 have minimum distance exactly 11. Taking h to be $m-2$ shows that all distance 23 codes of length at least 127 have minimum distance exactly 23, etc.

4. CONCLUSION

The first part of this paper presents theorems of Pele, which can be summarized as follows:

For each r-dimensional subspace V of $GF(q^m)$ there is a dual subspace U of dimension $m - r$. There is a unique polynomial $\bar{f}(x)$ which has all elements of V as roots and a unique polynomial $\bar{g}(x)$ which has the elements of U as roots. Furthermore $\bar{f}(x)$ maps $GF(q^m)$ onto U and $\bar{g}(x)$ maps $GF(q^m)$ onto V. Both $\bar{f}(x)$ and $\bar{g}(x)$ have non-zero terms only with degrees which are powers of q. Finally $\bar{f}(\bar{g}(x)) = \bar{g}(\bar{f}(x)) - x^{q^m} - x$.

In the second part of the paper these polynomials are used to construct minimum weight code words in certain BCH codes, establishing for additional infinite subclasses of BCH codes that the true minimum weight is given by the BCH bound.

Acknowledgement

These results are part of collaborative research of T. Kasami, S. Lin, W. Leahy, W. Peterson, and K. Rogers. Theorem 4 was discovered independently by E. Berlekamp.

References

1. Gorenstein, D., Peterson, W. W. and Zierler, N. "Two-Error Correcting Bose-Chaudhuri Codes are Quasi-Perfect," *Inf. and Control*, **3** (1960), 291-294.
2. Kasami, T., Shu Lin and Peterson, W. W. "Some Results on Cyclic Codes which are Invariant under the Affine Group," Report AFCRL-66-622, Department of Electrical Engineering, University of Hawaii, 1966. The portions of this report pertaining to minimum weight in BCH codes have been submitted to the *J. Inst. Elect. Comm. Eng. Japan*.
3. Pele, R. L. "Some Remarks on the Vector Subspaces of a Finite Field," Report AFCRL-66-477, Department of Electrical Engineering, University of Hawaii, 1966. Submitted to *Canad. J. Math*.
4. Peterson, W. W. *Error-Correcting Codes*, MIT Press, Cambridge, Mass., 1961, p. 169.

CHAPTER 20

Weight Distributions of Bose-Chaudhuri-Hocquenghem Codes

TADAO KASAMI[1], *Osaka University*

1. INTRODUCTION

The weight distribution problem of a code is to find the number of code vectors of each weight in the code. The weight distribution is one of the important properties of the structure of a code and gives the complete information on the probability of an undetected error when the code is used for error detection only.

To the author's knowledge, explicit formulas of weight distribution have been known only for the Hamming codes [13] and the Reed-Solomon codes [1, 3, 8]. The weight distributions of all cyclic codes of length 31 were computed by Prange [16] and a number of weight distributions for BCH codes and their dual codes of length 63 to 1023 found by digital computation have been tabulated by Peterson [14].

In this paper, several methods useful for finding the weight distributions of binary Bose-Chaudhuri-Hocquenghem codes [2]

[1] This work was supported in part by the Joint Services Electronics Program (U. S. Army, U. S. Navy, and U. S. Air Force) under Contract No. DA 28 043 AMC 00073(E), and in part by the National Science Foundation under Grant NSF GK-690, and in part by the Air Force Cambridge Research Laboratories under Contract AF 19 (628) 4379.

(BCH codes) of length $2^m - 1$ are presented. Explicit weight distribution formulas for several classes of BCH codes and some other cyclic codes are derived.

Let C be a binary linear code of length n and let k denote the number of information digits. Let a_j denote the number of vectors of weight j in C and b_j denote the number of vectors of weight j in the dual code of C. A series of identities by which each a_j can be calculated from the b_j's has been given by MacWilliams [9]. Thus it is enough to consider the case where $k \leq n - k$. The following power moment identities have been derived from MacWilliams identities by Pless [15].

$$(1) \qquad \sum_{j=0}^{n} j^l a_j = \sum_{j=0}^{n} (-1)^j b_j \left(\sum_{\nu=0}^{l} \nu! G_l^\nu 2^{k-\nu} \binom{n-j}{n-\nu} \right),$$

where G_l^ν is a Stirling number of the second kind [5].

Simple formulas for even j,

$$(2) \qquad ja_j = (n + 1 - j)a_{n+1-j}$$

$$(3) \qquad jb_j = (n + 1 - j)b_{n+1-j}$$

have been proved to hold for the BCH codes of length $2^m - 1$ by Prange and Peterson [14] as a simple consequence of the fact that it is possible to extend these codes by adding one more check digit in such a way that the extended code is invariant under a doubly transitive group of permutations on the components of a code vector.

For odd d, by a d-BCH code is meant a binary BCH code of length $2^m - 1$ which has $\beta, \beta^2, \ldots, \beta^{d-1}$ but not β^d as roots of its generator polynomial, where β is a primitive element of $GF(2^m)$. For even d, let a d-BCH code be a code consisting of the code vectors of even weight in a $(d - 1)$-BCH code. A t-error-correcting BCH code [2] is a $(2t + 1)$-BCH code.

A cyclic code of length $2^m - 1$ can be derived from the νth order Reed-Muller code of length 2^m [12] by deleting the first component of each code vector and permuting the remaining components suitably. The resulting code will be called the νth order modified Reed-Muller code. This code has been proved to be a subcode of a $(2^{m-\nu} - 1)$-BCH code [7, 8]. It is shown in section 3 that the possible values of weights of code vectors of the second order Reed-Muller code are very sparse. By using this fact as well as the power moment identities and the invariant property, explicit weight distribution formulas are obtained for the following subcodes of the second order modified Reed-Muller code: the dual code of every double-error-correcting BCH code, the dual code of triple-error-correcting BCH code for any odd $m \geq 5$ and several even m's, $(2^{m-1} - 2^{m/2-1})$-BCH codes for even $m \geq 4$, $(2^{m-1} - 2^{m/2})$-BCH codes for even $m \geq 4$, $(2^{m-1} - 2^{(m-1)/2})$-BCH codes for odd $m \geq 3$, $(2^{m-1} - 2^{(m+1)/2})$-BCH codes for odd $m \geq 5$, $(2^{m-1} - 2^{(m+3)/2})$-BCH codes for odd $m \geq 11$ and some other classes of cyclic codes.

2. INVARIANT PROPERTIES

Let $n = 2^m - 1$. The extended code of C is the code with an overall parity check added to C as the first digit. The first component in a code vector is numbered 0, and for $i > 1$ the ith component is numbered α^{i-2}, where α is a primitive element of $GF(2^m)$. Let v be a vector of the extended code. For a $(\neq 0)$ and b in $GF(2^m)$, permute the component of v in position X to position $aX + b$. Then, the resulting vector will be denoted by $\pi_{ab}v$. If the extended code of C is invariant under the doubly transitive group of permutations $\Pi = \{\pi_{ab} \mid a \neq 0, b \in GF(2^m)\}$, then C is a cyclic code by definition. Peterson [14] proved that the extended codes of $(2t + 1)$-BCH codes are invariant under permutation group Π.

Let i be a positive integer less than 2^m. Then i can be expressed in binary form:

$$i = \sum_{j=0}^{m-1} \delta_j 2^j .$$

Let $I(i)$ denote the set of all nonzero integers i' such that

$$i' = \sum_{j=0}^{m-1} \delta'_j 2^j ,$$

where $0 \leq \delta'_j \leq \delta_j$ for $0 \leq j < m$.

Theorem 1 [7, 8]. *Let C be a cyclic code of length $2^m - 1$ generated by a polynomial $g(X)$. The extended code of C is invariant under the permutation group Π if and only if* (1) $g(1) \neq 0$ *and* (2), *for every root α^i of $g(X)$, $g(\alpha^{i'}) = 0$ for i' in $I(i)$.*

Let C_0 be a cyclic code of length $2^m - 1$ generated by $g(X) = (X^{2^m-1} - 1)/(h_0(X) \ldots h_p(X))$, where $h_0(X) = X - 1$, $h_i(X)$ is an irreducible polynomial of degree m_i and $h_i(\alpha^{j_i}) = 0$ $(0 \leq i \leq p)$. Suppose that $g(X)$ satisfies the condition of Theorem 1. Let $v(X)$ be the polynomial representation [1] of a code vector of C_0. If $g(\alpha^j) = 0$, then $v(\alpha^j) = 0$. Obviously, $v(\alpha^{j_i}) \in GF(2^{m_i})$ $(0 \leq i \leq p)$. Conversely, for any set of β_i in $GF(2^{m_i})$ $(0 \leq i \leq p)$, there exists a unique code vector[2] $\boldsymbol{v}(\beta_0, \ldots, \beta_p\colon X)$ in C_0 such that $\boldsymbol{v}(\beta_0, \ldots, \beta_p\colon \alpha^{j_i}) = \beta_i$ $(0 \leq i \leq p)$ (Mattson and Solomon [10]). Let $\bar{\boldsymbol{v}}(\beta_0, \ldots, \beta_p)$ denote the vector with an overall parity added to code vector $\boldsymbol{v}(\beta_0, \ldots, \beta_p\colon X)$ as the first component. $\bar{\boldsymbol{v}}(\beta_0, \ldots, \beta_p)$ is a vector of the extended code C_{ex} of C_0. Let $X_1, \ldots, X_w$ be the location numbers of nonzero components of $\bar{\boldsymbol{v}}(\beta_0, \ldots, \beta_p)$. By definition, w is an even integer and

$$\sum_{f=1}^{w} X_f^{j_i} = \boldsymbol{v}(\beta_0, \ldots, \beta_p\colon \alpha^{j_i}) = \beta_i\,, \qquad (1 \leq i \leq p)\,.$$

If $g(\alpha^l) = 0$, then

$$(4) \qquad \sum_{f=1}^{w} X_f^l = 0\,.$$

Otherwise, $h_q(\alpha^l) = 0$ for some q and, consequently, $l \equiv j_q 2^\nu$ $(\text{mod } 2^{m_q} - 1)$ for some $0 \leq \nu < m_q$. Hence,

$$(5) \qquad \sum_{f=1}^{w} X_f^l = \beta_q^{2^\nu}\,.$$

For any a $(\neq 0)$ and b in $GF(2^m)$, there exists $\bar{\boldsymbol{v}}(\beta_0', \ldots, \beta_p')$ in C_{ex} such that

$$\bar{\boldsymbol{v}}(\beta_0', \ldots, \beta_p') = \pi_{ab}\bar{\boldsymbol{v}}(\beta_0, \ldots, \beta_p)\,.$$

By definition,

$$(6) \qquad \beta_i' = \sum_{f=1}^{w} (aX_f + b)^{j_i}\,, \qquad (1 \leq i \leq p)\,.$$

If

$$j = 2^{\sigma_1} + 2^{\sigma_2} + \ldots + 2^{\sigma_t} \qquad (0 \leq \sigma_1 < \sigma_2 < \ldots < \sigma_t < m)\,,$$

$$(7) \qquad (aX_f + b)^j = (a^{2^{\sigma_1}} X_f^{2^{\sigma_1}} + b^{2^{\sigma_1}}) \ldots (a^{2^{\sigma_t}} X_f^{2^{\sigma_t}} + b^{2^{\sigma_t}}) = \sum_{l \in I(j)} a^l X_f^l b^{j-l}\,.$$

It follows from (4) through (7) that

[2] Vector $\boldsymbol{v}(X)$ means the vector represented by polynomial $v(X)$.

$$\beta'_i = \sum_{q=1}^{p} \sum_{\nu \in E_{iq}} a^{j_q 2^\nu} \beta_q^{2^\nu} b^{j_i - j_q 2^\nu}, \qquad (1 \le i \le p) \tag{8}$$

where E_{iq} is the set of integer ν's such that the remainder of $j_q 2^\nu/(2^m - 1)$ is in $I(j_i)$ and that $0 \le \nu < m_q$.

Lemma 2. *Assume that, for given β_i and β'_i in $GF(2^{m_i})$ $(1 \le i \le p)$, there are a $(\neq 0)$ and b in $GF(2^m)$ which satisfy (8). Then, if the weight of $v(0, \beta_1, \ldots, \beta_p: X)$ is w, the weight of $v(0, \beta'_1, \ldots, \beta'_p: X)$ is either w or $n + 1 - w$.*

Proof. It follows from the assumption that there exists β'_0 in $GF(2)$ such that $\bar{v}(\beta'_0, \ldots, \beta'_p) = \pi_{ab}\bar{v}(0, \beta_1, \ldots, \beta_p)$. Obviously, the weights of $\bar{v}(0, \beta_1, \ldots, \beta_p)$ and $\bar{v}(\beta'_0, \ldots, \beta'_p)$ are equal to w. If $\beta'_0 = 0$, the weight of $v(0, \beta'_1, \ldots, \beta'_p: X)$ is w. If $\beta'_0 = 1$, the weight of $v(1, \beta'_1, \ldots, \beta'_p: X)$ is $w - 1$ by definition. C_0 contains the all-one vector $e(X) = 1 + X + \ldots + X^{2^m - 2}$. Since $e(\alpha^j) = \sum_{f=0}^{2^m-2} \alpha^{jf} = 0$ $(0 < j < 2^m - 1)$, $v(0, \beta'_1, \ldots, \beta'_p: X) = v(1, \beta'_1, \ldots, \beta'_p: X) + e(X)$. Therefore, the weight of $v(0, \beta'_1, \ldots, \beta'_p: X)$ is $n + 1 - w$.

Since C_0 contains the all-one vector $e = (1, \ldots, 1)$,

$$a_{n-j} = a_j, \qquad \text{for any } j.$$

Consequently, it is enough to consider the code C consisting of all the code vectors of even weight in C_0. In C, $\beta_0 = 0$. Since the symmetry property (2) holds for C_0 by Prange's Theorem [14, 16], it also holds for C. Hence, it is sufficient to find $a_j + a_{n+1-j}$ for even j $(0 < j \le (n + 1)/2)$. Thus the following power moments are convenient.

$$I_l = \sum_{j \neq 0} (j - [(n + 1)/2])^l a_j,$$

where $[x]$ denotes the integer part of x.

If n is odd and $b_1 = b_2 = 0$, then

$$I_2 = 2^{k-2}(n + 1) - 2^{-2}(n + 1)^2, \tag{9}$$

$$I_4 = 2^{k-4}[3(n + 1)^2 - 2(n + 1)] - 2^{-4}(n + 1)^4 + 3 \cdot 2^{k-1}(b_3 + b_4). \tag{10}$$

If n is odd and $b_i = 0$ $(1 \le i \le 4)$,

(11) $$I_6 = 15\cdot 2^{k-6}[(n+1)^3 - 2(n+1)^2] + 2^{k-2}(n+1) - 2^{-6}(n+1)^6 + 6!\,2^{k-6}(b_5 + b_6)\,.$$

The proof of (9), (10) and (11) is given in Appendix 1.

3. MODIFIED REED-MULLER CODES

Let V_j denote a j-dimensional vector space over $GF(2)$ and x_i $(1 \le i \le m)$ be a variable over $GF(2)$. For $1 \le \nu \le m$, let P_ν be the set of polynomials over $GF(2)$ of variables $x_1, \ldots, x_m$ of degree ν or less. For $0 \le j < 2^m - 1$, let

$$\alpha^j = \sum_{i=0}^{m-1} v_{ji}\alpha^i\,, \qquad v_{ji} \in GF(2)\,.$$

For $f(x_1, \ldots, x_m) \in P_\nu$, let $v(f)$ denote a vector in V_{2^m} of which the first component is $f(0, \ldots, 0)$ and the jth component $(j > 1)$ is $f(v_{j-2\,0}, v_{j-2\,1}, \ldots, v_{j-2\,m-1})$. Then the νth order Reed-Muller code of length 2^m is the set of vectors $\{v(f) \mid f \in P_\nu\}$.[3] Delete the first component of each vector of the νth order Reed-Muller code of length 2^m. Then the resulting set of vectors in V_{2^m-1} will be called the νth order modified Reed-Muller code. Let

$$y_j = u_{j0} + \sum_{i=1}^{m} u_{ji}x_i \in P_1 \qquad (1 \le j \le m)\,.$$

If vectors $(u_{j1}, u_{j2}, \ldots, u_{jm})$ $(1 \le j \le l)$ are linearly independent, $y_1, \ldots, y_l$ will be said to be independent. For $f(x_1, \ldots, x_m) \in P_\nu$, there is f' in P_ν such that $f(y_1, \ldots, y_m) = f'(x_1, \ldots, x_m)$. Therefore, if $v(f)$ is a code vector of the νth order Reed-Muller code, then $v(f')$ is also a code vector. It follows from this fact that a modified Reed-Muller code is cyclic [7, 8]. Let $w(j)$ denote the number of ones in the binary expression of j.

Theorem 3 [7, 8]. *Let $g(X)$ be the generator polynomial of the νth order modified Reed-Muller code of length $2^m - 1$. Then α^j is a root of $g(X)$ if and only if $0 < w(j) < m - \nu$.*

This theorem implies that the νth order modified Reed-Muller code is a subcode of a $(2^{m-\nu} - 1)$-BCH code.

For the polynomial $f(x_1, \ldots, x_m)$, let $|f|_m$ denote the number of m-tuple's $(v_1, \ldots, v_m)$ such that

[3] The order of the digit positions is different from the original one [12, 13].

$$f(v_1, \ldots, v_m) = 1 .$$

By definition, $|f|_m$ is the weight of vector $v(f)$. If the y_j's $(1 \leq j \leq m)$ in P_1 are independent and $f'(x_1, \ldots, x_m) = f(y_1, \ldots, y_m)$, then

$$|f'|_m = |f|_m . \tag{12}$$

This follows from the fact that $y_j = u_{j0} + \sum_{i=1}^{m} u_{ji}x_i$ $(1 \leq j \leq m)$ defines a one-to-one mapping from V_m onto itself.

Lemma 4. *Assume that* (1) $m \geq 2$, (2) $f(x_1, \ldots, x_m) \in P_2$, (3) f *does not depend on* x_i $(i < i_0 \leq m)$ *but on* x_{i_0}. *Then there exist independent* $y_j^{(i)}$'s $(1 \leq i \leq t; 1 \leq j \leq l_i)$ *in* P_1 *such that*

(1) $y_1^{(1)} = x_{i_0}$

(2) $f(x_1, \ldots, x_m) = u_0 + \sum_{i=1}^{t} (\sum_{j=1}^{l_i-1} y_j^{(i)} y_{j+1}^{(i)} + u_i y_{l_i}^{(i)})$,

where $u_i \in GF(2)$ $(0 \leq i \leq t)$.

Proof. If $m = 2$, it is easy to check that this lemma holds. Suppose that this lemma holds for $2 \leq m < m'$. Consider the case of $m = m'$. Let

$$f(x_1, \ldots, x_m) = F_0(x_2, \ldots, x_m) + x_1 F_1(x_2, \ldots, x_m) , \tag{13}$$

where $F_0 \in P_2$ and $F_1 \in P_1$. If $F_1 = 1$,

$$f(x_1, \ldots, x_m) = x_1 + F_0(x_2, \ldots, x_m) .$$

Then apply the induction hypothesis to $F_0(x_2, \ldots, x_m)$. Suppose that F_1 is not a constant. Since x_1 and $F_1(x_2, \ldots, x_m)$ are independent, there exist independent $y_1, \ldots, y_m$ such that

(1) $y_1 = x_1$,

(2) $y_2 = F_1(x_2, \ldots, x_m)$,

(3) $y_2, \ldots, y_m$ are polynomials of $x_2, \ldots, x_m$ of the first degree.

Then it follows from (13) that

$$f(x_1, \ldots, x_m) = y_1 y_2 + f'(y_2, \ldots, y_m) ,$$

where $f'(y_2, \dots, y_m) = F_0(x_2, \dots, x_m)$. Now apply the induction hypothesis to $f'(y_2, \dots, y_m)$.

Let $G_0(l)$ and $G_1(l)$ be defined by the following:

$$G_0(l) = |x_1x_2 + x_2x_3 + \dots + x_{l-1}x_l|_l\,, \qquad l \geq 2$$

$$G_0(1) = 0\,,$$

$$G_1(l) = |x_1x_2 + x_2x_3 + \dots + x_{l-1}x_l + x_l|_l\,, \qquad l \geq 1\,.$$

Note that

$$|f(x_1, \dots, x_m)|_m = |f(x_1, \dots, x_{m-1}, 0)|_{m-1} + |f(x_1, \dots, x_{m-1}, 1)|_{m-1} \tag{14}$$

$$|1 + f(x_1, \dots, x_m)|_m = 2^m - |f(x_1, \dots, x_m)|_m\,. \tag{15}$$

It is easy to check that, for $l \geq 2$,

$$\begin{aligned} G_0(l) &= |x_1x_2 + \dots + x_{l-2}x_{l-1}|_{l-1} \\ &\quad + |x_1x_2 + \dots + x_{l-2}x_{l-1} + x_{l-1}|_{l-1} \\ &= G_0(l-1) + G_1(l-1)\,, \end{aligned} \tag{16}$$

$$\begin{aligned} G_1(l) &= |x_1x_2 + \dots + x_{l-2}x_{l-1}|_{l-1} \\ &\quad + |x_1x_2 + \dots + x_{l-2}x_{l-1} + x_{l-1} + 1|_{l-1} \\ &= G_0(l-1) + 2^{l-1} - G_1(l-1)\,. \end{aligned} \tag{17}$$

By (16) and (17), for $l \geq 3$

$$\begin{aligned} G_0(l) &= G_0(l-2) + G_1(l-2) \\ &\quad + G_0(l-2) + 2^{l-2} - G_1(l-2) \\ &= 2G_0(l-2) + 2^{l-2}\,, \end{aligned}$$

$$\begin{aligned} G_1(l) &= G_0(l-2) + G_1(l-2) + 2^{l-1} \\ &\quad - (G_0(l-2) + 2^{l-2} - G_1(l-2)) \\ &= 2G_1(l-2) + 2^{l-2}\,. \end{aligned}$$

Hence,

$$2^{l-1} - G_i(l) = 2(2^{l-3} - G_i(l-2))\,, \qquad i = 0, 1\,. \tag{18}$$

On the other hand, it is easy to check that

$$G_i(2) = 1\,, \qquad i = 0, 1\,,$$

$$G_0(1) = 0\,,$$

$$G_1(1) = 1\,.$$

Therefore, it follows from (18) that, for even $l \geq 2$,

$$G_0(l) = G_1(l) = 2^{l-1} - 2^{l/2-1}\,, \tag{19}$$

and that, for odd $l \geq 1$,

$$G_0(l) = 2^{l-1} - 2^{(l-1)/2}\,, \tag{20}$$

$$G_1(l) = 2^{l-1}\,. \tag{21}$$

Lemma 5. *Suppose that* $m \geq 2$ *and* $f(x_1, \ldots, x_m) \in P_2$. *Then* $|f|_m$ *is of the form:*

$$2^{m-1} + \varepsilon 2^l\,,$$

where $m/2 - 1 \leq l \leq m - 1$ *and* ε *is either* 0, 1 *or* -1.

Proof. If $m = 2$, this lemma is obvious. Assume that this lemma holds for $2 \leq m < m'$ and consider the case of $m = m'$. By Lemma 4, there exist independent $y_1, \ldots, y_m$ in P_1 such that, for some h $(1 \leq h \leq m)$,

$$f(x_1, \ldots, x_m) = f_0(y_1, \ldots, y_h) + f_1(y_{h+1}, \ldots, y_m)\,,$$

$$f_0(y_1, \ldots, y_h) = y_1y_2 + y_2y_3 + \ldots + y_{h-1}y_h + uy_h\,,$$

where $u \in GF(2)$ and, if $h = m$, f_1 is a constant. If $h = m$, this lemma follows from (15), (19), (20) and (21). Otherwise, it follows from the induction hypothesis that

$$|f_0|_h = 2^{h-1} + \varepsilon_0 2^{l_0}\,, \tag{22}$$

$$|f_1|_{m-h} = 2^{m-h-1} + \varepsilon_1 2^{l_1}\,, \tag{23}$$

where ε_i $(i = 0, 1)$ is either 0, 1 or -1 and

$$h/2 - 1 \leq l_0 \leq h - 1 \tag{24}$$

$$(m - h)/2 - 1 \leq l_1 \leq (m - h) - 1\,. \tag{25}$$

It is easy to check that

$$|f|_m = |f_0|_h(2^{m-h} - |f_1|_{m-h}) + |f_1|_{m-h}(2^h - |f_0|_m) .$$

By (22) and (23),

$$\begin{aligned}|f|_m &= (2^{h-1} + \varepsilon_0 2^{l_0})(2^{m-h-1} - \varepsilon_1 2^{l_1}) \\ &\quad + (2^{h-1} - \varepsilon_0 2^{l_0})(2^{m-h-1} + \varepsilon_1 2^{l_1}) \\ &= 2^{m-h-1} - \varepsilon_0\varepsilon_1 2^{l_0+l_1+1} .\end{aligned}$$

By (24) and (25),

$$m/2 - 1 \leq l_0 + l_1 + 1 \leq m - 1 .$$

Since $\varepsilon_0\varepsilon_1$ is either 0, 1 or -1, the lemma holds.

The following theorem follows from the definition of Reed-Muller codes and Lemma 5.

Theorem 6. *The weight of a code vector of the second order Reed-Muller code of length* 2^m *is of the form*

$$2^{m-1} + \varepsilon 2^l ,$$

where $m/2 - 1 \leq l \leq m - 1$ *and* ε *is either* 0, 1 *or* -1.

4. SUBCODES OF THE SECOND-ORDER MODIFIED REED-MULLER CODES

In what follows, the weight distributions of subcodes of the second order modified Reed-Muller code will be considered.

Lemma 7. *A cyclic code C with overall parity check is a subcode of the second-order modified Reed-Muller code of length* $2^m - 1$, *if and only if the generator* $g(X)$ *is of the form*

$$g(X) = (X^{2^m-1} - 1)/(h_1(X) \ldots h_p(X)) ,$$

where $h_1(X), \ldots, h_p(X)$ *are different irreducible polynomials and there are integers* μ_i $(1 \leq i \leq p)$ *such that*

$$0 \leq \mu_1 < \mu_2 < \ldots < \mu_p \leq m/2 ,$$

$$h_i(\alpha^{-2^{\mu_i}-1}) = 0 .$$

Proof. It follows from Theorem 3 that if $h_i(\alpha^j) = 0$ $(0 < j < 2^m - 1)$, then $m - 2 \leq w(j) < m$. Hence, $j = 2^m - 1 - j'$, where $1 \leq w(j') \leq 2$. If $w(j') = 1$, let $\mu_i = 0$.

If $\mu_1 = 0$, the extended code of C is invariant under permutation group II by Theorem 1 and, consequently, the symmetry properties (2) and (3) hold for C. If $\mu_i = m/2$, the degree of $h_i(X)$ is $m/2$. Otherwise, the degree of $h_i(X)$ is m. (Refer to [6].) Hence, if $\mu_p = m/2$, then $k = (2p-1)m/2$, and otherwise $k = pm$.

Theorem 8. *Suppose that code C satisfies the condition of Lemma 7 and that $p \geq 2$. If $\mu_1 = 0$, let $\mu = \mu_2$. Otherwise, let $\mu = \mu_1$. Then the weight of a nonzero code vector of C is of the form*

$$2^{m-1} + \varepsilon 2^l ,$$

where $m/2 - 1 \leq l \leq m - 1 - \mu$ and ε is either 0, 1 or -1.

Proof. Let C_0 be the cyclic code of length $2^m - 1$ generated by $g(X)/(X-1)$. Then C is the set of the code vectors of even weight in C_0. It is easy to check that, for $1 \leq \mu_i \leq m/2$, $2^{m-1} - 2^{m-1-\mu_i} - 1$ is the smallest among the positive exponents of the roots of $h_i(X)$. Hence, the minimum distance of C_0 is at least $2^{m-1} - 2^{m-1-\mu} - 1$ by the BCH bound [2]. Since C_0 contains the all-one vector $\boldsymbol{e}$, there is no code vector of weight j with $2^{m-1} + 2^{m-1-\mu} < j < 2^m - 1$. Thus, this theorem follows from Theorem 6 and Lemma 7.

For $0 \leq i \leq [(m-1)/2]$, let

$$\bar{a}_i = a_{2^{m-1}-2^{[(m-1)/2]+i}} + a_{2^{m-1}+2^{[(m-1)/2]+i}} . \tag{26}$$

From Theorem 8, it follows that, for even l,

$$I_l = \sum_{i=0}^{[m/2]-\mu} 2^{l[(m-1)/2]+li} \bar{a}_i . \tag{27}$$

Lemma 9. *Let $m \geq 3$. If $\mu_1 = 0$, $\mu_i = [m/2] - p + i$ $(2 \leq i \leq p)$ and $[m/2] - [m/3] + 2 \geq p$, then the code C is a $(2^{m-1} - 2^{m-[m/2]+p-3})$-BCH code.*

Proof. For $0 < j < 2^m - 1$, let $j_{\min}$ denote the smallest exponent of the roots of the minimum polynomial of α^j. If $g(\alpha^j) = 0$ and $w(j) \leq m - 3$, then $j_{\min} \leq 2^m - 1 - (2^{m-1} + 2^{m-[m/3]-1} + 2^{[m/3]-1})$. Hence,

$$j_{\min} < 2^{m-1} - 2^{m-[m/3]-1} - 1 \leq 2^{m-1} - 2^{m-[m/2]+p-3} - 1 .$$

If $g(\alpha^j) = 0$ and $w(j) = m - 2$, then

$$j_{\min} \leq 2^m - 1 - (2^{m-1} + 2^{m-1-[m/2]+p-1}) < 2^{m-1} - 2^{m-[m/2]+p-3} - 1 .$$

On the other hand, if $g(\alpha^j) \neq 0$,

$$j_{\min} \geq 2^m - 1 - (2^{m-1} + 2^{m-[m/2]+p-3}) = 2^{m-1} - 2^{m-[m/2]+p-3} - 1 .$$

Thus, this lemma follows from the definition of $(2^{m-1} - 2^{m-[m/2]+p-3})$-BCH codes.

It is easy to check that $b_1 + b_2 = 0$, if and only if the exponent of $g(X)$ is equal to $2^m - 1$. Hereafter, this condition will be assumed. Let $(l_1, \ldots, l_f)$ denote the greatest common divisor of $l_1, \ldots$ and l_f.

Lemma 10. *Let* $\mu_1 = 0$. *Then* $b_3 + b_4 \neq 0$, *if and only if* $(m, \mu_2, \ldots, \mu_p) > 1$.

Proof. From (3)

$$4b_4 = (2^m - 4)b_{2^m-4} .$$

Since the dual code of C contains the all-one vector $(1, \ldots, 1)$,

$$b_{2^{m}-1-3} = b_3 .$$

Hence, $b_3 + b_4 \neq 0$ if and only if $b_3 \neq 0$. Assume that α^{j_1}, α^{j_2} and α^{j_3} are the location numbers of non-zero components of a code vector of weight 3 in C. Then,

(28) $\quad \alpha^{j_1} + \alpha^{j_2} = \alpha^{j_3}$,

(29) $\quad \alpha^{j_1(2^{\mu_i}+1)} + \alpha^{j_2(2^{\mu_i}+1)} = \alpha^{j_3(2^{\mu_i}+1)}$, $\quad (1 < i \leq p)$.

From (28),

(30) $\quad \alpha^{j_3(2^{\mu_i}+1)} = (\alpha^{j_1} + \alpha^{j_2})^{2^{\mu_i}+1}$

$$= \alpha^{j_1(2^{\mu_i}+1)} + \alpha^{j_1 2^{\mu_i}}\alpha^{j_2} + \alpha^{j_1}\alpha^{j_2 2^{\mu_i}} + \alpha^{j_2(2^{\mu_i}+1)} .$$

By subtracting (29) from (30),

$$\alpha^{j_1 2^{\mu_i}}\alpha^{j_2} + \alpha^{j_1}\alpha^{j_2 2^{\mu_i}} = 0 ,$$

$$\alpha^{(j_1-j_2)(2^{\mu_i}-1)} = 1 .$$

Thus, for $1 < i \leq p$,

$$(j_1 - j_2)(2^{\mu_i} - 1) \equiv 0 \pmod{2^m - 1} .$$

Since $j_1 - j_2 \not\equiv 0 \pmod{2^m - 1}$, the "only if part" of the lemma follows. The converse can be proved similarly.

5. WEIGHT DISTRIBUTION FORMULAS

Several cases will be considered in detail.

(a) $p = 2$ and $\mu_1 = 0$.

If $\mu_2 = m/2$, then $k = 3m/2$, and otherwise, $k = 2m$. For examples, $(2^{m-1} - 2^{m-[m/2]-1})$-BCH codes and the duals of double-error-correcting BCH codes belong to this case. Since the order of the permutation group Π is $2^m(2^m - 1)$ and the number of code vectors is 2^{2m} or $2^{3m/2}$, Lemma 2 is very useful. By using Lemma 2 and power moment identities (9) and (10), the weight distribution formula is derived for any μ_2.

Theorem 11. *Let $p = 2$ and $\mu_1 = 0$.*

(1) *If $(m, \mu_2) = (m, 2\mu_2) = c$, then*

$$a_{2^{m-1} \pm 2^{(m+c)/2-1}} = (2^{m-c-1} \mp 2^{(m-c)/2-1})(2^m - 1) ,$$

$$a_{2^{m-1}} = (2^m - 2^{m-c} + 1)(2^m - 1) ,$$

$$a_j = 0 , \quad \textit{for other nonzero } j.$$

(2) *If $2(m, \mu_2) = (m, 2\mu_2) = c$ and $c \neq m$, then*

$$a_{2^{m-1} \pm 2^{(m+c)/2-1}} = 2^{(m-c)/2-1}(2^{(m-c)/2} \mp 1)(2^m - 1)/(2^{c/2} + 1) ,$$

$$a_{2^{m-1} \pm 2^{m/2-1}} = 2^{(m+c)/2-1}(2^{m/2} \mp 1)(2^m - 1)/(2^{c/2} + 1) ,$$

$$a_{2^{m-1}} = ((2^{c/2} - 1)2^{m-c} + 1)(2^m - 1) ,$$

$$a_j = 0 , \quad \textit{for other nonzero } j.$$

(3) *If $2(m, \mu_2) = (m, 2\mu_2) = m$, then*

$$a_{2^{m-1} \pm 2^{m/2-1}} = (2^{m-1} \mp 2^{m/2-1})(2^{m/2} - 1) ,$$

$$a_{2^{m-1}} = 2^m - 1 ,$$

$$a_j = 0\,, \qquad \textit{for other nonzero } j.$$

The proof is given in [6].

The following theorem is due to Pless [15]:

Theorem 12. *If only u a_j's are unknown, and $b_1, b_2, \ldots, b_{u-1}$ are known, then a unique solution to (1) exists.*

(b) $m =$ **odd and** $k = 2m$.

Theorem 13. *Let C be any binary linear code for which $b_1 = b_2 = 0$, $n = 2^m - 1$ and $k = 2m$, where m is an odd integer.*

(i) *Let j_0 denote the smallest j such that*

$$a_j + a_{2^m - j} \neq 0\,, \qquad 0 < j < 2^{m-1}\,.$$

Then,

$$j_0 \leq 2^{m-1} - 2^{(m-1)/2}\,.$$

If j_0 is identical with the upperbound $2^{m-1} - 2^{(m-1)/2}$, the weight distribution is the same as the weight distribution of the dual code of a double-error-correcting BCH code:

$$a_{2^{m-1} \pm 2^{(m-1)/2}} = (2^{m-2} \mp 2^{(m-3)/2})(2^m - 1)\,,$$

$$a_{2^{m-1}} = (2^{m-1} + 1)(2^m - 1)\,,$$

$$a_j = 0\,, \qquad \textit{for other nonzero } j.$$

(ii) *If C is a subcode of the second order modified Reed-Muller code for which $b_3 = b_4 = 0$, C has the weight distribution mentioned above.*

Proof. By (9) and (10),

$$I_2 = 2^{2m-2}(2^m - 1)\,,$$

$$I_4 = 2^{3m-3}(2^m - 1) + 3 \cdot 2^{2m-1}(b_3 + b_4)\,.$$

Thus,

$$(31) \qquad I_4 - 2^{m-1} I_2 = \sum_{j_0 \leq j < 2^{m-1}} (2^{m-1} - j)^2[(2^{m-1} - j)^2 - 2^{m-1}](a_j + a_{2^m - j}) = 3 \cdot 2^{2m-1}(b_3 + b_4)\,.$$

Hence,

(32) $$(2^{m-1} - j_0)^2 \geq 2^{m-1} .$$

If $j_0 = 2^{m-1} - 2^{(m-1)/2}$, it follows from (31) that for $j_0 < j < 2^{m-1}$,

$$a_j + a_{2^m - j} = 0$$

and that

$$b_3 + b_4 = 0 .$$

Since only $a_{2^{m-1} \pm 2^{(m-1)/2}}$ and $a_{2^{m-1}}$ are unknown, part (i) follows from Theorem 12. The weight distribution of the dual code of a double-error-correcting BCH code is given by letting $\mu_2 = 1$ in Theorem 11 (1).

Consider part (ii). By Theorem 8, for $2^{m-1} - 2^{(m-1)/2} < j < 2^{m-1}$,

$$a_j + a_{2^m - j} = 0 .$$

Since $b_3 + b_4 = 0$, for $j \neq 0$, $2^{m-1} \pm 2^{(m-1)/2}$ and 2^{m-1},

$$a_j + a_{2^m - j} = 0 .$$

Thus part (ii) follows from Theorem 12.

The results in cases (a) and (b) can be applied to the cross-correlation problem of two maximum length sequences [4, 6].

(c) $(2^{m-1} - 2^{m/2})$-BCH codes for even $m \geq 4$.

Let $p = 3$, $\mu_1 = 0$, $\mu_2 = m/2 - 1$ and $\mu_3 = m/2$. Then $k = 5m/2$. Lemma 9 shows that this code is a $(2^{m-1} - 2^{m/2})$-BCH code. Therefore,

(33) $$\bar{a}_i = 0 \qquad (i \geq 2) .$$

Since $m/2 - 1$ is relatively prime to $m/2$, it follows from Lemma 10 that

$$b_3 + b_4 = 0 .$$

By (9) and (10),

$$I_2 = 2^{2m-2}(2^{3m/2} - 1) ,$$

$$I_4 = 2^{7m/2-4}(2^{m/2} - 1)(3 \cdot 2^{m/2} + 2) .$$

By solving (27) for $l = 2$ and 4,

$$\bar{a}_0 = 2^m(2^{m/2} - 1)(2^m + 2^{m/2+1} + 4)/3 ,$$

$$\bar{a}_1 = 2^{m+2}(2^{m/2+1} - 1)(2^m - 1)/3 .$$

By using (2), the following theorem is obtained.

Theorem 14. *For even $m \geq 4$, a $(2^{m-1} - 2^{m/2})$-BCH code has the following weight distribution:*

$$a_{2^{m-1} \pm 2^{m/2}} = 2^{m/2-2}(2^{m/2-1} \mp 1)(2^{m/2+1} - 1)(2^m - 1)/3$$

$$a_{2^{m-1} \pm 2^{m/2-1}} = 2^{m/2-1}(2^{m/2} \mp 1)(2^{m/2} - 1)(2^m + 2^{m/2+1} + 4)/3 ,$$

$$a_{2^{m-1}} = (2^{m/2} - 1)(2^{2m-1} + 2^{3m/2-2} - 2^{m-2} + 2^{m/2} + 1) ,$$

$$a_j = 0 , \quad \textit{for other nonzero } j.$$

(d) $p = 3$ **and** m = **odd** ≥ 5.

In this case, $k = 3m$. By (9), (10) and (11),

(34) $$I_2 = 2^{2m-2}(2^{2m} - 1) ,$$

(35) $$I_4 = 3 \cdot 2^{4m-4}(2^m - 1) ,$$

(36) $$I_6 = 2^{4m-5}(7 \cdot 2^m - 8)(2^m - 1) + 6!\, 2^{3m-6}(b_5 + b_6) .$$

(d1) Assume that $b_i = 0$ $(1 \leq i \leq 6)$. The dual code of a triple-error-correcting BCH code is this case.

By (27),

(37) $$2^{m-1}\bar{a}_0 + 2^{m-1} \sum_{i \geq 1} 2^{2i}\bar{a}_i = 2^{2m-2}(2^{2m} - 1) ,$$

(38) $$2^{2(m-1)}\bar{a}_0 + 2^{2(m-1)} \sum_{i \geq 1} 2^{4i}\bar{a}_i = (3 \cdot 2^{4m-4})(2^m - 1) ,$$

(39) $$2^{3(m-1)}\bar{a}_0 + 2^{3(m-1)} \sum_{i \geq 1} 2^{6i}\bar{a}_i = 2^{4m-5}(7 \cdot 2^m - 8)(2^m - 1) .$$

By subtracting 2^{m-1} times (37) from (38),

(40) $$2^{2(m-1)} \sum_{i \geq 1} 2^{2i}(2^{2i} - 1)\bar{a}_i = 2^{3m-3}(2^m - 1)(2^{m-1} - 1) .$$

By subtracting 2^{m-1} times (38) from (39),

(41) $$2^{3(m-1)} \sum_{i \geq 1} 2^{4i}(2^{2i} - 1)\bar{a}_i = 2^{4m-2}(2^m - 1)(2^{m-1} - 1) .$$

By subtracting 2^{m+1} times (40) from (41),

$$\sum_{i\geq 1} (2^{4i} - 2^{2i+2})(2^{2i} - 1)\bar{a}_i = 0 .$$

Since $(2^{4i} - 2^{2i+2})(2^{2i} - 1) > 0$ for $i > 1$,

$$\bar{a}_i = 0 , \qquad (i > 1) . \tag{42}$$

From (37) and (38),

$$\bar{a}_0 = 2^{m-1}(2^m - 1)(5\cdot 2^{m-1} + 4)/3 ,$$

$$\bar{a}_1 = a^{m-3}(2^m - 1)(2^{m-1} - 1)/3 ,$$

Since Equation (2) holds for the dual code of a triple-error-correcting BCH code, it follows from Equation (42) and Theorem 12 that Equation (2) holds for other cases. Hence the weight distribution can be calculated easily.

(d2) Now consider the case where $a_j + a_{2^m-j} = 0$ for $0 < j < 2^{m-1} - 2^{(m+1)/2}$ and $b_i = 0$ for $1 \leq i \leq 4$. For example, let $\mu_1 = 0$, $\mu_2 = (m-3)/2$ and $\mu_3 = (m-1)/2$. Lemma 9 shows that this code is a $(2^{m-1} - 2^{(m+1)/2})$-BCH code. Since $(m-1)/2$ is relatively prime to m, it follows from Lemma 10 that $b_i = 0$ for $1 \leq i \leq 4$.

Theorem 12 and Equation (42) imply that the weight distribution for case **(d2)** is the same as the one for case **(d1)**. Consequently, for $1 \leq i \leq 6$,

$$b_i = 0 .$$

Thus the following theorem holds.

Theorem 15. *Let m be an odd integer greater than 4.*

(i) *A $(2^{m-1} - 2^{(m+1)/2})$-BCH code and the dual code of a triple-error-correcting BCH code have the following weight distribution:*

$$a_{2^{m-1}\pm 2^{(m+1)/2}} = 2^{(m-5)/2}(2^{(m-3)/2} \mp 1)(2^m - 1)(2^{m-1} - 1)/3 ,$$

$$a_{2^{m-1}\pm 2^{(m-1)/2}} = 2^{(m-3)/2}(2^{(m-1)/2} \mp 1)(2^m - 1)(5\cdot 2^{m-1} + 4)/3 ,$$

$$a_{2^{m-1}} = (2^m - 1)(9\cdot 2^{2m-4} + 3\cdot 2^{m-3} + 1) ,$$

$$a_j = 0 , \qquad \textit{for other nonzero } j.$$

(ii) *These weight distribution formulas hold also for every subcode with $k = 3m$ of the second order modified Reed-Muller code that satisfies one of the following conditions:*

(1) $b_i = 0$ *for* $1 \le i \le 6$.

(2) $a_j + a_{2^m - j} = 0$ *for* $0 < j < 2^{m-1} - 2^{(m+1)/2}$
and $b_i = 0$ *for* $1 \le i \le 4$.

(e) $(2^{m-1} - 2^{(m+3)/2})$-BCH codes for odd $m \ge 11$.

Let $p = 4$, $\mu_1 = 0$, $\mu_2 = (m-5)/2$, $\mu_3 = (m-3)/2$ and $\mu_4 = (m-1)/2$. Then $k = 4m$. From Theorem 8, $\bar{a}_i = 0$ $(i > 3)$. Lemma 9 shows that, for $m \ge 11$, this code is a $(2^{m-1} - 2^{(m+3)/2})$-BCH code. The dual code is a subcode of the dual code of a $(2^{m-1} - 2^{(m+1)/2})$-BCH code, which has minimum weight 7 by Theorem 15. Consequently,

$$b_i = 0 \qquad (1 \le i \le 6) .$$

By solving (27) for $l = 2$, 4 and 6 and using symmetry property (2), the following theorem is obtained.

Theorem 16. (i) *For odd* $m \ge 7$, *let* $p = 4$, $\mu_1 = 0$, $\mu_2 = (m-5)/2$, $\mu_3 = (m-3)/2$ *and* $\mu_4 = (m-1)/2$. *Then,*

$$a_{2^{m-1} \pm 2^{(m-1)/2}} = (2^{m-1} \mp 2^{(m-1/2)}) \times (151 \cdot 2^{2m-3} + 25 \cdot 2^m + 2^5)(2^m - 1)/45 ,$$

$$a_{2^{m-1} \pm 2^{(m+1)/2}} = (2^{m-2} \mp 2^{(m-1)/2}) \times (23 \cdot 2^{m-5} + 1)(2^{m-1} - 1)(2^m - 1)/9 ,$$

$$a_{2^{m-1} \pm 2^{(m+3)/2}} = (2^{m-6} \mp 2^{(m-7)/2}) \times (2^{m-3} - 1)(2^{m-1} - 1)(2^m - 1)/45 ,$$

$$a_{2^{m-1}} = 2^{4m} - 1 - \sum_{j \ne 0, 2^{m-1}} a_j ,$$

$$a_j = 0 , \qquad \textit{for other nonzero } j.$$

(ii) *For* $m \ge 11$, *the code in* (i) *is a* $(2^{m-1} - 2^{(m+3)/2})$*-BCH code.*

(f) The dual codes of triple-error-correcting BCH codes for even $m \ge 6$.

Let $\mu_1 = 0$, $\mu_2 = 1$, and $\mu_3 = 2$. Then $k = 3m$. It is easy to check that this code is the dual code of a tripple-error-correcting BCH code. Hence, $b_i = 0$ $(1 \leq i \leq 6)$. From (27), (34), (35) and (36),

$$2^{m-2}\bar{a}_0 + 2^m\bar{a}_1 + 2^{m-2}\sum_{i\geq 2} 2^{2i}\bar{a}_i = 2^{2m-2}(2^{2m} - 1)\,, \tag{43}$$

$$2^{2m-4}\bar{a}_0 + 2^{2m}\bar{a}_1 + 2^{2m-4}\sum_{i\geq 2} 2^{4i}\bar{a}_i = 3\cdot 2^{4m-4}(2^m - 1)\,, \tag{44}$$

$$2^{3m-6}\bar{a}_0 + 2^{3m}\bar{a}_1 + 2^{3m-6}\sum_{i\geq 2} 2^{6i}\bar{a}_i = 2^{4m-5}(7\cdot 2^m - 8)(2^m - 1)\,. \tag{45}$$

By eliminating $\bar{a}_0$ and $\bar{a}_1$,

$$\bar{a}_2 + 28\bar{a}_3 + \sum_{i\geq 4} c_i\bar{a}_i = 2^{m-4}(2^{m-2} - 1)(2^m - 1)/15\,,$$

where $c_i > 28$.

On the other hand, it can be shown that, for $j \geq 3$, $\bar{a}_j$ is divisible by $2^{m-4}(2^m - 1)$. The proof is given in Appendix 2. Hence, it is easy to check that, for $6 \leq m \leq 10$, $\bar{a}_i = 0$ $(i \geq 3)$. Consequently, the weight distributions of the dual codes of triple-error-correcting BCH codes for $m = 6$, 8 and 10 can easily be found by solving Equations (43), (44) and (45) and using symmetry property (2).

APPENDIX 1

The Proof of (9), (10) and (11).

Assume that code C has no code vector with odd weight. Add the all-one vector $(1, \ldots, 1)$ to the basis of C, and let C' denote the resulting code with $k + 1$ information digits. Add an overall parity check to C' and denote the resulting code of length $n + 1$ by C''. Let a'_j (or a''_j) denote the number of code vectors of weight j of code C' (or C'') and let b'' denote the number of code vectors of weight j of the dual of code C''. Then, for even j

$$\begin{aligned} a'_j &= a'_{n-j} = a_j\,, \\ a''_j &= a'_j + a'_{j-1} = a_j + a_{n+1-j}\,, \qquad j \neq 0\,. \end{aligned} \tag{A1}$$

It is easy to check that for odd j

$$b''_j = 0\,, \tag{A2}$$

and that for even $j \neq 0$

(A3) $$b''_j = b_j + b_{j-1}\,.$$

By identity (1),

(A4) $$\sum_{j=0}^{n+1} j^l a''_j = \sum_{h=0}^{n+1} (-1)^h b''_h \left[\sum_{\nu=0}^{l} \nu!\, G_l^\nu 2^{k+1-\nu} \binom{n+1-h}{n+1-\nu} \right],$$

where G_l^ν is a Stirling number of the second kind and $\binom{n+1-h}{n+1-\nu} = 0$ for $h > \nu$. By (A1) through (A4),

(A5) $$\begin{aligned} I_p &= 2^{-1} \sum_{j=0}^{n+1} (j-(n+1)/2)^p a''_j - (n+1)^p 2^{-p} \\ &= 2^{-1} \sum_{l=0}^{p} (-1)^{p-l} \binom{p}{l} (n+1)^{p-l} 2^{-p+l} \sum_{h=0}^{p} (-1)^h b''_h \sum_{\nu=0}^{i} \nu! \\ &\qquad \times G_l^\nu 2^{k+1-\nu} \binom{n+1-h}{n+1-\nu} - (n+1)^p 2^{-p} \\ &= 2^{k-p} \sum_{h=0}^{p} (-1)^h b''_h J_{ph} - (n+1)^p 2^{-p}\,, \\ &= 2^{k-p} \sum_{h=0}^{p/2} (b_{2h} + b_{2h-1}) J_{p2h} - (n+1)^p 2^{-p}\,, \qquad \text{for even } p, \end{aligned}$$

where

$$J_{ph} = \sum_{l=0}^{p} (-1)^{p-l} \binom{p}{l} (n+1)^{p-l} 2^l \sum_{\nu=0}^{l} \nu!\, G_l^\nu 2^{-\nu} \binom{n+1-h}{n+1-\nu}\,.$$

By using the formula

$$(n+1)n \ldots (n-\nu+2) = \sum_{f=1}^{\nu} S_\nu^f (n+1)^f\,,$$

where S_ν^f is a Stirling number of the first kind [5], we have

$$J_{p0} = \sum_{l=0}^{p} (-1)^{p-l} \binom{p}{l} (n+1)^{p-l} 2^l \sum_{\nu=0}^{l} G_l^\nu 2^{-\nu} \sum_{f=0}^{\nu} S_\nu^f (n+1)^f\,.$$

By noting that $S_\nu^f = 0$ for $f > \nu$ and $G_l^\nu = 0$ for $\nu > l$,

(A6) $$\begin{aligned} J_{p0} &= \sum_{q=0}^{p} (n+1)^{p-q} \sum_{l=0}^{p} (-1)^l \binom{p}{l} 2^l \sum_{\nu=0} 2^{-\nu} G_l^\nu S_\nu^{l-q} \\ &= \sum_{q=0}^{p} (n+1)^{p-q} \sum_{i=0}^{p} 2^i \sum_{l=0}^{q} (-1)^l \binom{p}{l} G_l^{l-i} S_{l-i}^{l-q}\,. \end{aligned}$$

Since $G_l^l = 1$, we have

(A7) $$J_{pp} = p!\,.$$

By (A5), (A6), (A7) and a straightforward but tedious calculation, equations (9), (10) and (11) can be derived. Code C has been assumed to have no code vector with odd weight. However, it follows from identity (1) that the form of I_p depends on only n and k. Therefore, (9), (10) and (11) hold for the general case.

APPENDIX 2

The notations in section 2 will be used. Let $p = 3$, $j_1 = 2^m - 1 - 3$, $j_2 = 2^m - 1 - 5$ and $j_3 = 2^m - 1 - 1$. It is easy to check that (1) E_{iq} ($i \neq q$, $i = 1, 2$) is empty, (2) $E_{ii} = \{1\}$ ($1 \leq i \leq 3$) and (3) $E_{31} = \{0, m-1\}$ and $E_{32} = \{0, m-2\}$. From (8) it follows that

$$\beta_1' = a^{-3}\beta_1$$

$$\beta_2' = a^{-5}\beta_2$$

$$\beta_3' = a^{-3}\beta_1 b^2 + a^{-3\cdot 2^{m-1}}\beta_1^{2^{m-1}} b^{2^{m-1}} + a^{-5}\beta_2 b^4 + a^{-5\cdot 2^{m-2}}\beta_2^{2^{m-2}} b^{2^{m-2}} + a^{-1}\beta_3\,.$$

Let $Z = b^{2^{m-2}}$ and

$$f(Z) = a^{-5}\beta_2 Z^{2^4} + a^{-3}\beta_1 Z^{2^2} + a^{-3\cdot 2^{m-1}}\beta_1^{2^{m-1}} Z^2 + a^{-5\cdot 2^{m-2}}\beta_2^{2^{m-2}} Z\,.$$

If β_1 or β_2 is not equal to zero, the zeros of $f(Z)$ form a σ-dimensional subspace of $GF(2^m)$, where $1 \leq \sigma \leq 4$. Thus, for fixed a, β_1, β_2, and β_3, the number of elements of $\{\beta_3' | \beta_3' = f(Z) + a^{-1}\beta_3,\ Z = b^{2^{m-2}},\ b \in GF(2^m)\}$ is divisible by 2^{m-4}. Now, suppose that $v(0, \beta_1, \beta_2, \beta_3: X)$ has weight j with $2^{m/2+2} \leq |j - 2^{m-1}| < 2^{m-1}$. If β_1 or β_2 is equal to zero, $v(0, \beta_1, \beta_2, \beta_3: X)$ is a code vector of a code considered in case (**a**) and the weight of the nonzero vector $v(0, \beta_1, \beta_2, \beta_3: X)$ is greater than $2^{m-1} - 2^{m/2+1} - 1$ and smaller than $2^{m-1} + 2^{m/2+1} + 1$. Hence, β_1 and β_2 can not be zero.

Let l be a positive integer such that

$$3l \equiv 5l \equiv 0 \pmod{2^m - 1}.$$

Then,

$$2l \equiv 0 \pmod{2^m - 1}.$$

Hence, l must be a multiple of $2^m - 1$. Thus the number of pairs $\{(\beta_1', \beta_2') \mid \beta_1' = a^{-3}\beta_1, \beta_2' = a^{-5}\beta_2, a \neq 0, a \in GF(2^m)\}$ is equal to $2^m - 1$. Consequently, it follows from Lemma 2 that $a_j + a_{2^m-j}$ is divisible by $2^{m-4}(2^m - 1)$.

Acknowledgement

The author is grateful to Professor W. W. Peterson for many valuable suggestions and to Professors R. T. Chien and J. S. Lin for their helpful discussions.

References

1. Assmus, E. F., Mattson, H. F. and Turyn, R. "Cyclic Codes," Scientific Report, AFCRL-65-322, 1965, Air Force Cambridge Research Labs., Bedford, Mass.
2. Bose, R. C. and Ray-Chaudhuri, D. K. "On a Class of Error-Correcting Binary Group Codes," *Inf. and Control*, **3** (1960), 68-79.
3. Forney, G. D., Jr. *Concatenated Codes*, MIT Press, Cambridge, Mass., 1966.
4. Gold, R. and Kopitzka, E. "Study of Correlation Properties of Binary Sequences," Report of Magnavox Research Lab., Vol. 1-15, 1965.
5. Jordan, C. *Calculus of Finite Differences*, 2nd ed., Chelsea, New York, 1950.
6. Kasami, T. "Weight Distribution Formula for some Classes of Cyclic Codes," Report of Coordinated Science Lab., University of Illinois, 1966.
7. Kasami, T. and Lin, S. "Some Codes which Are Invariant under a Doubly-Transitive Permutation Group and their Connection with Balanced Incomplete Block Designs," Scientific Report AFCRL-66-142, Air Force Cambridge Research Labs., Bedford, Mass., 1966.
8. Kasami, T., Lin, S. and Peterson, W. W. "Some Results on Cyclic Codes which Are Invariant under the Affine Group," Scientific Report, AFCRL-66-622, Air Force Cambridge Research Labs., Bedford, Mass., 1966. Partly published in *Inf. and Con-*

trol, **11** (1967), 475–496, in *J. Inst. Elec. Commun. Engrs., Japan*, **50** (1967), 1617–1672, and in *IEEE Trans.*, **IT-14** (1968), 189–199.

9. MacWilliams, F. J. "A Theorem on the Distribution of Weights in a Systematic Code," *Bell System Tech. J.*, **42** (1962), 79–94.
10. Mattson, H. F. and Solomon, G. "A New Treatment of Bose-Chaudhuri Codes," *J. Soc. Indust. Appl. Math.*, **9** (1961), 654–669.
11. Mattson, H. F. and Assmus, E. F. "Research Program to Extend the Theory of Weight Distribution and Related Problems for Cyclic Error-Correcting Codes," Applied Research Lab., Sylvania Electronic Lab., Waltham, Mass., 1964.
12. Muller, D. E. "Application of Boolean Algebra to Switching Circuit Design and to Error Detection," *IRE Trans.*, **EC-3** (1954), 6–12.
13. Peterson, W. W. *Error Correcting Codes*, John Wiley and Sons, New York, 1961.
14. Peterson, W. W. "On the Weight Structure and Symmetry of BCH Codes," *J. Inst. Elec. Commun. Engrs., Japan*, **50** (1967), 1183–1190.
15. Pless, V. "Power Moment Identities on Weight Distributions in Error Correcting Codes," *Inf. and Control*, **6** (1963), 147–152.
16. Prange, E. Unpublished paper.

CHAPTER 21

Irreducible Polynomials, Synchronization Codes, Primitive Necklaces, and the Cyclotomic Algebra

SOLOMON W. GOLOMB, *University of Southern California*

1. INTRODUCTION

The number $T_q(n)$ of irreducible polynomials of degree n over $GF(q)$ is well-known [1] to be

$$T_q(n) = \frac{1}{n} \sum_{d|n} \mu(d) q^{n/d} \tag{1}$$

where μ is the Möbius function of elementary number theory.

The number of inequivalent necklaces (under cyclic permutation) consisting of n beads, where q colors of beads are available (q is now any positive integer) is also well-known [9] to be given by

$$N_q(n) = \frac{1}{n} \sum_{d|n} \phi(d) q^{n/d} \tag{2}$$

where ϕ is Euler's " totient function ". The number of these necklaces which have no periodic substructure is known [4] to be given by the expression (1) for $T_q(n)$. Conversely, the total number of irreducible polynomials in the factorization of $x^{q^n} - x$ over $GF(q)$, which is to say the total number of minimal polynomials for all the elements of $GF(q^n)$, is given by the expression (2) for $N_q(n)$.

If codewords of length n are formed from a q-symbol alphabet, the number of distinct classes of codewords under cyclic permutation is $N_q(n)$, by obvious isomorphism with the necklace problem. It has been shown [3] that $T_q(n)$ is an upper bound to the size of any *comma-free dictionary* consisting of words of length n over the q-symbol alphabet; and [5] that $T_q(n)$ is an *attainable* upper bound under the weaker conditions of a *bounded synchronization delay dictionary*; and [2] is also attainable for comma-free dictionaries if n is odd, but not [3] if n is even and q is large.

In this paper, it will be shown that there is a *natural* correspondence between irreducible polynomials and necklaces, and this correspondence will be exploited in connection with certain problems of coding theory.

2. FROM POLYNOMIALS TO NECKLACES

Let α be a primitive element of $GF(q^n)$; i.e., the powers $\alpha, \alpha^2, \alpha^3, \ldots, \alpha^{q^n-1}$ are all the distinct non-zero elements of $GF(q^n)$, with $\alpha^{q^n-1} = 1$. If $f(x)$ is any irreducible polynomial of degree n over $GF(q)$, then $f(x)$ factors into distinct linear factors over $GF(q^n)$, and the roots of $f(x)$ are $\alpha^m, \alpha^{qm}, \alpha^{q^2m}, \ldots, \alpha^{q^{n-1}m}$ for some integer m, corresponding to the fact that the automorphisms of $GF(q^n)$ over $GF(q)$ are the mappings $\sigma_i: \eta \to \eta^{q^i}$, $i = 0, 1, \ldots, n-1$. These mappings σ_i form the Galois Group for $f(x)$ over $GF(q)$, and there are $n = \deg f$ such mappings by the Fundamental Theorem of Galois Theory. (For all these facts, cf. [10].)

Thus, the roots of $f(x)$ correspond to a set of exponents $(m, qm, q^2m, \ldots, q^{n-1}m)$, where these exponents are considered modulo $q^n - 1$. Let us write m as an n-digit number in base q arithmetic, viz. $m = m_1m_2 \ldots m_n$, $0 \leq m_i \leq q-1$. Then, working modulo $q^n - 1$, we have

$$\begin{aligned} m &= m_1m_2 \ldots m_{n-1}m_n \\ qm &= m_2m_3 \ldots m_nm_1 \\ q^2m &= m_3m_4 \ldots m_1m_2 \\ &\vdots \\ q^{n-1}m &= m_nm_1 \ldots m_{n-2}m_{n-1} \,. \end{aligned}$$

That is, as n-digit numbers in base q notation, the n exponents associated with the roots of an irreducible polynomial of degree n over $GF(q)$ are simply cyclic shifts of one another, and correspond to a single necklace composed of n beads (digits) in q colors (symbols).

Conversely, let ω be any element of $GF(q^n)$. We form the *field polynomial* [11] for ω as follows:

$$(3) \qquad F_\omega(x) = \prod_{i=1}^{n} (x - \sigma_i \omega)$$

where the product is extended over the Galois group. If and only if all automorphisms of ω are distinct, we get an irreducible field polynomial, so that the number of irreducible polynomials of degree n over $GF(q)$ equals the number of equivalence classes of n-digit integers in base q notation for which all cyclic shifts are distinct. That is, the number of irreducible (n, q) polynomials equals the number of primitive (n, q) necklaces.

Every element of $GF(q^n)$ satisfies the equation $x^{q^n} - x = 0$, and the polynomial $x^{q^n} - x$ is the product of the distinct *minimal* polynomials for the elements of $GF(q^n)$. The minimal polynomials $f(x)$ for the elements of $GF(q^n)$ are in one-to-one correspondence with the field polynomials $F(x)$ for the elements of $GF(q^n)$ relative to $GF(q)$, with $F(x) = (f(x))^d$, where $d = n/(\deg f)$. Thus:

(total number of irreducible factors of $x^{q^n} - x$)
= (# of distinct field polynomials for elements of $GF(q^n)$ over $GF(q)$)
= (# of irreducible polynomials over $GF(q)$ of degrees which divide n)
= (total number of necklaces of n beads in q colors)
= (# of primitive necklaces in q colors with a length dividing n).

Actually, with $b = q - 1$, the exponents $00 \ldots 0$ and $bb \ldots b$ are equivalent modulo $q^n - 1$, and doubly represent the class of $\alpha^0 = 1 \in GF(q^n)$; but this is precisely offset by the fact that the element $0 \in GF(q^n)$ corresponds to no exponent on α.

3. AN ILLUSTRATIVE EXAMPLE

We will consider $GF(2^4)$ over $GF(2)$. The factorization over $GF(2)$ of $x^{2^4} - x$ is:

$$x^{16} - x = x(x+1)(x^2 + x + 1)(x^4 + x + 1) \\ \times (x^4 + x^3 + 1)(x^4 + x^3 + x^2 + x + 1) .$$

Let α be a root of $x^4 + x + 1 = 0$. Then the correspondence between factors and roots, exponents and their binary expressions, is:

Factors	Roots	Exponents	Binary
x^4+x+1	α, α^2, α^4, α^8	1, 2, 4, 8	0001, 0010, 0100, 1000
x^4+x^3+1	α^{14}, α^{13}, α^{11}, α^7	14, 13, 11, 7	1110, 1101, 1011, 0111
$x^4+x^3+x^2+x+1$	α^3, α^6, α^{12}, α^9	3, 6, 12, 9	0011, 0110, 1100, 1001
x^2+x+1	α^5, α^{10}	5, 10	0101, 1010
$x+1$	α^0, or α^{15}	0 or 15	0000 or 1111
x	0	none	none

The binary expressions clearly correspond to the necklaces. One could artificially assign the " necklace " 11 . . . 1 to the polynomial x, leaving the " necklace " 00 . . . 0 for the polynomial $x + 1$, to complete the one-to-one nature of the correspondence.

The distinction between the minimal polynomials and the field polynomials is mirrored in the distinction between primitive necklaces of lengths dividing n and all necklaces of length n. Specifically, we can set up the chart:

Minimal Polynomial	Primitive Necklace	Field Polynomial	General Necklace
x^4+x+1	(0001)	x^4+x+1	(0001)
x^4+x^3+1	(1110)	x^4+x^3+1	(1110)
$x^4+x^3+x^2+x+1$	(0011)	$x^4+x^3+x^2+x+1$	(0011)
x^2+x+1	(01)	x^4+x^2+1	(0101)
$x+1$	(0)	x^4+1	(0000)
x	(1)	x^4	(1111)

The two-bead necklace (01) and the four-bead necklace (0101) are certainly different, but they both generate the same periodic sequence . . . 0101010101010101 . . . .

4. FROM NECKLACES TO POLYNOMIALS

Let α be a primitive $(q^n - 1)$st root of unity over $GF(q)$, and let $f(x)$ be the field polynomial of degree n for α^m. If we write $m = \sum_{i=0}^{n-1} c_i q^i$, then the coefficients $\{c_i\}$ in base q arithmetic are the " beads of the necklace " corresponding to the polynomial $f(x)$. The polynomial $g(x) = x^n f(1/x)$ has as its roots the reciprocals of the roots of $f(x)$, and thus corresponds to an exponent $m' = (q^n - 1) - m = \sum_{i=0}^{n-1} (q - 1 - c_i)q^i$. Thus there is a " reciprocal necklace " whose ith bead has the complementary color, $q - 1 - c_i$, to the original ith bead c_i.

It is natural to consider also the " transposed necklace " corresponding to the exponent $m'' = \sum_{i=0}^{n-1} c_{n-i-1} q^i$, i.e., with the beads in reverse order. The question of which polynomials are " paired up " under this transposing operation is quite different from the pairing $f(x) \leftrightarrow g(x) = x^n f(1/x)$ for " reciprocal necklaces ". Specifically, the pairing of polynomials corresponding to transpose pair necklaces is not an intrinsic property of the polynomials, but depends on the specific designation of a primitive root α. Thus over $GF(2)$, whichever primitive polynomial has α as a root is mapped into itself by the transpose operation; while α^{11} and α^{13} correspond to transpose-pair polynomials (1011 $\leftrightarrow$ 1101). It is, of course, possible to define transpose classes of polynomials if two polynomials are in the same class whenever there is *some* choice of the root α for which they are transpose pairs. (It is not difficult to define these " classes " so that the reflexive, symmetric, and transitive laws hold.)

5. CODING APPLICATIONS

In the study of the structure and correlation properties of linear shift register sequences a key role is played by the " cyclotomic cosets " (cf. [6]). These are closely related to Gauss's " periods of the cyclotomic equation " [10], and correspond precisely to the necklaces of this paper.[1]

As a new approach to the study of synchronization codes, we may seek a bounded synchronization delay code (or even a comma-free code) as illustrated herewith for wordlength 5 over the binary alphabet:

(**a**) We list all the powers of a primitive root mod $2^5 - 1 = 31$, where the rows correspond to cyclotomic cosets:

[1] The author has just become aware of a recent paper by H. B. Mann [7] in which the necklace isomorphism is used in the decoding of BCH codes.

1	16	8	4	2
3	17	24	12	6
9	20	10	5	18
27	29	30	15	23
19	25	28	14	7
26	13	22	11	21

(**b**) We rewrite these in base-2 notation, so that each row exhibits the five views of a necklace:

00001	10000	01000	00100	00010
00011	10001	11000	01100	**00110**
01001	10100	01010	**00101**	10010
11011	11101	**11110**	01111	10111
10011	**11001**	11100	01110	00111
11010	01101	10110	01011	10101

(**c**) We select a uniformly spaced subset of these objects as codewords. The words in boldtype are the binary representations of 3^{5i} (mod 31), $i = 0, 1, 2, 3, 4, 5$. In this case, the resulting dictionary:

00001
11010
11001
11110
00101
00110

is maximum-sized and comma-free. (This can be quickly checked by noting that no digram or tetragram ever occurs in both initial and final position.)

It must be mentioned that this example is purely heuristic, and has not been properly generalized as yet.

6. A RATIONAL ALGORITHM FOR MARSH'S CUBIC TRANSFORMATION

If $f(x)$ is the irreducible polynomial over $GF(2)$ for α, we can obtain the field polynomial $f^*(x)$ of the same degree for α^3

by Marsh's "cubic transformation" [8]. That is, Marsh's procedure obtains the polynomial corresponding to "tripling" the value of the necklace. (Doubling the value does not change the polynomial, over $GF(2)$.) In this section we present a "rational algorithm" for Marsh's cubic transformation, as follows:

Marsh introduced the "cubic transformation"

$$(4) \qquad M:\ f(x) \to f(x^{1/3})f(\omega x^{1/3})f(\omega^2 x^{1/3}) = f^*(x)$$

where $\omega^3 = 1$. It is easily seen that the roots of $f^*(x)$ are the cubes of the roots of $f(x)$. In particular, for odd degree n, $2^n - 1$ is not divisible by 3, and the transformation M preserves not only irreducibility, but also the degree of primitivity of the roots. In a variety of cases, iteration of M enables one to generate *all* irreducible polynomials of degree n from a given one. (These degrees include $n = 3, 5, 7, 13, 17$, and 19.)

The procedure given here is "rational" in the sense that ω and ω^2, which do not appear in the final result $f^*(x)$, do not occur in the intermediate computations either.

Algorithm. Divide the exponents of the terms in $f(x)$ into three classes, A, B, and C, according to the residue class of the exponent modulo 3. We produce the set of exponents for $f^*(x)$ from those for $f(x)$ by the following 3 steps:

(1) Copy the exponents of $f(x)$.

(2) Adjoin all numbers $(2u_1 + u_2)/3$ where u_1 and u_2 are distinct exponents of $f(x)$ in the same residue class modulo 3.

(3) Adjoin all numbers $(a + b + c)/3$ where $a \in A$, $b \in B$, and $c \in C$.

Any exponent for $f^*(x)$ which is produced an *even* number of times by these operations must be discarded; if an *odd* number of times, it should be retained once. If any of the three categories A, B, C is empty, then Step (3) is vacuous. If a category has fewer than two members, it does not contribute to Step (2).

Example 1. Let $f(x) = x^5 + x^2 + 1$. Then the categories are

A	B	C
0		2, 5

To produce $f^*(x)$, we follow the three steps:

	A	B	C	
Step 1	0		2, 5	copy
Step 2	3	4		$\frac{2\times 2+5}{3}, \frac{2\times 5+2}{3}$
Step 3				vacuous
mod 2 sum	0, 3	4	2, 5	

Thus $f^*(x)$ has the exponents 0, 2, 3, 4, 5, and $f^*(x) = x^5 + x^4 + x^3 + x^2 + 1$.

Example 2. We iterate the transformation, this time starting with $f^*(x) = x^5 + x^4 + x^3 + x^2 + 1$. To form $f^{**}(x)$, we follow the three steps:

	A	B	C	
Step 1	0, 3	4	2, 5	copy
Step 2	3	1, 4	2	$\frac{2\cdot 0+3}{3}, \frac{2\cdot 3+0}{3}, \frac{2\cdot 2+5}{3}, \frac{2\cdot 5+2}{3}$
Step 3	3, 3	4	2	$\frac{0+4+2}{3}, \frac{0+4+5}{3}, \frac{3+4+2}{3}, \frac{3+4+5}{3}$
mod 2 sum	0	1, 4	2, 5	

Thus $f^{**}(x) = x^5 + x^4 + x^2 + x + 1$.

The reader is invited to verify $f^{***}(x) = x^5 + x^3 + 1$.

Proof of Algorithm. We wish to show that $f^*(y^3) = f(y)\cdot f(\omega y)f(\omega^2 y)$ can be obtained in the manner just described, where $y = x^{1/3}$. Write $f(y) = f_0(y) + f_1(y) + f_2(y)$, where $f_i(y)$ contains precisely those terms of $f(y)$ with exponent congruent to i modulo 3. Then

$$f(\omega y) = f_0(y) + \omega f_1(y) + \omega^2 f_2(y) ,$$

and
$$f(\omega^2 y) = f_0(y) + \omega^2 f_1(y) + \omega f_2(y) .$$

Thus

$$\begin{aligned} f^*(y^3) &= (f_0 + f_1 + f_2)(f_0 + \omega f_1 + \omega^2 f_2)(f_0 + \omega^2 f_1 + \omega f_2) \\ &= (f_0^3 + f_1^3 + f_2^3) + (1 + \omega + \omega^2)(f_0 f_1^2 + f_1 f_2^2 + f_2 f_0^2 \\ &\quad + f_0^2 f_1 + f_1^2 f_2 + f_2^2 f_0) + \begin{vmatrix} 1 & 1 & 1 \\ 1 & \omega & \omega^2 \\ 1 & \omega^2 & \omega \end{vmatrix} f_0 f_1 f_2 \\ &= (f_0^3 + f_1^3 + f_2^3) + f_0 f_1 f_2 , \end{aligned}$$

since $1 + \omega + \omega^2 = 0$, while the determinant is 1 (the matrix is non-singular by linear independence of the rows, and the determinant is " rational " by symmetry in ω and ω^2).

The exponents in f_0, f_1, and f_2 are those in the classes A, B, and C, respectively. The exponents of f_i^3 are (1) the triples of the exponents of f_i, and (2) the sums $2u_1 + u_2$ where u_1 and u_2 are distinct exponents of f_i. The exponents of $f_0 f_1 f_2$ are (3) all sums of the form $a + b + c$, with $a \in A$, $b \in B$, and $c \in C$. Allowing for $y^3 = x$, these are the three steps in the algorithm for finding $f^*(x)$ from $f(x)$.

Similar algorithms can be given for quintic and higher order transformations on the roots, although the amount of computation increases rapidly. These are discussed in the next section.

6. GENERALIZATION TO pth POWER TRANSFORMATIONS

We may generalize the cubic transformation of the previous section to the pth power transformation on the roots of $f(x)$ for any prime $p > 2$, as follows:

Let $f(x) = f_0(x) + f_1(x) + \dots + f_{p-1}(x)$, where $f_k(x)$ consists of those terms of $f(x)$ with exponents congruent to k modulo p. Let y be an indeterminate with $y^p = x$, and let $\rho^p = 1$. Then the pth power transformation P_p on $f(x)$, which gives rise to the polynomial $f^*(x)$, whose roots are the pth powers of the roots of $f(x)$, is given by

$$(5) \quad P_p\colon\ f(x) \to f^*(x) = f^*(y^p) = \prod_{i=0}^{p-1} f(\rho^i y) = \prod_{i=0}^{p-1} \sum_{k=0}^{p-1} \rho^{ik} f_k(y) .$$

This expression may be regarded as a product of a complete set of "Fourier coefficients", and is known to equal the determinant of a circulant matrix.

Thus far, the discussion in this section applies to pth power transformations over any field. When we specialize to $GF(2)$, every distinct term in the determinant occurs with coefficient 1 or 0. These terms are precisely those of the form:

$$(6) \qquad \prod_{i=0}^{p-1} f_i^{a_i}, \qquad \sum_{i=0}^{p-1} a_i = p, \qquad \sum_{i=0}^{p-1} i a_i \equiv 0 \pmod{p}.$$

The total number of such terms is readily found to be $1 + (1/p)\left[\binom{2p-1}{p} - 1\right]$. The term $f_0 f_1 \ldots f_{p-1}$ always occurs; and the remaining terms occur in sets of p at a time, under cyclic permutation of the f_i's. (Note that the two side-conditions in (6) are invariant under cyclic permutation of the subscripts.) This furnishes a combinatorial proof of the congruence

$$(7) \qquad \binom{2p-1}{p} \equiv 1 \pmod{p^2}.$$

For $p > 3$, it is actually known by number theory methods that the congruence holds modulo p^3.

Examples. Let P_p be the pth power transformation.

1. $P_3: \quad f(x) \to f_0^3(y) + f_1^3(y) + f_2^3(y) + f_0(y)f_1(y)f_2(y)$,

with $1 + (1/3)\left[\binom{5}{3} - 1\right] = 4$ terms, in agreement with the previous section.

2. $$P_5: \quad f(x) \to f_0^5 + f_1^5 + f_2^5 + f_3^5 + f_4^5$$
$$+ f_0^3(f_1 f_4 + f_2 f_3) + f_1^3(f_0 f_2 + f_3 f_4) + f_2^3(f_0 f_4 + f_1 f_3)$$
$$+ f_3^3(f_0 f_1 + f_2 f_4) + f_4^3(f_0 f_3 + f_1 f_2) + f_0(f_1^2 f_4^2 + f_2^2 f_3^2)$$
$$+ f_1(f_0^2 f_2^2 + f_3^2 f_4^2) + f_2(f_0^2 f_4^2 + f_1^2 f_3^2) + f_3(f_0^2 f_1^2 + f_2^2 f_4^2)$$
$$+ f_4(f_0^2 f_3^2 + f_1^2 f_2^2) + f_0 f_1 f_2 f_3 f_4 .$$

The number of terms is indeed $1 + (1/5)\left[\binom{9}{5} - 1\right] = 26$. All 26 of these terms do occur, i.e. with coefficient 1. Moreover, as a practical matter, this transformation is not so difficult to apply as one might suppose. Thus, with $f(x) = x^6 + x + 1$, so that $f_0 = 1$ and $f_1 = y^6 + y$, while $f_2 = f_3 = f_4 = 0$, we find

$$f^*(x) = f_0^5 + f_1^5 = 1 + [y(x+1)]^5 = x^6 + x^5 + x^2 + x + 1\,,$$

since only two of the 26 terms actually appeared. Iterating, we find

$$f^*(x) = (x^5 + 1) + (x^6 + x) + (x^2)\,,$$

so that

$$f_0^* = y^5 + 1\,, \quad f_1^* = y^6 + y\,, \quad f_2^* = y^2\,, \quad f_3^* = f_4^* = 0\,.$$

Thus

$$\begin{aligned} f^{**}(x) &= f_0^{*5} + f_1^{*5} + f_2^{*5} + f_1^{*3} f_0^* f_2^* + f_1^* f_0^{*2} f_2^{*2} \\ &= (x+1)^5 + x(x+1)^5 + x^2 + y^3(x+1)^3 (x+1) y^2 \\ &\qquad + y(x+1)(x+1)^2 y^4 \\ &= x^6 + x^5 + x^3 + x^2 + 1\,. \end{aligned}$$

In this case, only 5 of the 26 terms actually appeared.

The terms for P_5 come from the following simple table:

subscripts:	0	1	2	3	4
cyclic sets of exponents	1	1	1	1	1
	1	2	0	0	2
	1	0	2	2	0
	3	1	0	0	1
	3	0	1	1	0
	5	0	0	0	0

Each row is a set of exponents which can be used with the subscripts appearing as column headings. Moreover, these rows can be permuted cyclically (necklace fashion) to generate *all* valid sets of exponents. Note further that rows 2 and 3 are " complementary ", and that rows 4 and 5 arise from rows 2 and 3, respectively, upon multiplication by 3, modulo 5. Rows 1 and 6 are of course " trivial ".

3. P_7 involves $1 + (1/7)\left[\binom{13}{7} - 1\right] = 246$ terms. However, in the case $p = 7$, some of the terms have coefficient 0 rather than 1. Also, in any *specific* case, only those terms satisfying the conditions (6) need be computed. Thus, to apply P_7 to

$f(x) = x^{20} + x^3 + 1$, we have $f_0 = 1$, $f_3 = y^3$, $f_6 = y^{20}$, with $f_1 = f_2 = f_4 = f_5 = 0$. Then,

$$\begin{aligned} f^*(x) &= f_0^7 + f_3^7 + f_6^7 + c_1 f_0 f_3^5 f_6 + c_2 f_0^3 f_3 f_6^3 + c_3 f_0^2 f_3^3 f_6^2 \\ &= 1 + x^3 + x^{20} + 1 \cdot x^5 + 1 \cdot x^9 + 0 \cdot x^7 \\ &= x^{20} + x^9 + x^5 + x^3 + 1 , \end{aligned}$$

where we have used $c_1 = c_2 = 1$, $c_3 = 0$. It would not be difficult to make a table indicating which types of terms have coefficient 1, and which types have coefficient 0. The exponent patterns corresponding to coefficient 1 could then be listed in a table similar to the one given for P_5. (Which patterns have coefficient 1 and which have coefficient 0 depends of course on the choice of ρ as either a root of $x^3 + x + 1$ or of $x^3 + x^2 + 1$.) The set of all 36 exponent patterns which satisfy (6) for $p = 7$ appears in the following table, where $36 = 1 + (1/7^2)\left[\binom{13}{7} - 1\right]$.

subscripts:	0	1	2	3	4	5	6
cyclic sets of exponents	1	1	1	1	1	1	1
	2	2	1	0	1	0	1
	2	1	2	0	1	1	0
	2	1	1	2	0	1	0
	2	0	0	2	1	1	1
	2	2	0	1	1	1	0
	2	1	2	1	0	0	1
	1	2	2	0	2	0	0
	1	0	0	2	0	2	2
	4	1	1	0	1	0	0
	4	0	0	1	0	1	1
	1	3	0	0	0	0	3
	1	0	3	0	0	3	0
	1	0	0	3	3	0	0
	3	2	0	0	0	0	2
	3	0	2	0	0	2	0
	3	0	0	2	2	0	0
	3	0	1	1	1	1	0
	3	1	0	1	1	0	1
	3	1	1	0	0	1	1

subscripts:	0	1	2	3	4	5	6
cyclic sets of exponents	5	1	0	0	0	0	1
	5	0	1	0	0	1	0
	5	0	0	1	1	0	0
	3	2	1	1	0	0	0
	3	0	0	0	1	1	2
	3	0	2	0	1	0	1
	3	1	0	1	0	2	0
	3	0	1	2	0	0	1
	3	1	0	0	2	1	0
	4	2	0	0	0	1	0
	4	0	1	0	0	0	2
	4	0	2	1	0	0	0
	4	0	0	0	1	2	0
	4	1	0	2	0	0	0
	4	0	0	0	2	0	1
	7	0	0	0	0	0	0

References

1. Albert, A. A. *Fundamental Concepts of Higher Algebra*, University of Chicago Press, 1956.
2. Eastman, W. L. "On the Construction of Comma-Free Codes," *IEEE Trans.*, **IT-11** (1965), 263–267.
3. Golomb, S. W., Gordon, B. and Welch, L. R. "Comma-Free Codes," *Canad. J. Math.*, **10** (1958), 202–209.
4. Golomb, S. W. "A Mathematical Theory of Discrete Classification," in *Proceedings of the Fourth London Symposium on Information Theory*, Butterworths, London, 1961.
5. Golomb, S. W. and Gordon, B. "Codes with Bounded Synchronization Delay," *Info. and Control*, **8** (1965), 355–372.
6. Golomb, S. W. *Shift Register Sequences*, Holden-Day, 1967.
7. Mann, H. B. "On the Number of Information Symbols in Bose-Chaudhuri Codes," *Info. and Control*, **5** (1962), 153–162.
8. Marsh, R. W. *Table of Irreducible Polynomials Over GF(2) Through Degree 19*, U. S. Department of Commerce, Office of Technical Services, Washington, D. C., 1957.
9. Riordan, John. *An Introduction to Combinatorial Analysis*, John Wiley and Sons, New York, 1958.
10. van der Waerden, B. L. *Modern Algebra*, Vol. 1, English translation, Frederick Ungar Co., 1949.
11. Zariski, O. and Samuel, P. *Commutative Algebra*, Van Nostrand, 1958.

CHAPTER 22

A Proof of Some Properties of Reed-Muller Codes by Means of the Normal Basis Theorem

PAUL CAMION, *Toulouse University, France*

1. NOTATION

We shall denote by F_{q^ν} the Galois field with q^ν elements, i.e. the extension of degree ν of the field F_q with q elements. A denotes the F_q-algebra $F_q[X]/(X^m - 1)F_q[X]$ and B the F_{q^ν}-algebra $F_{q^\nu}[Y]/(Y^m - 1)F_{q^\nu}[Y]$, where F_{q^ν} is the splitting field of $X^m - 1$. $\bar{X}$, $\bar{Y}$ denote respectively the residue classes of X and Y in A and B. We shall write $\cdot C$ for the K-algebra C when we shall only have to consider it as a vector space over the field K. The mapping $\sigma: h(\bar{X}) \to h(\bar{Y})$ is the isomorphism of A into B that could be simply defined by $\sigma\bar{X} = \bar{Y}$.

By a we denote a primitive root of $X^m - 1$; $f(X)$ will denote a monic factor of degree r of $X^m - 1$; $X^m - 1/f(X)$ will be denoted by $g(X)$ and $L_h = \{j: h(a^j) \neq 0\}$.

We shall write S_h for the generic element of the algebra (isomorphic to C) of the endomorphisms of $\cdot C$. It is defined by $S_h l = h \cdot l$, for all $l \in C$. We shall write $S_h^* = h(S_{\bar{X}}^*)$ for the matrix of the endomorphism $S_h = h(S_X)$ of $\cdot A$ in the canonical basis $\{\bar{1}, \bar{X}, \ldots, \bar{X}^{m-1}\}$ of $\cdot A$. $\underline{Z}$ will denote the ring of relative integers.

2. THE FUNDAMENTAL ISOMORPHISM

It is known that the ideals of A (i.e., the cyclic codes), considered as vector spaces over F_q, are the invariant subspaces under S_X of $\cdot A$. Each of those ideals is of the type $f(\bar{X})A$, where $f(X)$ is a monic factor of $X^m - 1$.

Let us recall some properties which are derived in [3].

We have $\sigma(f(\bar{X})A) = f(\bar{Y})\sigma A \subset f(\bar{Y})B$, and since $f(\bar{Y})B$ is the direct sum of the $\mathrm{Ker}(S_Y - a^i)$, where i runs over L_f, one can state

Property 1. *The isomorphism σ of A into B sends every invariant subspace $\cdot f(\bar{X})A$ under S_X of A into the direct sum of the $Ker(S_Y - a^i)$, where i runs over L_f.*

We identify $\cdot f(\bar{Y})B$ with the vector subspace $F_{q^\nu}^m$ consisting of the m-tuples whose components of each element are the coefficients of the terms of the polynomial with smallest degree in a class of $\cdot f(\bar{Y})B$, and we write $[t]$ for the smallest integer in the class of $\underline{Z}/(m)$ containing t.

Let $F_{q^\nu}[u]$ be the F_{q^ν}-algebra of the polynomial functions defined on F_{q^ν} and with values in F_{q^ν}. Each element of $F_{q^\nu}[u]$ may be represented as a polynomial in u with degree less than m and $F_{q^\nu}[u]$ is isomorphic with B.

To each primitive root c of $X^m - 1$, with $c^{m'} = a$, corresponds a $1-1$ mapping $\lambda_{m'}$ of B into $F_{q^\nu}[u]$: for all $y = y_k \in B$, $0 \leq k < m$, $\lambda_{m'}y = T_{m'}(u)$, where $T_{m'}(u)$ is the (unique) element of $F_{q^\nu}[u]$ such that $y_k = T_{m'}(c^{m-k-1})$, $0 \leq k < m$. One then deduces from Property 1:

Statement 1. *The restriction of $\lambda_{m'}$ to $\cdot f(\bar{Y})B$ is an isomorphism over F_{q^ν} of $\cdot f(\bar{Y})B$ onto the vector space of polynomials in $F_{q^\nu}[u]$ of which the terms with degree $[m'j]$, $j \in L_g$ are zero. The restriction of $\lambda_{m'}$ to $f(\bar{Y})\sigma A$ is an isomorphism over F_q of $\cdot f(\bar{Y})\sigma A$ onto the set of all polynomials $T_{m'}(u)$ in $\lambda_{m'}f(\bar{Y})B$ such that $T_{m'}(a^j) \in F_q$, for all j. Moreover, for every j in L_f, there exists a polynomial $T_{m'}(u)$ in $\lambda_{m'}f(\bar{Y})\sigma A$ whose term of degree $[m'j]$ is not zero.*

We shall give here some more details about the third assertion by describing explicitly the polynomials in $\lambda_{m'}f(\bar{Y})\sigma A$.

Let $\prod_{i \in I} g_i(X) = g(X) = X^m - 1/f(X)$ be the factorization of

$g(X)$ into polynomials irreducible in $F_q[X]$. Let $f_i(X) = X^m - 1/g_i(X)$, for all $i \in I$.

Since $\cdot f(\bar{X})A$ is the direct sum of the $\cdot f_i(\bar{X})A$, where i runs over I, $\lambda_{m'} f(\bar{Y})\sigma A$ is the direct sum of the $\cdot \lambda_{m'} f_i(Y)\sigma A$, where i runs over I and the support of an element in $\lambda_{m'} f_{i'}(\bar{Y}) \cdot \sigma A$ is disjoint with that of an element in $\lambda_{m'} f_{i''}(\bar{Y})\sigma A$, whenever $i' \neq i''$, since $L_{f_{i'}}$ is disjoint with $L_{f_{i''}}$. $\cdot \lambda_{m'} f(\bar{Y})\sigma A$ will be thus perfectly described when we are able to designate the polynomials in $\lambda_{m'} f(\bar{Y})\sigma A$ in the case where $g(X)$ is irreducible in $F_q[X]$. $g(X)$ has then the form $\prod_{s<\nu/\delta} (X - a^{tq^s})$ and $T_{m'}(u) \in \lambda_{m'} f(\bar{Y})\sigma A$ has the form $\sum_{s<\nu/\delta} x_s u^{tq^s}$. But since $T_{m'}(u)$ has its range in F_q, $(T_{m'}(u))^{q^s} = T_{m'}(u)$, for all s; thus $x_s = x_{s-1}^q$ (s mod ν/δ), and $T_{m'}(u)$ is determined by any one of its coefficients.

When $T_{m'}(u)$ runs over $\lambda_{m'} f(\bar{Y})\sigma A$, its first coefficient runs over a subspace E with dimension ν/δ of the vector space F_{q^ν} over F_q. On the other hand, those elements of E belong to the fixed field in the subgroup $(\alpha^{\nu/\delta})$ of the Galois group (α) over F_q of F_{q^ν}. The dimension over F_q of this fixed field is the index of $(\alpha^{\nu/\delta})$ in (α), i.e., ν/δ; thus E is the subfield of F_{q^ν} isomorphic with $F_{q^{\nu/\delta}}$. $T_{m'}(u)$ is now completely defined and we have

Statement 2. *The image under $\lambda_{m'}\sigma$ of an ideal $f(\bar{X})A$ of A is the direct sum of subspaces over F_q of $\cdot F_{q^\nu}[u]$. Each term of this sum has the form $\{\sum_{s<\nu/\delta} (bu^t)^{q^s}\}_{b \in F_{q^{\nu/\delta}}}$ where t is a fixed element in L_f and where $F_{q^{\nu/\delta}}$ is the subfield of F_{q^ν} with order $q^{\nu/\delta}$.*

Remark. When $\nu = 1$, $\lambda\sigma A = F_q[u]$, and if $\lambda\sigma E$ is the subspace of $F_q[u]$ consisting of all polynomials with degree less than k, E is a *Reed-Solomon code.* In that case, one can affirm that the *Bose-Chaudhuri* estimation of the distance is correct, because at least one of the polynomials in $\lambda\sigma E$ does split in F_q. We shall see in the following that this affirmation may be done for a larger class of BCH codes.

We have more closely investigated in [3] the vector space

$$\mathcal{F} = \{\sum_{s<\nu} a^i u^{q^s}\}_{0 \le i < m}$$

which is the vector space of linear forms of F_{q^ν} over F_q. $\mathcal{F}$ is the image under $\lambda\sigma = \mu^{-1}$ of $g(\bar{X})A$, where $f(X) = X^m - 1/g(X)$

is a primitive irreducible factor ($\in F_q[X]$) of $X^m - 1$, $m = q^\nu - 1$; i.e. μ: $h(u) \to \sum_{j<m} h(\alpha^{m-j-1})X^j$.

We stated that the algebra of polynomial functions generated by $\mathcal{F}$ is $\lambda\sigma A$; now let us recall more precisely the main result of this paper.

Let us consider $\cdot g(\bar{X})A$ as a subspace of F_q^m and let $\mathcal{L}_0$ be a system $\{x_0, \ldots, x_{\nu-1}\}$ of linearly independent m-tuples in $g(\bar{X})A$. Let $\mathcal{L}_1$ be the set

$$\{x_0, x_0 \times x_0 = x_0^2, \ldots, x_0^{q-1}, \ldots, x_{\nu-1}, \ldots, x_{\nu-1}^{q-1}\}$$

where $\times$ is the product in the algebra F_q^m, considered as the sum of algebras F_q. $\mathcal{L}_r$ is the set of products i by i of elements in $\mathcal{L}_1$, for all $i \leq r$. Let $\cdot C_r$ be the vector space over F_q spanned by $\mathcal{L}_r$. We then have the

Theorem. *C_r is an ideal $h_r(\bar{X})A$ in A. If $r < \nu/2$, C_r is contained in its orthogonal vector space, i.e., $l_r(X) = X^m - 1/h_r(X)$ has no factor which is proportional to its reciprocal polynomial. If $r \leq \nu$, the dimension of $\cdot C_r$ is at least $\sum_{1 \leq i \leq r} (q-1)^i \binom{\nu}{i}$. The C_r are the Reed-Muller ideals.*

Main line of the proof. To prove the theorem, we choose $\mathcal{L}_0$ so that $\mu^{-1}\mathcal{L}_0$ is the set $\{\sum_{s<\nu} \alpha^s a_i u^{q^s}\}$, $1 \leq i \leq \nu$, where $\{a_i\}_{0<i\leq\nu}$ is a *normal basis* of F_{q^ν}. Let $\mu^{-1}x_i$ be the polynomial $\sum_{s<\nu} \alpha^s a_i u^{q^s}$.

The essential point of the proof consists in stating that to each set of ν or fewer distinct integers $\{s_1, \ldots, s_r\}$ corresponds a set of integers $1 \leq i_1, \ldots, i_r \leq \nu$ such that for every $k \leq (q-1)r$ and for all $j_1, \ldots, j_r$ with $\sum_l j_l = k$ and $j_l < q$, there exists a product

$$y = \mu^{-1}(x_{i_1}^{j_1'} \ldots x_{i_r}^{j_r'})$$

with $\sum_l j_l' = k$ which has a non zero term with degree $j_1 q^{s_1} + \ldots + j_r q^{s_r}$. Let $p(x)$ denote the Hamming norm.

One also obtains

Corollary 1 (Reed and Muller). *If $q = 2$, the dimension of $\cdot C_r$ equals $\sum_{1 \leq i \leq r} \binom{\nu}{i}$, there does exist an element x in $\cdot C_r$ with $p(x) = 2^{\nu-r}$, and for all $y \in \cdot C_r$, $p(y) \geq 2^{\nu-r}$. If $\cdot C_r' = Ker(S_X - 1)$*

$+ \cdot h_r(\bar{X})A$, *dim* $C'_r = \sum_{i \le r} \binom{\nu}{i}$, *and for all* $y \in C'_r$, $p(y) \ge 2^{\nu-r} - 1$.

Corollary 2. *If* $q = 2$, $\mu^{-1}C_r$ *is the set of all polynomial functions with range in* F_q *whose terms with degrees* $\sum_{l<\nu} j_l 2^l$, $(j_l)_{l<\nu} \in L_{l_r}$ *are zero, where* L_{l_r} *is the set of* ν*-tuples with components in* $\{0, 1\}$ *such that* $\sum_l j_l > r$ *or* $\sum_l j_l = 0$.

C_r being an ideal, the corollary immediately follows from Statement 2, Corollary 1, and from the argument of the proof of the theorem that we have just recalled.

3. SOME CONCLUSIONS ABOUT CYCLIC CODES

Let us consider the Bose-Chaudhuri-Hocquenghem code whose generating polynomial is the least common multiple of the minimal polynomials of the $2t$ first powers of a and 1.

Let $b = a^{-1}$; thus

$$L_g \supset \{m - 2t, m - 2t + 1, \ldots, m - 1, 0\}.$$

If $t = 2^{\nu-r-1} - 1$, $2t + 1 = 2^{\nu-r} - 1$ and

$$\begin{aligned} m - 2t - 1 &= m - 2^{\nu-r} + 1 \\ &= 2^{\nu} - 2^{\nu-r} \\ &= 2^{\nu-r} + \ldots + 2^{\nu-1}. \end{aligned}$$

Let B_r be this cyclic ideal.

Every integer $\sum_{l<\nu} j_l 2^l$, with $\sum_{l<\nu} j_l \le r$ belongs to L_f, thus $L_{l_r} \subset L_f$. B_r thus contains an x with $p(x) = 2^{\nu-r}$, and for these codes we may be assured that the BCH estimation is not pessimistic.

$L_f \cup \{0\}$ is defined by the set of ν-tuples $(j_l)_{l<\nu}$ which have at most $r - 1$ consecutive components (circularly) equal to 1 augmented with the set of the $(j_l)_{l<\nu}$ having r nonzero consecutive components (circularly). $\{m - 2t, m - 2t + 1, \ldots, m - 1, 0\}$ does represent the longest circular sequence of integers $< m$ not belonging to L_f, because it is impossible that every polynomial in $\mu^{-1}B_r$ has less than $2^{\nu} - 2^{\nu-r} - 1$ roots, since $\mu^{-1}x$ belongs to $\mu^{-1}B_r$.

We also observe that if $g(X)$ is the least common multiple

of the minimal polynomials of the a^s, $s \in \{m'(m-2t), m'(m-2t+1), \ldots, m'(m-1)\}$ where $(m', m) = 1$, the longest sequence of integers not belonging to $L_{g'}$ *equals* $2t$.

We may then verify, starting from the given definition of $L_f \cup \{0\}$, by considering the least r such that $t \geq 2^{\nu-r-1} - 1$, that if the ratio t/m is kept fixed, the rate k/m in the class of the cyclic codes considered *actually approaches zero* as ν becomes very large. In conclusion, the problem of finding interesting cyclic codes reduces to the problem of determining spaces of polynomial functions described in Statement 2 whose splitting fields all differ from F_{q^ν}. Such spaces do exist, as is shown by the Golay codes.

References

1. Bose, R. C. and Kuebler, R. R., Jr. "On the Construction of a Class of Error Correcting Binary Signaling Codes," *Ann. Math. Statist.*, **31** (1960), 113–139.
2. Bose, R. C. and Ray-Chaudhuri, D. K. "On a Class of Error Correcting Binary Group Codes," *Info. and Control*, **3** (1960), 68–79.
3. Camion, P. "Codes Correcteurs d'erreurs," Revue du Cethedec, 3eme Annee, 3eme trim., 1966, Numero Special.
4. Mann, H. B. "On the Number of Information Symbols in Bose-Chaudhuri Codes," *Info. and Control*, **5** (1962), 153–162.
5. Mattson, H. F., Jr. and Solomon, G. "A New Treatment of Bose-Chaudhuri Codes'" *J. Soc. Indust. Appl. Math.*, **9** (1961), 654–669.
6. Menon, P. K. "On a Combinatorial Problem of Good," *Arch. Math.*, **16** (1965), 37–46.
7. Neveu, J. "Etude generale du codage," Document du Laboratoire Central de Telecommunications, Paris.
8. Peterson, W. W. *Error-Correcting Codes*, MIT Press, Cambridge, 1961.
9. Reed, I. S. "A Class of Multiple-Error-Correcting Codes and the Decoding Scheme," *IRE Trans.*, **IT-4** (1954), 38–49.

CHAPTER 23

Euclidean Geometry Cyclic Codes

E. J. WELDON, Jr., *University of Hawaii and ADTECH, Inc.*

1. INTRODUCTION

On channels which can be closely approximated by a memoryless channel, the use of random-error-correcting codes to improve the reliability of digital communication appears to hold great promise. For example, it is well-known that classes of codes exist for which the probability of decoding erroneously on a memoryless channel decreases exponentially with increasing code length. Similar, although less dramatic, results hold if the channel is " almost " memoryless.

The improvement in reliability attainable with coding is paid for in two ways. First, the effective rate at which data can be transmitted over the channel is reduced. Second, the equipment necessary to implement the codes tends to be costly if a long, powerful code is employed. These two costs, and especially the latter, have often caused communication system designers to eschew coding and to seek other ways of improving reliability, e.g., repeating signals, improving channels. Thus the promise intimated by coding theory is, in the all-important practical sense, largely unfulfilled.

Of all the classes of random-error-correcting codes proposed to date, perhaps the most widely known are the Bose-Chaudhuri [2]—Hocquenghem [3] (BCH) cyclic codes. For practical values of code length, these codes are highly efficient random-error-correctors. Also a decoding algorithm, which can be implemented with a reasonable amount of equipment, has been devised for these codes. A second, less well-known, class of codes for which a simple decoding algorithm has also been devised is the Reed-Muller (RM) codes [7, 10]. These codes, recently proved to be cyclic [4], are slightly inferior to the BCH codes in terms of error-correcting ability but in some cases seem to be easier to implement.

In this paper, a third class of random-error-correcting cyclic codes is investigated[1]. These codes, first discovered by L. D. Rudolph [11], contain the RM codes as a proper subclass. Since the name " generalized Reed-Muller codes " has been employed elsewhere [5] (with respect to a non-binary generalization of the binary RM codes), the codes discussed herein will be called *Euclidean geometry* (EG) *codes*. In Section 2, the class of codes is defined and it is proved that the null spaces of the codes contain certain Euclidean subspaces. As a consequence of the structure of their null spaces the codes are rather powerful random-error-correctors. In Section 3 it is shown that with a minor modification the simply implemented Reed [7] algorithm can be used to decode these codes. The final section summarizes the paper.

2. THE CODES

The reader will be assumed to be familiar with the material in Peterson [9] on cyclic codes. Where possible the notation and conventions employed therein will be used here.

It is necessary to state a few well-known facts about Euclidean geometries. The points of a Euclidean geometry of dimension D over $GF(2^s)$, i.e., $EG(D, 2^s)$, from a field. In particular, they form $GF(2^m)$ where $m = sD$. Thus every point in the geometry can be associated uniquely with some binary m-tuple and, except for the all-zero m-tuple (geometrically, the point at the origin), with some power of a primitive element, α, of $GF(2^m)$. A line (1-flat) consists of all points linearly dependent over $GF(2^s)$ on a single point. In other words a line through the point α^{e_0} consists of the 2^s points α^j such that

[1] Because of their relative importance, only binary codes are considered here. The generalization to the non-binary case is immediate, however, and should cause the interested reader little difficulty.

(1) $\alpha^j = \alpha^{e_0} + \beta^{i_1}\alpha^{e_1}$, $i_1 = 0, 1, \ldots, 2^s - 2, \infty$,

where β is a primitive element of $GF(2^s)$. The symbol β^∞ represents the element 0 in $GF(2^s)$. More generally, a b-flat consists of the 2^{sb} points

(2) $\alpha^j = \alpha^{e_0} + \beta^{i_1}\alpha^{e_1} + \beta^{i_2}\alpha^{e_2} + \ldots + \beta^{i_b}\alpha^{e_b}$,

where the α^{e_i}, $1 \le i \le b$, are elements of $GF(2^m)$ which are chosen to be linearly independent over $GF(2^s)$.

Before defining the Euclidean geometry codes it is necessary, unfortunately, to introduce more notation. Consider an element of $GF(2^m)$, say α^t. The s-weight of the integer t, $0 \le t \le 2^m - 2$, is defined as follows. Let

(3) $t = c_{m-1}2^{m-1} + c_{m-2}2^{m-2} + \ldots + c_0 2^0$

be the binary expansion of t, where $c_i = 0, 1$. The binary m-tuple associated with t in this manner may contain one or more disjoint binary representations of multiples of $2^s - 1$. The s-*weight* of t, denoted $w_s(t)$, is defined as the maximum number of such disjoint multiples. An example should clarify this simple but complicated-sounding point. For $s = 2$ and $m = 6$, the 2-weights of various values of t are given in the accompanying table:

t	Binary representation	$w_2(t)$
3	000011	1
5	000101	0
7	000111	1
9	001001	1
15	001111	2
21	010101	1
27	011011	2

For $s = 1$, $w_s(t)$ clearly reduces to the usual definition of weight.

The concept of s-weight is basic to the definition of the *EG* codes. The (ν, s)th-*order Euclidean geometry code* is the augmented cyclic code[2] whose parity check polynomial, $h(x)$ contains all roots α^t such that $w_s(t) \le \nu$. The reader familiar with the Reed-Muller codes will note that for $s = 1$ this is just one

[2] A cyclic code with an overall parity check added. In the sequel both the cyclic and augmented codes will be referred to as *EG* codes—context will make clear which is meant in a particular situation.

definition of the latter codes. (See [5] for details.)

The Euclidean geometry codes defined as above can be shown to have several useful properties as a result of the following theorem.

Theorem 1. *The null space of the EG code of order* (ν, s) *and length* 2^m *contains every* $(\nu + 1)$*-flat of* $EG(D, 2^s)$*, where* $D = m/s$.

Proof. Let $f(x)$ denote the polynomial representation of an arbitrary $(\nu + 1)$-flat of $EG(D, 2^s)$. If α^t is a root of $f(x)$ for every t such that $w_s(t) \leq \nu$, then $h(x)$ must divide $f(x)$ and the theorem is proved.

It follows from (2) that

$$f(x) = \sum_i x^{j_i} \tag{4}$$

where the j_i are the $2^{s(\nu+1)}$ values of j such that (2) is satisfied. Substituting the root α^t into this expression gives

$$f(\alpha)^t = \sum_i (\alpha^{e_0} + \beta^{i_1}\alpha^{e_1} + \beta^{i_2}\alpha^{e_2} + \ldots + \beta^{i_{\nu+1}}\alpha^{e_{\nu+1}})^t \tag{5}$$

where again the sum is over the $2^{(\nu+1)s}$ points of the flat. Expanding this result in a multinomial expansion yields

$$f(\alpha^t) = \sum_i \sum_j \frac{t!}{j_0!\, j_1! \ldots j_{\nu+1}!} (\alpha^{e_0})^{j_0} (\beta^{i_1}\alpha^{e_1})^{j_1} \ldots (\beta^{i_{\nu+1}}\alpha^{\nu+1})^{j_{\nu+1}} , \tag{6}$$

where

$$\sum_{v=0}^{v+1} j_v = t . \tag{7}$$

Equation (6) can be simplified considerably with the help of the following result, proved by Kasami, Lin and Peterson [5].

Lemma. *Let* β *be an element of* $GF(2^s)$*. Then*

$$\sum_{i=0}^{2^s-1, \infty} \beta^{ij} = 1 ; \qquad j = k(2^s - 1), \quad j \neq 0 , \tag{8}$$
$$= 0 ; \qquad \textit{otherwise} ,$$

where the field element 0 *is denoted by* β^∞.

Proof. Since $\beta^{2^s-1} = 1$, clearly

$$(9)\qquad \sum_{i=0}^{2^s-1,\infty} \beta^{ij} = 1\,; \qquad j = k(2^s-1),\quad j \neq 0\,,$$
$$= 0\,; \qquad j = 0\,,$$

where $(\beta^\infty)^0 = (0)^0 = 1$. If on the other hand, $GCD(j, 2^s - 1) = a < 2^s - 1$, the order of α^j, say n, divides $2^s - 1$. Then

$$(10)\qquad \sum_{i=0}^{2^s-1,\infty} \beta^{ij} = \left(\frac{2^s-1}{n}\right) \sum_{i=0}^{n-1} \beta^{ij}\,.$$

But this last sum is just the coefficient of x^{n-1} in the product of all the factors $(x + \beta^{ij})$, that is, the coefficient of x^{n-1} in the polynomial $x^n + 1$, or in other words, zero. This proves the lemma.

Upon applying this result to (6) with the order of summation interchanged, we find that $f(\alpha^t) = 0$ unless for every j_v, $1 \leq v \leq \nu + 1$,

$$(11)\qquad j_v = k_v(2^s - 1)\,,$$

where k_v is greater than zero. Inserting (11) into (6) gives

$$(12)\qquad f(\alpha^t) = \sum_k \frac{t!}{j_0!(k_1(2^s-1))!\,(k_2(2^s-1))!\,\ldots\,(k_{\nu+1}(2^s-1))!}$$
$$\times\ \alpha^{e_0 j_0}\alpha^{e_1(k_1(2^s-1))}\ \ldots\ \alpha^{e_{\nu+1}(k_{\nu+1}(2^s-1))}\,.$$

Now it is known [5, 12] that for the multinomial coefficient in (12) to be congruent to 1 modulo 2 it is necessary and sufficient that

$$(13)\qquad w_1(t) = w_1(j_0) + \sum_{v=1}^{\nu+1} w_1(k_v(2^s-1))\,.$$

But by definition $w_s(t) \leq \nu + 1$ so t cannot be of the form required by (13). Consequently every coefficient in (12) is zero and the theorem is proved.

This result provides a means of determining k, the number of information symbols in a Euclidean geometry code. It would be desirable to be able to translate Theorem 1 into some sort of combinatorial expression for k or, equivalently, $n - k$. This has not yet been done and it appears that any such general expression will necessarily be rather involved. However in a few special cases a simple formula for k has been obtained.

For $s = 1$ the Euclidean geometry codes of length $2^m - 1$

reduce to the cyclic Reed-Muller codes. In this case it is known that

$$k = \sum_{i=0}^{\nu} \binom{m}{i} . \tag{14}$$

Another practically important case which can be analyzed without undue difficulty is the case of $D = 2$.

Corollary. *The* (0, *s*)th *order Euclidean geometry code of length* $n = 2^{2s} - 1$ *has* $3^s - 1$ *parity checks, that is* $k = 4^s - 3^s$.

Proof. Since $w_s(t)$ is always less than D by definition the generator polynomial of the (0, *s*)th order code contains only those roots α^t for which $w_s(t) = 1$. Let $t = t_0 + t_1(2^s - 1)$, $t_1 > 0$, and let

$$\begin{aligned} t_1(2^s - 1) &= a_0 2^0 + a_1 2^1 + \dots + a_{2s-1} 2^{2s-1} \\ &= (a_0 2^0 + a_1 2^1 + \dots + a_{s-1} 2^{s-1}) \\ &\quad + (a_s 2^0 + a_{s+1} 2^1 + \dots + a_{2s-1} 2^{s-1}) 2^s \end{aligned} \tag{15}$$

$$\begin{aligned} &= (a_0 + a_s) 2^0 + (a_1 + a_{s+1}) 2^1 \\ &\quad + \dots + (a_{s-1} + a_{2s-1}) 2^{s-1} \\ &\quad + (2^s - 1)(a_s 2^0 + a_{s+1} 2^1 + \dots + a_{2s-1} 2^{s-1}) . \end{aligned} \tag{16}$$

Now each of the s columns of the array of (15) must contain exactly a single 1. Clearly every such array corresponds to some value of t_1. Since there are 2^s such arrays and 2^s non-zero multiples of $2^s - 1$ less than $2^{2s} - 1$, every multiple of $2^s - 1$ corresponds to a different array. Therefore the array of every value of t for which $w_s(t) = 1$ contains at least a single 1 in each column and conversely. Since there are s columns and three ways of choosing each, there are 3^s arrays in which each column contains one or two 1's. But one of these is the all-1's array which does not correspond to a multiple of $2^s - 1$ less than $2^{2s} - 1$. The stated result follows directly.

At this point we turn our attention to the problem of determining the minimum distance of the Euclidean geometry codes. By Theorem 1 the (ν, *s*)th order code contains all ($\nu + 1$)-flats of $EG(D, 2^s)$ in its null space. There are exactly.

$$\frac{2^{sD} - 2^{s\nu}}{2^{s(\nu+1)} - 2^{s\nu}} = 2^{s(D-\nu-1)} + 2^{s(D-\nu-2)} + \dots + 2^s + 1 = J \tag{17}$$

$(\nu + 1)$-flats in $EG(D, 2^s)$ which intersect only in a given ν-flat. That it is always possible to construct J such flats follows from the definition of a $(\nu + 1)$-flat: It consists of the points linearly dependent on the linearly independent points of a ν-flat and another point not in the ν-flat. After constructing a number of $(\nu + 1)$-flats intersecting only on a given ν-flat, either all the points of the geometry are employed, in which case (19) holds, or they are not, in which case another $(\nu + 1)$-flat can be constructed with the ν-flat and one of the remaining points.

The parity check equations are said to be orthogonal in the sense of Massey [6]. Clearly it is possible to construct a set of J orthogonal equations for every ν-flat in $EG(D, 2^s)$. As was originally realized by Reed for the Muller codes ($s = 1$), if $[J/2]$ or fewer errors occur the parity check sum corresponding to any flat is correctly determined by the majority of the parity check sums orthogonal on it. For if the ν-flat sum is 1, then there are at most $[J/2] - 1$ errors in the orthogonal $(\nu + 1)$-flats which are not in the ν-flat. Consequently at least $J + 1 - [J/2]$ (a majority) of the orthogonal sums must be 1. Similarly, if the ν-flat is 0 then at most $[J/2]$ of the orthogonal sums can be 1. But $J - [J/2]$ constitutes a majority, so the assertion that every ν-flat check sum is known when $[J/2]$ or fewer errors occur is proved.

Now given these check sums it is possible to determine the $(\nu - 1)$-flat check sums in the same manner. For the number of ν-flats orthogonal on an $(\nu - 1)$-flat is greater than J; therefore, the argument given above holds here as well. In general the number of orthogonal flats increases as their dimensionality decreases (see (17)) so it is finally possible to determine the 0-flats (the binary symbols of code word) and thus to correct up to and including $[J/2]$ errors.

If, as is usual, one wishes to consider only the cyclic code of length $2^m - 1$ contained in the (ν, s)th order Euclidean geometry code, the number of orthogonal flats is reduced by one. Then since it is always true that $d \leq 2t + 1$, where t is the maximum number of errors which can always be corrected, (17) gives

$$(18) \qquad d \geq 2^{s(D-\nu-1)} + 2^{s(D-\nu-2)} + \ldots + 2^s + 1 .$$

For $s = 1$ this result is known to hold with equality [5], although the general case is an open question at present.

The Euclidean geometry codes are closely related to the

BCH codes. For example, for $n = 15$ and 31 every Euclidean geometry code is a BCH code. The codes of length 63 are compared in Table 1. Although the Euclidean geometry codes deteriorate with increasing code length faster than do the BCH codes, there are nonetheless many interesting and practically attractive long codes in the former class.

Table 1. A Comparison of Some BCH Codes and the Euclidean Geometry Codes of Length 63.

d	BCH Codes (n, k)	*EG* Codes (n, k)	*EG* Codes (ν, s)
3	(63,57)	(63,57)	(4,1)
5	(63,51)	(63,48)	(1,2)
7	(63,45)	(63,42)	(3,1)
9	(63,39)	(63,37)	(0,3)
15	(63,24)	(63,22)	(2,1)
21	(63,18)	(63,13)	(0,2)
31	(63, 7)	(63, 7)	(1,1)
63	(63, 1)	(63, 1)	(0,1)

The attractiveness of these codes for use in error control systems stems from the fact that they can be decoded rather simply. This decoding procedure is explained in the following section.

3. DECODING

The argument employed in lower-bounding the minimum distance of the Euclidean geometry codes suggests how to construct their decoders, at least in theory. The procedure explained below is essentially the Reed [10] algorithm modified in a trivial way to account for the fact that s may be greater than 1.

Given the generator polynomial of a particular code, its encoder and syndrome calculator can be constructed readily (see Peterson [9], for example). The second step in decoding, translating the syndrome into an error pattern, is ordinarily the difficult one. For the Euclidean geometry codes this is accomplished as follows.

Since every $(\nu + 1)$-flat is in the null space of the order-(ν, s) code, the corresponding check-sum can be formed by adding appropriate digits of the syndrome. Taking the majority vote of the $d-1$ check sums orthogonal on a given ν-flat produces the check sum associated with that flat. In like manner the check sum corresponding to a $(\nu-1)$-flat is produced at the output of a majority element whose $d-1$ inputs are the ν-flat check sums orthogonal on the $(\nu-1)$-flat. Repeating this process a total of $\nu+1$ times produces the 0-flats (points, noise bits) and error correction is accomplished by adding these to the appropriate information bits.

The above decoding procedure has been referred to as $(\nu+1)$-step orthogonalization [6]. We employ this terminology in stating the result as

Theorem 2. *The Euclidean geometry codes of order* (ν, s) *can be* $(\nu+1)$*-step orthogonalized.*

It is clear from the above description that the complexity of the decoder decreases with decreasing ν. For example, the codes with $\nu=0$ can be decoded with a single majority gate. The most interesting codes in this class are those for which s is as large as possible. In particular if m is even and $s=m/2$ (see the corollary to Theorem 2) one obtains the affine plane codes considered by Rudolph.

A simplification of this decoding procedure is given in Reference 14.

4. SUMMARY AND CONCLUSIONS

In this paper we have considered a class of majority-logic decodable cyclic codes. This class of codes, which was first discovered by L. D. Rudolph [11], contains the Reed-Muller codes as a proper subclass. These Euclidean geometry codes, as they are called, are shown to be capable of correcting relatively large numbers of random errors. In fact for values of code length in the region of interest, the codes are nearly as powerful as the BCH codes. (See Table 1, for example.) Because of their geometric structure, the codes can be decoded very simply using majority-logic decoding or, more accurately, the Reed decoding algorithm. In conclusion it can be said that because of their relatively large minimum distances and the simplicity of their decoders, these Euclidean geometry codes seem attractive for use in error control systems requiring the correction of multiple random errors.

APPENDIX

In this paper we have defined a class of codes in terms of its generator polynomial and shown, as a direct consequence of the definition, that the null space of the code contained all flats of a given type. One can pose the question in reverse and seek to find the largest cyclic code which contains all flats of a given type in its null space. The minimum distance of a code obtained in this manner will be bounded by (18) and the code will have at least as many information symbols as the corresponding *EG* codes. It seems likely that the two classes of codes are, in fact, identical but the author has been unable to prove this in general. However when $\nu = 0$, it can be proved using the argument of Graham and MacWilliams [3].

If there exists a code whose parity check polynomial contains a root α^t for which $w_s(t) > 0$ and whose null space contains all the 1-flats of $EG(D, 2^s)$, then the two classes of codes are not identical. For $\nu = 0$, (12) reduces to

$$\text{(A.1)} \qquad f(\alpha^t) = \sum_{k=1}^{[t/(2^s-1)]} \binom{t}{k(2^s-1)} \alpha^{e_1 k(2^s-1)} \alpha^{e_0(t-k(2^s-1))}$$

$$= \alpha^{e_0 t} \sum_{k=1}^{[t/(2^s-1)]} \binom{t}{k(2^s-1)} \beta^{k(e_1-e_0)}$$

where $[x]$ denotes the greatest integer contained in x and where $\beta = \alpha^{2^s-1}$. Now if $w_s(t) \geq 1$ then some of the binomial coefficients in this expression are non-zero. In order for this sum to be zero for all lines in $EG(D, 2^s)$ it must hold for all values of the difference $e_1 - e_0$ greater than zero and less than $2^{sD} - 1$. But this says that the polynomial

$$\text{(A.2)} \qquad \sum_{k=1}^{[t/(2^s-1)]} c_k x^k$$

has $(2^{sD} - 1)/(2^s - 1)$ non-zero roots. Since $t < 2^{sD} - 1$ this is impossible. Thus the EG codes of order $(0, s)$ are the largest codes having all 1-flats of $EG(D, 2^s)$ in their null spaces.

References

1. Berlekamp, E. R. *Algebraic Coding Theory*, McGraw-Hill, 1968.

2. Bose, R. C. and Ray-Chaudhuri, D. K. "On a Class of Error Correcting Binary Group Codes," *Inf. and Control*, **3** (1960), 68–79.
3. Graham, R. L. and Jessie MacWilliams. "On the Number of Parity Checks in Difference-Set Cyclic Codes," *BSTJ*, **XLV**, 7, 1966.
4. Hocquenghem, A. "Codes correcteurs d'erreurs," *Chiffres*, **2** (1959), 147–156.
5. Kasami, T. and S. Lin. "Some Codes which Are Invariant under a Doubly-Transitive Permutation Group and their Connection with Balanced Incomplete Block Designs," Scientific Report, AFCRL-66-142, Air Force Cambridge Research Laboratory, Bedford, Mass., 1966.
6. Kasami, T., S. Lin and W. W. Peterson. "Some Results on Cyclic Codes which Are Invariant under the Affine Group," Scientific Report, AFCRL-66-622, Air Force Cambridge Research Laboratory, Bedford, Mass., 1966.
7. Massey, J. L. *Threshold Decoding*, MIT Press, Cambridge, Mass., 1963.
8. Muller, D. E. "Applications of Boolean Algebra to Switching Circuit Design and to Error Detection," *IRE Trans.*, **EC-3** (1954), 6–12.
9. Peterson, W. W. "Encoding and Error-Correction Procedures for the Bose-Chaudhuri Codes," *IRE Trans.*, **IT-6** (1960), 459–470.
10. Peterson, W. W. *Error-Correcting Codes*, John Wiley and Sons, New York, 1961.
11. Reed, I. S. "A Class of Multiple-Error-Correcting Codes and the Decoding Scheme," *IRE Trans.*, **IT-4** (1954), 38–49.
12. Rudolph, L. D. "Geometric Configuration and Majority Logic Decodable Codes," M. E. E thesis, University of Oklahoma, Norman, Oklahoma, 1964.
13. Weldon, E. J., Jr. "New Generalizations of the Reed-Muller Codes—Part II, Nonprimitive Codes," *IEEE Trans.*, **IT-14** (1968), 199–205.
14. Weldon, E. J., Jr. "Some Results on Majority-Logic Decoding," *Proceedings of Symposium on Error Correcting Codes*, Mathematics Research Center, Madison, Wisconsin, May, 1968.

CHAPTER 24

Some Codes Which are Invariant under a Transitive Permutation Group and Their Connection with Balanced Incomplete Block Designs

SHU LIN[1], *University of Hawaii and ADTECH, Inc.*

1. INVARIANCE UNDER A TRANSITIVE PERMUTATION GROUP

The set of all n-tuples of field elements is a vector space. A set of such vectors is called a linear code if it is a subspace of the space of all n-tuples [5]. In this paper, codes are assumed to have symbols from the binary field $GF(2)$.

Let α be a primitive element of the Galois field $GF(2^m)$. A t-error-correcting binary *primitive Bose-Chaudhuri-Hocquenghem* code (PBCH) is obtained by requiring

$$(1) \qquad \alpha, \alpha^2, \ldots, \alpha^{2t}$$

to be roots of every code polynomial $f(X) = a_0 + a_1X + \ldots +$

[1] This work was supported by the Air Force Cambridge Research Labs., Office of Aerospace Research (USAF), Laurence G. Hanscom Field, Bedford, Mass., under contract No. AF19(826)-4379.

$a_{n-1}X^{n-1}$. This polynomial corresponds to the code vector $(a_0, a_1, \ldots, a_{n-1})$, where $a_i \in GF(2)$ and $n = 2^m - 1$ [5]. An *extended* PBCH code is defined as the null space of the matrix

$$(2) \qquad H = \begin{pmatrix} 1 & 1 & 1 & \ldots & 1 \\ 0 & 1 & \alpha & & \alpha^{n-1} \\ 0 & 1 & \alpha^2 & & (\alpha^2)^{n-1} \\ \cdot & \cdot & \cdot & & \cdot \\ \cdot & \cdot & \cdot & & \cdot \\ \cdot & \cdot & \cdot & & \cdot \\ 0 & 1 & \alpha^{2t} & \ldots & (\alpha^{2t})^{n-1} \end{pmatrix}$$

where $n = 2^m - 1$. This is a PBCH code with an overall parity check added as the first digit. Let us number each digit in a code vector with the Galois field element that appears in the corresponding position in the second row of $\boldsymbol{H}$, i.e., the first symbol in each code vector is numbered 0, and for $i > 1$, the ith digit is numbered α^{i-2} [5], [6].

An affine transformation with parameters $a, b \in GF(2^m)$, $a \neq 0$, is a permutation which carries the symbol in position X to the position $aX + b$. Such a transformation can be applied to any extended code associated with a primitive element of $GF(2^m)$. A code will be called invariant under the affine group if every affine permutation carries every code vector into another code vector. It has been shown that the extended PBCH codes are invariant under the affine group of permutations [6].

We now show that the Reed-Muller codes are invariant under a transitive permutation group. In this paper we shall describe this class of codes as follows: Consider a matrix

$$(3) \qquad G = \begin{pmatrix} 1 & 1 & 1 & 1 & \ldots & 1 \\ 0 & 1 & \alpha & \alpha^2 & \ldots & \alpha^{n-1} \end{pmatrix}$$

where α is a primitive element in $GF(2^m)$ and $n = 2^m - 1$. Consider the basis $1, \alpha, \alpha^2, \ldots, \alpha^{m-1}$ of $GF(2^m)$. Any field element α^j in $GF(2^m)$ can be expressed as

$$(4) \qquad \alpha^j = \sum_{i=1}^{m} a_{ij}\alpha^{i-1}, \qquad 0 \leq j \leq 2^m - 2,$$

where $a_{ij} \in GF(2)$. If each α^j in $\boldsymbol{G}$ is represented as a column

$(a_{1j}, a_{2j}, \ldots, a_{mj})^T$ of m binary digits, $\boldsymbol{G}$ is then an $(m+1) \times 2^m$ matrix of zeros and ones:

$$\begin{pmatrix} \boldsymbol{v}_0 \\ \boldsymbol{v}_1 \\ \boldsymbol{v}_2 \\ \cdot \\ \cdot \\ \cdot \\ \boldsymbol{v}_m \end{pmatrix} = \begin{pmatrix} 1 & 1 & 1 & 1 & \ldots & 1 \\ 0 & 1 & a_{11} & a_{12} & \ldots & a_{1,2^m-2} \\ 0 & 0 & a_{21} & a_{22} & \ldots & a_{2,2^m-2} \\ \cdot & \cdot & \cdot & & & \\ \cdot & \cdot & \cdot & & & \\ \cdot & \cdot & \cdot & & & \\ 0 & 0 & a_{m1} & a_{m2} & \ldots & a_{m,2^m-2} \end{pmatrix} . \tag{5}$$

Now, let us define the vector product of two vectors as follows [6]:

$$\begin{aligned} \boldsymbol{u} &= (a_1, a_2, \ldots, a_n) \\ \boldsymbol{v} &= (b_1, b_2, \ldots, b_n)\,, \\ \boldsymbol{uv} &= (a_1b_1, a_2b_2, \ldots, a_nb_n)\,. \end{aligned} \tag{6}$$

Then, the νth order Reed-Muller code is formed by using as a basis the vectors $\boldsymbol{v}_0, \boldsymbol{v}_1, \boldsymbol{v}_2, \ldots, \boldsymbol{v}_m$ of Eq. (5) and all vector products of these elementary vectors ν or fewer at a time, where $\nu < m$. That is the basis

$$B_\nu = \{\boldsymbol{v}_0^{l_0}\boldsymbol{v}_1^{l_1} \ldots \boldsymbol{v}_m^{l_m} : 0 \leqq l_i \leqq 1 \text{ and } \sum_{i=1}^{m} l_i \leqq \nu\} \tag{7}$$

is employed. Therefore, each code vector of the νth order Reed-Muller code is a linear combination of $\boldsymbol{v}_0, \boldsymbol{v}_1, \ldots, \boldsymbol{v}_m$ and their vector products. We can express every code vector as a polynomial of degree ν or less in $\boldsymbol{v}_1, \boldsymbol{v}_2, \ldots, \boldsymbol{v}_m$:

$$\boldsymbol{u} = f(\boldsymbol{v}_1, \boldsymbol{v}_2, \ldots, \boldsymbol{v}_m) \tag{8}$$

with coefficients in the binary field $GF(2)$. Conversely, for any polynomial in $\boldsymbol{v}_1, \boldsymbol{v}_2, \ldots, \boldsymbol{v}_m$ of degree ν or less, there is a code vector in the νth-order Reed-Muller code.

For each code vector,

$$\begin{aligned} \boldsymbol{u} &= (u_1, u_2, \ldots, u_n) \\ &= f(\boldsymbol{v}_1, \boldsymbol{v}_2, \ldots, \boldsymbol{v}_m) \end{aligned} \tag{9}$$

the first component u_1 can be expressed as

$$u_1 = f(0, 0, \ldots, 0) \tag{10}$$

and the jth component, for $j > 1$, can be expressed as

$$u_j = f(a_{1,j-2}, a_{2,j-2}, \ldots, a_{m,j-2}) \tag{11}$$

where $(a_{1,j-2}, a_{2,j-2}, \ldots, a_{m,j-2})$ corresponds to the field element α^{j-2}.

Let $x_1, x_2, \ldots, x_m$ be m binary variables over $GF(2)$, V_m be the m-dimensional vector space of m-tuples over $GF(2)$ and P_ν be the set of all polynomials $f(x_1, x_2, \ldots, x_m)$ of degree ν or less in variables $x_1, x_2, \ldots, x_m$ with coefficients in $GF(2)$. Let $\boldsymbol{X} = (x_1, x_2, \ldots, x_m)$, then $f(\boldsymbol{X}) = f(x_1, x_2, \ldots, x_m)$ is a binary function over V_m. Each $f(\boldsymbol{X})$ uniquely determines a code vector as in Eq. (10) and Eq. (11). Thus, we have the following lemma:

Lemma 1. *There exists a one-to-one correspondence between a code vector in the* νth *order Reed-Muller code and a polynomial* $f(\boldsymbol{X}) \in P_\nu$.

Let $\boldsymbol{A}$ be an $m \times m$ nonsingular matrix over $GF(2)$ and $\boldsymbol{B}$ be a vector in V_m. Then, for any $f(\boldsymbol{X}) \in P_\nu$, there is $f'(\boldsymbol{X})$ in P_ν such that

$$f'(\boldsymbol{X}) = f(\boldsymbol{X}') = f(\boldsymbol{XA} + \boldsymbol{B}) \tag{12}$$

where $\boldsymbol{X}' = \boldsymbol{XA} + \boldsymbol{B}$ and $\boldsymbol{X} \in V_m$. Let $\boldsymbol{v}_f = (v_1, v_2, \ldots, v_{2^m})$ be the vector determined by $f(\boldsymbol{X}) \in P_\nu$, i.e.,

$$\begin{aligned} &v_1 = f(\boldsymbol{X}_1) = f(0, 0, \ldots, 0)\,, \\ &v_j = f(\boldsymbol{X}_j) = f(a_{1,j-2}, a_{2,j-2}, \ldots, a_{m,j-2})\,, \qquad j > 1\,, \end{aligned} \tag{13}$$

where $2 \leq j \leq 2^m$.

Consider the permutation π_{AB} on the components of this vector such that, for any $f(\boldsymbol{X}) \in P_\nu$

$$\pi_{AB}\boldsymbol{v}_f = \boldsymbol{v}_{f'} \tag{14}$$

where $f'(\boldsymbol{X}) = f(\boldsymbol{XA}^{-1} - \boldsymbol{BA}^{-1})$, i.e., *the component which corresponds to* $\boldsymbol{X}_j$ is permuted to a position which corresponds to $\boldsymbol{X}_j\boldsymbol{A} + \boldsymbol{B}$.

Let Π denote the transitive permutation group

$$\{\pi_{AB}: \ \boldsymbol{A} \text{ is a nonsingular } m \times m \text{ matrix over } GF(2) \text{ and } \boldsymbol{B} \in V_m\} .$$

From Eq. (12) and Lemma 1, we obtain the following theorem:

Theorem 2. *The class of Reed-Muller codes are invariant under the transitive permutation group Π.*

Let us again number each component in a code vector with elements of $GF(2^m)$ such that the first component in each code vector is numbered 0, and for $j > 1$, the jth component is numbered α^{j-2}. From Theorem 2, it is obvious that the Reed-Muller codes are invariant under the doubly-transitive affine group of permutations $aX + b$ with $a \neq 0$ and b in $GF(2^m)$.

Let α be a primitive element of $GF(2^m)$. Consider the permutation αX. This permutation will leave the first component of each code vector unpermuted, but shift cyclically the remaining $2^m - 1$ digits by one position. Since the code is invariant under the permutation αX, the shortened code obtained by deleting the first component of each code vector of a Reed-Muller code is cyclic [2]. We shall call this shortened code as modified Reed-Muller code (mRM).

All the previous results can be generalized to the non-binary case [3].

Theorem 3. *Binary Reed-Muller codes are invariant under a triply-transitive permutation group.*

Proof. From Theorem 2, it is sufficient to prove that, for any different vectors $\boldsymbol{X}_i$, $\boldsymbol{X}_j$, $\boldsymbol{X}_k$ in V_m, and any different vectors $\boldsymbol{Y}_i$, $\boldsymbol{Y}_j$, $\boldsymbol{Y}_k$ in V_m, there is a nonsingular $m \times m$ binary matrix $\boldsymbol{A}$ and a vector $\boldsymbol{B}$ in V_m such that

$$(15) \qquad \boldsymbol{Y}_l = \boldsymbol{X}_l\boldsymbol{A} + \boldsymbol{B}$$

for $l = i, j, k$.

From a fundamental theorem of algebra, Eq. (15) is obvious except for the case where $\boldsymbol{X}_i + \boldsymbol{X}_j = \boldsymbol{X}_k$ and $\boldsymbol{Y}_i + \boldsymbol{Y}_j \neq \boldsymbol{Y}_k$ (or $\boldsymbol{X}_i + \boldsymbol{X}_j \neq \boldsymbol{X}_k$ and $\boldsymbol{Y}_i + \boldsymbol{Y}_j = \boldsymbol{Y}_k$). Suppose that $\boldsymbol{X}_i + \boldsymbol{X}_j = \boldsymbol{X}_k$ and $\boldsymbol{Y}_i + \boldsymbol{Y}_j \neq \boldsymbol{Y}_k$. Let

$$(16) \qquad \boldsymbol{Y}_i + \boldsymbol{Y}_j + \boldsymbol{Y}_k = \boldsymbol{B} ,$$

where $\boldsymbol{B} \in V_m$. Since $\boldsymbol{X}_i \neq \boldsymbol{X}_j$ and $\boldsymbol{Y}_i + \boldsymbol{B} \neq \boldsymbol{Y}_j + \boldsymbol{B}$, there is a nonsingular $m \times m$ matrix $\boldsymbol{A}$ such that

$$Y_i + B = X_i A\ ,$$
(17)
$$Y_j + B = X_j A\ .$$

Hence

(18)
$$B + Y_k = Y_i + Y_j = (X_i + X_j)A$$
$$= X_k A\ .$$

By a similar argument, the other case can be proved.

2. ON BALANCED INCOMPLETE BLOCK DESIGNS

A balanced incomplete block design (BIBD) [4], [8] is an arrangement of v objects in b sets satisfying the following conditions:

(1) Each set contains exactly k different objects,
(2) Each object occurs in exactly r different sets,
(3) Any pair of objects occurs in exactly λ different sets.

The sets are called blocks. The parameters v, b, r, k, λ, satisfy

$$bk = vr$$
(19)
$$\lambda(v-1) = r(k-1)\ .$$

With any block design, we may associate a $b \times v$ incidence matrix A of zeros and ones.

Definition. *If D is a block design with blocks $B_1, B_2, \ldots, B_b$ and objects $O_1, O_2, \ldots, O_v$, the incidence matrix $A = [a_{ij}]$, where $i = 1, 2, \ldots, b$ and $j = 1, 2, \ldots\ v$, of D is defined by the rules*

$$a_{ij} = 1 \quad \text{if} \quad O_j \in B_i\ ,$$
(20)
$$a_{ij} = 0 \quad \text{if} \quad O_j \notin B_i\ .$$

Clearly, A is a $b \times v$ matrix of zeros and ones such that each row contains k ones, each column contains r ones, any two columns have ones in corresponding positions exactly λ times. Conversely, the existence of a matrix A with these properties is equivalent to the existence of a balanced incomplete design.

Consider a binary code V of length n which is invariant

under a doubly-transitive permutation group. Let D_k be the set of all code vectors of weight k. Since a permutation does not affect code word weight, D_k is also invariant under the doubly-transitive permutation group. Let N_k be the number of vectors in D_k. If the code vectors in D_k are arranged as a $N_k \times n$ matrix A, then we have the following theorems:

Theorem 4. *The number of ones in each column of A is constant and is equal to*

$$r = kN_k/n \tag{21}$$

where n is the length of the code V.

Proof. Because the permutation is transitive, for every l, there exists a permutation that carries column 1 into column l. This permutation leaves the rows of A unchanged except it rearranges the rows. It follows that column 1 and column l of A have the same number of ones which is equal to

$$r = kN_k/n\ . \tag{22}$$

Theorem 5. *Any two columns of the matrix A have ones in corresponding positions exactly λ times, where λ is given by*

$$\lambda = k(k-1)N_k/n(n-1)\ . \tag{23}$$

Proof. Because the permutation group is doubly transitive, there exists a permutation which will permute the ith column of A to the first column of A, and the jth column to the second column. Since the permutation leaves the rows of A invariant (except it rearranges the rows), we can rearrange the rows to obtain the original matrix A. It follows that the number of ones in corresponding positions between the ith and jth columns is exactly equal to the number of ones in corresponding positions between the first and second columns. This implies the stated theorem.

The matrix A is thus a $N_k \times n$ matrix with zeros and ones, such that each row has exactly k ones, each column has exactly $r = kN_k/n$ ones, and any two columns have ones in corresponding positions for exactly $\lambda = r(k-1)/(n-1)$ positions. Therefore, A is the incidence matrix of a balanced incomplete block design with parameters

$$v = n\,,\quad b = N_k\,,\quad k\,,\quad r = kN_k/n\,,\quad \lambda = r(k-1)/n\ .$$

From Theorem 4 and Theorem 5, we have observed that if a binary code is invariant under a doubly-transitive permutation group Π, then the set of all code vectors of weight k forms a balanced incomplete block design. In this report, we have shown that extended binary BCH codes of length 2^m and Reed-Muller codes are invariant under a doubly-transitive permutation group; thus BIB designs can be derived from them. The extended quadratic residue codes have also been proven to be invariant under a doubly-transitive permutation group [7]. The connection between perfect codes and Steiner systems has been discussed by Assmus, Mattson and Turyn [1].

Since a binary Reed-Muller code has been proved to be invariant under a triply-transitive permutation group, the modified Reed-Muller obtained by deleting the first component of each code vector is then invariant under a doubly-transitive permutation group. Therefore, BIB designs can also be derived from mRM codes. Since mRM codes are proved to be cyclic, the block designs obtained from this class of codes are cyclic.

It has been shown that the minimum weight of a modified νth order Reed-Muller code of length $2^m - 1$ is $2^{m-\nu} - 1$. The weight distributions of this class of codes have been partly solved [3]. We shall summarize the results without proofs.

Let the location numbers of the non-zero components of a minimum weight code vector be $X_1, X_2 \ldots, X_{2^h-1}$ where $h = m - \nu$ and $X_i \in GF(2^m)$. It has been shown that $X_1, X_2, \ldots, X_{2^h-1}$ are the non-zero elements of an $(m-\nu)$-dimensional subspace of $GF(2^m)$ [3].

Theorem 6 [3]. *There exists a one-to-one correspondence between the location numbers of non-zero components of a minimum code vector in a* νth-*order modified Reed-Muller code and a* $(m-\nu)$-*dimensional subspace of the field* $GF(2^m)$.

Since there are

$$\prod_{i=0}^{m-\nu-1} \left(\frac{2^m - 2^i}{2^{m-\nu} - 2^i}\right) \tag{26}$$

distinct $(m-\nu)$-dimensional subspaces, therefore, the number of minimum weight code vectors of a νth order mRM code is

$$N_{2^h-1} = \prod_{i=0}^{m-\nu-1} \left(\frac{2^m - 2^i}{2^{m-\nu} - 2^i}\right). \tag{27}$$

The block design derived from the set of minimum weight code vectors has the following parameters

$$
\begin{aligned}
v &= 2^m - 1\,, \\
b &= \prod_{i=0}^{m-\nu-1} \left(\frac{2^m - 2^i}{2^{m-\nu} - 2^i} \right), \\
r &= \prod_{i=1}^{m-\nu-1} \left(\frac{2^m - 2^i}{2^{m-\nu} - 2^i} \right), \\
k &= 2^{m-\nu} - 1\,, \\
\lambda &= \begin{cases} \displaystyle\prod_{i=2}^{m-\nu-1} \left(\frac{2^m - 2^i}{2^{m-\nu} - 2^i} \right), \\ 1 \quad m - \nu = 2\,. \end{cases}
\end{aligned}
\tag{28}
$$

It is interesting to note that the above designs is exactly the BIB designs obtained from the finite projective geometry $PG(m-1, 2)$ [4].

The weight next to the minimum weight in a νth order modified RM code is $2^{m-\nu}$. The number of code vectors of this weight can be found from the following theorem [3].

Theorem 7. *For a* νth *order modified binary Reed-Muller code, there exists a one-to-one correspondence between the location numbers of the non-zero components of a code vector of weight* $2^{m-\nu}$ *and a coset of a* $(m-\nu)$*-dimensional subspace of* $GF(2^m)$ *(excluding the subspaces).*

Therefore, the number of the code vectors of weight $2^{m-\nu}$ in a νth order RM code is

$$
N_{2^h} = \left\{ \prod_{i=0}^{m-\nu-1} \left(\frac{2^m - 2^i}{2^{m-\nu} - 2^i} \right) \right\} \cdot (2^\nu - 1)\,. \tag{29}
$$

The BIB design derived from the set of code vectors of weight $2^{m-\nu}$ has the following parameters

$$
\begin{aligned}
v &= 2^m - 1\,, \\
b &= \left\{ \prod_{i=0}^{m-\nu-1} \left(\frac{2^m - 2^i}{2^{m-\nu} - 2^i} \right) \right\} \cdot (2^\nu - 1)\,, \\
r &= \left(\frac{2^m - 2^{m-\nu}}{2^{m-\nu} - 1} \right) \cdot \left\{ \prod_{i=1}^{m-\nu-1} \left(\frac{2^m - 2^i}{2^{m-\nu} - 2^i} \right) \right\},
\end{aligned}
\tag{30}
$$

$$k = 2^{m-\nu},$$

$$\lambda = \left(\frac{2^m - 2^{m-\nu}}{2^{m-\nu} - 2}\right) \cdot \left\{ \prod_{i=1}^{m-\nu-1} \left(\frac{2^m - 2^i}{2^{m-\nu} - 2^i}\right) \right\}.$$

For a νth-order normal Reed-Muller code of length 2^m, the minimum weight is $2^{m-\nu}$ and the number of the minimum code vectors is equal to the number of all distinct $(m - \nu)$-dimensional subspaces and their cosets, i.e.,

$$N'_{2^h} = \left\{ \prod_{i=0}^{m-\nu-1} \left(\frac{2^m - 2^i}{2^{m-\nu} - 2^i}\right) \right\} \cdot 2^\nu. \tag{31}$$

Thus, the BIB design derived from the set of minimum weight code vectors of a νth order normal RM code has the following parameters

$$\nu = 2^m,$$

$$b = 2^\nu \left\{ \prod_{i=0}^{m-\nu-1} \left(\frac{2^m - 2^i}{2^{m-\nu} - 2^i}\right) \right\},$$

$$k = 2^{m-\nu}, \tag{32}$$

$$r = \prod_{i=0}^{m-\nu-1} \left(\frac{2^m - 2^i}{2^{m-\nu} - 2^i}\right),$$

$$\lambda = \prod_{i=1}^{m-\nu-1} \left(\frac{2^m - 2^i}{2^{m-\nu} - 2^i}\right).$$

3. EXAMPLES

Example 1. Consider the extended 2-error correcting NBCH code with $m = 4$. The weight distribution is as follows:

Weight	*Number of code vectors*
0	1
6	48
8	30
10	48
16	1

Three BIB designs can be derived from this code with parameters as follows:

(a) $v = 16, \quad b = 48, \quad r = 18, \quad k = 6, \quad \lambda = 6.$

(b) $v = 16, \quad b = 30, \quad r = 15, \quad k = 8, \quad \lambda = 7.$

(c) $v = 16, \quad b = 48, \quad r = 30, \quad k = 10, \quad \lambda = 18.$

Example 2. Consider the extended NBCH 5-error correcting code with $m = 5$. The weight distribution is

Weight	*Number of code vectors*
0	1
12	496
16	1054
20	496
32	1

Three BIB designs can be derived from this code with parameters as follows:

(a) $v = 32, \quad b = 496, \quad r = 186, \quad k = 12, \quad \lambda = 66.$

(b) $v = 32, \quad b = 1054, \quad r = 527, \quad k = 16, \quad \lambda = 255.$

(c) $v = 32, \quad b = 496, \quad r = 310, \quad k = 20, \quad \lambda = 190.$

Example 3. Consider the second order Reed-Muller code with $m = 4$. The weight distribution of this code is as follows:

Weight	*Number of code vectors*
0	1
4	140
6	448
8	870
10	448
12	140
16	1

Five block designs can be derived from this code; they are

(a) $v = 16, \quad b = 140, \quad r = 35, \quad k = 4, \quad \lambda = 7.$

(b) $v = 16, \quad b = 448, \quad r = 168, \quad k = 6, \quad \lambda = 56.$

(c) $v = 16, \quad b = 870, \quad r = 435, \quad k = 8, \quad \lambda = 203.$

(d) $v = 16, \quad b = 448, \quad r = 280, \quad k = 10, \quad \lambda = 168.$

(e) $v = 16, \quad b = 140, \quad r = 105, \quad k = 12, \quad \lambda = 77.$

Example 4. Consider the extended quadratic residue (18, 9) code with weight distribution

Weight	*Number of code vectors*
0	1
6	6×17
8	9×17
10	9×17
12	6×17
18	1

Four block designs can be derived from this code with parameters as follows:

(a) $v = 18,\quad b = 102,\quad r = 34,\quad k = 6,\quad \lambda = 10.$

(b) $v = 18,\quad b = 153,\quad r = 68,\quad k = 8,\quad \lambda = 28.$

(c) $v = 18,\quad b = 153,\quad r = 85,\quad k = 10,\quad \lambda = 45.$

(d) $v = 18,\quad b = 102,\quad r = 68,\quad k = 12,\quad \lambda = 44.$

4. CONCLUSION

We have shown that if a binary code is invariant under a doubly-transitive permutation group, then the set of all code vectors of weight k forms the incidence matrix of a balanced incomplete block design. The extended PBCH codes and the normal Reed-Muller codes have been proved to be invariant under a doubly-transitive permutation group. Thus, BIB designs can be derived from these classes of codes. Normal Reed-Muller codes have also been proved invariant under a triply-transitive permutation group. Therefore, a modified RM code obtained by deleting the first component from each code vector of a normal Reed-Muller code is invariant under a doubly-transitive permutation group. Since the modified RM codes are cyclic, the BIB designs derived from this class of codes are thus cyclic.

Acknowlegment

The author wishes to thank Professor T. Kasami of Osaka University, Japan and Professor W. W. Peterson of the University of Hawaii for their comments and help. They have made substantial contributions to this work.

References

1. Assmus, E. F., Mattson, H. F. and Turyn, R. "Cyclic Codes," Scientific Report AFCRL-65-332, Air Force Cambridge Research Laboratory, Bedford, Mass., 1965.
2. Kasami, T. and Lin, S. "Some Codes which Are Invariant under a Doubly-Transitive Permutation Group and Their Connection with Balanced Incomplete Block Designs," Scientific Report AFCRL-66-142, Air Force Cambridge Research Laboratory, Bedford, Mass., 1966.
3. Kasami, T., Lin, S. and Peterson, W. W. "Some Results on Cyclic Codes which Are Invariant under the Affine Group," Scientific Report AFCRL-66-622, Air Force Cambridge Research Laboratory, Bedford, Mass., 1966.
4. Mann, H. B. *Analysis and Design of Experiments*, Dover, New York, 1949.
5. Peterson, W. W. *Error-Correcting Codes*, John Wiley and Sons, New York, 1961.
6. Peterson, W. E. "On the Weight Structure and Symmetry of BCH Codes," Scientific Report, Air Force Cambridge Research Laboratory, Bedford, Mass., 1965.
7. Prange, E. Private Communication.
8. Ryser, H. J. *Combinatorial Mathematics*, John Wiley and Sons, New York, 1963.

Discussion on Professor Lin's Paper

DR. T. A. DOWLING: Certainly the most significant results contained in Professor Lin's paper are his theorems on the invariance of Reed-Muller codes under a doubly-transitive permutation group in general, and under a triply transitive permutation group in the binary case. The application of these results to derive balanced incomplete block designs is a by-product of the invariance property, and, while it does not appear to yield many new block designs, the designs which do result point up the connection of R-M codes with finite projective and affine geometrys. This connection was mentioned by Professor Peterson in his book and also by Professor Weldon earlier this evening.

As the speaker has noted, the design with $v = 2^m - 1$, $k = 2^{m-\nu} - 1$ based on the minimum weight $(2^{m-\nu} - 1)$ code vectors of a νth order modified R-M code is equivalent to the design obtained by taking as treatments the points of $PG(m-1, 2)$ and as blocks the $(m-\nu-1)$-flats of the geometry.

The design with $v = 2^m$, $k = 2^{m-\nu}$ based on the minimum weight $(2^{m-\nu})$ code words of a normal R-M code is equivalent to the design based on the $(m-\nu)$-flats of $EG(m, 2)$. Actually in this case every triple occurs in the same number of blocks since any three points in $EG(m, 2)$ determine a plane. This fact of course is a consequence of the invariance of binary R-M codes under a triply-transitive group.

The design with $v = 2^{m-\nu} - 1$ and $k = 2^{m-\nu}$ based on the code words of weight $2^{m-\nu}$ in a modified R-M code also has an interesting geometric interpretation. If we delete a point from $EG(m, 2)$ together with the $(m-\nu)$-flats which contain it, then the retained points and $(m-\nu)$-flats give this design. The fact that every pair of retained points (i.e., lines) is contained in the same number of retained flats is a consequence of the original design, based on $EG(m, 2)$, having the property that every *triple* is contained in the same number of blocks.

Part IV

FINITE GEOMETRIES

CHAPTER 25

Classification of Finite Projective Planes

A. BARLOTTI, *University of Palermo, Italy*

1. INTRODUCTION

In recent years some ideas have been developed to classify projective planes. We shall present here a brief exposition of those ideas which also work in the finite case[1].

2. THE CLASSIFICATION BASED ON (C, a)-TRANSITIVITIES

A projective plane π is said to be (C, a)-transitive (see Baer [1]) if it contains a point C and a line a such that for each pair of points P, P' with $P, P' \notin a$, $P, P' \neq C$ and $CP = CP'$ there is a (C, a)-perspectivity (i.e. a perspectivity with center C and axis a) which takes P into P'.

It is well known that the existence of (C, a)-transitivities in π is closely related to certain properties of a ternary ring

[1] As first attempts to classify projective planes we can mention those made by Andre (see [1]) (who gave a classification for translation planes) and by Naumann (see [1]). The reader interested in infinite planes is referred to Salzmann [3].

which is suitably chosen in π (see Gingerich [1], Hall [1], [2], Pickert [1], [6]).

The problem arises of classifying projective planes according to the set of (C, a)-transitivities which they have. The point-line pairs (C, a) for which a plane is (C, a)-transitive form a figure $F = \{(C, a) : C$ a point, a a line, Π is (C, a)-transitive$\}$. Lenz [1] studied the subset $\{(C, a) \in F : C \in a\}$. Later Barlotti gave a refinement of the classification obtained by Lenz by removing the restriction $C \in a$. Both authors obtained upper bounds for the number of various types of possible figures. Results of other authors (Skornyakov [1], San Soucie [1], Ostrom [1], [2], Pickert [3], Spencer [1]) allow us now to rule out some of those types, and we can state the following theorem:

Theorem. *Let F be the figure formed by all the point-line pairs (C, a) for which a projective plane is (C, a)-transitive. Then the following exhaust the possible cases* (see Figure 1)*:*

Class I.

I.1) *F is the empty set.*

I.2) *F contains a single pair (C, a), with $C \notin a$.*

I.3) *F contains two pairs (C_i, a_i), $i = 1, 2$, with $C_i \in a_j$ if and only if $i \neq j$.*

I.4) *F contains three pairs (C_i, a_i), $i = 1, 2, 3$, with $C_i \in a_j$ if and only if $i \neq j$.*

I.5) [formerly I.6] *The plane contains an incident point-line pair (R, r) and a one-to-one mapping φ of the points of $r - \{R\}$ onto the lines passing through R and different from r. F contains all the pairs $(C, \varphi(C))$ with $C \in r - \{R\}$.*

Class II.

II.1) *F contains a single pair (C, a), with $C \in a$.*

II.2) *F contains two pairs (C_i, a_i), $i = 1, 2$, with $C_i \in a_1$, $i = 1, 2$, $C_2 \notin a_2$, $C_1 \neq C_2$, $a_1 \neq a_2$.*

Class III.

III.1) *There exist in the plane a line r and a point $R \notin r$ such that the figure F contains all the pairs (C, CR) for all $C \in r$.*

III.2) *F contains all the pairs of the figure of type* III.1) *together with the single additional pair (R, r).*

Class IV_a.

IV_a.1) *There exists a line r in the plane such that F contains all the pairs (C, r) for all $C \in r$.*

IV_a.2) *There exist in the plane a line r and two points A and B, $A, B \in r$, such that F contains all the following pairs:*
(C, r) for all $C \in r$,
(A, b) for all $b \ni B$,
(B, a) for all $a \ni A$.

IV_a.3) *There exist in the plane a line r and an involutory permutation φ of the points of r, which fixes no point of r, such that F contains all the pairs (C, a) with $C \in r$, $a \ni \varphi(C)$.*

Class IV_b.

Types IV_b1)–3) *are the duals to* IV_a.1)–3).

Class V.

V.1) *There exist in the plane a line r and a point $R \in r$, such that F contains all the following pairs:*
(C, r) with $C \in r$,
(R, a) with $a \ni R$.

Class VI. [formerly Class VII]

VI.1) *F contains all the pairs (C, a) with $C \in a$.*

VI.2) *F contains all the point-line pairs.*

Note. The existence of planes of Type I.5) is still undetermined. Examples of planes of every other type have been found.

3. THE FINITE CASE

Infinite planes provide examples for the existence of different types of planes. Planes of finite order are considered here.

Type I.1) The Hughes planes (Hughes [2]) are of this type as is shown by the results of Zappa [1], Rosati [1] and Jonsson [1]. Professor Ostrom has informed me that the Fryxell planes and some other semi-translation planes (Ostrom [3], [6]) also belong to this class.

Type I.2) No example is known.

Type I.3) No example is known.

Type I.4) No example is known. However, a number of

properties that planes of this class would have, should they exist, are given by Hughes [1].

Type I.5) Finite planes of this type do not exist. The problem was first approached by Pickert [4] and Jónsson [1]. The nonexistence for odd order has been demonstrated by Lüneburg [1]. Afterwards Cofman [2] showed that if there exists a finite plane of this type, it is of order $n \equiv 0 \pmod 8$. The final proof of nonexistence is given by Yaqub [3].

Type II.1) The only known examples are:

a) The Ostrom-Rosati planes derived from the Hughes planes (Ostrom [4], Rosati [2]).
b) The planes derived from the dual Lüneburg planes (Ostrom [7]).

Suitably chosen coordinate systems for these planes will furnish finite cartesian number systems which satisfy neither the left nor the right distributive law. A direct construction of a class of such systems has been given by Panella [1]. Bartolozzi [1] proved that projective planes constructed on the finite systems given by Panella are Ostrom-Rosati planes (see also Fryxell [1] and Pickert [7]).

Type II.2) No example is known.

Type III.1) No example is known. Lüneburg made the conjecture that finite planes of this type do not exist. It has been proved (Lüneburg [2], Hering (unpublished), and Cofman [1]) that if there exists a plane of Type III.1), it is of order $n \equiv 1 \pmod 8$, n not a prime.

Type III.2) Attempts to prove the nonexistence of planes of this type have been made by Zappa [2] and Lüneburg [2], [3]. Nonexistence has been proved very recently by Mrs. Yaqub [4].

Type IV_a.1) Examples are given by the Hall planes of order greater than nine (Jonsson [1]).

Type IV_a.2) Planes of order greater than nine over a near-field belong to this type (André [1]).

Type IV_a.3) There exists only one plane of this type: the translation plane of order nine (see André [1], where this plane is called the "exceptional plane").

Type IV_b.1)-3): are the duals to IV_a.1)-3).

Type V.1) Planes over a distributive quasifield belong to this type. Examples of such structures are given by Dickson [1] and Albert [1] (see also Magari [1]).

Type VI.1) Finite planes of this type do not exist. In fact, in a plane of Type VI.1) the little Desargues theorem holds, and if the plane is a finite one this theorem is equivalent to Desargues' theorem (i.e. every finite alternative division ring is a field: see Zorn [1] and Levi [1]).

Type VI.2) The only planes of this type are the Galois planes (finite Desarguesian planes).

From the previous results we state the following theorem (Pickert [5], [6]):

Theorem. *If a finite plane is (C, a)-transitive for all pairs (C, a) of a set M and if M is not a subset of a figure of one of the following types:*

$$\text{I.4)}; \quad \text{IV}_a.3); \quad \text{IV}_b.3); \quad \text{V.1)}; \quad \text{III.1)};$$

then the plane is Desarguesian.

Many theorems have been proved which have the following structure:

If a finite projective plane is (C, a)-transitive for at least all pairs (C, a) of a figure F of a type W.x), *and if the plane satisfies some additional properties, then the plane is (at least) of type* Z.y).

In many cases Z.y) is VI.2) and we obtain in this way characterizations of Desarguesian planes.

See for example: Cofman [1], [2], Lüneburg [1], [2], [3], Pickert [4]. Examples for the infinite case are given in Salzmann [2].

4. THE CLASSIFICATION OBTAINED BY USING (C, a, μ)-HOMOGENEITIES.

Jónsson [1] extended the concept of (C, a)-homogeneity (see Bear [1]) to what he called (C, a, μ)-homogeneity.

Definition. *A projective plane is (C, a, μ)-homogeneous if:*

(1) *It is (C, a)-transitive;*

(2) *There is a correlation τ whose square is a (C, a)-perspectivity and μ is the mapping of the lines through C onto the points of a induced by τ.*

In a given plane let S be the figure formed by all the point-line pairs (C, a) for each one of which there is a mapping

μ such that the plane is (C, a, μ)-homogeneous. Jónsson [1], [2], obtained an upper bound for all the possible types of the figures S and deduced from his results a refinement of the classification we have seen in Section 2.

Many questions regarding this refinement in the finite case are still unanswered.

5. OTHER POSSIBLE REFINEMENTS OF THE CLASSIFICATION OF SECTION 2

We now look at the problem of obtaining further refinements of the classification in Section 2.

Consider first the following problem: To determine for each type of plane shown in Section 2 all the possible figures formed by all the point-line pairs (C, a) for which in the plane there is at least a (C, a)-perspectivity.

The solution requires theorems having the following structure:

(A) If the plane possesses a certain set of perspectivities, it follows that in the plane there is a larger set of perspectivities.

(B) From the knowledge that the plane possesses a certain set of perspectivities, some properties of the set of these perspectivities can be deduced.

A theorem of Type (B) is the following (Ostrom [3]):

Theorem. *Let π be a projective plane of order n. If for some line l, π admits a group G of elations with axis l, where the order of G is greater than n, then every point P on l is the center of a nontrivial elation with axis l.*

If this refinement is too fine to be obtained, we can try to use one of the modifications of the notion of (C, a)-transitivity introduced with the following definitions:

(i) *A projective plane is called (C, a)-halftransitive* (André [4]) *if there is a line l through C and different from a, such that under the group of the (C, a)-perspectivities the set of points of l, which are different from C and $a \cap l$, splits in exactly two orbits.*

(ii) *Let π_0 be a subplane of the projective plane π. Let a be a fixed line of π_0 and let C be a fixed point of π_0. Let P and Q be any two points of π_0 that are collinear with C,*

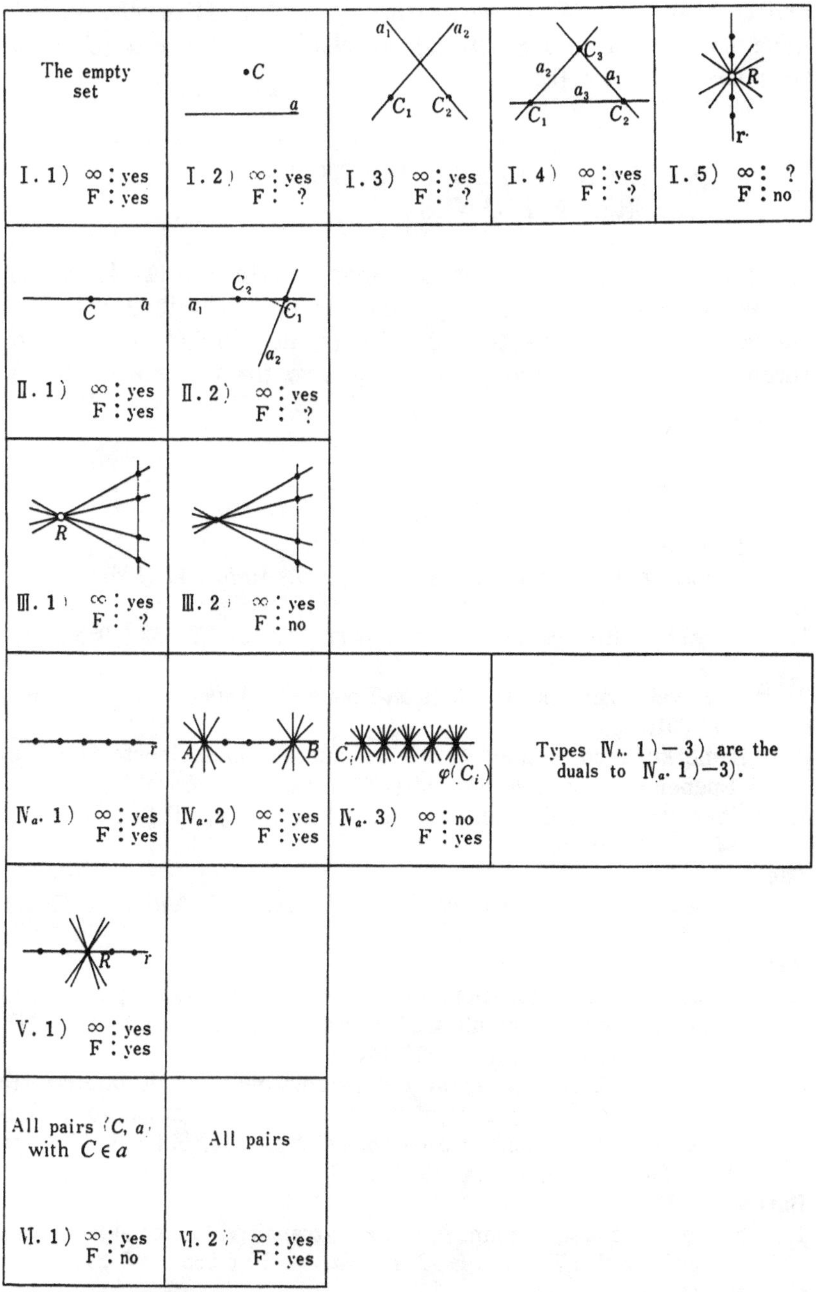

Figure 1. The various classes of projective planes and known results concerning existence in the finite (*F*) and infinite (∞) cases.

distinct from C, and not on a. If, for each such choice of P and Q, there is a (C, a)-collineation of π *that* (1) *carries* π_0 *into itself and* (2) *carries P into Q, we shall say that* π *is* (C, a, π_0)-*transitive* (Ostrom [5]).

6. NOTE ON A CLASSIFICATION OF INVERSIVE PLANES

We want to point out a paper of Hering [1] in which a classification of inversive planes has been obtained which is similar to that of Section 2. Questions similar to some of those we have indicated above arise also for this classification.

References

Albert, A. A.:
1. " Generalized Twisted Fields ", *Pacific J. Math.*, **11** (1961), 1–8.

André, J.:
1. " Projektive Ebenen über Fastkörpern ", *Math. Z.*, **62** (1955), 137–160.
2. " Über verallgemeinerte Moulton-Ebenen ", *Arch. Math.*, **13** (1962), 290–301.
3. " Bemerkung zu meiner Arbeit « Über verallgemeinerte Moulton-Ebenen » ", *Arch. Math.*, **14** (1963), 359–360.
4. " Über projektive Ebenen vom Lenz-Barlotti-Typ III 2 ", *Math. Z.*, **84** (1964), 316–328.

Baer, R.:
1. " Homogeneity of Projective Planes ", *Amer. J. Math.*, **64** (1942), 137–152.

Barlotti, A.:
1. " Le possibili configurazioni dei sistemi delle coppie punto-retta (A, a) per cui un piano grafico risulta (A, a)-transitivo ", *Boll. Un. Mat. Ital.*, **12** (1957), 212–226.
2. " Sulle possibili configurazioni del sistema delle coppie punto-retta (A, a) per cui un piano grafico risulta (A, a)-transitivo ", *Convegno internaz. Reticoli e Geometrie proiettive*, Palermo-Messina 1957, Cremonese, Roma, 1958, 75–78.

Bartolozzi, F.:
1. " Sopra una classe di piani finiti (R, r)-transitivi ", *Atti Accad. Naz. Lincei Rend. Cl. Sci. Fis. Mat. Natur.*, **39** (1966), 245–248.

Cofman, J.:
1. " On a Characterization of Finite Desarguesian Projective Planes ", *Arch. Math.*, **17** (1966), 200–205.

2. " On the Nonexistence of Finite Projective Planes of Lenz-Barlotti Type I_6 ", *J. Algebra*, **4** (1966), 64–70.

Fryxell, R. C.:

1. Ph. D. Thesis, Washington State University, 1961.

Gingerich, H. F.:

1. *Generalized Fields and Desargues Configurations*, Abstr. of Thesis, Univ. of Illinois, Urbana, Ill., 1945.

Hall, M. Jr.:

1. " Projective Planes ", *Trans. Amer. Math. Soc.*, **54** (1943), 229–277.
2. *The Theory of Groups*, Macmillan, New York, 1959.

Hering, C.:

1. " Eine Klassifikation der Möbius-Ebenen ", *Math. Z.*, **87** (1965), 252–262.

Hughes, D. H.:

1. " Planar Division Neo-Rings ", *Trans. Amer. Math. Soc.*, **80** (1955), 502–527.
2. "A Class of Non-Desarguesian Projective Planes ", *Canad. J. Math.*, **9** (1957), 378–388.

Jónsson, W. J.:

1. " Transitivität und Homogenität projektiver Ebenen ", *Math. Z.*, **80** (1963), 269–292.
2. " (C, γ, μ)-homogeneity of Projective Planes and Polarities ", *Canad. J. Math.*, **17** (1965), 331–334.

Lenz, H.:

1. " Kleiner Desarguesscher Satz und Dualität in projektiven Ebenen ", *Jahresbericht der Deutschen Math.-Ver.*, **57** (1954), 20–31.

Levi, F. W.:

1. *Finite Geometrical Systems*, Calcutta, 1942.

Lüneburg, H.:

1. " Endliche projektive Ebenen von Lenz-Barlotti Typ I-6 ", *Abh. Math. Sem. Univ. Hamburg*, **27** (1964), 75–79.
2. " Charakterisierungen der endlichen desarguesschen projektiven Ebenen ", *Math. Z.*, **85** (1964), 419–450.
3. " Zur Frage der Existenz von endlichen projektiven Ebenen vom Lenz-Barlotti-Typ III-2 ", *J. Reine Angew. Math.*, **220** (1965), 63–67.
4. " Über projektive Ebenen, in denen jede Fahne von einer nichttrivialen Elation invariant gelassen wird ", *Abh. Math. Sem. Univ. Hamburg*, **29** (1965), 37–76.

Magari, R.:

1. " Sul gruppo delle collineazioni di un piano grafico di ordine 27 introdotto da O. Veblen e J. H. Maclagan Wedderburn ", *Boll. Un. Mat. Ital.*, **14** (1959), 190–199.

Naumann, H.:

1. " Stufen der Begründung der Ebenen affinen Geometrie ", *Math. Z.*, **60** (1954), 120–141.

Ostrom, T. G.:
1. "Transitivities in Projective Planes", *Canad. J. Math.*, **9** (1957), 389–399.
2. "Correction to « Transitivities in Projective Planes »", *Canad. J. Math.*, **10** (1958), 507–512.
3. "Semi-Translation Planes", *Trans. Amer. Math. Soc.*, **111** (1964), 1–18.
4. "Finite Planes with a Single (p, L)-Transitivity", *Arch. Math.*, **15** (1964), 378–384; "Correction to...", *Arch. Math.*, **17** (1966), 480.
5. "A Characterization of the Hughes Planes", *Canad. J. Math.*, **17** (1965), 916–922.
6. "Collineation Groups of Semi-Translation Planes", *Pacific J. Math.*, **15** (1965), 273–279.
7. "The Dual Lüneburg Planes", *Math. Z.*, **92** (1966), 201–209.

Panella, G.:
1. "Una classe di sistemi cartesiani", *Atti Accad. Naz. Lincei Rend. Cl. Sci. Fis. Mat. Natur.*, **38** (1965), 480–485.

Pickert, G.:
1. *Projektive Ebenen*, Springer, Berlin-Göttingen-Heidelberg, 1955.
2. "Projektive Ebenen über Neokörpern", *Wiss. Zeitschr. Univ. Jena*, 1955–1956, 131–135.
3. "Gemeinsame Kennzeichnung zweier projektiver Ebenen der Ordnung 9", *Abh. Math. Sem. Univ. Hamburg*, **23** (1959), 69–74.
4. "Eine Kennzeichnung desarguesscher Ebenen", *Math. Z.*, **71** (1959), 99–108.
5. "Kennzeichnung desarguesscher Ebenen", *Bull. de la Soc. des math. et phys. de la R.P. de Serbie*, **14** (1962), 157–160.
6. *Lectures on Projective Planes*, mimeographed notes, Saskatoon, 1963.
7. "Die cartesischen Gruppen der Ostrom-Rosati-Ebenen", *Abh. Math. Sem. Univ. Hamburg*, **30** (1967), 106–117.

Rosati, L. A.:
1. "I gruppi di collineazioni dei piani di Hughes", *Boll. Un. Mat. Ital.*, **13** (1958), 505–513.
2. "Su una nuova classe di piani grafici", *Ric. Mat.*, **13** (1964), 39–55.

San Soucie, R. L.:
1. "Right Alternative Division Rings of Characteristic Two", *Proc. Amer. Math. Soc.*, **6** (1955), 291–296.

Salzmann, H.:
1. "Topologische projektive Ebenen", *Math. Z.*, **67** (1957), 436–466.
2. "Kompakte zweidimensionale projektive Ebenen", *Math. Ann.*, **145** (1962), 401–428.
3. "Zur Klassifikation topologischer Ebenen", *Math. Ann.*, **150** (1963), 226–241.

4. " Zur Klassifikation topologischer Ebenen II ", *Abh. Math. Sem. Univ. Hamburg*, **27** (1964), 145–166.
5. " Zur Klassifikation topologischer Ebenen III ", *Abh. Math. Sem. Univ. Hamburg*, **28** (1965), 250–261.

Segre, B.:
1. *Lectures on Modern Geometry*, with an appendix by L. Lombardo Radice, Cremonese Rome 1961.

Skornyakov, L. A.:
1. " Right Alternative Fields ", *Bull. Acad. Sci. URSS, Sér. Math.*, **15** (1951), 177–184.

Spencer, J. C. D.:
1. " On the Lenz-Barlotti Classification of Projective Planes ", *Quart. J. Math. Oxford*, **11** (1960), 241–257.

Yaqub, J. C. D. S.:
1. " On Projective Planes of Class III ", *Arch. Math.*, **12** (1961), 146–150.
2. " The Existence of Projective Planes of Class I 3 ", *Arch. Math.*, **12** (1961), 374–381.
3. " On Projective Planes of Lenz-Barlotti Class I 6 ", *Math. Z.*, **95** (1967), 60–70.
4. " The Non-Existence of Finite Planes of Class III 2 ", to appear in *Arch. Math.*

Zappa, G.:
1. " Sui gruppi di collineazioni dei piani di Hughes ", *Boll. Un. Mat. Ital.*, **12** (1957), 507–516.
2. " Sui piani quasi di traslazione secondo Lingenberg ", *Rend. Mat. e Appl.*, **23** (1964), 124–127.

Zorn, M.:
1. " Theorie der alternativen Ringe ", *Abh. Math. Sem. Univ. Hamburg*, **8** (1931), 123–147.

CHAPTER 26

Translation Planes of Order $\mathbf{p}^2$

T. G. OSTROM[1], *Washington State University and The University of Frankfurt, Germany*

We shall attempt to give a discussion of methods of constructing translation planes of order p^2, where p is a prime. Since there is only one such plane of order 4, we shall assume that p is odd. The notions we shall discuss have broader applications. They can be modified in one direction to apply to translation planes whose order is not a square; they can be modified in another direction to apply to planes of square order which are not translation planes. We achieve a certain simplicity by our restriction on the order.

André [1] and Bruck and Bose [2] have shown that every translation plane may be represented in terms of a "spread". Let V be a vector space of dimension $2r$ over the Galois field of order p. A spread (over V) is a class of $p^r + 1$ mutually independent vector subspaces of dimension r. These subspaces will be called the components of the spread.

The points of the translation plane are the elements of V. The lines are the translates of the components. The order of the plane is the number of points on a line and is here equal to p^r.

We are restricting ourselves to the case where $r = 2$. Let

[1] This research was supported in part by National Science Foundation Grant GP 5036.

π be the Desarguesian affine plane of order p^2. Then π is a particular kind of translation plane and may be interpreted in the above manner. We may choose a point 0 as reference point to be identified with the zero element of V. The lines through 0 are two-dimensional subspaces of V. There will be other two-dimensional subspaces. It is not difficult to verify that they are the affine subplanes of order p which contain 0. The other affine subplanes of order p are translates of these vector subspaces.

Thus if π' is another translation plane of order p^2, then each line of π' may be interpreted either (1) as a line of π or (2) as a subplane of π.

We now wish to introduce the idea of a "replaceable net". There is a close connection between replaceable nets and what Bruck and Bose call switching sets [4]. Indeed, we shall be looking at a particular kind of replaceable net which is equivalent to a particular kind of switching set.

A net is a set of "points" together with a class of designated subsets, called "lines", such that:

1. The lines occur in "parallel classes" with the property that each point belongs to exactly one line of each parallel class.

2. Lines of different parallel classes have exactly one point in common.

The order of the net is the number of points on a line. (This parameter is independent of the choice of line.) We impose the nontriviality conditions that the order be greater than one and that there be at least three parallel classes. It is well known that a net with k parallel classes is equivalent to a set of $k-2$ mutually orthogonal latin squares.

A net N is said to be replaceable if there is another net N' defined on the same points such that two points are on a line of N if and only if they are on a line of N' [8].

A (finite) affine plane is a net of order n with $n+1$ parallel classes. A net N of order n is said to be embedded in an affine plane π of order n if they are defined on the same points and every line of N is also a line of π. Thus we can get a net embedded in an affine plane by merely choosing some subset of the parallel classes of π. The lines of the designated parallel classes are the lines of the embedded net. The lines of π which are not in N form another net M which is said to be the complement of N. We then write $\pi = N \cup M$. (Note that we are not talking about the set theoretic union of the points in the two nets. They are defined on the same points.)

Our basic idea is that if $\pi = N \cup M$ and N is replaceable, then M is also complementary to the " replacing net " N'—i.e., $N' \cup M$ is an affine plane π'. The planes π and π' can be isomorphic; in most cases they are not.

We shall say that two nets embedded in a plane are " disjoint " if they have no parallel classes in common. In a similar spirit, the union of two nets N_1 and N_2 embedded in a plane is the smallest net embedded in the plane which includes all of the parallel classes of N_1 and N_2. If N_1 and N_2 are disjoint replaceable nets embedded in a plane, we may replace N_1 by its replacing net N_1', or N_2' can replace N_2, or we may replace $N_1 \cup N_2$ by $N_1' \cup N_2'$, thus possibly obtaining three different planes from the original plane.

It so happens that Desarguesian planes not of prime order contain many replaceable nets. Let π be the Desarguesian plane coordinatized by the field K of order p^2, where K is a quadratic extension of a field F of order p. The vector space interpretation mentioned earlier arises out of the fact that the additive group in K is a two dimensional vector space over F. Each point (x, y) may be interpreted as an element of a four dimensional vector space V over F.

If a and b are elements of K, the set of points (x, y) such that $y = ax + bx^p$ will constitute a two dimensional subspace of V. If $b = 0$, this subspace is a line; otherwise, it is a subplane of order p. Adding a constant is equivalent to taking a translate. Thus, if $b \neq 0$, $y = ax + bx^p + c$ is the equation of a subplane of order p. (in fact, the only subplanes of order p which cannot be represented in this way are those which intersect some line $x =$ constant in more than one point.)

The slope of the line connecting two points of the subplane $y = ax + bx^p + c$ is always of the form $a + bx^{p-1}$ for some value of x. If $m = a + bx^{p-1}$, then $(m - a)^{p+1} = b^{p+1}$. Conversely, if $(m - a)^{p+1} = b^{p+1}$, then $m = a + bx^{p-1}$ for some value of x. That is, the subplane is embedded in the net N consisting of those parallel classes for which the slope m satisfies $(m - a)^{p+1} = b^{p+1}$.

The multiplicative group in K is of order $p^2 - 1$; the elements of F are the $(p - 1)$st roots of unity. It follows that b^{p+1} is an element of F. Furthermore, if $e^{p+1} = b^{p+1}$, then the subplane $y = ax + ex^p + d$ is embedded in the same net. Counting the subplanes embedded in N and using Lemma 9 and

Theorem 6 of [6], it doesn't require very much checking to verify that N is replaceable by a net N', where the lines of N' are the subplanes of order p embedded in N.

We are now dealing with replaceable nets of a rather special type. We call them "derivable nets" [7]. Specifically, a derivable net is a net which can be embedded in a Desarguesian plane and can be replaced by a net whose lines are subplanes of the original net. A derivable net must be of square order but need not be of order p^2. For the remainder of this paper, the term should always be understood to refer to derivable nets of order p^2.

Clearly, a Desarguesian plane of order p^2 will contain many derivable nets—including some which contain the parallel class $x =$ constant. We hope it is also clear that (unless p is too small) it will be possible to pick out a number of pairwise disjoint derivable nets. Note that the total number of parallel classes in an affine plane of order p^2 is equal to $p^2 + 1$ while the number of parallel classes in a derivable net is equal to $p + 1$ (the number of parallel classes in a plane of order p). Thus we connot have more than $p - 1$ mutually disjoint derivable nets. The total number of parallel classes in the union will be $p^2 - 1$ in this maximum case. Hence, given any set of mutually disjoint derivable nets, there will always be at least two parallel classes which are not included in any derivable net of the given set. Without loss of generality, we can assume that the parallel classes $x =$ constant and $y =$ constant are not included in any of the nets in our given set.

We can summarize the implications of the preceding discussion in the following Theorems:

Theorem 1. *Let a be any element of K and let ρ be any non-zero element of F. Then the net $N(a, \rho)$ consisting of those parallel classes for which the slope m satisfies the condition $(m - a)^{p+1} = \rho$ is a derivable net. Every derivable net which does not include the parallel classes $x =$ constant and $y =$ constant can be represented in this way with the additional restriction $a^{p+1} \neq \rho$.*

Theorem 2. *Let $a_1, a_2, \ldots, a_k$ be a set of elements of K and let $\rho_1, \rho_2, \ldots, \rho_k$ be a set of non-zero elements of F. Suppose that these parameters satisfy the following conditions:*

(1) *For each i, $i = 1, \ldots, k$, $a_i^{p+1} \neq \rho_i$;*

(2) *If $(m - a_i)^{p+1} = \rho_i$ and $(m - a_j)^{p+1} = \rho_j$ for any value of m, then $i = j$.*

Then the nets $N(a_i, \rho_i)$ form a set of k pairwise disjoint derivable nets. Without loss of generality, each set of k pairwise disjoint derivable nets can be represented in this fashion.

Theorem 3. *Under the conditions of Theorem 2, if we replace each of the nets $N(a_i, \rho_i)$, we obtain a translation plane π' in which the lines are represented by equations of the following types:*

(1) $x = c$.

(2) $y = a_i x + (m - a_i)x^p + c$ *if* $(m - a_i)^{p+1} = \rho_i$, $i = 1, \ldots, k$.

(3) $y = xm + c$ *if* m *satisfies none of the conditions* (2).

There is one line for each choice of m and c in K.

Proof. In view of the preceding discussion, we only need to prove that π' is a translation plane. This follows immediately from the fact that the lines through the origin constitute a spread and that the other lines are translates of the lines through the origin.

Comment. This reduces to a case of a construction due to André [1] when all of the a_i are chosen to be zero. Here the ρ_i will have to be distinct. If we choose all of the a_i to be zero and let $k = p - 1$, the "new" plane turns out to be Desarguesian.

Up to an isomorphism—and with a finite number of exceptions—all of the known translation planes of order p^2 can be constructed by these methods.

Possible exceptions are the planes coordinatized by the special nearfields of Zassenhaus [9] which are not of order 9. There are six of these nearfields. There are two more possible exceptions which are discussed below. By way of contrast, there are several planes in the class we have been discussing for each value of $p > 3$ and one of order 3^2. Hence this class contains an infinite number of members.

A word on isomorphisms: Let N be any replaceable net embedded in a plane. Suppose that the plane admits a collineation carrying N into another net M. Then the plane (or planes) constructed by replacing M will be isomorphic to the plane (or planes) constructed by replacing N. This includes the case where N is a union of disjoint derivable nets. Planes constructed by derivation can be isomorphic in other situations. See the example mentioned above where a Desarguesian plane is converted into a (isomorphic) Desarguesian plane.

It is an open question as to whether all translation planes of order p^2 can be constructed by replacing a set of mutually disjoint derivable nets in a Desarguesian plane. A related question: Let π' be a plane constructed in this way. Can we replace derivable nets in π' to obtain a plane not isomorphic to a plane gotten directly from the Desarguesian plane by replacing disjoint derivable nets? We suspect that the answer is "no" to both questions.

Foulser [5] has constructed two translation planes of order 25 which do not seem to be obtainable directly from a Desarguesian plane by replacing derivable nets. Foulser's construction may be looked at in the following way:

Bose [2] has shown that every Desarguesian affine plane of order n admits a cyclic collineation of order $n^2 - 1$. One point is fixed, all of the other points are in a single orbit. Let U be the fixed point and let ω be the collineation. Let $P \neq U$ be some reference point. We may represent the image of P under the ith power of ω by the integer i. In our case, we take $n = p^2$, and i is one of the residues mod $(p^4 - 1)$.

We may also think of our Desarguesian plane in terms of a four dimensional vector space over the field of order p, with U as the zero vector. Let λ denote ω raised to the power $(p^2 + 1)(p + 1)$. Then λ is a collineation of order $p - 1$. It can be shown that λ is a homology with center U which fixes all one dimensional vector subspaces. This implies that the non-zero vectors in a one-space consist of a set of $p - 1$ values of i congruent to each other mod $(p^2 + 1)(p + 1)$.

Consider the class of subplanes of order p which contain the one-space in which the residues are congruent to 1 mod $(p^2 + 1)(p + 1)$. Now two distinct subplanes of order p can have at most p points in common. Every one of the $p^4 - p^2$ points not on the line containing the given one-space will belong to exactly one subplane of our class. Each such subplane contains $p^2 - p$ points not in the given one-space. Hence the number of subplanes in our class is equal to $(p^4 - p^2)$ divided by $(p^2 - p)$, i.e., to $p(p + 1)$.

Now let σ be ω raised to the power $2(p + 1)$. Since λ is a power of σ, the orbit of σ which contains i also contains all of the points in the same one-space with i. Let S be the orbit of σ which contains the points congruent to 1 mod $2(p + 1)$. The number of points in S is $\frac{1}{2}(p^2 + 1)(p - 1)$. The number of subplanes in the class considered above which contain additional points of S can be at most equal to $\frac{1}{2}(p^2 + 1)$, since if a subplane contains one of these points, it must contain at least $p - 1$ of them. Since $p(p+1) > \frac{1}{2}(p^2+1)$, there is at least one subplane which contains the points congruent to $1 \bmod (p^2+1)(p+1)$ and no other points in the same orbit with them under σ. Let π_0 be one of these subplanes.

Then σ raised to the power $\frac{1}{2}(p^2+1)$ induces a collineation of π_0. The image of π_0 under any smaller power of σ will intersect π_0 only in the point U. The subplane π_0 and its $\frac{1}{2}(p^2+1)$ distinct images may be interpreted as a set of $\frac{1}{2}(p^2+1)$ independent vector spaces of dimension two.

Suppose that we can find another subplane π_1 which contains U and satisfies the following: (1) π_1 is independent of π_0 and its images (2) π_1, like π_0, is carried into itself by the $\frac{1}{2}(p^2+1)$ power of σ and is independent of its images under smaller powers of σ.

Then π_1 and its images will be independent of π_0 and its images. Altogether, we have a set of p^2+1 mutually independent vector spaces of dimension two. That is, we will have a spread and a translation plane. Foulser constructed two planes of order 25 by doing what amounts to making appropriate choices for π_0 and π_1.

We shall conclude with a few remarks about other applications of replaceable nets to the construction of planes. Most of what we have said about derivable nets remains valid if the prime p is replaced by a power of a prime. The known planes of square order which are not translation planes (or their duals) are, again up to a duality, representable in affine form as planes which contain derivable nets. Indeed, many of them were constructed from other planes in this class by replacement of derivable nets [6].

If K is a field of order q, where q is any power of the prime p but not equal to p, and if π is the Desarguesian affine plane coordinatized by K, π will contain replaceable nets. Let σ be a power of p such that $x \to x^\sigma$ is a nontrival automorphism of K. Let i be the smallest integer such that $(\sigma - 1) \equiv 0 \bmod q - 1$. Let a be any element of K and let ρ be any nonzero element of the subfield left elementwise fixed by σ. It is straightforward to verify that the net consisting of those parallel classes for which the slope m satisfies the condition $(m-a)^i = \rho$ is replaceable. The lines of the replacing net can be represented by equations of the form $y = xa + (m-a)x^\sigma + c$. Details for the case $a = 0$ are given in [8]. Each such net with $a \neq 0$ will be isomorphic to one in which $a = 0$. Here we are again very close to the work of André [1] and to the switching sets of Bruck and Bose [4].

A complete classification of all translation planes seems to be hopeless at the moment. We might be well on the road to giving a complete classification of all translation planes of order p^2 if we could answer the following questions:

1. Can all translation planes of order p^2 be constructed from Desarguesian planes by replacement of derivable nets? If not, can we give a finite list of exceptions?

2. Can we list in systematic fashion all possible ways of picking out a set of mutually disjoint derivable nets in a Desarguesian plane of order p^2?

There are other related questions raised earlier in this paper. There are not easy questions, but it is conceivable that they can be answered. A less ambitious program would be to try to answer them for some particular values of p. Question 2 may well be within the range of computer attack for reasonably small values of p.

References

1. André, J. "Über nicht-Desarguesche Ebenen mit transitiver Translations Gruppe," *Math. Z.*, **60** (1954), 156–186.
2. Bose, R. C. "An Affine Analogue of Singer's Theorem," *J. Indian Math. Soc.*, **6** (1942), 1-15.
3. Bruck, R. H. and Bose, R. C. "The Construction of Translation Planes from Projective Spaces," *J. Algebra*, **1** (1964), 85–102.
4. Bruck, R. H. and Bose, R. C. "Linear Representations of Projective Planes in Projective Spaces," *J. Algebra*, **4** (1966), 117–172.
5. Foulser, David. "Solvable Flag Transitive Groups," *Math. Z.*, **86** (1964), 191-204.
6. Ostrom, T. G. "Semi-Translation Planes," *Trans. Amer. Math. Soc.*, **111** (1964), 1-18.
7. Ostrom, T. G. "Derivable Nets," *Canad. Math. Bull.*, **8** (1965), 601–613.
8. Ostrom, T. G. "Replaceable Nets, Net Collineations, and Net Extensions," *Canad. J. Math.*, **18** (1966), 666–672.
9. Zassenhaus, H. "Über endliche Fastkörper," *Abh. Math. Sem., Univ. Hamburg*, **11** (1936), 187-220.

Discussion on Professor Ostrom's Paper

DR. R. LASKAR: Professor Ostrom has discussed methods of constructing translation planes of order p^2. The presentation of the paper is lucid, clear and is quite suggestive of new ideas.

As pointed out by Professor Ostrom, first André (1954), then Bruck and Bose (1964) have shown that every translation plane may be represented in terms of a "spread", though the constructions by André and Bruck-Bose are different. The former has used a vector-space of even dimension, whereas the latter authors have used a projective space of even dimension and hence a vector space of odd dimension. Professor Ostrom uses here the vector-space representation of André, and shows how from a given Desarguesian affine plane of order p^2, and hence from a particular kind of translation plane of order p^2, other translation planes of the same order can be constructed by replacing appropriate "nets" by the "replacing nets". He also shows that most of the known translation planes of order p^2 can be constructed by the same method.

At this point, it will not be inappropriate to look at the simplest case of the Bruck-Bose construction and show certain properties geometrically, which Professor Ostrom discusses in his paper.

For $r = 2$, the Bruck-Bose construction is as follows. Define

Σ_4: projective 4-space over $GF(p)$.

Σ_3: a fixed projective 3-space contained in Σ_4.

S : spread of Σ_3, i.e., a collection of $p^2 + 1$ mutually skew lines.

π-points: points of $\Sigma_4 - \Sigma_3$

π-lines : planes through the lines of the spread not wholly contained in Σ_3.

Incidence relation is induced by that of Σ_4. π is an affine translation plane of order p^2. It can be made projective by adjoining Σ_3 as a π-line at infinity and lines of the spread as points at infinity.

A method of constructing a spread which will be regular is as follows:

It is known that the Plucker co-ordinates of lines of a projective 3-space, say Σ_3, determine a point in a projective 5-space, say S_5, and the condition that a point in S_5 is determined by a line in Σ_3 is that these points in S_5 lie on a quadric surface Q_5 in S_5. With a suitably chosen 3-space, say S_3, contained in S_5, we can have the section $Q_5 \cap S_5 = Q_3$ an elliptie quadric. The $p^2 + 1$ points of Q_3 are such that they are mutually non-conjugate to each other. The $p^2 + 1$ lines in S_3 corresponding to the $p^2 + 1$ points of Q_3 will be mutually skew to one another and form a spread S.

Suppose S is a spread constructed by the method described above in the Bruck-Bose space at infinity Σ_3. It can be shown that S is regular. For consider any three lines A, B, C, of S. The lines A, B, C correspond to three points a, b, c in Q_5 which determine a plane π. Let π^* be the polar plane of π with respect to Q_5. Then each of the sections $S_1 = \pi \cap Q_3$ and $\pi^* \cap Q_5$ contains $p + 1$ points, such that each point of one section is conjugate to every point of the other section and vice versa; also points belonging to one section are non-conjugate to each other. The $p + 1$ lines, including A, B, C of S, corresponding to the $p + 1$ points of S_1, and the $p + 1$ lines corresponding to the $p + 1$ points of S_2, form two conjugate reguli. It could also be shown that except for two lines, say L, M, in the spread S, the remaining $p^2 - 1$ lines of S can be partitioned into $p - 1$ disjoint reguli, each regulus consisting of $p + 1$ lines. To see this, consider a line l in S_3, which is a non-intersector to Q_3. Each of the sections of Q_3 by planes π through l will be disjoint, each section containing $p + 1$ points. Two planes through l will be tangent planes to Q_3 at two different points, corresponding to the lines L, M in S. These $(p - 1)$ disjoint reguli are what Professor Ostrom calls mutually disjoint derivable nets.

PROFESSOR OSTROM replied as follows:

I am glad that the interpretation of derivation in terms of quadrics in S_5 was included in the discussion. To me it seems more complicated than my own way of looking at the construction but different interpretations often lead to different ways of generalization. Because of the connection between vector spaces and affine spaces, I insist that my interpretation is also geometrical.

CHAPTER 27

Construction Problems of Finite Projective Planes

R. H. BRUCK[1], *University of Wisconsin*

Let q be a prime. (In the paper, q is a prime-power but the statements which follow require space-consuming modifications.) Let $PG(3, q)$, $IP(q)$ denote the projective 3-space and the inversive plane, respectively, over the field $GF(q)$. The isomorphism classes of the André planes of order q^2 are in one-to-one correspondence with the orbits of the subregular spreads of linear type of $PG(3, q)$ under the group of all collineations of $PG(3, q)$. These orbits are in one-to-one correspondence with the orbits of the set of at most $(q-1)/2$ "concentric" circles of $IP(q)$ under the group of all collineations of $IP(q)$—subject to a small correction which we shall ignore here. Subregular spreads of $PG(3, q)$ correspond to (not necessarily "concentric") sets of disjoint circles of $IP(q)$ and also to projective planes of order q^2 belonging to a class called subregular. This class contains the class of André planes of order q^2 as a much smaller sub-class. Both the André planes of order q^2 and the "non-

[1] This research was supported as follows: During 1963-64 at The University of North Carolina at Chapel Hill, by the National Science Foundation, Grant No. GP 16-60 and by the Mathematics Division of the Air Force Office of Scientific Research, grant No. 84-63. Since June, 1964, at the University of Wisconsin, Madison, by the National Science Foundation, Grants No. GP-3993 and GP-7465.

concentric" triples of disjoint circles of $IP(q)$ are completely classified. The paper represents a first step in a proposed attempt at constructing projective planes of order $q(q+1)$ where q is a prime-power.

1. INTRODUCTION

If F_0 is any field such that the polynomial ring $F_0[x]$ has an irreducible quadratic, then $F_0 \subset F \subset K$ for superfields F, K such that

(i) K is 2-dimensional over F;

(ii) every quadratic polynomial in $F[x]$ splits into linear factors in $K[x]$.

Thus pairs F, K of fields with properties (i), (ii) are not at all rare. When F is finite they have an additional property which, I suspect, is always false in the infinite case, namely:

(iii) every element of F is the norm (relative to F) of some element of K.

In the present paper we restrict attention to the finite examples: $F = GF(q)$, $K = GF(q^2)$, where q is an arbitrary prime-power. This is because the results are dedicated to problems in the theory of finite projective planes, and it would be foolish not to avail ourselves of simple, revealing combinatorial arguments when these happen to be on hand. Nevertheless, much of the present paper goes through for the general pair F, K with properties (i), (ii), most differences in the theory being ascribable to the lack of property (iii).

I mention this because of something which I have not done —and perhaps should have done—in the present paper. If F, K are the field of reals and the field of complex numbers, respectively, there is a very rich theory of classical geometry over F from which we could probably draw many results of deep significance in the present context. At most, I have been content to recall some familiar theorems of real geometry and to shape them to my ends, when I could, in the finite case. Only now, when this paper is about to leave my hands, has it become clear that much more should be attempted along this line. And what should be done? I believe that the problem of "translation" should be settled at one stroke by a proof of the following *Conjecture: Any true proposition of real geometry concerning a finite number of objects (points, lines, planes, conics, quadrics, or the like) and specified incidences or non-*

incidences between them can be illustrated by a model over a suitably large subring A *of the field of reals such that* (a) A *consists of algebraic integers;* (b) *the quotient field of* A *has finite dimension over the field of rationals and* (c) *at least one finite homomorphic image (a finite field) of* A *inherits the model.* —For all I know to the contrary, there may be a proof of this conjecture in the literature.

In Section 2 we propose a method of constructing finite projective planes which, if successful, would produce a plane of order $q(q+1)$ where q is a prime-power. The construction requires the existence of suitable collections of spreads and packings of $PG(3, q)$; that is, of projective 3-space over a finite field $GF(q)$. In infinitely many cases—for example, when the prime-power q is a square—a proof that the construction is impossible would probably require a very deep knowledge of the theory of spreads of $PG(3, q)$.

In the rest of the paper we restrict attention to the theory of spreads of $PG(3, q)$. To show that the study is worthwhile, we discuss (in Section 3) the connection between the isomorphism classes of translation planes of order q^2, on the one hand, and the orbits of the spreads of $PG(3, q)$ under the group of all collineations of $PG(3, q)$, on the other. Later (in Section 8) we give a simple and explicit prescription for determining (at least when q is a prime) the number of non-isomorphic André planes of order q^2. We also give a table for $q \leq 11$.

Actually we restrict attention to a large class of spreads called subregular. (In Appendix II we exhibit a type of spread which is not subregular; this exists only for $q = 2^{2t-1} \geq 8$, and corresponds to the Suzuki group of order $(q^2+1)q^2(q-1)$. No other examples seem to be known at present.) The essential content of Sections 4, 5, 6 and Appendix I is the following: For $0 \leq k < (q-1)/2$, there is a one-to-one correspondence between the orbits of subregular spreads of $PG(3, q)$ of index k and the orbits of sets of k disjoint circles of $IP(q)$.—Here $IP(q)$ is an inversive plane which we shall describe in a moment.— Consequently, in a very long section (Section 7) we abandon the study of subregular spreads in favor of the study of sets of disjoint circles of $IP(q)$. Then in Section 8, we come back to the connection with subregular spreads and show that the André planes of order q^2 can be imbedded in a much larger class of subregular planes of order q^2.

Suppose that F, K are two fields with the properties (i), (ii) mentioned in our opening paragraph. Let Σ_1 be the projective line over K. Then each unordered triple of distinct points P, Q, R can be imbedded uniquely in a point-set (P, Q, R) consisting of the points of the projective sub-line over F which contains P, Q and R. If we call (P, Q, R) the *circle* through P, Q, R, then Σ_1 becomes the *inversive plane* $IP(F)$. Inversion with respect to the circle (P, Q, R) is the unique collineation of Σ_1 of order two which fixes each of P, Q, R. When F is the field of reals, $IP(F)$ may be obtained from the real Euclidean plane by adjoining a single point at infinity, ∞, interpreting straight lines as circles through ∞, and performing inversion with respect to a circle in the classical manner. The center of a circle may be replaced by any pair of points conjugate with respect to the circle. When F is a finite field $GF(q)$, $IP(F)$ is the inversive plane $IP(q)$ studied in Section 7. We shall mention some facts about $IP(F)$ of special pertinence to the present inquiry, with references to proofs for the special case $F = GF(q)$.

Two disjoint circles of $IP(F)$ have a unique common pair of conjugate points (Lemma 7.4). A set of disjoint circles of $IP(F)$ is called *linear* provided the circles of the set have a common pair of conjugate points. (This is the analog of a set of concentric circles of the real Euclidean plane.) Linear sets of disjoint circles correspond to André planes (Section 8). Such linear sets are completely discussed in Theorem 7.5. Nonlinear triples of disjoint circles of $IP(F)$ exist when F has at least 5 elements. To each such nonlinear triple corresponds a unique *common orthogonal circle* containing the three pairs of common conjugate points (Theorem 7.14). The nonlinear triples are completely classified following Theorem 7.14. The existence of nonlinear sets of $k > 3$ disjoint circles is shown in considerable variety in Lemma 7.23, but no complete classification is attempted. Behind Lemma 7.23 is a simple proposition about circles: Let P, Q, Q', S be four distinct points of $IP(F)$, let C be the unique circle through S with P, Q as conjugate points, let C' be the unique circle through S with P, Q' as conjugate points, and let D be the unique circle through P, Q, Q'; then C, C' are tangent if and only if D contains S (Lemma 7.22). When F is the field of reals and $P = \infty$, then C, C' are circles through S with centers Q, Q' respectively, and D is the line through these centers.—There is no doubt in my mind that other equally simple and familiar propositions about elementary geometry have a profound significance in the area of the present paper.

2. A METHOD OF DECOMPOSING OR CONSTRUCTING PROJECTIVE PLANES

Let π be a projective plane of finite order n. Assume that π can be decomposed into two disjoint partial planes G, G' subject to the following axioms:

(i) Each of G, G' contains at least one point and at least one line of π.

(ii) No point of G is incident in π with a line of G'.

(iii) For some positive integer k, each line of G is incident in π with exactly k points of G.

Then the following dual of (iii) also holds:

(iii′) For some positive integer k', each point of G' is incident in π with exactly k' lines of G'.

In addition, if we regard the points and lines of G as the treatments and blocks, respectively, of a design with parameters v, b, k, r, λ and, at the same time, regard (dually) the lines and points of G' as the treatments and blocks, respectively, of a design with parameters v', b', k', r', λ', we may verify the following relations:

$$n = kk' \geq 2 \tag{2.1}$$

$$\begin{aligned} v &= k(n+1) - n & v' &= k'(n+1) - n \\ b &= (n+1)^2 - k'(n+1) & b' &= (n+1)^2 - k(n+1) \\ k &= k & k' &= k' \\ r &= n+1 & r' &= n+1 \\ \lambda &= 1 & \lambda' &= 1 \end{aligned}$$

$$v + b' = v' + b = n^2 + n + 1 \,. \tag{2.2}$$

In the opposite direction, we may start from a partial plane π_0 which, when points are identified with treatments and lines with blocks, is a design with parameters v, b, k, r, λ subject to

$$\lambda = 1 \,, \qquad r \geq 3 \,, \qquad k \text{ divides } v \,. \tag{2.3}$$

Then, if we define n and k' by $n = r - \lambda = kk'$, the parameters of π_0 are related as in (2.1). The significance of n and the dual set of parameters becomes clear if we assume that π_0 can be extended to a projective plane π subject to the following condition:

(*) The only lines of π incident in π with a point P of π_0 are the lines of π_0 incident in π_0 with P.

In this case, π has order n and is partitioned into disjoint partial planes G, G' with $G = \pi_0$.

If, with $G = \pi_0$ as in the last paragraph, we wish to characterize G' intrinsically in terms of G, we need two definitions:

Definition 2.1. *A spread, $\mathcal{S}$, of π_0 is a collection of lines of π_0 which, as point sets, partition the points of π_0.*

Definition 2.2. *A packing, $\mathcal{P}$, of π_0 is a collection of spreads of π_0, which, as line sets, partition the lines of π_0.*

Now we may note the following: Each point P of G' corresponds to a unique spread of π_0, consisting of all the lines of $G = \pi_0$ incident in π with P. Each line L of G' corresponds to a unique packing of π_0, made up of the spreads corresponding to the points of G' incident in G' with L. Thus, if π and G' exist, there exists a collection C_s of spreads of π_0 and a collection C_p, of packings of π_0 subject to the following conditions:

(a) If $\mathcal{P}$ is a packing in C_p, each spread in $\mathcal{P}$ is a spread in C_s.

(b) If A, B are any two distinct lines of π_0 not incident in π_0 with a point of π_0, there exists one and only one spread in C_s containing A and B.

(c) If $\mathcal{S}_1$, $\mathcal{S}_2$ are any two disjoint spreads in C_s, there exists one and only one packing in C_p containing $\mathcal{S}_1$ and $\mathcal{S}_2$.

Moreover, conditions (a), (b), (c) imply that π_0 can be extended to a projective plane π subject to (*). In particular they imply.

(d) Two distinct packings in C_p have one and only one common spread.

Designs π_0 subject to (2.3) are exceedingly common and, for a few of them, the problem of constructing π subject to (*) is familiar. We shall limit our attention to the only type relévant to the rest of the paper.

Let $\Sigma = PG(3, q)$ be projective 3-space over a finite field $GF(q)$, and let π_0 be the partial plane consisting of the points and lines of Σ under the natural incidence. Then π_0 has type (2.3), and (2.1), (2.2) hold with $k = q + 1$, $k' = q$ and hence

$$n = q(q + 1)\ . \tag{2.4}$$

Thus, if π_0 can be extended to a projective plane π of order n subject to (*), (2.4) holds; in particular the order of π is not a prime-power. If we try to show non-existence of π by use of the Bruck-Ryser Theorem then, since q is a prime-power, we are only successful in the following cases:

Cases of non-existence

(1) $q = 2$

(2) $q \equiv 1 \bmod 4$ but $q + 1$ is not a sum of two integral squares.

All other cases are still undecided. Specifically, these are:

Undecided cases

(3) $q \equiv 0$ or $3 \bmod 4$

(4) $q \equiv 1 \bmod 4$ and $q + 1$ is a sum of two integral squares.

In particular, if $q \equiv 5 \bmod 8$, we have case (2), whereas, if q is an odd square, we have case (4).

By a spread of $\Sigma = PG(3, q)$ we mean a spread of the natural partial plane π_0. Thus a spread of Σ is a set of $q^2 + 1$ skew lines of Σ. Similarly, a packing of Σ is a set of $q^2 + q + 1$ disjoint spreads of Σ. As a first step in the extension problem, we need to study the structure of the spreads of Σ. The rest of this paper will be devoted to the theory of a broad class of spreads called subregular, and to one important application of that theory.

3. BOSE'S CONSTRUCTION

I need to repeat here the simplest case of Bose's construction in [2].

Let $\Sigma' = PG(4, q)$ be projective 4-space over a finite field $GF(q)$, let $\Sigma = PG(3, q)$ be a fixed hyperplane of Σ', and let $\mathcal{S}$ be a spread of Σ. Then we may define an affine plane $\pi = \pi(\mathcal{S})$ as follows: The points of π are the points of Σ' which are not in Σ. The lines of π are the planes of Σ' containing a line of $\mathcal{S}$ and not entirely contained in Σ. The incidence relation of π is induced by that of Σ'.

The construction yields an affine translation plane π of order q^2. And π may be extended to a projective plane by adjoining Σ (or $\mathcal{S}$) as a line at infinity and the lines in $\mathcal{S}$ as points at infinity. We shall be interested in the following uniqueness condition $U(q)$:

$U(q)$: *For every choice of a spread* $\mathcal{S}$ *of* $\Sigma = PG(3, q)$, *if* P, Q, R *are any three distinct points of* $\pi(\mathcal{S})$ *not lying on a line of* $\pi(\mathcal{S})$, *there is at most one Desarguesian affine subplane of order* q *of* $\pi(\mathcal{S})$ *which contains* P, Q, R *(and has the same line at infinity as* $\pi(\mathcal{S})$*).*

Theorem 3.1. *Assume that* $U(q)$ *is true for the given prime-power* q. *Let* $\mathcal{S}, \mathcal{S}'$ *be spreads of* $\Sigma = PG(3, q)$. *A necessary and sufficient condition that there exist an isomorphism* θ *of the affine plane* $\pi(\mathcal{S})$ *upon the affine plane* $\pi(\mathcal{S}')$ *is that there exist a collineation* ϕ *of* Σ *which maps* $\mathcal{S}$ *upon* $\mathcal{S}'$.

Remark. The proof of Theorem 3.1 will show that $U(q)$ is true if q is a prime·

Proof. First we note that a collineation ϕ of Σ can be extended (in many ways) to a collineation ϕ' of Σ'. In this case, if ϕ maps $\mathcal{S}$ upon $\mathcal{S}'$, ϕ' induces an isomorphism θ of $\pi(\mathcal{S})$ upon $\pi(\mathcal{S}')$.

Next suppose that there exists an isomorphism θ of $\pi(\mathcal{S})$ upon $\pi(\mathcal{S}')$. Then $\pi(\mathcal{S})$ and $\pi(\mathcal{S}')$ have the same points, and θ permutes these points in a one-to-one manner. Also θ maps the lines of $\pi(\mathcal{S})$ upon the lines of $\pi(\mathcal{S}')$ in a one-to-one manner. Thus θ maps certain planes upon other planes, namely planes (not in Σ) containing a line of $\mathcal{S}$ upon planes (not in Σ) containing a line of $\mathcal{S}'$. In addition, obviously, θ maps affine subplanes of $\pi(\mathcal{S})$ upon affine subplanes of $\pi(\mathcal{S}')$, preserving both order and the Desarguesian property.

Now consider a projective plane Σ_0 of Σ' (not in Σ). If Σ_0 is a line of $\pi(\mathcal{S})$, then θ maps Σ_0 upon a plane, namely a line of $\pi(\mathcal{S}')$. If Σ_0 is not a line of $\pi(\mathcal{S})$, then Σ_0 meets Σ in a line L, not in $\mathcal{S}$, and L determines a unique subset, $\mathcal{R}$, of $\mathcal{S}$ consisting of $q+1$ lines of $\mathcal{S}$, one through each point of L. It is now easy to see that the affine plane obtained from Σ_0 by deleting L and its points is isomorphic to a (Desarguesian) affine subplane of $\pi(\mathcal{S})$ of order q, consisting of the affine points of Σ_0 and the lines of $\pi(\mathcal{S})$ containing a line of $\mathcal{R}$ and at least one affine point of Σ_0. Now let P, Q, R be three distinct affine non-collinear points of Σ_0. Then θ must map the set of all affine points of Σ_0 upon the points of a Desarguesian affine subplane of order q of $\pi(\mathcal{S}')$ which contains the points $P\theta, Q\theta, R\theta$. One such subplane consists of the affine points of the unique plane Σ_0' containing $P\theta, Q\theta, R\theta$. If there is a second such subplane, it must intersect Σ_0' in an affine subplane whose order properly divides q.—Thus $U(q)$ holds if q is a prime.—In any case, if $U(q)$ holds, θ maps Σ_0 upon Σ_0'.

It is now easy to see that θ maps lines of Σ' (not in Σ) upon lines of the same sort. Hence θ may be extended uniquely to a collineation θ' of Σ', and θ' induces a collineation of Σ which maps $\mathcal{S}$ upon $\mathcal{S}'$. This completes the proof of Theorem 3.1.

$U(q)$ can be rephrased as follows in terms of Veblen-Wedderburn systems:

$U'(q)$. *For every Veblen-Wedderburn system R of order q^2 containing a field F of order q such that*

$$f(x+y) = fx + fy\,, \qquad f(xy) = (fx)y \tag{3.1}$$

for all f in F and x, y of R, the only subfield of R of order q is F.

If it should turn out that $U(q)$ is false for some q, this fact will be annoying rather than serious: we will merely be forced to classify translation planes of order q^2 in terms of Bose's construction in higher dimensions.

4. DOUBLY-RULED QUADRICS; REGULAR SPREADS

We shall begin with a rapid review of some well-known properties of $\Sigma = PG(3, q)$. If one or two of our remarks should prove unfamiliar, their proofs will be obvious from the sequel.

By an ordered double-triple

$$(A, B, C\,;\; A', B', C') \tag{4.1}$$

of Σ we mean an ordered set of 6 distinct lines of Σ made up of a triple of skew lines A, B, C and a triple of skew lines A', B', C' such that each line of one triple is a transversal to the other triple (that is, meets each line of the other triple in a point). Given A, B, C we may choose A', B', C' as any three distinct transversals of A, B, C. The projective linear group (of collineations) of Σ is strictly transitive on the ordered double-triples.

Given a double-triple (4.1), a fourth transversal, D, to A', B', C' and a fourth transversal, D', to A, B, C, then D meets D' in a point. Hence the set

$$\mathfrak{R} = \mathfrak{R}(A, B, C) \tag{4.2}$$

consisting of the $q + 1$ transversals (including A, B, C) to A', B', C' depends only on A, B, C and not on the choice of A', B', C'. We call $\mathfrak{R}$ the *regulus* containing A, B, C. If A_1, B_1, C_1 are any three distinct lines of $\mathfrak{R}$, then $\mathfrak{R} = \mathfrak{R}(A_1, B_1, C_1)$. By the *opposite regulus*, $\mathfrak{R}'$, we mean

$$\mathfrak{R}' = \mathfrak{R}(A', B', C') . \tag{4.3}$$

Clearly, $(\mathfrak{R}')' = \mathfrak{R}$.

By the *doubly-ruled quadric*

$$Q = Q(A, B, C) = Q(A', B', C') = Q(\mathfrak{R}) = Q(\mathfrak{R}') \tag{4.4}$$

we mean the set of $(q + 1)^2$ distinct points of Σ lying on the lines of $\mathfrak{R}$ and $\mathfrak{R}'$. A line, L, of Σ may be classified relative to Q according to the number, n, of distinct points of Q contained in L: a *ruling* $(n = q + 1)$; an *ordinary tangent* $(n = 1)$; a *secant* $(n = 2)$; or a *nonsecant* $(n = 0)$. There are $2(q + 1)$ rulings (the lines of $\mathfrak{R}$ and $\mathfrak{R}'$), $(q + 1)^2(q - 1)$ ordinary tangents, $(q + 1)^2q^2/2$ secants and $q^2(q - 1)^2/2$ nonsecants. A point P of Q lies on exactly two rulings, one in $\mathfrak{R}$ and one in $\mathfrak{R}'$, and these rulings lie in the *tangent plane* to Q at P. A plane of Σ either is a tangent plane to Q at a unique point of Q or contains no rulings.

By a *polarity* of Σ we mean a one-to-one, incidence-preserving mapping of Σ of order two which maps points, lines, planes upon planes, lines, points respectively. To each doubly-ruled quadric Q of Σ there corresponds a polarity ρ of Σ uniquely defined by the requirement that, for every point P of Q, $P\rho$ is the tangent plane to Q at P. A line L of Σ is called *self-conjugate* (with respect to Q) if $L\rho = L$; otherwise $L, L\rho$ form a *conjugate pair*. The secants split into conjugate pairs of skew lines; and so do the nonsecants; the rulings are self-conjugate. If q is even, the ordinary tangents are self-conjugate; if q is odd, they split into conjugate pairs of intersecting lines.

One cautionary remark is perhaps in order at this point. Let Q be a doubly-ruled quadric with polarity ρ. If q is odd, the lines L such that $L\rho = L$ are simply the rulings of Q, whence Q is uniquely determined by ρ. However (as we shall point out more explicitly in Appendix I) if q is even, there always exist several doubly-ruled quadrics distinct from Q with the same polarity ρ.

By an *involution* of Σ we mean a collineation of Σ of order two.

We recall that a *spread*, $\mathcal{S}$, of Σ is a set of q^2+1 skew lines of Σ, one through each point of Σ. We call a spread, $\mathcal{S}$, *regular* provided that, for every line L of Σ which is not in $\mathcal{S}$, the $q+1$ lines of $\mathcal{S}$ which meet L form a regulus of Σ. Equivalently, for every three distinct lines A, B, C of $\mathcal{S}$, $\mathcal{S}$ contains all the lines of the regulus $\mathcal{R}(A, B, C)$. We note that if a spread $\mathcal{S}$ contains a regulus $\mathcal{R}$, then $\mathcal{S}$ also contains q^2-q distinct nonsecants of the quadric $Q(\mathcal{R})$. Now we may state the following theorems, the proofs of which are given in Appendix I:

Theorem 4.1. *The projective linear group $PL(\Sigma)$ of $\Sigma = PG(3, q)$ is strictly transitive on the ordered double-triples* (4.1).

Corollary. *$PL(\Sigma)$ is transitive on the doubly-ruled quadrics of Σ.*

Theorem 4.2. *Let Q be a doubly-ruled quadric of $\Sigma = PG(3, q)$ with reguli $\mathcal{R}$ and $\mathcal{R}'$. Let $K = K(Q)$ be the subgroup of $PL(\Sigma)$ which maps Q upon Q and let $K_0 = K_0(Q)$ be the subgroup of K which maps each of $\mathcal{R}, \mathcal{R}'$ upon itself. Then K_0 is a (normal) subgroup of index* 2 *in K, and $K = K_1 \otimes K_2$ where $K_1(K_2)$ is the subgroup of K which maps $\mathcal{R}(\mathcal{R}')$ upon itself and fixes every line of $\mathcal{R}'(\mathcal{R})$. In particular, each of K_1, K_2 is isomorphic to the projective linear group of the line $PG(1, q)$; that is, to $LF(2, q)$.*

Theorem 4.3. *Let $\mathcal{R}$ be a regulus of $\Sigma = PG(3, q)$ and let D be a nonsecant of the doubly-ruled quadric $Q = Q(\mathcal{R})$. Then there exists one and only one regular spread $\mathcal{S}$ of Σ containing $\mathcal{R}$ and D. The q^2-q lines of $\mathcal{S}$ which are not in $\mathcal{R}$ split into $(q^2-q)/2$ pairs of conjugate nonsecants of Q.*

Corollary. *There are exactly $(q^2-q)/2$ regular spreads of $\Sigma = PG(3, q)$ containing a given regulus $\mathcal{R}$. Each two of these have only $\mathcal{R}$ in common.*

Theorem 4.4. *The projective linear group $PL(\Sigma)$ of $\Sigma = PG(3, q)$ is transitive on the regular spreads of Σ.*

Theorem 4.5. *Let $\mathcal{S}$ be a regular spread of $\Sigma = PG(3, q)$. Let $G = G(\mathcal{S})$ be the subgroup of $PL(\Sigma)$ which maps $\mathcal{S}$ upon $\mathcal{S}$, and let $N = N(\mathcal{S})$ be the (normal) subgroup of G which fixes each line of $\mathcal{S}$. Then:*

(i) *N is cyclic of order $q + 1$. Moreover, for each line, A, of $\mathcal{S}$, N is (strictly) transitive on the points of A.*

(ii) *G contains a unique (normal) subgroup $G_0 = G_0(\mathcal{S})$ of index 2 which is strictly transitive on the ordered quadruples*

$$(A, B, C;\ T) \tag{4.5}$$

consisting of an ordered triple A, B, C of three distinct (hence skew) lines of $\mathcal{S}$ together with a transversal, T, to A, B, C.

(iii) *Given such a quadruple (4.5), let $\mathcal{R}$ be the regulus $\mathcal{R}(A, B, C)$ and let ρ be the polarity defined by the doubly-ruled quadric $Q = Q(\mathcal{R})$. There exists a unique involution θ in G which fixes the ordered quadruple (4.5); and θ induces ρ on $\mathcal{S}$. Moreover, the coset $N\theta$ is the set of all collineations in G inducing ρ on $\mathcal{S}$.*

(iv) *Let $PG(1, q^2)$ be a projective line over $GF(q^2)$, and let $IP(q)$ be the inversive plane consisting of the points of $PG(1, q^2)$ and of the sub-lines of $PG(1, q^2)$ of order q, regarded as circles, under the natural incidence. Then there exists an isomorphism α of $\mathcal{S}$ upon $IP(q)$ which maps lines upon points, reguli upon circles and preserves incidence. The group $\alpha^{-1}N\alpha$ is the set of all elements of $\alpha^{-1}G\alpha$ inducing the identity on $IP(q)$. The group $\alpha^{-1}G_0\alpha$ is the projective linear group of $PG(1, q^2)$. In the notation of (iii), $\alpha^{-1}\theta\alpha$ is the inversion with respect to the circle $\mathcal{R}\alpha$; that is, the unique involution of $PG(1, q^2)$ fixing every point of the sub-line $\mathcal{R}\alpha$.*

Theorem 4.6. *Let Q be a doubly-ruled quadric of $\Sigma = PG(3, q)$ with reguli $\mathcal{R}, \mathcal{R}'$. Let $\mathcal{S}, \mathcal{S}'$ be regular spreads of Σ containing $\mathcal{R}, \mathcal{R}'$ respectively. In the notation of Theorem 4.5, let $N = N(\mathcal{S})$, $N' = N(\mathcal{S}')$. Then:*

(i) *$NN' = N \otimes N'$ is an abelian group of order $(q + 1)^2$.*

(ii) *$\mathcal{S}, \mathcal{S}'$ have exactly two common lines, namely a pair of conjugate nonsecants L, L' of Q.*

(iii) *The points of Σ split into $q + 1$ orbits under NN', namely the $q + 1$ points of L, the $q + 1$ points of L', and the*

$(q+1)^2$ *points of each of* $q-1$ *(disjoint) quadrics* Q_i, *one of which is* Q. *Moreover, for each* $i = 1, 2, \ldots, q-1$.

(iv) L, L' *are conjugate nonsecants of* Q_i.

(v) *One regulus,* $\mathcal{R}_i$, *of* Q_i *is in* S *and the opposite regulus,* $\mathcal{R}_i'$, *is in* S'. *Finally,*

(vi) *The lines of* S *split into* $q+1$ *distinct orbits under* N', *namely the fixed line* L, *the fixed line* L' *and the* $q-1$ *reguli* $\mathcal{R}_i$. *Similarly:*

(vii) *The orbits of the lines of* S' *under* N *are* L, L' *and the* $\mathcal{R}_i'$.

Theorem 4.7. *Let* S *be a regular spread of* $\Sigma = PG(3, q)$ *and let* A, B *be two distinct lines of* S. *Then the* $q^2 - 1$ *lines of* S *distinct from* A, B *may be partitioned into* $q-1$ *disjoint reguli* $\mathcal{R}_i$ *uniquely defined, apart from order, by the requirement that, for each* i, A, B *are conjugate nonsecants of the doubly ruled quadric* $Q_i = Q(\mathcal{R}_i)$. *The line-set,* S', *obtained from* S *by replacing each of the* $q-1$ *reguli* $\mathcal{R}_i$ *by its opposite regulus* $\mathcal{R}_i'$, *is a regular spread of* Σ.

As stated above, the proofs of these theorems will be given in Appendix I.

5. $PG(3, q)$ IMBEDDED IN $PG(3, q^2)$

Projective 3-space $\Sigma = PG(3, q)$ can always be imbedded as a sub-3-space of order q in $\Sigma^* = PG(3, q^2)$. An analysis of this imbedding will throw additional light on the material of the preceding section.

Lemma 5.1. *Let* L *be a line of* $\Sigma^* = PG(3, q^2)$ *which is not in the sub-3-space* $\Sigma = PG(3, q)$. *Then exactly one of the following is true:*

(a) L *contains exactly one point of* Σ *and* L *lies in exactly one plane of* Σ.

(b) L *contains no point of* Σ *and lies in no plane of* Σ.

Proof. Since L is not in Σ, L contains at most one point of Σ and L lies in at most one plane of Σ.

First assume that L contains a (unique) point P of Σ. If A is a line of Σ through P, then $A \neq L$, whence there exists a unique plane (A, L) of Σ^* containing A and L. If no plane of Σ contains L, then, as A ranges over the q^2+q+1 lines of Σ through P, (A, L) ranges over q^2+q+1 distinct planes of Σ^* containing L. But L is contained in precisely q^2+1 distinct planes of Σ^*, a contradiction. Hence L is contained in a (unique) plane of Σ.

A dual argument shows that if L is contained in a (unique) plane of Σ, then L contains a (unique) point of Σ. This proves Lemma 5.1.

Lemma 5.2. *Let P be a point of $\Sigma^* = PG(3, q^2)$ which is not in the sub-3-space $\Sigma = PG(3, q)$. Then P lies on*

(i) *exactly one line of Σ;*

(ii) *exactly q^3+q^2 lines containing one point of Σ;*

(iii) *exactly q^4-q^3 lines containing no points of Σ.*

Proof. Since P is not in Σ, P lies on at most one line of Σ. Hence P lies on a line L which contains exactly one point of Σ. By Lemma 4.1, L lies in a unique plane, π, of Σ. Of the lines through P which are not in π, at least one, say L', contains exactly one point of Σ, and L' lies in a unique plane, π', of Σ. The planes π, π' are distinct and intersect in a line, A, of Σ through P. The line A contains $q+1$ points of Σ. The remaining q^3+q^2 points of Σ lie, one each, on lines through P. Thus, since P lies on exactly q^4+q^2+1 lines of Σ^*, the proof of Lemma 4.2 is complete.

Theorem 5.3. *Let $\Sigma = PG(3, q)$ be imbedded as a sub-3-space of $\Sigma^* = PG(3, q^2)$. Let τ be the unique involution of Σ^* which fixes every point of Σ. Let L be any line of Σ^* which contains no point of Σ. For each such line, L, let $\mathcal{S}(L)$ denote the set of all lines of Σ which meet L. Then:*

(i) *$\mathcal{S}(L) = \mathcal{S}(L\tau)$ is a regular spread of Σ. Every regular spread of Σ can be represented in this manner for a unique pair of lines L, $L\tau$.*

(ii) *If P, Q, R are three distinct points of L, and if A, B, C are the unique lines of Σ through P, Q, R, respectively, then the $q+1$ lines of the regulus $\mathcal{R}(A, B, C)$ of Σ meet L in the points of the unique projective sub-line of order q of L which contains P, Q, R.*

(iii) *If two distinct lines A, B of $\mathcal{S}(L)$ meet L in points P, Q respectively, and if M is the line of Σ^* containing P and $Q\tau$, then M contains no points of Σ. The regular spreads $\mathcal{S}(L)$, $\mathcal{S}(M)$ of Σ have the following properties:* (a) *A, B are the* *only common lines of* $\mathcal{S}(L)$, $\mathcal{S}(M)$. (b) *To each point* R *of* Σ *which is not on* A *or* B *there corresponds a unique doubly-ruled quadric* Q *of* Σ *containing* R *such that one regulus of* Q *is in* $\mathcal{S}(L)$ *and the other is in* $\mathcal{S}(M)$. *And* A, B *are conjugate nonsecants (in* Σ*) of* Q.

Remark. The proof of Theorem 5.3 is independent of the theorems of Section 4, except that the proof of the last sentence of (i) and the last sentence of (iii) depend on Theorem 4.3.

Proof. (i) Let P, Q be two distinct points of L. By Lemma 5.2, P lies on a unique line, A, of Σ and Q lies on a unique line, B, of Σ. If A meets B, then A, B lie in a plane of Σ containing L, in contradiction to Lemma 5.1. Hence A, B are skew. Thus $\mathcal{S}(L)$ consists of q^2+1 skew lines of Σ, and therefore is a spread of Σ. Now let A, B, C be three distinct lines of $\mathcal{S}(L)$, A', B', C' be three distinct transversals to A, B, C in Σ, and D be a fourth transversal to A', B', C' in Σ. Then, viewed in Σ^*, L is a fourth transversal to A, B, C so L meets D. That is, every line of the regulus $\mathcal{R}(A, B, C)$ of Σ is in $\mathcal{S}(L)$. Hence $\mathcal{S}(L)$ is regular.

Conversely, let $\mathcal{S}$ be a regular spread of Σ, let $\mathcal{R}$ be a regulus in $\mathcal{S}$, and let D be a line of $\mathcal{S}$ which is not in $\mathcal{R}$. Thus D is a nonsecant of the quadric $Q = Q(\mathcal{R})$. Let $\mathcal{R}^*$ be the unique regulus of Σ^* containing the regulus $\mathcal{R}$ of Σ, so that the opposite regulus of $\mathcal{R}^*$ is the unique regulus $(\mathcal{R}')^*$ of Σ^* containing $\mathcal{R}'$. Then D meets the quadric $Q^* = Q(\mathcal{R}^*)$ in two distinct points $P, P\tau$ not in Σ. The lines $L, L\tau$ of $(\mathcal{R}')^*$ through $P, P\tau$ respectively can contain no points of Σ—else they would be in $\mathcal{R}'$ and would meet D in points of Σ. Hence $\mathcal{S}(L) = \mathcal{S}(L\tau)$ is a regular spread of Σ containing $\mathcal{R}$ and D. Assuming Theorem 4.3, we conclude that $\mathcal{S} = \mathcal{S}(L) = \mathcal{S}(L\tau)$. This proves (i).

(ii) This becomes clear when we make a suitable introduction of coordinates. The details will be omitted.

(iii) If M contains at least one point of Σ, then, by Lemma 5.1, M lies in at least one plane π of Σ. But then π contains $P, Q\tau$, hence also $P\tau, Q$, and therefore the lines $L, L\tau$; a contradiction to Lemma 5.1. Therefore M (and $M\tau$) contains no point of Σ. For like reasons, the lines $L, L\tau$ are skew; and the lines $M, M\tau$ are skew. The lines L, M intersect in P and hence lie in a unique plane LM of Σ^*. Since LM contains L, then LM is not in Σ; hence (by the space dual of Lemma 5.2) LM contains a unique line of Σ—and this we recognize to be B. The only other line of Σ meeting both of L, M is that through P, namely A. This proves (a). Now consider a point R of Σ, not on A or B, and let C, C' be the unique lines of $\mathcal{S}(L), \mathcal{S}(M)$ respectively through R. Then

$$(5.1) \qquad (L, L\tau, C'; M, M\tau, C)$$

is an ordered double-triple of Σ^*. Consider any point R_1 of Σ which is on C', and let K be the unique transversal to $L, L\tau, C'$ through R_1. Then $K\tau$ is also a transversal to $L, L\tau, C'$ through R_1, whence $K\tau = K$. Thus, if K meets L in R_2, then K meets $L\tau$ in $R_2\tau$, whence we see that K must be the unique line of Σ through R_2. By this and by symmetry, we see that the set, $\mathcal{R}$, of all transversals to $L, L\tau, C'$ which meet C' in a point of Σ, is a regulus of Σ contained in $\mathcal{S}(L)$, that the opposite regulus $\mathcal{R}'$ consists of all transversals to $M, M\tau, C$ which meet C in a point of Σ, and that $\mathcal{R}'$ is contained in $\mathcal{S}(M)$. By appeal to Theorem 4.3 we see that A, B are conjugate nonsecants of the doubly-ruled quadric $Q = Q(\mathcal{R}) = Q(\mathcal{R}')$ of Σ. This proves (b) and completes the proof of Theorem 5.3.

If, in the notation of Theorem 5.3, $\mathcal{S} = \mathcal{S}(L)$, various parts of Theorem 4.5 can be explained simply as follows: $G = G(\mathcal{S})$ is (isomorphic to) that subgroup of $PL(\Sigma^*)$ which permutes the ordered double-triples belonging to the types

$$(5.2) \qquad (A, B, C;\ T, L, L\tau)\ ,$$

$$(5.3) \qquad (A, B, C;\ T, L\tau, L)\ ,$$

where, in each case, A, B, C are three distinct lines of $\mathcal{S}$ and T is a transversal, in Σ, to A, B, C. G_0 is the subgroup of G which permutes the double-triples of type (5.2) among themselves. In (iv) of Theorem 4.5, $PG(1, q^2)$ can be taken to be L, with α the mapping which sends a line of $\mathcal{S}$ upon its intersection with L. In (iii), θ is that collineation in $PL(\Sigma^*)$ which maps a given ordered double-triple (5.2) upon the corresponding ordered double-triple (5.3).—Similarly, Theorems 4.6, 4.7 are intimately related to Theorem 5.3 (iii).

6. SUBREGULAR SPREADS

If $\mathcal{S}$ is a spread of $\Sigma = PG(3, q)$ containing a regulus $\mathcal{R}$, and if $\mathcal{S}'$ is the line-set obtained from $\mathcal{S}$ by replacing $\mathcal{R}$ by its opposite regulus $\mathcal{R}'$, we shall say more briefly that $\mathcal{S}'$ is obtained from $\mathcal{S}$ *by reversing the regulus* $\mathcal{R}$. We note that $\mathcal{S}'$ is a spread. Similarly, we may speak of (simultaneously) reversing several disjoint reguli of $\mathcal{S}$.

By a *subregular sequence of length k* we mean a sequence

$$\mathcal{S}_0, \mathcal{S}_1, \mathcal{S}_2, \ldots, \mathcal{S}_k \qquad (k \geq 0) \tag{6.1}$$

of spreads $\mathcal{S}_i$ of Σ such that $\mathcal{S}_0$ is regular and, for $1 \leq i \leq k$, $\mathcal{S}_i$ is obtained from $\mathcal{S}_{i-1}$ by reversing a regulus. We call a spread $\mathcal{S}$ of Σ *subregular of index k* if there exists a subregular sequence of length k ending in $\mathcal{S}$ but none of shorter length. For example, the regular spreads are the subregular spreads of index zero. Two questions of interest are these: (a) For given q, what indices k occur? (b) If in (6.1), $\mathcal{S}_k$ has index k, what conditions will insure that $\mathcal{S}_0$ is uniquely determined?

For $q = 2$, questions (a), (b) have an obvious answer: every spread is regular. For $q = 3$, every spread is subregular of index k where $k = 0$ or 1. The spreads with $k = 1$ are all equivalent (for $q = 3$) under the projective linear group, but each can be obtained from 10 different regular spreads by reversing a regulus.

The following notation will be convenient: a set of k disjoint reguli

$$\mathcal{R}_i \qquad (1 \leq i \leq k) \tag{6.2}$$

of a regular spread $\mathcal{S}$ will be called *linear* if there exist two distinct lines A, B in $\mathcal{S}$ which are conjugate nonsecants of each of the k doubly-ruled quadrics $Q(\mathcal{R}_i)$. Such a linear set will be called *complete* if $k = q - 1$. By Theorem 4.6, 4.7 or 5.3, a regular spread of Σ has complete linear sets of disjoint reguli. We shall see later that, for $q > 4$, a regular spread also has non-linear sets of three or more disjoint reguli.

Lemma 6.1. *Let* $\mathcal{S}$ *be a regular spread of* $\Sigma = PG(3, q)$ *and, for some positive integer k, let* (6.2) *be a set of k disjoint reguli of* $\mathcal{S}$. *Let* $\mathcal{S}'$ *be obtained from* $\mathcal{S}$ *by reversing the k reguli* (6.2). *Then:*

(i) $1 \leq k \leq q - 1$.

(ii) *If* $k < q - 1$, *then* $\mathcal{S}'$ *is not regular. More precisely:*

(iii) *A necessary and sufficient condition that* $\mathcal{S}'$ *be regular is that* (6.2) *be a complete linear set of disjoint reguli of* $\mathcal{S}$.

This implies that $k = q - 1$.

(iv) *If* $2k < q - 1$, *every regulus* $\mathcal{R}$ *of* $\mathcal{S}'$ *distinct from the* k *reguli*

$$\mathcal{R}'_i \qquad (1 \leq i \leq k) \tag{6.3}$$

is also a regulus of $\mathcal{S}$. *Hence* $\mathcal{R}$ *is disjoint from the reguli* (6.2) *and the reguli* (6.3).

Proof. (i) Since $\mathcal{S}$ has exactly

$$q^2 + 1 = (q - 1)(q + 1) + 2$$

lines and since each of the k disjoint reguli (6.2) consists of $q + 1$ lines of $\mathcal{S}$, then $k \leq q - 1$, with equality precisely when $\mathcal{S}$ has exactly two lines which are in none of the reguli (6.2). This proves (i).

(ii) Note that if $q = 2$, we cannot have $1 \leq k < q - 1$. We proceed to (iii).

(iii) If (6.2) is a complete linear set then $\mathcal{S}'$ is regular by Theorem 4.7. Now we assume, conversely, that $\mathcal{S}'$ is regular. Since $k \geq 1$, the hypotheses of Theorem 4.6 are fulfilled with $\mathcal{R} = \mathcal{R}_1$. Hence (6.2) is a complete linear set, and therefore, also, $k = q - 1$. This proves (iii).

(iv) Now assume $2k \leq q - 2$, and let $\mathcal{R}$ be a regulus of $\mathcal{S}'$ distinct from the k reguli (6.3). If $\mathcal{R}$ has three or more distinct lines in common with some $\mathcal{R}'_i$, then $\mathcal{R}$ coincides with $\mathcal{R}'_i$, a contradiction. Hence $\mathcal{R}$ contains at least

$$(q + 1) - 2k \geq 3$$

distinct lines common to $\mathcal{S}$ and $\mathcal{S}'$. Since $\mathcal{S}$ is regular, $\mathcal{R}$ is contained in $\mathcal{S}$. Hence $\mathcal{R}$ is common to $\mathcal{S}$ and $\mathcal{S}'$. Since no regulus in (6.2) has a line in common with a regulus in (6.3), $\mathcal{R}$ is disjoint from (6.2) and (6.3). This proves (iv) and completes the proof of Lemma 6.1.

Lemma 6.2. *Let* $\mathcal{S}_1, \mathcal{S}_2$ *be two distinct regular spreads of* $\Sigma = PG(3, q)$. *Let* $\mathcal{S}'$ *be a spread of* Σ *which can be obtained both from* $\mathcal{S}_1$ *by reversing a set* (6.2) *of* k *disjoint reguli of* $\mathcal{S}_1$ *and from* $\mathcal{S}_2$ *by reversing a set*

$$\mathcal{A}_j \qquad (j = 1, 2, \ldots, t) \tag{6.4}$$

of t *disjoint reguli of* $\mathcal{S}_2$, *where*

(6.5) $$1 \leq k \leq t \leq q-2\,.$$

Let w be the number of distinct lines in $\mathcal{S}_1 \cap \mathcal{S}_2 \cap \mathcal{S}'$. Of the t opposite reguli

(6.6) $$\mathcal{R}'_j \qquad (j = 1, 2, \ldots, t)$$

let c be included among the k opposite reguli (6.3) *and d be disjoint from the reguli* (6.3). *Then there are only the following possibilities:*

(i) $c = 1,\ \ d = w = 0,\ \ k = t = q-2,\ \ q \geq 9\,;$

(ii) $c = d = 0,\ \ 3 \leq w,\ \ q-3 \leq k \leq t \leq q-2 < 2k\,;$

(iii) $c = d = 0,\ \ k = t = q-2,\ \ 0 \leq 2 - w \leq q-5\,;$

(iv) $c = 0,\ \ 1 \leq d \leq k-1 < t,\ \ k + d = q-2,\ \ w \leq 2\,;$

(v) $c = 0,\ \ 1 \leq d \leq k,\ \ d < t,\ \ k + d = q-1,\ \ w \leq 2\,;$

(vi) $c = 0,\ \ d = t,\ \ k + t = q-1,\ \ w = 2\,.$

Case (vi) *occurs precisely when the reguli* (6.2), (6.6) *are distinct and form a complete linear set of $q-1$ disjoint reguli of $\mathcal{S}_1$, which yield $\mathcal{S}_2$ upon reversal. Moreover:*

(a) *If $2k < q-1$, case* (vi) *occurs.*

(b) *If q is odd and $2k = q-1$, either Case* (vi) *occurs or we have a special case of Case* (ii):

(6.7) $$q = 3,\ \ k = t = 1,\ \ c = d = 0,\ \ w = 4\,;$$

or we have a special case of Case (v):

(6.8) $$q > 3,\ \ 2k = q-1,\ \ c = 0,\ \ d = k,\ \ t = k+1,\ \ w = 0\,.$$

(c) *If $2t \leq q+1$ and $k < t$, then either Case* (vi) *occurs or we have* (6.8). *Thus either* (6.6) *is linear or* (6.6) *becomes linear on removal of a single regulus.*

Remark. Case (vi) may be regarded as normal, the other five cases as abnormal. The only abnormal example I know is (6.7). It would be of considerable interest to settle the existence of (6.8), particularly in view of (II. 1) in Lemma 7.23 (see the next section).

Proof. We begin by modifying the proof of Lemma 6.1 (iv). Suppose that $t > c + d$ and let $\mathcal{R}$ be a regulus in (6.4) such that $\mathcal{A}'$ is distinct from the reguli (6.3) but not disjoint from them. Now $\mathcal{A}'$ is disjoint from c of the reguli (6.3), namely those in (6.6), and has at most two lines in common with each of the $k - c$ others. Also $\mathcal{A}'$ is not entirely in $\mathcal{S}_1$, so $\mathcal{A}'$ has at most two lines in none of the reguli (6.3). Therefore

$$2(k - c) + 2 \geq q + 1\,, \quad \text{or} \quad 2(k - c) \geq q - 1\,,$$

with equality precisely when each of the counts of two is exact. In particular,

$$(6.9) \qquad t = c + d \quad \text{if} \quad 2(k - c) > q - 1\,.$$

Returning to the case $t > c + d$, we see that the $t - (c + d)$ reguli $\mathcal{A}'$ have at most

$$2(t - c - d)$$

distinct lines not in the union of (6.3). This count is exact if $2(k - c) = q - 1$ and also if $t = c + d$. None of these lines is in $\mathcal{S}_2$, and hence none is in $\mathcal{S}_1 \cap \mathcal{S}_2 \cap \mathcal{S}$. Adding to the above the number of lines in the reguli (6.3), in the d reguli of (6.6) disjoint from (6.3), and in $\mathcal{S}_1 \cap \mathcal{S}_2 \cap \mathcal{S}$, we get

$$2(t - c - d) + (k + d)(q - 1) + w \geq q^2 + 1$$

or, equivalently,

$$(6.10) \qquad (q - 1)\{(q - 1) - (k + d)\} \leq w - 2 + 2\{(k + t - c) - (q - 1)\}$$

with equality if $2(k - c) = q - 1$ *or if* $t = c + d$.

Now assume $c > 0$. Then (6.3), (6.6) have a regulus in common. Thus (6.2), (6.4) have a regulus in common. That is, $\mathcal{S}_1, \mathcal{S}_2$ have a common regulus. Since $\mathcal{S}_1, \mathcal{S}_2$ are distinct and regular, they have exactly one common regulus and no other lines in common. Thus $c = 1$, $w = 0$. If we assume that $d > 0$, we must conclude that $\mathcal{S}_2$ has a regulus whose opposite regulus is in $\mathcal{S}_1$, which implies that $\mathcal{S}_1, \mathcal{S}_2$ have exactly two common lines, a contradiction. Thus $d = 0$. From (6.10) with $c = 1$, $d = w = 0$ we get

$$(6.11) \qquad (q - 1)(q - 1 - k) \leq 2(k + t - q - 1) \quad \text{if} \quad c = 1\,.$$

By (6.5), the right-hand side of (6.11) is at most

$$2(q-5) < 2(q-1) \, .$$

Hence $q-1-k \leq 1$, whence, by (6.5), $k = t = q-2$. From this in (6.11), $q-1 \leq 2(q-5)$ or $q \geq 9$. Thus we have (i).

Until further notice we assume $c = 0$.

If $w \geq 3$, then $\mathcal{S}_1, \mathcal{S}_2$ have a common regulus. As above, this forces $d = 0$, which conflicts with (6.9). Thus $2k \geq q-1$. Also, $w \leq q+1$. Then, from (6.10), (6.5),

$$\begin{aligned}(q-1)(q-1-k) &\leq w-2+2(k+t-q+1) \\ &\leq q-1+2(q-3) < 3(q-1) \, .\end{aligned}$$

Hence $q-3 \leq k \leq t \leq q-2 < 2k$, which gives (ii). If we add the assumption $2k = q-1$ and use equality in (6.10), we get only (6.7).

If $w \leq 2$ and $d = 0$, whence $2k \geq q-1$, then (6.10), (6.5) yield

$$\begin{aligned}(q-1)(q-1-k) &\leq w-2+2(k+t-q+1) \\ &\leq 2(q-3) < 2(q-1) \, ,\end{aligned}$$

whence $q-2 \leq k \leq t \leq q-2$. Setting $k = t = q-2$ in (6.10), we get $2-w \leq q-5$. This gives (iii). The added assumption $2k = q-1$ yields $q = 3$ and $2-w = -2$, a contradiction.

In the remaining cases, $d > 0$. This implies that $\mathcal{S}_2$ has a regulus opposite to one in $\mathcal{S}_1$, whence $w \leq 2$. Now (6.10) yields

$$\begin{aligned}(q-1)\{q-1-(k+d)\} &\leq w-2+2\{(k+t)-(q-1)\} \\ &\leq 2\{(k+t)-(q-1)\} \\ &\qquad \text{if} \quad c=0,\ d>0 \, .\end{aligned} \tag{6.12}$$

Also, from (6.9) and the discussion of (i),

$$c = 0, \quad d = 1 \quad \text{if} \quad 2k < q-1 \, . \tag{6.13}$$

Our next two cases come by assuming $d < t$ and hence, also, $2k \geq q-1$. From (6.5), the right-most term of (6.12) is at most $2(q-3) < 2(q-1)$. Hence $q-1-(k+d) < 2$ and

$$q-2 \leq k+d \leq q-1 \, ,$$

the last inequality following since $\mathcal{S}'$ has $k+d$ disjoint reguli. Thus either $k+d = q-2 < 2k$, which gives (iv), or $k+d = q-1 \leq 2k$ and $d < t$, which gives (v). If we try $2k = q-1$

in connection with (iv), we get a contradiction, whereas (v) yields (6.8).

Finally we take $d = t$. Then, clearly, all three parts of (6.12) must be zero. This gives (vi). Then (6.6), (6.2) together yield a set of $t + k = q - 1$ distinct and disjoint reguli of $\mathcal{S}_1$, which on reversal yield $\mathcal{S}_2$. Therefore by Lemma 6.1 (iii), these reguli form a (complete) linear set of disjoint reguli of $\mathcal{S}_1$.

We have already proved (a), (b). As for (c), if $k < t$ and $2t \leq q + 1$ then $2k \leq 2(t - 1) \leq q - 1$. Thus, by (a), (b) we either have (vi) or (6.7), (6.8). However, (6.7) conflicts with our assumption that $k < t$. If (6.7) holds, the $d = t - 1 \geq 2$ reguli in (6.6) disjoint from (6.3) are in $\mathcal{S}_1$, and the corresponding opposite reguli are in $\mathcal{S}_2$. Hence the latter form a linear set derived from (6.4) by deleting one regulus, though we do not know that (6.4) is linear.

This completes the proof of Lemma 6.2.

At this point let us introduce a simple concept which will perhaps clarify our thinking. Let $\mathcal{S}_1, \mathcal{S}_2$ be any two spreads of $\Sigma = PG(3, q)$, not necessarily distinct. We define

$$\Theta = \Theta(\mathcal{S}_1, \mathcal{S}_2) \tag{6.14}$$

to be the set of all reguli $\mathcal{R}$ of Σ such that $\mathcal{R}$ is contained in $\mathcal{S}_1$ and the opposite regulus, $\mathcal{R}'$, is contained in $\mathcal{S}_2$. Thus

$$\Theta' = \Theta(\mathcal{S}_2, \mathcal{S}_1) \tag{6.15}$$

is obtained from Θ by reversing all the reguli in Θ.

Lemma 6.3. *Let $\mathcal{S}_1, \mathcal{S}_2$ be spreads of $\Sigma = PG(3, q)$, and let k be the number of reguli in $\Theta = \Theta(\mathcal{S}_1, \mathcal{S}_2)$, v be the number of lines in $\mathcal{S}_1 \cap \mathcal{S}_2$. Then Θ consists of k disjoint reguli of $\mathcal{S}_1$, where $0 \leq k \leq q - 1$, and*

$$v - 2 \leq (q + 1)(q - 1 - k)\,, \tag{6.16}$$

with equality precisely when $\mathcal{S}_2$ comes from $\mathcal{S}_1$ by replacing Θ by $\Theta' = \Theta(\mathcal{S}_2, \mathcal{S}_1)$.

Proof. Let L be a line of $\mathcal{S}_1$ and let $\mathcal{R}$ be a regulus of Σ containing L. Then $\mathcal{R}'$ consists of $q + 1$ skew lines, one through each point of L. If L is also in $\mathcal{S}_2$, no line of $\mathcal{R}'$ is in $\mathcal{S}_2$, whence $\mathcal{R}$ is not in Θ. If L is not in $\mathcal{S}_2$ then L meets a unique set, $\mathcal{A}$, of $q + 1$ distinct lines of $\mathcal{S}_2$, one through each point of L. Thus $\mathcal{R}$ is in Θ if and only if (i) $\mathcal{A}$ is a regulus of Σ, (ii) the opposite regulus, $\mathcal{A}'$, is contained in $\mathcal{S}_2$ and (iii) $\mathcal{R} = \mathcal{A}'$. (Note that, if $\mathcal{S}_2$ is regular, condition (i) is trivial.) Thus Θ consists of k disjoint reguli of $\mathcal{S}_1$, for some integer $k \geq 0$. If w is the number of lines of $\mathcal{S}_1$ which are neither in $\mathcal{S}_2$ nor in a regulus of Θ, then

$$v + w + k(q + 1) = q^2 + 1 = 2 + (q - 1)(q + 1)$$

and

$$v - 2 + w = (q + 1)(q - 1 - k) .$$

This proves Lemma 6.3.

Lemma 6.4. *Let $\mathcal{S}_0$, $\mathcal{S}'$ be spreads of $\Sigma = PG(3, q)$, $q > 2$. Assume that $\mathcal{S}_0$ is regular and that there exists a subregular sequence of length t leading from $\mathcal{S}_0$ to $\mathcal{S}'$, but none of shorter length. Assume further that $2t \leq q$. Then*

(i) *$\Theta = \Theta(\mathcal{S}_0, \mathcal{S}')$ consists of t disjoint reguli of $\mathcal{S}_0$ whose reversal yields $\mathcal{S}'$.*

(ii) *If $2t \leq q - 1$, then $\mathcal{S}'$ has index t.*

(iii) *If $2t = q$, then $\mathcal{S}'$ has index $t - 1$ or t according to whether Θ is linear or nonlinear.*

Remark. That the possibilities envisaged in Lemma 6.4 can all occur will be shown in Section 7, which follows. In particular, that we can have $2t = q$ with Θ nonlinear follows from I. 2 in Lemma 7.23 below.

Proof. We prove (i) by induction on t. Certainly (i) is true for $t = 0$. If $t > 0$, set $k = t - 1$, so that $2k \leq q - 2$. We may suppose that (6.1) is a subregular sequence of length k leading from $\mathcal{S}_0$ to $\mathcal{S}_k$, and that $\mathcal{S}'$ comes from $\mathcal{S}_k$ by reversing a regulus $\mathcal{R}$ of $\mathcal{S}_k$. By the minimality of t, there is no shorter subregular sequence from $\mathcal{S}_0$ to $\mathcal{S}_k$. Hence we may assume inductively that $\Theta(\mathcal{S}_0, \mathcal{S}_k)$ consists of k disjoint reguli whose reversal gives $\mathcal{S}_k$. Thus $\mathcal{S}_k$ consists of $\mathcal{S} \cap \mathcal{S}_k$ and the set of k reguli $\Theta(\mathcal{S}_k, \mathcal{S}_0)$. If $\mathcal{R}$ is in $\Theta(\mathcal{S}_k, \mathcal{S}_0)$, clearly we can get a shorter sequence from $\mathcal{S}_0$ to $\mathcal{S}'$ in which $\mathcal{R}$ and $\mathcal{R}'$ are not interchanged. Hence $\mathcal{R}$ is not in $\Theta(\mathcal{S}_k, \mathcal{S}_0)$. Since $2k < q - 1$, we conclude that $\mathcal{R}$ is in $\mathcal{S}_0$ and is disjoint from $\Theta(\mathcal{S}_0, \mathcal{S}_k)$. Thus $\Theta = \Theta(\mathcal{S}_0, \mathcal{S}')$ is the union of $\Theta(\mathcal{S}_0, \mathcal{S}_k)$ and $\mathcal{R}$. This is enough for the proof of (i).

Now assume that $\mathcal{S}'$ is subregular of index k for some integer k with $k < t$. Then, by (a) of Lemma 6.2, with $\mathcal{S}_0$ taking the role of $\mathcal{S}_2$, $k + t = q - 1$, and Θ is linear. If $2t \leq q - 1 = k + t$, then $t \leq k$, a contradiction. This proves (ii). If $2t = q = k + t + 1$, then $k = t - 1$. In this case, we have a contradiction if Θ is nonlinear. On the other hand, if Θ is linear, then, since $2t = q > 2$ and hence $t > 1$, there is a unique regular spread containing Θ', namely that containing one regulus in Θ' and any line of a second regulus. With this spread as $\mathcal{S}_2$, we get the situation of Lemma 6.2 with $k = t - 1$. This proves (ii) and completes the proof of Lemma 6.4.

Lemma 6.4 tells us that, in order to characterize the subregular spreads of $PG(3, q)$ of index not exceeding $q/2$, we need not concern ourselves with the whole space $PG(3, q)$ but only with the structural properties of a single regular spread $\mathcal{S}$ and, in particular, with sets of disjoint reguli of $\mathcal{S}$. Moreover, by Theorem 5.3 (ii) or Theorem 4.5 (iv), the lines and reguli of $\mathcal{S}$ constitute (isomorphically speaking) the points and circles, respectively, of the inversive plane $IP(q)$ formed from the projective line $PG(1, q^2)$. In addition, Theorem 4.5 (iv) dictates the appropriate group of collineations of $IP(q)$ with which to concern ourselves. Accordingly, in the section which follows, we make an intensive study of the sets of disjoint circles of $IP(q)$, classifying them with respect to the appropriate group.

7. THE INVERSIVE PLANE $IP(q)$

By an *inversive plane* (German: *Möbiusebene*) we mean a set, M, of objects called points and a collection of subsets of M, called circles such that (i) every three distinct points of M lie on exactly one circle of M; (ii) given a circle C of M, a point P on C, and a point Q of M which is not on C, there exists exactly one circle, C', of M such that C' contains P, Q and has only P in common with C; (iii) every circle of M is non-empty and there exist four points of M not lying on any circle of M.

Let M be an inversive plane. We say M has *order* q (where q is a positive integer) if some circle of M has exactly $q + 1$ distinct points. If C, C' are distinct circles of M, we call C, C' *secant*, *tangent* or *disjoint* according as they have two, one or no common points. If P is a point of M, we define $M(P)$ to be the system consisting of the points of M distinct from P

and the circles of M through P, considered as lines. By a *collineation* of M we mean a permutation of the points of M which maps circles upon circles. An *involution* is a collineation of order two. If C is a circle of M and if θ is a collineation of M which fixes every point of C but is not the identity collineation, we call θ an *inversion* in C. For the next two lemmas we refer to Lüneburg [4].

Lemma 7.1. *If M is an inversive plane then $M(P)$ is an affine plane for every point P of M. If, in addition, M has finite order q then $q \geq 2$ and:*

(a) *M has exactly $q^2 + 1$ distinct points. Each circle of M contains exactly $q + 1$ distinct points. Every three distinct points of M lie on exactly one circle of M.*

(b) *Every two distinct points of M lie on exactly $q + 1$ distinct circles of M. Every point of M lies on exactly $q^2 + q$ distinct circles of M. The number of distinct circles of M is $q^3 + q$.*

(c) *If C is a circle of M and if P, Q are points of M with P on C, Q not on C, then, of the $q + 1$ distinct circles of M through P and Q, q are secants of C and one is tangent to C.*

(d) *If C is a circle of M and P is a point of M on C, then the tangent pencil, consisting of C and the circles through P tangent to C, is a set of q distinct, mutually tangent circles, exactly one through each point of M distinct from P.*

(e) *If C is a circle of M and if Q is a point of M not on C, then, of the $q^2 + q$ distinct circles of M through Q, $(q^2 + q)/2$ are secants of C, $q + 1$ are tangents to C, and $(q^2 - q)/2$ are disjoint from C.*

(f) *If C is a circle of M, then, of the $q^3 + q - 1$ distinct circles of M other than C, $(q^3 + q^2)/2$ are secants of C, $q^2 - 1$ are tangents to C, and $q(q - 1)(q - 2)/2$ are disjoint from C.*

Remark. If we assume (a) for some integer $q \geq 2$, we can prove (b) through (f) by elementary counting arguments. In this case, by (a), (c) and the last sentence of (b), M is an inversive plane.

Lemma 7.2. *Let M be an inversive plane and let C be a circle of M. Then:*

(i) *If θ is an inversion of M in C then (a) θ is an involution and (b) θ fixes no points of M except the points of C.*

(ii) *There is at most one inversion, of M in C.*

Now we are ready to discuss $IP(q)$, where q is a prime-power. The projective line $L = PG(1, q^2)$ may be represented as a 2-dimensional vector space V over the finite field $GF(q^2)$. The points of L are the one-dimensional vector subspaces of V over $GF(q^2)$. If P, Q, R are three distinct points of L and if e is a basis vector for P over $GF(q^2)$, there exists one and only one basis vector, e', for Q over $GF(q^2)$ such that $e + e'$ is a basis vector for R over $GF(q^2)$. We define

$$(7.1) \qquad P(\infty) = \{e\} , \qquad P(x) = \{xe + e'\}$$

for every x in $GF(q^2)$, where $\{w\}$ denotes the subspace of V over $GF(q^2)$ spanned by the vector w. Thus

$$(7.2) \qquad P(\infty) = P , \qquad P(0) = Q , \qquad P(1) = R .$$

Also, as x ranges over $GF(q^2)$, $P(x)$ ranges over the q^2 points of L distinct from P. Replacement of e by ke, where k is a non-zero element of $GF(q^2)$, will preserve (7.2) provided e' is replaced by ke'. Moreover, the unique projective sub-line of L of order q which contains P, Q and R is the set of points $P(\infty)$ and $P(x)$ where x is restricted to range over $GF(q)$. In affine terms, we may represent this sub-line as

$$(7.3) \qquad C_0 : \infty \quad \text{and} \quad GF(q)$$

and call it a circle. Now let $IP(q)$ consist of the $q^2 + 1$ points of $L = PG(1, q^2)$ and of the projective sublines of order q, regarded as circles. Then (a) of Lemma 7.1 holds. Hence $IP(q)$ is an inversive plane of order q. Next consider the mapping θ of our vector space $V = \{e, e'\}$ defined by

$$(7.4) \qquad (xe + ye')\theta = x^q e + y^q e'$$

for all x, y in $GF(q^2)$. Then θ is a one-to-one semi-linear transformation of V upon V, and θ has order two. Moreover

$$(7.5) \qquad P(\infty)\theta = P(\infty) , \qquad P(x)\theta = P(x^q)$$

for all x in $GF(q^2)$. Hence the involuntary automorphism

$$(7.6) \qquad x \to x^q$$

of $GF(q^2)$, regarded as fixing ∞, is the unique inversion of $IP(q)$ in the circle C_0, given by (7.3). We note further that the projective linear group of $L = PG(1, q^2)$—that is, the group induced on the points of L by the non-singular linear transformations of $V = \{e, e'\}$ over $GF(q^2)$—is strictly transitive on the ordered triples of three distinct points P, Q, R of L.

If a, b, c, d are elements of $GF(q^2)$ such that

$$ad - bc \neq 0 \tag{7.7}$$

then the linear fractional transformation

$$x \to f(x) = (ax + b)/(cx + d), \quad \text{for every} \quad x \in \infty \cup GF(q^2) \tag{7.8}$$

is to be interpreted as follows:

$$\begin{aligned} P(f(\infty)) &= \{ae + ce'\} \\ P(f(x)) &= \{(ax + b)e + (cx + d)e'\} \\ &\quad \text{for every} \quad x \in GF(q^2) . \end{aligned} \tag{7.9}$$

The inverse linear fractional transformation is given by

$$x \to f^{-1}(x) = (dx - b)/(-cx + a), \quad \text{for every} \quad x \in \infty \cup GF(q^2) . \tag{7.10}$$

Again, if λ, α, β are elements of $GF(q^2)$ such that

$$\lambda^{q+1} + \alpha\beta \neq 0 , \tag{7.11}$$

then the semi-linear fractional transformation

$$x \to h(x) = (\lambda x^q + \alpha)/(\beta x^q - \lambda^q), \quad \text{for every} \quad x \in \infty \cup GF(q^2) \tag{7.12}$$

is to be interpreted as follows:

$$\begin{aligned} P(h(\infty)) &= \{\lambda e + \beta e'\} ; \\ P(h(x)) &= \{(\lambda x^q + \alpha)e + (\beta x^q - \lambda^q)e'\} , \\ &\quad \text{for every} \quad x \in GF(q^2) . \end{aligned} \tag{7.13}$$

Clearly the mapping $P(x) \to P(f(x))$ or $P(x) \to P(h(x))$ can be induced by a linear or semi-linear transformation of V over $GF(q)$, respectively. Since the multiplicative group of $GF(q^2)$ is cyclic, we can choose (in $q - 1$ ways) an element ε of $GF(q^2)$ such that

(7.14) $$\varepsilon^{q-1} = -1 , \qquad \varepsilon^q = -\varepsilon \neq 0 .$$

We shall be interested in the case that f, h (given by (7.8), (7.12) respectively) are related by

(7.15)
$$\begin{aligned} \varepsilon\lambda &= ad^q - bc^q , \\ \varepsilon\alpha &= -ab^q + a^q b , \\ \varepsilon\beta &= cd^q - c^q d , \end{aligned}$$

for a suitable choice of ε subject to (7.14).

Lemma 7.3. *Consider $\infty \cup GF(q^2)$ as the set of points of $IP(q)$. Then inversion with respect to a circle C of $IP(q)$ takes the form $x \to h(x)$ where h is given by* (7.11), (7.12) *and satisfies*

(7.16) $$\lambda \in GF(q^2) , \qquad \alpha, \beta \in GF(q) .$$

Furthermore:

(i) *If*

(7.17) $$C = f(C_0)$$

where f is given by (7.7), (7.8), *then λ, α, β have form* (7.15) *aside from a proportionality factor.*

(ii) *If h is given by* (7.11), (7.12), *subject only to* (7.16), *and if C is the set of all x in $\infty \cup GF(q^2)$ such that $h(x) = x$, then C is a circle of $IP(q)$ and inversion with respect to C is given by $x \to h(x)$. In addition:*

(iia) *If $\beta = 0$, then* (7.17) *holds with*

(7.18) $$f(x) = \varepsilon\lambda x + b , \qquad \text{for every} \quad x \in \infty \cup GF(q^2) ,$$

for a suitable choise of b, where ε satisfies (7.14).

(iib) *If $\beta \neq 0$ we may assume $\beta = 1$. If $\beta = 1$ then* (7.17) *holds with*

(7.19) $$f(x) - \lambda = k(x - u^q)/(x - u) , \qquad \text{for every} \quad x \in \infty \cup GF(q^2) ,$$

for a suitable choice of $k \neq 0$ in $GF(q^2)$ and for any u in $GF(q^2)$ which is not in $GF(q)$.

Proof. First suppose that (7.17) holds where f is given by (7.8), (7.9). Then the inversion mapping $x \to h(x)$ of C must map $f(x)$ upon $f(x^q)$ for every x in $\infty \cup GF(q^2)$. Equivalently,

(7.20) $$h(x) = f(f^{-1}(x)^q)\,, \qquad \text{for every} \quad x \in \infty \cup GF(q^2)\,.$$

From (7.20) we see that

$$h(x) = (Ax + B)/(Cx + A^q)$$

where

$$A = ad^q - bc^q\,, \qquad B = -ab^q + a^q b\,, \qquad C = cd^q - c^q d$$

and hence

$$B^q = -B\,, \qquad C^q = -C\,.$$

Thus if ε satisfies (7.14) and if we define λ, α, β by

$$\varepsilon\lambda = A\,, \qquad \varepsilon\alpha = B\,, \qquad \varepsilon\beta = C$$

we will have

$$\varepsilon(-\lambda^q) = A^q\,, \qquad \alpha^q = \alpha\,, \qquad \beta^q = \beta\,.$$

This shows that h has form (7.11), (7.12) subject to (7.16). If $\beta = 0$ we try f of form (7.18); thus $c = 0$, $d = 1$. Then (7.15) holds provided

(7.21) $$b = \lambda s \qquad \text{where} \qquad s + s^q = -(\alpha/\lambda^{q+1})\,.$$

We can always find s in $GF(q^2)$; in fact we can add the condition $s^{q+1} = \gamma$, for an arbitrary γ in $GF(q)$, and obtain s as a solution of the equation

$$x^2 - (\alpha/\lambda^{q+1})x + \gamma = 0\,.$$

This proves (iia). If $\beta \neq 0$, then, since β is in $GF(q)$, we can replace λ, α, β by λ/β, α/β, 1 respectively without changing h or the conditions (7.16). Thus we may assume that $\beta = 1$. If $\beta = 1$, we try f of form (7.19). This means that (7.7), (7.8) holds with

$$a = k + \lambda\,, \qquad b = -(ku^q + \lambda u)\,, \qquad c = 1\,, \qquad d = -u\,.$$

Then, from (7.20), after choosing $\varepsilon = u^q - u \neq 0$, so that (7.19) holds, we get the condition

(7.22) $$k^{q+1} = \lambda^{q+1} + \alpha \neq 0\,.$$

We can certainly find a $k \neq 0$ in $GF(q^2)$ subject to (7.22). This proves (iib) and completes the proof of Lemma 7.3.

Let us call two distinct points P, Q of $IP(q)$ *conjugate* (or inverse) with respect to a circle C of $IP(q)$ provided that inversion with respect to C interchanges P and Q. By a *pair of conjugate points of a circle* C we mean a pair of distinct points conjugate with respect to C.

Lemma 7.4. *Let C, C' be two distinct circles of $IP(q)$. Then:*

(i) *If C, C' are disjoint, they have one and only one pair P, Q of conjugate points in common.*

(ii) *If C, C' have a common pair of conjugate points, then C, C' are disjoint.*

Proof. Since C, C' are distinct, they have at most two common points. Thus C has at least $(q+1)-2=q-1$ points which are not in C'. Hence we may assume without loss of generality that $C=C_0$ and that C' does not contain ∞. Then, in the sense of Lemma 7.3, inversion with respect to $C=C_0$ is given by $x \to x^q$ and inversion with respect to C' is given by $x \to h(x)$ for h of form (7.11), (7.12) subject to (7.16). Since C' does not contain ∞, then $h(\infty) \neq \infty$; that is

$$\beta \neq 0 . \tag{7.23}$$

A conjugate pair of C_0 has the form x, x^q where x is in $GF(q^2)$ but not in $GF(q)$, so that $x \neq x^q$. This will be a conjugate pair of C' provided $h(x^q)=x$ or

$$\beta x^2-(\lambda+\lambda^q)x-\alpha=0 . \tag{7.24}$$

An element of C_0, other than ∞, is an element x of $GF(q^2)$ which is in $GF(q)$ and therefore satisfies $x^q=x$. This will be an element of $C' \cap C_0$ if $h(x)=x$ and hence $h(x^q)=x$. Thus C, C' will have a common point if and only if (7.24) holds subject to

$$x^q=x . \tag{7.25}$$

By (7.23), equation (7.24) is a non-degenerate quadratic with coefficients in $GF(q)$. Therefore (7.24) has solutions in $GF(q^2)$. If C, C' are disjoint, then (7.24) has no solutions in $GF(q)$; in this case it has a unique pair of distinct solutions x, x^q. This proves (i). If C, C' are not disjoint, then (7.24) has either a single solution, which is in $GF(q)$, or two distinct solutions, both in $GF(q)$. In this case, C, C' have no common pair of conjugate points. This proves (ii) and completes the proof of Lemma 7.4.

Let K be a non-empty set of circles of $IP(q)$. We shall say that K is *disjoint* if every two distinct circles in K are disjoint. We shall say that K is *complete* if K is disjoint and if the number of distinct circles in K is $q-1$. We shall say that K is *linear* if K is disjoint (though possibly not complete) and if the circles in K all have a common pair P, Q of conjugate points.

Before we go on, let us refer briefly to the real inversive plane, A^*, whose points are the points of the real Euclidean affine plane, A, together with a single additional point, ∞, and whose circles are the circles of A together with the lines of A. (The latter are understood to contain ∞ in A^*.) There is, of course, no special significance to the point ∞; any other point can assume its role. If C is a circle of A, then, in A^*, the center of C has ∞ as its conjugate with respect to C. Thus the analogue of Lemma 7.4 (i) for A^* is the following: If C, C' are any two disjoint circles of A^*, then ∞ can be chosen in exactly two ways so as to make C, C' concentric circles of A. In the same sense, the theorem which follows may be compared with the theory of a complete set of concentric circles of A; and the proof also discusses the lines through the common center. However, the analogy must not be relied upon too heavily. In the real plane, A, circles have positive radii, the squared radii being non-zero norms of complex numbers. Again in $IP(q)$, "squared radii" may be thought of as non-zero norms of elements of $GF(q^2)$ relative to $GF(q)$; but such norms can assume every non-zero value in $GF(q)$. This makes considerable difference!

Theorem 7.5. *Assume $q > 2$. Let P, Q be two distinct points of $IP(q)$ and let $K = K(P, Q)$ be the set of all circles of $IP(q)$ with P, Q as conjugate points. Let G be that subgroup of the collineation group of $IP(q)$ generated by the inversions and the collineations induced by the projective linear group of the line $PG(1, q^2)$. Let H be the subgroup of G which maps K upon K. Let A be the group generated by the inversions with respect to circles in K, and let B be the group generated by the inversions with respect to circles containing P, Q. Then:*

(i) *K is a complete linear set of ($q-1$ disjoint) circles* *of IP(q), one through each point distinct from P, Q.*

(ii) *The orders of G, H, A, B are as follows:*

(7.26) $$|G| = 2q^2(q^4 - 1)\,, \qquad |H| = 4(q^2 - 1)\,,$$
$$|A| = 2(q - 1)\,, \qquad |B| = 2(q + 1)\,.$$

(iii) *A, B are normal subgroups of H.*

(iv) *AB has index 1 or 2 in H according as q is even or odd.*

(v) *Every element of A commutes with every element of B.*

(vi) *B consists of all collineations in H which map every circle in K upon itself. And B has a cyclic normal subgroup of order $q+1$, index 2 which acts transitively (and hence regularly) on the points of each circle in K.*

(vii) *H/B has order $2(q-1)$. The $q-1$ circles of K can be enumerated as*

(7.27) $$C(i)\,, \qquad i \text{ reduced} \bmod q - 1\,,$$

in such a way that H/B acts isomorphically on these circles according to the group of permutations

(7.28) $$C(i) \to C(si + t)\,, \qquad \text{for every } i \bmod q - 1$$

where $s = \pm 1$ and where t is an integer $\bmod q - 1$.

(viii) *A induces on the circles* (7.26) *the permutations* (7.28) *with t even.*

Remark. In the excluded case ($q = 2$), K consists of a single circle through the three points distinct from P, Q. In this case, (7.28) is the identity permutation; G/B does not act isomorphically on the (unique) circle of K.

Proof. We may assume without loss of generality that $P = \infty$, $Q = 0$. For all x in $\infty \cup GF(q^2)$ define

(7.29) $$h_i(x) = \omega^i / x^q\,, \qquad 0 \leq i < q - 1\,,$$

(7.30) $$H_j(x) = U^j x^q\,, \qquad 0 \leq j < q + 1$$

where ω is a multiplicative generator of $GF(q)$ (and hence is of order $q - 1$) and where U is an element of $GF(q^2)$ of multiplicative order $q + 1$.

By Lemma 7.3, mapping $x \to h(x)$ is an inversion with respect to some circle C of $IP(q)$ if and only if h has form (7.12) subject to (7.11), (7.16). The circle C has ∞, 0 as conjugate points if and only if $h(\infty) = 0$, $h(0) = \infty$. This requires $\lambda = 0$, $\alpha\beta \neq 0$. Hence, without loss of generality, we may assume $\lambda = 0$, $\beta = 1$, $\alpha = \omega^i$ for some i reduced mod $q - 1$. That is, $h = h_i$ for some i. Hence the circles in K correspond to the $q - 1$ inversions (7.29). On the other hand, the circle C with inversion h contains ∞ and 0 if and only if $h(\infty) = 0$, $h(0) = 0$. This requires $\alpha = \beta = 0$, $\lambda \neq 0$. In this case, $h(x) = kx^q$ where $k = -(\lambda^{-1})^{q-1} = (\varepsilon\lambda^{-1})^{q-1}$, with ε satisfying (7.14). Thus $k = U^j$ for some j reduced mod $q + 1$. Hence the circles through $P = \infty$, $Q = 0$ correspond to the $q + 1$ inversions (7.30).

Consider a point c distinct from ∞, 0. That is, c is in $GF(q^2)$ and $c \neq 0$. Then $h_i(c) = 0$ if and only if $c^{q+1} = \omega^i$; and this is true for a unique i mod $q - 1$. This proves (i). Similarly, $H_i(c) = c$ if and only if $U^j = (c^{-1})^{q-1}$; and this is true for a unique j reduced mod $q + 1$. In other words, c lies on a unique circle through ∞, 0—which we knew in advance!

In view of Lemma 7.3, we see readily that G consists of all mappings $x \to f(x)$ of form

$$f(x) = (ax^n + b)/(cx^n + d)\,, \qquad ad - bc \neq 0\,, \quad n = 1 \text{ or } q\,, \tag{7.31}$$

where (for all x in $\infty \cup GF(q^2)$) a, b, c, d are in $GF(q^2)$ and either $n = 1$ or $n = q$. For each choice of n, the number of distinct mappings is

$$(q^4 - 1)(q^4 - q^2)/(q^2 - 1) = q^2(q^4 - 1)\,,$$

as we see by computing the number of ordered bases of a two-dimensional vector space over $GF(q^2)$ and dividing by $q^2 - 1$. Thus we get the order of G: $|G| = 2q^2(q^4 - 1)$.

The mapping f will map K upon K if and only if either (1) $f(\infty) = \infty$, $f(0) = 0$ or (2) $f(\infty) = 0$, $f(0) = \infty$. For case (1) we have $b = c = 0$, $ad \neq 0$, whence $f(x) = kx^n$ with $k \neq 0$, $n = 1$ or q. For case (2) we have $a = d = 0$, $bc \neq 0$, whence $f(x) = k/x^n$. To sum up, the mapping $x \to f(x)$ is in H if and only if

$$f(x) = kx^m\,, \qquad k \neq 0\,, \qquad m = \pm 1,\ \pm q\,, \tag{7.32}$$

and k is in $GF(q^2)$. Thus $|H| = 4(q^2 - 1)$.

Since A is generated by the h_i (given by (7.29)) and since

$$h_i(h_0(x)) = \omega^i x\,,$$

we see readily that f, given by (7.32), is in A if and only if $k = \omega^i$ for some i reduced mod $q-1$ and $m = 1$ or $-q$. In particular $|A| = 2(q-1)$. Similarly f, given by (7.32), is in B if and only if $k = U^j$ for some j reduced mod $q+1$ and $m = 1$ or q. In particular, $|B| = 2(q+1)$. This proves (ii). We may add that $A \cap B$ consists of those f of form $f(x) = kx$ where the multiplicative order of k divides $(q-1, q+1) = (2, q-1)$. That is,

$$|A \cap B| = (2, q-1)\,, \tag{7.33}$$

and hence is 1 or 2 according as q is even or odd.

For f given by (7.32),

$$\begin{aligned} f^{-1}(x) &= k^{-m}x^m\,, \\ f(h_i(f^{-1}(x))) &= k^{q+1}\omega^{im}x^{-q}\,, \\ f(H_j(f^{-1}(x))) &= k^{1-q}\omega^{jm}x^q\,. \end{aligned} \tag{7.34}$$

Since k^{q+1}, k^{1-q} are powers of ω, U respectively, we see that A, B are normal in H. This proves (iii). Also

$$h_i(H_j(x)) = H_j(h_i(x)) = \omega^i U^j x\,, \tag{7.35}$$

which proves (v). For (iv) we note, for example, in view of (i), (ii) and (7.33), that

$$|AB| = |A|\,|B|/|A \cap B| = |H|/(2, q-1)$$

so that AB is a subgroup of index $(2, q-1)$. Or we may use (7.35) and note that every non-zero element of $GF(q^2)$ has form $\omega^i U^j$ precisely when $(2, q-1) = 1$.

For f given by (7.32), the mapping $x \to f(x)$ will map each circle of K upon itself if and only if

$$f(h_i(f^{-1}(x))) = h_i(x)$$

for all x in $\infty \cup GF(q^2)$ and for every i mod $q-1$. In view of (7.34), we see that this is true precisely when $k^{q+1} = 1$ and $m = 1$ or q; that is, precisely when f is in B. Thus we have proved the first part of (vi). If f is in B, so that $k = U^j$ and $m = 1$ or q, and if c is a non-zero element of $GF(q^2)$, then

$$f(c) = cU^jc^{m-1}$$

and c^{m-1} $(= 1$ or $c^{q-1})$ is a power of U. In particular, those f

in B with $m = 1$ form a cyclic subgroup of order $q + 1$, index 2, which acts transitively on the points cU^j of the circle in K containing c. This completes the proof of (vi).

For (viii), let $C(i)$ denote the circle with inversion h_i. Thus x is in $C(i)$ if and only if $x^{q+1} = \omega^i$. If f is given by (7.31) and if x is in $C(i)$ then

$$f(x)^{q+1} = k^{q+1}\omega^{mi} = \omega^{si+t}$$

where $k^{q+1} = \omega^t$ and where $\omega^m = \omega^s$. Here we take $s = 1$ if $m = 1$ or q; and $s = -1$ if $m = -1$ or $-q$. Thus f maps $C(i)$ into $C(si + t)$ for all i. This proves (vii). For (viii) we have the case that $k = \omega^u$ for some integer u and $m = 1$ or $-q$; this gives $t = 2u$ but places no restriction on s. Now the proof of Theorem 7.5 is complete.

The two corollaries which follow are vacuously true for $q = 2$. For $q > 2$ they follow from Lemma 7.4 and Theorem 7.5.

Corollary 7.6. *If C, C' are two disjoint circles of $IP(q)$, then C, C' can be imbedded in exactly one complete linear set of ($q - 1$ disjoint) circles of $IP(q)$.*

Corollary 7.7. *If L is a linear set of two or more disjoint circles of $IP(q)$, then L can be imbedded in exactly one complete linear set of circles of $IP(q)$.*

If we were only interested in characterizing the so-called André planes, we could stop our investigation of $IP(q)$ at this point. In what follows, our interest is centered on nonlinear sets of disjoint cycles. For our next lemma we contemplate a complete disjoint set K (linear or not) and a disjoint set L having a "large" intersection with K.

Lemma 7.8. *Let K be a complete set (possibly non-linear) of $q - 1$ disjoint circles of $IP(q)$. Let L be a set of t disjoint circles of $IP(q)$, where $2t > q - 1$. Assume that K, L have precisely s distinct circles in common. Then:*

(i) *If $2s > q - 1$, then $s = t$; that is, L is a subset of K.*

(ii) *If $2s > q - 3$, and $s < t$, then one of the following must be true:*

(iia) *q is odd, $s = (q - 1)/2$, $t = s + 1 = (q + 1)/2$*

(iib) *q is even, $s = (q - 2)/2$ and either $t = s + 1 = q/2$ or $t = s + 2 = (q + 2)/2$.*

Corollary 7.9. *A set of more than $(q-1)/2$ disjoint circles of $IP(q)$ can be imbedded in at most one complete set of $q-1$ disjoint circles of $IP(q)$.*

Corollary 7.10. *If $q < 5$ every set of disjoint circles of $IP(q)$ is linear.*

Proof. First we note that the $q-1$ circles in K use up $(q-1)(q+1) = q^2 - 1$ of the q^2+1 points of $IP(q)$. Thus there are precisely two distinct points, call them P, Q, which lie on no circle of K. Each of the $t-s$ circles of L which is not in K is disjoint from the s circles of K which are in L and has at most 2 points in common with each of the remaining $q-1-s$ circles of K. Consequently, the $t-s$ circles of L which are not in K contain at least

$$(t-s)\{q+1-2(q-1-s)\} = (t-s)(2s+3-q)$$

distinct points of $IP(q)$ which are not on any circle of K. Thus

$$2 \geq (t-s)(2s+3-q) \tag{7.36}$$

with equality precisely when the following conditions hold:

(a) $t > s$.

(b) Each of P, Q lies on some circle in L.

(c) Each of the $t-s$ circles of L which are not in K meets each of the $q-1-s$ circles of K which are not in L in two distinct points.

If $t > s$ and $2s > q-1$, the right-hand side of (7.36) exceeds 2. Hence $2s > q-1$ requires $t = s$. This proves (i).

If $t > s$ and $2s > q-3$, both factors on the right of (7.36) are positive integers. Thus we must have one of the following:

(α) $t-s=1$, $2s+3-q=1$

(β) $t-s=2$, $2s+3-q=1$

(γ) $t-s=1$, $2s+3-q=2$.

In cases (α), (β), q is even, $s=(q-2)/2$, and $t=s+1$ or $s+2$. This gives (iib). In case (γ), q is odd, $s=(q-1)/2$, and $t=s+1$. This gives (iia) and completes the proof of Lemma 7.8.

Note that (β), (γ) require equality in (7.36). We shall say more about these cases later.

Corollary 7.9 follows directly from (i) of Lemma 7.8. For

if K, L are complete linear sets with $2s > q-1$, then $s = t = q-1$.

For Corollary 7.10, let L be a set of t disjoint circles of $IP(q)$, $t > 0$. Then, certainly, $t \leq q-1$. Also L is linear if $t \leq 2$ and, in particular, if $q \leq 3$. Next suppose that $t = 3 = q-1$. Choose some two disjoint circles of L and let K be the unique complete linear set containing these two circles. If K, L have s common circles, then $s \geq 2$, $2s \geq 4 > q-1$ and hence $s = t = 3 = q-1$. That is, $L = K$. Therefore, if $q < 5$ every set of disjoint circle of $IP(q)$ is linear. This proves Corollary 7.10.

At this point we may mention an unsolved problem: we have been unable to decide whether a complete set of disjoint circles of $IP(q)$ is necessarily linear for every prime-power q.

Now we shall take a closer look at inversion.

Lemma 7.11. *Assume $q > 2$. Let P, Q, P', Q' be four distinct points of $IP(q)$. Then:*

(a) *Each of the following statements implies the other:*

(i) *There exists a circle C of $IP(q)$ with P, Q and P', Q' as pairs of conjugate points.*

(ii) *There exists a circle D of $IP(q)$ containing the four points P, Q, P', Q'.*

(b) *If the circles C, D (of statements (i), (ii)) exist, they are unique.*

(c) *Let C, D exist. If q is even, C and D are tangent. If q is odd, C and D are secant or disjoint; both cases can occur.*

Proof. We may assume without loss of generality that $P = \infty$, $Q = 0$, $P' = 1$, $Q' = k$ where k is an element of $GF(q^2)$ distinct from $0, 1$.

If C is a circle with $\infty, 0$ as conjugate points, then inversion $x \rightarrow h_i(x)$ with respect to C is given by (7.29) for some i mod $q-1$. The points $1, k$ are also conjugate with respect to C if and only if $h_i(1) = k$; that is, $\omega^i = k$. This requires that k be in $GF(q)$. When k is in $GF(q)$, i is determined uniquely mod $q-1$. Thus the circle C of statement (i) exists precisely when k is in $GF(q)$; and C is then unique.

If D is the unique circle containing $\infty, 0, 1$, then $D = C_0$. And C_0 contains k if and only if k is in $GF(q)$. Thus the circle D of statement (ii) exists precisely when k is in $GF(q)$; and D is then unique. This proves (a) and (b).

Now suppose that C, D exist. Thus k is in $GF(q)$, $D = C_0$, and inversion with respect to C is given by $x \to kx^{-q}$. Then a common point of C, D (if one exists) is a non-zero element x of $GF(q)$ such that $x = kx^{-q}$ and hence

$$(7.37) \qquad x^2 = k\ .$$

If q is even then (7.37) has a unique solution x in $GF(q)$. Hence C, D are tangent. If q is odd then (7.37) has two distinct solutions x in $GF(q)$ or no solutions, according as the element k of $GF(q)$ is a square or non-square in $GF(q)$; both cases occur. Hence C, D are secant or disjoint according as k is a square or a non-square. This proves (c) and completes the proof of Lemma 7.11.

We have used the hypothesis $q > 2$ in Lemma 7.11 only to ensure that each circle of $IP(q)$ has at least 4 points. If $q = 2$, the circles C, D of the lemma cannot exist.

Lemma 7.2. *Let C, C' be two distinct circles of $IP(q)$ and let θ, ϕ be the inversions with respect to C, C' respectively. Then:*

(a) *Each of the following statements implies all the others:*

(i) $\theta\phi = \phi\theta$.

(ii) *C' is fixed under inversion with respect to C.*

(ii′) *C is fixed under inversion with respect to C'.*

(iii) *C' contains a pair of conjugate points of C.*

(iii′) *C contains a pair of conjugate points of C'.*

(b) *Assume that the equivalent statements (i)-(iii′) are true. If q is even, then C and C' are tangent. If q is odd, then C and C' are secant or disjoint; both cases can occur.*

Proof of (a).

(i)→(ii) and (i)→(ii′). Assume (i). If P is a point of C' then $(P\theta)\phi = (P\phi)\theta = P\theta$, so $P\theta$ is in C'. That is, $C'\theta = C'$. Hence (ii) holds. Similarly (ii′) holds. Thus (i)→(ii), (ii′).

(ii)→(iii) and (ii′)→(iii′). Assume (ii). Since C' is distinct from C, C' must contain at least one point P which is not in C. Since P is not in C, $P\theta = Q \neq P$. Thus P, Q are conjugate points of C. Since $C' = C'\theta$ contains P and $P\theta = Q$, then (iii) holds. Thus (ii)→(iii). Similarly, (ii′)→(iii′).

(iii)→(i) and (iii′)→(i). Assume (iii). Thus C' contains a conjugate pair P, Q of C. Therefore by Theorem 7.5 (v) (the proof of which did not require $q > 2$), $\theta\phi = \phi\theta$. Thus (iii)→(i) and, similarly, (iii′)→(i). This completes the proof of (a).

Proof of (b). By assumption, the five equivalent statements (i)–(iii′) are true. By (iii′), C contains a pair P, Q of conjugate points of C'. We may assume without loss of generality that $P = \infty$, $Q = 0$ and $C = C_0$. If $x \to h(x)$ is inversion with respect to C' then $h(\infty) = 0$, $h(0) = \infty$ and hence $h = h_i$, given by (7.29), for some integer $i \bmod q - 1$. A point common to $C = C_0$ and C' must be a non-zero element x of $GF(q)$ such that $h_i(x) = x$ and hence

$$x^2 = \omega^i \,. \tag{7.38}$$

Then—just as in the treatment of (7.37)—we see that, if q is even, C, C' are tangent whereas, if q is odd, C, C' are secant or disjoint according as the integer $i \bmod q - 1$ is even or odd. This proves (b) and completes the proof of Lemma 7.12.

Lemma 7.13. *Let C, C' be two disjoint circles of $IP(q)$ with common conjugate pair P, Q and let D be a circle of $IP(q)$ for which P, Q is not a pair of conjugate points. Then a necessary and sufficient condition that both of C, C' be fixed under inversion with respect to D is that D contain P, Q.*

Proof. Let θ be the inversion with respect to D. If D contains P, Q, then, by Lemma 7.12, θ fixes both of C, C'. If θ fixes both of C, C', then $P\theta, Q\theta$ must be the unique common conjugate pair of C, C'. Thus either (a) $P\theta = Q$, $Q\theta = P$ or (b) $P\theta = P$, $Q\theta = Q$. We have excluded (a) by hypothesis; and (b) means that D contains P and Q. This completes the proof of Lemma 7.13.

In the theorem which follows we show the existence (and give a general method for construction) of nonlinear triples of disjoint circles of $IP(q)$ for $q > 4$. That the hypothesis $q > 4$ is necessary follows from Corollary 7.10.

Theorem 7.14. *Assume $q > 4$. Then $IP(q)$ has at least one nonlinear triple*

$$C_1, C_2, C_3 \tag{7.39}$$

of disjoint circles. Given such a triple, let points P_i, Q_i be defined for $i = 1, 2, 3$ such that (with subscripts reduced mod 3)

(7.40) *P_i, Q_i is the common conjugate pair of C_{i+1}, C_{i+2}.*

Then:

(a) *The six points* P_i, Q_i $(i = 1, 2, 3)$, *are six distinct points of a circle,* D, *of* $IP(q)$.

(b) *No circle of* $IP(q)$ *has all three pairs* P_i, Q_i *as pairs of conjugate points.*

Moreover, given any circle D *of* $IP(q)$, *six points* P_i, Q_i *of* D *can always be found subject to* (a) *and* (b). *Each such set of points determines a unique nonlinear disjoint triple* (7.39) *such that* (7.40) *holds. If* q *is even,* D *is a common tangent to the three circles* (7.39). *If* q *is odd,* D *and* C_i *must be secants or disjoint for each* i; *there are various possibilities. In any case, each of the circles* (7.39) *is fixed under inversion with respect to* D.

Remark. In the context of Theorem 7.14, it will be convenient to refer to D as the *common orthogonal circle* of the nonlinear disjoint triple (7.39).

Proof. First assume that (7.39) is a nonlinear disjoint triple. Let the points P_i, Q_i be defined by (7.40) for $i = 1, 2, 3$. By nonlinearity, the three circles C_i have no common conjugate pair. Hence the three two-element sets $\{P_i, Q_i\}$ are distinct. Since, for example, P_1, Q_1 and P_2, Q_2 are two distinct pairs of conjugate points of C_3, then the four points P_1, Q_1, P_2, Q_2 are distinct. By symmetry, all six points are distinct. By Lemma 7.11 there exists, for each $i \bmod 3$, a unique circle D_i containing the four points $P_{i+1}, Q_{i+1}, P_{i+2}, Q_{i+2}$. To prove (a), we must show that

(7.14) $$D_1 = D_2 = D_3 = D\,.$$

Note that if, for some i, D_i contains P_i, Q_i, then (7.41) will be true. Hence, if (7.41) fails, D_i does not contain P_i, Q_i $(i = 1, 2, 3)$. By Lemma 7.12, since D_i contains a common conjugate pair P_{i+1}, Q_{i+1} of C_{i+2}, C_i and a common conjugate pair P_{i+2}, Q_{i+2} of C_i, C_{i+1}, then each of the circles (7.39) is fixed under inversion with respect to D_i. Hence, by Lemma 7.13, if D_i does not contain the common conjugate pair P_i, Q_i of C_{i+1}, C_{i+2}, then D_i has P_i, Q_i as a conjugate pair. Thus, if (7.41) is false, then, for each i,

(7.42) $\quad D_i$ has P_i, Q_i as a conjugate pair, $\quad i = 1, 2, 3$.

In order to dispense of (7.42) we may assume without loss

of generality that $P_1 = \infty$, $Q_1 = 0$, $P_2 = 1$. Then $D_3 = C_0$. Hence $Q_2 = \alpha$ is an element of $GF(q)$. Also, by (7.42), $P_3 = u$, $Q_3 = u^q$ for some u in $GF(q^2)$ which is not in $GF(q)$. By (7.42) with $i = 1$, D_1 must have an inversion of form

$$x \to kx^{-q} .$$

Since D_1 contains u, u^q, then $k = u^{q+1}$. Since D_1 contains 1 and α, then $k = 1 = \alpha^2$. Also $\alpha \neq 1$, so $\alpha = -1$ (and q is odd). Now we have $P_1 = \infty$, $Q_1 = 0$, $P_2 = 1$, $Q_2 = -1$, $P_3 = u$, $Q_3 = u^q = u^{-1}$. Since D_2 contains $\infty, 0$, D_2 must have an inversion of form

$$x \to tx^q .$$

Since D_2 has conjugate points $1, -1$, then $t = -1$. Since D_2 contains u, then $t = u^{1-q} = u^2$. Hence $u^2 = -1$. But now we note that the circle C with inversion

$$x \to -x^{-q}$$

has all three of the pairs P_i, Q_i as conjugate points. This would contradict (b). Hence—provided we prove (b) without deciding between (7.41), (7.42)—we see that (a) must be true.

We prove (b) as follows: since D_1 contains the four distinct points P_2, Q_2, P_3, Q_3, then, by Lemma 7.11, there is one and only one circle with P_2, Q_2 and P_3, Q_3 as pairs of conjugate points; and this is C_1. Similarly, C_2 is the unique circle with P_3, Q_3 and P_1, Q_1 as conjugate pairs. Since $C_1 \neq C_2$, (b) must be true.

Next let us assume (a), (b) for some circle D. By Lemma 7.11, there exists for each i mod 3 a unique circle C_i with P_{i+1}, Q_{i+1} and P_{i+2}, Q_{i+2} as pairs of conjugate points. If, for some i, C_i has P_i, Q_i as a conjugate pair, then $C_1 = C_2 = C_3 = C$, contradicting (b). Hence the three circles C_i are distinct. Since, for each i, C_i, C_{i+1} are distinct circles with a common conjugate pair P_{i+2}, Q_{i+2}, then, by Lemma 7.4 (ii), C_i, C_{i+1} are disjoint. Thus (7.39) is a disjoint triple. In view of (a), (7.39) is also nonlinear.

Next we wish to show that, given the circle D, we can always find six points P_i, Q_i subject to (a), (b). First let us treat the special case $q = 5$.

If $q = 5$, let P_1, Q_1, P_2 be any three distinct points of D. There exists (see Theorem 7.5) a unique circle C_2 which contains P_2 and has P_1, Q_1 as a conjugate pair. Since $q = 5$ is odd, then, by Lemma 7.12, C_2 meets D in a second point Q_2. The remaining $5+1-4=2$ points of D must be a conjugate pair, P_3, Q_3, of C_2. Since C_2 contains the distinct points P_2, Q_2 and has P_1, Q_1 and P_3, Q_3 as two distinct conjugate pairs, then (a) and (b) are satisfied. It will be clear from the discussion for the case $q > 5$ that the indicated pattern for $q = 5$ is the only possible one.

Now assume (when necessary) that $q > 5$. Let P_1, Q_1, P_2, Q_2 be any four distinct points of D. By Lemma 7.11, there exists a unique circle C_3 with P_1, Q_1 and P_2, Q_2 as pairs of conjugate points. By Lemma 7.12, D is fixed under inversion with respect to C_3. Also, a necessary and sufficient condition that C_3 and D be tangent is that q be even.

It q is even, C_3 has a unique (fifth) point R in common with D, and the remaining $q+1-5=q-4$ points of D break up into $(q-4)/2$ conjugate pairs with respect to C_3. We can either take one of P_3, Q_3 to be R and pick the other in $q-4$ ways; or we take P_3, Q_3 distinct from R and pick the unordered pairs P_3, Q_3 in $(q-4)(q-6)/2$ ways. This makes a total of $(q-4)^2/2$ admissible choices for $\{P_3, Q_3\}$.

If q is odd and C_3, D are disjoint, the $q-3$ points of D distinct from P_1, Q_1, P_2, Q_3 break up into $(q-3)/2$ conjugate pairs with respect to D. Thus there are $(q-3)(q-5)/2$ admissible choices for $\{P_3, Q_3\}$. (No choice if $q = 5$.)

If q is odd and C_3, D are secant, let R, S be the two distinct common points of C_3 and D. The $q-5$ points of D distinct from P_1, Q_1, P_2, Q_2, R, S break up into $(q-5)/2$ conjugate pairs of C_3. We can enumerate the number of admissible choices of the unordered pair P_3, Q_3 as follows: (i) $\{P_3, Q_3\} = \{R, S\}$: one choice. (ii) $\{P_3, Q_3\}$, $\{R, S\}$ have one common point: $2(q-5)$ choices. (No choice if $q = 5$). (iii) $\{P_3, Q_3\}$ disjoint from $\{R, S\}$: $(q-5)(q-7)/2$ choices. (No choice if $q = 5$ or 7.) Total: $\{1+(q-4)^2\}/2$ choices. (One choice if $q = 5$.)

Since the last two sentences of Theorem 7.14 should be clear from the foregoing discussion, the proof of Theorem 7.14 is now complete.

Further examination of the situation described in Theorem 7.14 can be facilitated slightly by the use of the concept of *cross-ratio*. The cross-ratio, $(P, Q; R, S)$, of four distinct points P, Q, R, S of $IP(q)$ in given order, is that unique element r of $GF(q^2)$ such that there exists a collineation in the projective linear group of $IP(q)$ which maps P, Q, R, S upon $\infty, 0, 1, r$ respectively. (In particular, $r \neq 0, 1$.) As a consequence, if $r = (P, Q; R, S)$ is in $GF(q)$, then

$$(7.43) \qquad (P\theta, Q\theta; R\theta, S\theta) = (P, Q; R, S)\,, \qquad \text{for every} \quad \theta \in G\,,$$

where G is the collineation group of $IP(q)$ defined in Theorem 7.5. As is well-known and easily checked, an unordered set of four distinct points gives rise to only $4!/4 = 6$ cross-ratios (some of which may be equal). Specifically,

$$(7.44) \qquad \begin{aligned} r &= (P, Q; R, S) = (Q, P; S, R) \\ &\qquad = (R, S; P, Q) = (S, R; Q, P)\,; \\ r^{-1} &= (P, Q; S, R) = (Q, P; R, S) \\ &\qquad = (R, S; Q, P) = (S, R; P, Q)\,; \end{aligned}$$

and (to give a sufficient sample)

$$(7.45) \qquad \begin{aligned} r &= (P, Q; R, S)\,, \qquad & (1 - r)^{-1} &= (P, R; S, Q)\,, \\ & & 1 - r^{-1} &= (P, S; Q, R)\,, \\ r^{-1} &= (P, Q; S, R)\,, & 1 - r &= (P, R; Q, S)\,, \\ & & r(r - 1)^{-1} &= (P, S; R, Q)\,. \end{aligned}$$

We note that four distinct points P, Q, P', Q' of $IP(q)$ satisfy the equivalent conditions (i), (ii) of Lemma 7.11 if and only if $(P, Q; P', Q')$ is in $GF(q)$.

With six distinct points P_i, Q_i $(i = 1, 2, 3)$ satisfying the conditions (a), (b) of Theorem 7.14 we associate an ordered triple

$$(7.46) \qquad (P_1, Q_1; P_2, Q_2; P_3, Q_3) = (\lambda, \mu, \nu)$$

of elements λ, μ, ν of $GF(q)$ defined by

$$(7.47) \qquad \lambda = (P_2, Q_2; P_3, Q_3)\,, \qquad \mu = (P_3, Q_3; P_1, Q_1)\,, \qquad \nu = (P_1, Q_1; P_2, Q_2)\,.$$

If P_i', Q_i' $(i = 1, 2, 3)$ are the same six points, and the same three pairs, in one of the $2^3 \cdot 2! = 48$ arrangements, the triple

$$(P_1', Q_1'; P_2', Q_2'; P_3', Q_3')$$

has one of 24 forms, some of which may be equal. Indeed, assuming (7.46), (7.47), then, clearly,

$$(7.48) \qquad (P_2, Q_2; P_3, Q_3; P_1, Q_1) = (\mu, \nu, \lambda)\,.$$

Moreover, by (7.45),

(7.49) $$(Q_1, P_1; Q_2, P_2; Q_3, P_3) = (\lambda, \mu, \nu)$$

and

(7.50)
$$\begin{aligned}
(P_1, Q_1; P_2, Q_2; P_3, Q_3) &= (\lambda, \mu, \nu)\,, \\
(P_1, Q_1; P_2, Q_2; Q_3, P_3) &= (\lambda^{-1}, \mu^{-1}, \nu)\,, \\
(P_1, Q_1; Q_2, P_2; P_3, Q_3) &= (\lambda^{-1}, \mu, \nu^{-1})\,, \\
(P_1, Q_1; Q_2, P_2; Q_3, P_3) &= (\lambda, \mu^{-1}, \nu^{-1})\,, \\
(P_1, Q_1; P_3, Q_3; P_2, Q_2) &= (\lambda, \nu, \mu)\,, \\
(P_1, Q_1; P_3, Q_3; Q_2, P_2) &= (\lambda^{-1}, \nu^{-1}, \mu)\,, \\
(P_1, Q_1; Q_3, P_3; P_2, Q_2) &= (\lambda^{-1}, \nu, \mu^{-1})\,, \\
(P_1, Q_1; Q_3, P_3; Q_2, P_2) &= (\lambda, \nu^{-1}, \mu^{-1})\,.
\end{aligned}$$

Thus there are the 8 triples (7.50) and their cyclic permutations. Also

(7.51) $$\lambda, \mu, \nu \in GF(q)\,, \qquad \lambda \neq 0, 1\,;\ \mu \neq 0, 1\,;\ \nu \neq 0, 1\,.$$

The conditions (7.51) merely ensure that the four points of every two pairs lie on a circle, not that all six points lie on a circle. We need to add, for example, the condition that

(7.52) $$(P_1, Q_1; P_2, Q_3) = \gamma \neq 0, 1\,, \qquad \gamma \in GF(q)\,.$$

Then (a) of Theorem 7.14 holds. For further investigation we may assume without loss of generality that

(7.53) $$D = C_0 = \infty \cup GF(q)\,,$$

(7.54) $$P_1 = \infty\,, \qquad Q_1 = 0\,;\ P_2 = 1\,, \qquad Q_2 = \alpha\,;\ P_3 = \beta\,, \qquad Q_3 = \gamma\,,$$

where

(7.55) $0, 1, \alpha, \beta, \gamma$ are distinct elements of $GF(q)$.

This is consistent with (7.52). Also, by (7.47), (7.54),

(7.56) $$\lambda = (1-\beta)(1-\gamma)^{-1}(\alpha-\gamma)(\alpha-\beta)^{-1}\,, \qquad \mu = \gamma\beta^{-1}\,, \qquad \nu = \alpha\,.$$

If we eliminate α, β from (7.56), we may write the result in form

$$(1-\mu)(1-\nu)(1-\lambda)^{-1} = \mu\nu + 1 - f(\gamma) \tag{7.57}$$

where

$$f(\gamma) = \gamma + \mu\nu\gamma^{-1} . \tag{7.58}$$

It may also be verified that

$$(\gamma - \mu\nu\gamma^{-1})^2 = f(\gamma)^2 - 4\mu\nu = (1-\lambda)^{-2}\lambda\mu\nu F(\lambda, \mu, \nu) \tag{7.59}$$

where

$$\begin{aligned} F(\lambda, \mu, \nu) = {} & \lambda\mu\nu + \lambda\mu^{-1}\nu^{-1} + \lambda^{-1}\mu\nu^{-1} + \lambda^{-1}\mu^{-1}\nu \\ & - 2(\lambda + \lambda^{-1} + \mu + \mu^{-1} + \nu + \nu^{-1}) + 8 . \end{aligned} \tag{7.60}$$

We note that $F(\lambda, \mu, \nu)$ has appropriate symmetries:

$$F(\lambda, \mu, \nu) = F(\mu, \lambda, \nu) = F(\mu, \nu, \lambda) = F(\lambda, \mu^{-1}, \nu^{-1}) , \tag{7.61}$$

whereas

$$F(\lambda, \mu, \nu) - F(\lambda^{-1}, \mu^{-1}, \nu^{-1}) = (\lambda - \lambda^{-1})(\mu - \mu^{-1})(\nu - \nu^{-1}) . \tag{7.62}$$

If we suppose that λ, μ, ν are given elements of $GF(q)$, distinct from 0 and 1, the conditions upon the solutions γ of (7.57), (7.58) which ensure that the six points (7.54) satisfy conditions (a), (b) of Theorem 7.14 may be expressed entirely in terms of λ, μ, ν and $F(\lambda, \mu, \nu)$—with differences according as q is odd or even. For example, if q is odd, the relevant necessary and sufficient condition is that $\lambda\mu\nu F(\lambda, \mu, \nu)$ be a non-zero square in $GF(q)$. However, this approach does not seem to be very helpful. Instead, we shall consider μ, ν as given and treat (7.57), (7.58) as expressing λ as a function of γ. Then, by examining the admissible choices for γ, we can determine the number of ways of choosing λ.

Before proceeding, we need to notice the condition

$$\alpha \neq \beta\gamma , \tag{7.63}$$

which corresponds to condition (b) of Theorem 7.14. Indeed, in view of (7.54), the circles C_2, C_3 are given by their inversions as follows:

$$\begin{aligned} C_2 &: \quad x \to \beta\gamma x^{-q} \,, \\ C_3 &: \quad x \to \alpha x^{-q} \,. \end{aligned} \tag{7.64}$$

Hence (7.63) says that C_2, C_3 are distinct and therefore that condition (b) holds. We note from (7.64) that a point x of $IP(q)$ is common to $D = C_0$ and C_3 if and only if x is in $GF(q)$ and $x^2 = \alpha$. Since $\alpha = \nu$, we see the following: *A necessary and sufficient condition that C_3 be disjoint from D is that ν be a non-square in $GF(q)$.* Similarly, C_1 (or C_2) is disjoint from D if and only if λ (or μ) is a non-square in $GF(q)$.

Condition (7.63) can be given another interpretation, as follows: let G be the group defined in Theorem 7.5. Then the only collineations in G which map $D = C_0$ upon itself and interchange $\infty, 0$ are those of form

$$x \to kx^{-1} \,, \qquad x \to kx^{-q} \,, \qquad k \neq 0 \,,$$

where k is in $GF(q)$. For each $k \neq 0$ in $GF(q)$, the two collineations have the same effect on $D = C_0$; they interchange $1, \alpha$ if $k = \alpha$; and they interchange β, γ if $k = \beta\gamma$. Since $\alpha \neq \beta\gamma$, we see the following: *No collineation in G can induce the permutation*

$$(P_1, Q_1)(P_2, Q_2)(P_3, Q_3) \,.$$

Since, by applying the collineation $x \to \alpha x^{-1}$, we get

$$\begin{aligned} (\lambda, \mu, \nu) = (\infty, 0; 1, \alpha; \beta, \gamma) &= (0, \infty; \alpha, 1; \gamma', \beta') \\ &= (\infty, 0; 1, \alpha; \beta', \gamma') \end{aligned}$$

where $\gamma' = \alpha\beta^{-1}$, $\beta' = \alpha\gamma^{-1}$, we know that

$$\gamma' = \alpha\beta^{-1} = \mu\nu\gamma^{-1} \tag{7.65}$$

is distinct from γ, and that (as is obvious) (7.57), (7.58) remain true with γ replaced by γ'. We now can list the restrictions on γ as follows:

$$\gamma \neq 0, 1, \mu, \nu, \mu\nu \qquad \text{and} \quad \gamma^2 \neq \mu\nu \,. \tag{7.66}$$

Indeed, since $\gamma, \gamma' \neq 1$, then $\gamma, \gamma' \neq \mu\nu$; since $\gamma, \gamma' \neq \alpha$ and $\alpha = \nu$, then $\gamma, \gamma' \neq \mu\nu$; since $\gamma \neq \gamma'$, then $\gamma^2 \neq \mu\nu$. And, of course, $\gamma \neq 0$. Furthermore, with (7.56) understood, and with $0, 1, \alpha, \beta$ distinct elements of $GF(q)$, if the element γ of $GF(q)$ satisfies (7.66), then (7.55), (7.63) are true.

Now suppose that μ, ν are elements of $GF(q)$, distinct from 0 and 1. Then, depending on μ, ν, (7.66) says that the element γ of $GF(q)$ must be taken distinct from a certain number, $n = n(\mu, \nu)$, of elements of $GF(q)$. The remaining $q - n$ elements of $GF(q)$ yield $(q - n)/2$ distinct values for $f(\gamma)$, given by (7.58), and a like number of values for λ, given by (7.57). The facts are as follows:

Lemma 7.15. *Let μ, ν be elements of $GF(q)$ distinct from 0 and 1. Let λ be given by* (7.57), (7.58), *where γ ranges over $GF(q)$ subject to* (7.66). *Then the number of distinct values of λ is as follows, according to the case:*

(i) *q even. $(q - 6)/2$ values if $\mu \neq \nu$ and $\mu\nu \neq 1$. Otherwise, $(q - 4)/2$ values.*

(ii) *q odd. $(q - 3)/2$ values if $\mu = \nu = -1$. And $(q - 7)/2$ values if $\mu \neq \nu$, $\mu\nu \neq 1$and $\mu\nu$ is a square in $GF(q)$. Finally, $(q - 5)/2$ values in all other cases.*

Proof. First assume that q is even. Then $\mu\nu = r^2$ for a unique r in $GF(q)$. Also $r = 0, 1, \mu, \nu$ or $\mu\nu$ respectively according as $\mu\nu = 0, 1, \mu^2, \nu^2$ or $\mu^2\nu^2$ respectively. Hence: $r \neq 0$. Also

$$r = 1 \Longleftrightarrow \mu\nu = 1 \Longleftrightarrow r = \mu\nu\ ;$$

$$r = \mu \Longleftrightarrow \mu = \nu \Longleftrightarrow r = \nu\ .$$

Thus (in the notation of the paragraph preceding the statement of Lemma 7.15) $n(\mu, \nu) = 6$ if $\mu \neq \nu$ and $\mu\nu \neq 1$; and $n(\mu, \nu) = 4$ otherwise. This proves (i).

Next assume that q is odd. Then the equation $\mu\nu = r^2$ has two distinct solutions $r, -r$ in $GF(q^2)$. First take the case that r is in $GF(q)$. This time we find that

$$r = \pm 1 \Longleftrightarrow \mu\nu = 1 \Longleftrightarrow r = \pm\mu\nu\ ;$$

$$r = \pm\mu \Longleftrightarrow \mu = \nu \Longleftrightarrow r = \pm\nu\ .$$

Thus, if $\mu = \nu = -1$, then $n(\mu, \nu) = 3$; if $\mu \neq \nu$ and $\mu\nu \neq 1$, then $n(\mu, \nu) = 7$; and if either $\mu = \nu \neq -1$ or $\mu \neq \nu$, $\mu\nu = 1$ then $n(\mu, \nu) = 5$. Next assume that r is not in $GF(q)$; then $\mu\nu \neq 1$ and $\mu\nu \neq \nu^2$, whence $\mu \neq \nu$. Hence, again, $n(\mu, \nu) = 5$. This proves (ii) and completes the proof of Lemma 7.15.

Now we are ready to begin a classification of the nonlinear disjoint triples (7.39) of Theorem 7.14. We shall say that a triple (7.39) has *type* $\{t\}$ if there are just t distinct ordered cross-ratio triples (λ, μ, ν) among the 24 associated with the pairs P_i, Q_i. Furthermore, if q is odd, we shall partition the triples (7.39) of type $\{t\}$ into those of type $\{t, d\}$, $d = 0, 1, 2$ or 3, where d is the number of circles among C_1, C_2, C_3 which are *disjoint* from D. Equivalently, if (λ, μ, ν) is any one of the t cross-ratio triples, d is the number of components λ, μ, ν which are non-squares in $GF(q)$. Note that, if $d = 3$, we have a nonlinear disjoint quadruple.

We also wish to partition the triples (7.39) into equivalence classes under the group G of Theorem 7.5. To this end, we note that if a cross-ratio triple (λ, μ, ν) is associated with a nonlinear disjoint triple (7.39), then we may assume that (7.46) and (7.53), (7.54) hold, for some α, β, γ. Then $\alpha = \nu$. Moreover, by (7.57), (7.58), there exist two distinct elements $r, r' = \mu\nu r^{-1}$ in $GF(q)$ such that either

$$\gamma = r\,, \qquad \beta = r\mu^{-1}$$

or

$$\gamma = r'\,, \qquad \beta = r'\mu^{-1}\,.$$

The two cases are equivalent under G: we use the mapping

$$x \to \nu x^{-1}\,.$$

Hence our classification problem reduces to a mere counting. For the most part, Lemma 7.15 suffices, but there are difficulties in connection with finer classification into types $\{t, d\}$.

It will be convenient—and perhaps instructive—to begin by disposing of the cases $q = 5$, $q = 7$.

By Lemma 7.15, with $q = 5$, an admissible cross-ratio triple (λ, μ, ν) must have $\mu = \nu = -1$. Hence, also, by cyclic permutation, $\lambda = -1$. Since -1 is a square in $GF(5)$, we have:

(7.67) If $q = 5$, the nonlinear disjoint triples (7.39) belong to a single equivalence class under G. This class has type $\{1, 0\}$.

By Lemma 7.15 for the case $q = 7$, there are exactly two cross-ratio triples of form $(\lambda, -1, -1)$. By (7.66) with $q = 7$, and $\mu = \nu = -1$; and by (7.57); we find that either (a) $\gamma = 2$ or -3, and $\lambda \equiv 2 \bmod 7$ or (b) $\gamma \equiv 3$ or -2, and $\lambda \equiv -3$. The complete set of triples corresponding to (a) is

$$(2, -1, -1)\,, \quad (-1, 2, -1)\,, \quad (-1, -1, 2)$$
$$(-3, -1, -1)\,, \quad (-1, -3, -1)\,, \quad (-1, -1, -3)\,,$$

and this includes the triples for (b). Hence we have an equivalence class of type $\{6\}$. Since -1 is a non-square and 2 is a square in $GF(7)$, the class has type $\{6, 2\}$. By Lemma 7.15 again, there can be no triple (λ, μ, ν) such that $\mu \neq \nu$, $\mu\nu \neq 1$ and $\mu\nu$ is a square in $GF(7)$. In all other cases (aside from the case $\mu = \nu = -1$) there is just one triple (λ, μ, ν) for given μ, ν. Taking $\mu = \nu = 2$, we find $\gamma \equiv -1$ or 3, $\lambda \equiv 3 \bmod 7$. The corresponding complete set of triples consists of 12 triples, namely

$$(3, 2, 2)\,, \quad (3, -3, -3)\,, \quad (-2, 2, -3)\,, \quad (-2, -3, 2)\,,$$

$$(2, 3, 2)\,, \quad (-3, 3, -3)\,, \quad (2, -2, -3)\,, \quad (-3, -2, 2)\,,$$

$$(2, 2, 3)\,, \quad (-3, -3, 3)\,, \quad (2, -3, -2)\,, \quad (-3, 2, -2)\,,$$

and gives a class of type $\{12, 1\}$. Taking $\mu = \nu = -2$, we find $\gamma \equiv -1$ or 3, $\lambda \equiv -2$. The corresponding complete set is

$$(-2, -2, -2)\,, \quad (-2, 3, 3)\,, \quad (3, -2, 3)\,, \quad (3, 3, -2)\,,$$

of type $\{4, 3\}$. The only unordered pairs (μ, ν), with $\mu, \nu \not\equiv 0, 1$ mod 7, which have not appeared above are $(-1, -2)$ and $(-1, 3)$. Each of these corresponds to the excluded case: $\mu \not\equiv \nu$, $\mu\nu \not\equiv 1$, $\mu\nu$ is a square mod 7. Hence our enumeration is complete. Thus:

(7.68) If $q = 7$, the nonlinear disjoint triples (7.39) split into 3 equivalence classes under G, one for each of the types $\{4, 3\}$, $\{6, 2\}$ and $\{12, 1\}$.

Note that type $\{4, 3\}$ yields a nonlinear set of 4 disjoint circles; each 3 of which also form a nonlinear disjoint triple.

In view of Theorem 7.14 and statements (7.67), (7.68), there will be no loss of generality in the sequel if we assume $q > 7$ when this proves convenient.

Next we wish to be a little more explicit about the subgroup of G which preserves a given nonlinear triple (7.39). Let P_i, Q_i $(i = 1, 2, 3)$ be the three pairs of common conjugate points corresponding to (7.39), and let P_i', Q_i' $(i = 1, 2, 3)$ be the same three pairs in one of the 48 possible arrangements. We have already noted the following: If

(7.69) $$(P_1, Q_1;\ P_2, Q_2;\ P_3, Q_3) = (P_1', Q_1';\ P_2', Q_2';\ P_3', Q_3')\,,$$

then either (a) there exists a collineation in G mapping P_i

upon P_i' and Q_i upon Q_i' for $i = 1, 2, 3$ or (b) there exists a collineation in G mapping P_i upon Q_i' and Q_i upon P_i' for $i = 1, 2, 3$. Both cannot be true, since no element of G can induce the permutation

$$(7.70) \qquad (P_1, Q_1)(P_2, Q_2)(P_3, Q_3) \, .$$

Conversely, (a) or (b) implies (7.69).

We can say a good deal more. For $i = 1, 2, 3$, let θ_i be the inversion with respect to the circle C_i, and let ϕ be the inversion with respect to the common orthogonal circle D of the triple (7.39). Let Z be the group of order two consisting of ϕ and the identity collineation of $IP(q)$. We note that G (the group described in Theorem 7.5) is in fact G_1Z where G_1 is the projective linear group of $IP(q)$. If ψ is any collineation of $IP(q)$ which maps D upon D, then $\psi^{-1}\phi\psi$ is an involution which fixes every point of D, and hence, by Lemma 7.2, $\psi^{-1}\phi\psi = \phi$. In particular, θ_i commutes with ϕ for each i. If ψ is in G and fixes at least three points of D, then ψ is in Z. If ψ is in G and interchanges P_2 with Q_2 and P_3 with Q_3, then $\psi\theta_1$ is in G and fixes each of the four points; hence $\psi = \theta_1$ or $\theta_1\phi$.

Now let K be the subgroup of G which maps each of the three circles C_i upon itself. Then K maps each of the three unordered point-pairs P_i, Q_i upon itself. Hence, as we shall see, one of the following must be true:

(7.71) (a) $K = Z$. Here $|K/Z| = 1$.
(b) K is the four-group generated by ϕ and one of the θ_i. Here $|K/Z| = 2$.
(c) K is an elementary abelian group of order 3 generated by $\theta_1, \theta_2, \theta_3$; and $\phi = \theta_1\theta_2\theta_3$. Here $|K/Z| = 4$.

In fact, K certainly contains Z. If K has an element ψ which is not in Z, then ψ must fix at least two and at most two of the six points P_i, Q_i. Hence $\psi = \theta_i$ or $\theta_i\phi$ for some i. Suppose that K contains θ_1. Then θ_1 must induce the permutation

$$(7.72) \qquad (P_1)(Q_1)(P_2, Q_2)(P_3, Q_3) \, .$$

Since θ_1 fixes P_1 and Q_1, and since P_1, Q_1 is the common pair of conjugate points of C_2, C_3, then θ_1 commutes with both of θ_2, θ_3. Since P_1, Q_1 is not a pair of conjugate points of C_1, then, conversely, if θ_1 commutes with both of θ_2, θ_3, it must be true that θ_1 induces (7.72) and therefore is in K. If K contains θ_1 but is larger than the four-group $\{\theta_1, \phi\}$, then K must contain, say, θ_2. In this case $\theta_1\theta_2\theta_3$ fixes each of P_1, Q_1, P_3, Q_3 but is not in the projective linear group; hence $\theta_1\theta_2\theta_3 = \phi$. This proves (a), (b), (c) of (7.71) and completes our preliminary analysis of K.

Finally, let H be the subgroup of G which permutes the circles of the unordered triple (7.39) among themselves. Then H is that subgroup of G which permutes the three unordered point-pairs P_i, Q_i among themselves. Furthermore, Z is contained in the center of H; and K is a normal subgroup of H; and H/K is isomorphic to a subgroup of the symmetric group of degree 3, order 6. Moreover, if the nonlinear triple (7.39) has type $\{t\}$, then the integer t must satisfy

(7.73) $$24/t = |H/Z| = |K/Z|\cdot|H/K| .$$

Indeed, each equality (7.69) among the 24 cross-ratio triples associated with (7.39) corresponds to an element of H (unique modulo Z) which maps the unordered pair P_i, Q_i upon the unordered pair P_i', Q_i' for $i = 1, 2, 3$. Thus we gain information about t by discussing the structure of K/Z and H/K (and conversely). We shall prove the following statements simultaneously:

(7.74) θ_1 is in K if and only if (7.46) holds with $\mu = \nu = -1$.

(7.75) $|K/Z| = 4$ if and only if (7.46) holds with $\lambda = \mu = \nu = -1$.

(7.76) If q is even, then $|K/Z| = 1$.

(7.77) If $|K/Z| = 2$, then $|H/K| = 2$ and H/Z is a four-group.

(7.78) If $|K/Z| = 4$, then $|H/K| = 6$ and $q \equiv 1 \bmod 4$.

(7.79) A nonlinear triple (7.39) of type $\{1\}$ $(t = 1)$ exists if and only if $q \equiv 1 \bmod 4$. If $q \equiv 1 \bmod 4$, the triples (7.39) of type $\{1\}$ have type $\{1, 0\}$ and form a single equivalence class under G.

In essence, we need only prove (7.74), (7.77), (7.78) and give an existence proof in connection with (7.79). Indeed by (7.71) and by symmetry, (7.75) follows from (7.74), and (7.76)

follows from (7.74) and the fact that a cross-ratio r must satisfy $r \neq 1$. Again, by (7.73), (7.78), $t = 1$ is equivalent to $|K/Z| = 4$ and requires $q \equiv 1 \bmod 4$. If we note that $q \equiv 1 \bmod 4$ makes -1 a square, we see from (7.75) that (7.79) now depends only on (7.74), (7.78) and an existence proof connected with $\lambda = \mu = \nu = -1$. Now we begin the proof of (7.74), (7.77), (7.78).

If θ_1 is in K, then θ_1 induces the permutation (7.72). In this case, by (7.43), (7.46), (7.50) we get

$$(\lambda, \mu, \nu) = (P_1, Q_1; Q_2, P_2; Q_3, P_3) = (\lambda, \mu^{-1}, \nu^{-1}) .$$

Thus $\mu = \mu^{-1}$. Since $\mu \neq 0, 1$, we see that $\mu = -1$. Similarly, $\nu = -1$. (And, of course, q is odd, $-1 \neq 1$.) Conversely, if $\mu = \nu = -1$, then we have (7.69) with $P_1' = P_1$, $Q_1' = Q_1$, $P_2' = Q_2$, $Q_2' = P_2$, $P_3' = Q_3$, $Q_3' = P_2$. Since no element of G can fix the four points P_2, Q_2, P_3, Q_3 and interchange P_1, Q_1, then H has an element inducing the permutation (7.72); and this must be θ_1. Hence we have proved (7.74) and therefore (7.75), (7.76). Now consider (7.77). If $|K/Z| = 2$ we may assume without loss of generality that K is the four-group generated by θ_1 and ϕ. If ψ is any element of H, then $\psi^{-1}\theta_1\psi$ is in K, is an inversion, and is distinct from ϕ. Since $\theta_1\phi$ and the identity, 1, are not inversions, we see that $\psi^{-1}\theta_1\psi = \theta_1$. Hence θ_1 lies in the center of H. Since, among the six points P_i, Q_i, the only points fixed by θ_1 are P_1, Q_1, we see that H must map the unordered pair P_1, Q_1 upon itself (and hence map the other four points upon themselves). In particular, H/K must have order 1 or 2. Suppose that ψ is in H but not in K. Then ψ^2 is both in H and in the linear projective group, which implies that $\psi^2 = 1$ or $\theta_1\phi$. We must rule out the second alternative. If $\psi^2 = \theta_1\phi$, then, since θ_1 induces (7.72), we may assume (after replacing ψ by ψ^{-1} if necessary) that ψ induces the cycle

$$(P_2, P_3, Q_2, Q_3)$$

and either $(P_1)(Q_1)$ or (P_1, Q_1). Thus either

$$(\lambda, \mu, \nu) = (P_1, Q_1; P_3, Q_3; Q_2, P_2) = (\lambda^{-1}, \nu^{-1}, \mu)$$

or

$$(\lambda, \mu, \nu) = (Q_1, P_1; P_3, Q_3; Q_3, P_2) = (\lambda^{-1}, \nu, \mu^{-1}) .$$

In either case, $\lambda = \lambda^{-1}$ and $\mu = \nu = \nu^{-1}$, whence (without any use of (7.74)) we get $\lambda = \mu = \nu = -1$. (We may remark at this point that if $\lambda = \mu = \nu = -1$, then H does indeed have an element ϕ satisfying $\phi^2 = \theta_1\phi$; more specifically, the Sylow 2-subgroups of H/Z are dihedral.) To go on, we have arrived at a contradiction to (7.75). Hence $\phi^2 = 1$ for every ϕ in H. We still must show that $|H/Z| \neq 1$.

For this it is perhaps simplest to use the representation (7.54). Consistent with (7.74), we shall assume $\mu = \nu = -1$ but, in order to prove (7.78) at the same time, we shall leave λ arbitrary. From (7.56) with $\mu = \nu = -1$ we get $\alpha = -1$, $\beta = -\gamma$. Then the mapping $x \to \beta x^{-1}$ is in G and induces the permutation

$$(P_1, Q_1)(P_2, Q_3)(P_3, Q_2) .$$

Hence H has an element (of order two) which is not in K. This proves (7.74). Now consider the inversion θ_2, namely the mapping $x \to \beta\gamma x^{-q}$. This inversion will be in H if and only if it fixes $P_2 = 1$ and $Q_2 = -1$; that is, if and only if $\beta\gamma = 1$. Since $\beta = -\gamma$, an equivalent condition is $\gamma^2 = -1$; and this may be satisfied precisely when $q \equiv 1 \bmod 4$. However, from (7.57) and (7.58) with $\mu = \nu = -1$ and $\gamma^2 = -1$ we get $\lambda = -1$, as predicted by (7.75). This supplies the desired existence proof in connection with (7.79). Finally, consider the mapping $x \to h(x)$ where

$$h(x) = (x + \beta)/(x - \beta) .$$

Since $\alpha = -1$, $\beta = -\gamma$ and $\gamma^2 = -1$, a simple calculation shows that $x \to h(x)$ induces the permutation

$$(P_1, P_2, P_3)(Q_1, Q_2, Q_3)$$

and hence is in H. Then, since H/K contains elements of order two and three respectively, $|H/K| = 6$. Now the proof of (7.74)–(7.79) is complete.

The next group of statements will complete our chracterization of H/Z. In three of these statements we use the letters "WLG" (standing for "without loss of generality") to mean "when the three point-pairs P_i, Q_i are suitably numbered and when the order in each pair is suitably chosen". In other words: "provided the complete set of 24 ordered cross-ratio triples contains a triple as indicated".

(7.80) $|H/K| = 1, 2$, or 6.

(7.81) $|H/Z| = 6$ if and only if (WLG) (7.46) holds with $\lambda = \mu = \nu = \rho$ where $\rho \neq -1$.

(7.82) $|H/Z| = 2$ if and only if (WLG) (7.46) holds with $\mu = \rho$, $\nu = \rho^{-1}$, $\lambda \neq \rho, \rho^{-1}$, and $\rho \neq -1$.

(7.83) $|H/Z| = 1$ if and only if (WLG) (7.46) holds with $\lambda, \mu, \mu^{-1}, \nu, \nu^{-1}$ all distinct and either (a) $\lambda \neq \lambda^{-1}$ or (b) $\lambda = -1$.

To prove (7.80), we must show that if H/K has an element of order 3, then $|H/K| = 6$. If H/K has an element of order 3, then, since K is a 2-group, H must have an element ψ of order 3. Clearly ψ can fix no unordered pair P_i, Q_i. Hence we may assume, without loss of generality, that ψ induces the permutation

$$(P_1, P_2, P_3)(Q_1, Q_2, Q_3) .$$

Then, by comparing (7.46), (7.48) and using (7.43), we get

$$\lambda = \mu = \nu = \rho , \quad \text{say.} \tag{7.84}$$

If, conversely, (7.84) holds, then, in view of (7.46), (7.47), (7.50), H induces the symmetric group of degree 3 on the unordered point-pairs P_i, Q_i. Thus $|H/K| = 6$. This proves (7.80).

In view of (7.77), (7.78), $|H/Z| = 2$ if and only if $|K/Z| = 1$ and $|H/K| = 2$. Let us begin by assuming merely that H has an element ψ, not in K, such that ψ^2 is in Z (and hence $\psi^2 = 1$). Then ψ must fix one of the unordered point-pairs, say P_1, Q_1, and (without loss of generality) permute the other four points according to $(P_2, P_3)(Q_2, Q_3)$. Then, unless ψ fixes both of P_1, Q_1, ψ must induce the permutation

$$(P_1, Q_1)(P_2, P_3)(Q_2, Q_3) . \tag{7.85}$$

To see that (7.85) is correct, we may use the representation (7.54) and assume that ψ is the mapping $x \rightarrow h(x)$. In particular, $h(1) = \beta$, $h(\beta) = 1$, $h(\alpha) = \gamma$, $h(\gamma) = \alpha$. If $h(\infty) = \infty$ and $h(0) = h(\infty)$ then $h(x) = sx$ or sx^q where $s = h(1) = \beta$ and hence $\alpha = h(\gamma) = \beta\gamma$, contradicting (7.63). Hence (7.85) is correct and we can deduce that

$$\mu = \rho , \qquad \nu = \rho^{-1} \tag{7.86}$$

for a suitable ρ. Indeed, by continuing our computation, $h(x) = sx^{-1}$ or sx^{-q} where $s = h(1) = \beta$; and hence everything is all right provided $\gamma = h(\alpha) = \beta\alpha^{-1}$. Since $\alpha = \nu$ and $\gamma = \beta\mu$, this gives $\mu\nu = 1$ or (7.86). So (7.86) is equivalent to the existence of an element of H which induces (7.85). Assuming (7.86), then, by (7.75), (7.74) and by symmetry, we will have $|K/Z| = 1$ precisely when $\rho \neq -1$. If $\lambda = \rho$ or ρ^{-1}, but not otherwise) then one of the 24 triples will have equal components, whence H/K will have an element of order 3. Thus, to ensure that $|H/Z| = 1$ on the assumption of (7.86), it is necessary and sufficient that we add the conditions $\rho \neq -1$ and $\lambda \neq \rho, \rho^{-1}$. This proves (7.82).

Now consider the method of forming the set of 24 cross-ratio triples from a given cross-ratio triple (λ, μ, ν): we may permute the components λ, μ, ν arbitrarily and replace any two components by their inverses. Thus, if an element from one of three pairs

$$\{\lambda, \lambda^{-1}\}\ , \qquad \{\mu, \mu^{-1}\}\ , \qquad \{\nu, \nu^{-1}\} \tag{7.87}$$

is equal to an element from one of the other pairs (and not otherwise)the set of 24 triples contains one of form (7.86). But, as just pointed out, (7.86) holds precisely when H has an element inducing (7.85). Thus, by (7.80), $|H/K| = 1$ if and only if no two of the pairs (7.87) have a common element. In addition, we know that if $|H/K| = 1$, then $|K/Z| = 1$ and hence $|H/Z| = 1$. Suppose that the pair $\{\lambda, \lambda^{-1}\}$ consists of a single element: $\lambda = \lambda^{-1}$. Then, since $\lambda \neq 0, 1$, we must have q odd and $\lambda = -1$. Hence we see that $|H/Z| = 1$ if and only if either the six elements listed in (7.87) are distinct or q is odd, five of the elements in (7.87) are distinct and one of the pairs (without loss of generality, the pair $\{\lambda, \lambda^{-1}\}$) consists of -1 repeated. This proves (7.83).

At this point, let us sum up our knowledge of the types $\{t\}$ which can exist:

(7.88) If q is even, $t = 4$, 12 or 24.

(7.89) If q is odd, $t = 1$, 4, 6, 12 or 24.

For (7.88) we need only (7.73), (7.76) and (7.80); for (7.89) we add (7.77), (7.78).—In (7.79) we have the complete story on $t = 1$; in (7.67), (7.68), the full details for $q = 5$, $q = 7$ respectively.

For q odd, a close examination of the types $\{t\}$ and, more particularly, of the types $\{t, d\}$ shows that much depends on the nature of -1; for example, on whether it is a fourth power, merely a square, or a non-square in $GF(q)$. Thus, for the types $\{12\}$ and $\{24\}$, corresponding to $|H/Z| = 2$ and 1 respectively, $\lambda \neq -1$ or $\lambda = -1$. Accordingly we split each of these types, and the corresponding types $\{t, d\}$, ($t = 12$ or 24; $d = 0, 1, 2, 3$), into two pieces

$$\{t\} = \{t\}^* \cup \{t\}'; \quad \{t, d\} = \{t, d\}^* \cup \{t, d\}' \qquad (q \text{ odd}; \ t = 12 \text{ or } 24; \ d = 0, 1, 2, 3). \tag{7.90}$$

The pieces may be indicated sufficiently by the nature of one of the cross-ratio triples. For the types $\{t\}^*$, $\{t\}'$ we have:

$$\begin{aligned}
&\{12\}^*: \quad \mu = \rho, \quad \nu = \rho^{-1}; \quad \rho \neq -1; \quad \lambda \neq -1, \rho, \rho^{-1}. \\
&\{12\}': \quad \mu = \rho, \quad \nu = \rho^{-1}; \quad \rho \neq -1; \quad \lambda = -1. \\
&\{24\}^*: \quad \lambda, \lambda^{-1}, \mu, \mu^{-1}, \nu, \nu^{-1} \text{ all distinct.} \\
&\{24\}': \quad \mu, \mu^{-1}, \nu, \nu^{-1} \text{ distinct}; \ \lambda = -1, \qquad (q \text{ odd})
\end{aligned} \tag{7.91}$$

That is, for a disjoint triple (7.39) having one of the types in (7.91), one or more of the 12 or 24 cross-ratio triples (λ, μ, ν) has the indicated form. The description of the types $\{t, d\}^*$, $\{t, d\}'$ should now be clear, since d is the number of non-squares among λ, μ, ν.

For any prime-power q, let $N_t(q)$ be the number of equivalence classes under G (the group defined in Theorem 7.5) among the nonlinear triples (7.39) of type $\{t\}$. For q odd, let $N_{t,d}(q)$ be the corresponding number for type $\{t, d\}$; and let $N_{t,d}^*(q)$, $N_{t,d}'(q)$ be the corresponding numbers for the types in (7.91). Naturally

$$\begin{aligned}
&N_t(q) = \sum_{d=0}^{3} N_{t,d}(q); \\
&N_{t,d}(q) = N_{t,d}^*(q) + N_{t,d}'(q) \text{ if } t = 12 \text{ or } 24 \quad (q \text{ odd}).
\end{aligned} \tag{7.92}$$

When $t = 1$ or 4, H is transitive on the three circles C_i of (7.39). Thus

$$N_1(q) = N_{1,0}(q), \quad N_4(q) = N_{4,0}(q) + N_{4,3}(q) \qquad (q \text{ odd}). \tag{7.93}$$

We can give explicit information about types $\{1\}$ and $\{6\}$:

Lemma 7.16. *Types $\{1\}$ and $\{6\}$ are empty except as follows:*

(i) *If* $q \equiv 1 \bmod 4$, *then* $N_1(q) = N_{1,0}(q) = 1$ *and*

$$N_6(q) = N_{6,0}(q) = (q-5)/4 .$$

(ii) *If* $q \equiv 3 \bmod 4$ *then*

$$N_6(q) = N_{6,2}(q) = (q-3)/4 .$$

Proof. From (7.57) and (7.58) with $\mu = \nu = -1$ we get $\lambda = s^2$ where $s = (\gamma+1)/(\gamma-1)$ and hence $\gamma = (s+1)/(s-1)$. By (7.66), $\gamma \neq 0, 1, -1$. Hence $\lambda = s^2$ can be any square in $GF(q)$, except that $\lambda \neq 0, 1$. For type $\{1\}$ we must have $\lambda = -1$. Each equivalence class of type $\{6\}$ has a pair of cross-ratio triples $(\lambda, -1, -1)$ and $(\lambda^{-1}, -1, -1)$ where $\lambda \neq -1$ and hence $\lambda \neq \lambda^{-1}$. This proves Lemma 7.16. Note that the results check for $q = 5, 7$ and also (trivially) for $q = 3$.

A similar method may be applied to the consideration of types $\{4\}$ and $\{12\}$. Here the situation is more complex and the results are not complete. We prove the next two lemmas together.

Lemma 7.17.

(i) *If* q *is even,*

$$N_4(q) + N_{12}(q) = (q-2)(q-4)/4 .$$

(ii) *If* q *is odd,*

$$N_4(q) + N_{12}(q) = (q-3)(q-5)/4 .$$

Lemma 7.18.

(i) *If* q *is even,* $N_4(q) = (q-3-\varepsilon)/2$ *where* $\varepsilon = \left(\dfrac{q}{3}\right)$. *That is,* $\varepsilon = 1$ *if* q *is a power of* 4, *and* $\varepsilon = -1$ *otherwise.*

(ii) *If* q *is odd,* $N_4(q) = (q-5-2s)/2$ *where* $s = 1$ *if* $q \equiv 1 \bmod 12$, $s = -1$ *if* $q \equiv -1$ *or* $3 \bmod 12$, *and* $s = 0$ *in the remaining cases.*

Proof. The proof of Lemmas 7.17, 7.18 may be discussed rather nicely in terms of the function F defined by

$$(7.94) \qquad F(x) = (1-x)(1-x^{-1}) = 2 - (x + x^{-1}) = F(x^{-1}) ,$$

We note that an equivalence class of nonlinear disjoint triples in class $\{4\}$ or $\{12\}$ has a cross-ratio triple of form $(\lambda, \rho, \rho^{-1})$

where $\lambda \neq \rho, \rho^{-1}$ for type {12} and where we may suppose that $\lambda = \rho^{-1}$ for type {4}. In either case,

(7.95) $$\rho \neq 0, 1, -1 .$$

Setting $\mu = \rho$, $\nu = \rho^{-1}$ in (7.57), (7.58), we find that

(7.96) $$1 - \lambda = F(\rho)/F(\gamma) .$$

The conditions (7.66) on γ can be stated as follows: γ satisfies the same conditions (7.95) as ρ; in addition, the unordered pair γ, γ^{-1} is distinct from the unordered pair ρ, ρ^{-1}. Replacement of ρ by ρ^{-1} in (7.96) does not change λ; and this as it should be. If q is odd, there are $(q-3)/2$ unordered pairs ρ, ρ^{-1} in $GF(q)$ subject to (7.95); and we may choose γ, γ^{-1} as any one of the remaining $(q-5)/2$ pairs. This proves (ii) of Lemma 7.17. If q is even, there are, instead, $(q-2)/2$ unordered pairs ρ, ρ^{-1} subject to (7.95); this completes the proof of Lemma 7.17.

To treat type {4}, we may suppose that $\lambda = \rho^{-1}$ in (7.96). This gives

$$1 - \rho = F(\gamma)$$

or

(7.97) $$\rho = \gamma + \gamma^{-1} - 1 .$$

Here the conditions (7.95) for ρ and γ, and the conditions that the unordered set γ, γ^{-1} be distinct from the unordered set ρ, ρ^{-1}, reduce to the following, when expressed as conditions on γ:

(a) $\gamma \neq 0, 1, -1$;
(b) $\gamma^2 \neq -1$;
(c) $\gamma^2 - \gamma + 1 \neq 0$.

Thus $N_4(q) = (q-k)/2$ where k is the number of values for γ in $GF(q)$ excluded by (a), (b), (c). If q is even, conditions (a), (b) contribute 2 to k; and (c) contributes 2 or 0 according as $q \equiv 1$ or $-1 \bmod 3$. This proves (i) of Lemma 7.18. If q is a power of 3, (a) and (c) contribute 3 to k, and (b) contributes 2 or 0 according as $q \equiv 1 \bmod 4$ or $q \equiv -1 \bmod 4$. Thus, if q is a power of 3, (ii) of Lemma 7.18 holds with $s = 0$ or -1 according as $q \equiv 9$ or $3 \bmod 12$. If q is a power of a prime other than 2 or 3, then (a) contributes 3 to k; (b), (c) contribute 4 if $q \equiv 1 \bmod 12$, 0 if $q \equiv -1 \bmod 12$, and 2 in the remaining cases. This completes the proof of Lemma 7.18.

We can also single out type $\{12\}'$ by insisting that $\lambda = -1$ in (7.96). However, we have not gained any special insight by this approach. Again, when q is odd, we have had little success determining $N_{4,d}(q)$ for $d = 0, 3$, for general q, let alone $N'_{12,d}(q)$ and $N^*_{12,d}(q)$, though (as the proof of the following lemmas will show) partial results are easy to obtain. Two special cases may be mentioned. First, if q is a power of 3, (7.97) may be written as

$$\gamma\rho = (\gamma - 1)^2 ,$$

which shows that γ, ρ are both squares or both non-squares. In this case, counting is easy. For example, when $q = 9$, conditions (a), (b), (c) rule out the squares in $GF(9)$. Hence

(7.98a) $$N_4(9) = N_{4,3}(9) = 2 .$$

Secondly, we shall prove the following:

(7.98b) *For* $q > 9$ (q odd), $\{4, 0\}$ *and* $\{4, 3\}$ *are both non-empty.*

To begin with, we note that γ and $1 - \gamma$ will both satisfy the conditions (a), (b), (c) (in connection with (7.97)) provided γ satisfies (a), (b), (c) and also

(d) $\gamma \neq 2$, (e) $(1 - \gamma)^2 \neq -1$.

Furthermore, if ρ' has form (7.97) with γ replaced by $1 - \gamma$, then

$$\rho/\rho' = \gamma^{-1} - 1 .$$

Hence if we can choose γ in $GF(q)$, subject to (a), (b), (c), (d), (e), so that $\gamma^{-1} - 1$ is a non-square, then one of ρ, ρ' will be a square and the other will be a non-square, proving (7.98). Conditions (a), (d) tell us that $\gamma^{-1} - 1$ cannot express the elements $-1, 0, -2, -1/2$; of these, at most three are non-squares, namely $-1, -2, -1/2$. If $GF(q)$ has an element ω such that

$$\omega^2 - \omega + 1 = 0$$

then $$(\omega^{-1} - 1)\omega^2 = 1 ;$$

hence condition (c) can be ignored. Thus if $q \equiv 3 \bmod 4$, so that (b), (e) can be ignored, the proof goes through for $q > 7$.

On the other hand, if $q \equiv 1 \bmod 4$, so that $GF(q)$ has an element i such that $i^2 = -1$, the only possible non-squares not expressible in form $\gamma^{-1} - 1$ (subject to the stated conditions) are the following three pairs of inverses:

$$-2, -1/2; \ \theta, \theta^{-1}; \ \ \theta/2, 2\theta^{-1},$$

where $\theta = -1 + i$, $\theta^2 = -2i$. Since the product of $-2, \theta^{-1}$ and $\theta/2$ is -1, a square, at most four of the six elements can be non-squares. Hence the proof goes through for $q > 9$. This proves (7.98b).—Note that (7.98b) is false for $q = 5$ or 7, as well as for $q = 9$; indeed, $N_4(5) = 0$, $N_4(7) = 1$.

In our next two lemmas we use the familiar method of "counting in two ways" to evaluate N_{24}.

Lemma 7.19.

(i) *If* $q \equiv 1 \bmod 4$, *then*

$$N'_{12}(q) + 2N'_{24}(q) = (q-5)^2/8 ;$$

(ii) *If* $q \equiv 3 \bmod 4$, *then*

$$N'_{12}(q) + 2N'_{24}(q) = [(q-5)^2 - 4]/8 .$$

Lemma 7.20.

(i) *If* q *is even,*

$$N_{12}(q) + 3N_{24}(q) = (q-2)[(q-5)^2 - 1]/16 .$$

(ii) *If* $q \equiv 1 \bmod 4$,

$$N_{12}(q) + 3N_{24}(q) = (q-2)(q-5)^2/16 .$$

(iii) *If* $q \equiv 3 \bmod 4$,

$$N_{12}(q) + 3N_{24}(q) = ((q-2)[(q-5)^2 - 4]/16) + ((q-3)/4) .$$

Proof. We shall prove Lemmas 7.19, 7.20 in detail for the case $q \equiv 1 \bmod 4$, actually deriving more than is claimed in the lemmas. The proof for $q \equiv 3 \bmod 4$ is quite similar. The proof for q even (of Lemma 7.20 only) is much simpler; roughly speaking, five steps are replaced by a single one.

First we note that, since -1 is a square for $q \equiv 1 \bmod 4$,

(7.99) $N'_{12,1}(q) = N'_{12,3}(q) = N'_{24,3}(q) = 0 \qquad (q \equiv 1 \bmod 4)$.

Next we count in two ways the total number, x, of ordered cross-ratio triples (λ, μ, ν) with $\mu = -1$ and with ν a square $\neq 0, 1, -1$. On the one hand, given $\mu = -1$, we can choose a square ν distinct from the squares $0, 1, -1$ in $(q-5)/2$ ways; then, given $\mu = -1$ and the square ν (so that $\mu \neq \nu$, $\mu\nu \neq 1$, $\mu\nu =$ square) Lemma 7.15 tells that we can choose λ in $(q-7)/2$ ways. Therefore

$$x = (q-5)(q-7)/4 \, .$$

On the other hand, for example, among the 6 ordered cross-ratio triples belonging to a triple (7.39) of disjoint circles of type $\{6, 0\}$, exactly two have form $(\lambda, -1, \nu)$ where ν is a square $\neq 0, 1, -1$. Similarly, "frequency numbers" for each type may be given as follows:

$$\{6, 0\} : 2 ; \quad \{12, 0\}' : 4 ; \quad \{24, 0\}' : 8 ; \quad \{24, 1\}' : 4 \, ,$$

and 0 for all other types. Hence

$$x = 2N_{6,0}(q) + 4N'_{12,0}(q) + 8N'_{24,0}(q) + 4N'_{24,1}(q) \, .$$

By equating the two values for x, we get

$$\begin{aligned} (7.100) \quad & N_{6,0}(q) + 2N'_{12,0}(q) + 4N'_{24,0}(q) + 2N'_{24,1}(q) \\ & \qquad = (q-5)(q-7)/8 \qquad (q \equiv 1 \bmod 4) \, . \end{aligned}$$

Again, by counting in two ways the number, y, of ordered cross-ratio triples $(\lambda, -1, \nu)$ where ν is a non-square, we get

$$\begin{aligned} (7.101) \quad & 2N'_{12,2}(q) + 4N'_{24,2}(q) + 2N'_{24,1}(q) = (q-1)(q-5)/8 \\ & \qquad\qquad (q \equiv 1 \bmod 4) \, . \end{aligned}$$

From (7.100), (7.101), by addition, in view of (7.99),

$$N_{6,0}(q) + 2N'_{12}(q) + 4N'_{24}(q) = (q-4)(q-5)/4 \, .$$

Eliminating $N_{6,0}(q)$ by use of Lemma 7.16 (i), and dividing by two, we get

$$(7.102) \quad N'_{12}(q) + 2N'_{24}(q) = (q-5)^2/8 \qquad (q \equiv 1 \bmod 4) \, ,$$

as claimed in Lemma 7.19 (i).

Next we consider the ordered cross-ratio triples (λ, μ, ν) subject to the requirement that $0, 1, -1, \mu, \mu^{-1}, \nu, \nu^{-1}$ are all distinct and to one of the following additional requirements: (a)

μ and ν are squares; (b) μ is a square, ν is a non-square; (c) μ and ν are non-squares. Counting in two ways subject to (a) (b), (c) in turn, we get

$$\begin{aligned}
&N^*_{12,0}(q) + N'_{24,0}(q) + 3N^*_{24,0}(q) + N^*_{24,1}(q) \\
&\qquad = (q-5)(q-7)(q-9)/64\ , \\
&N^*_{12,1}(q) + N^*_{12,2}(q) + N'_{24,1}(q) + 2N^*_{24,1}(q) + 2N^*_{24,2}(q) \\
&\qquad = (q-1)(q-5)^2/32\ , \\
&N^*_{12,3}(q) + N'_{24,2}(q) + 3N^*_{24,3}(q) + N^*_{24,2}(q) \\
&\qquad = (q-1)(q-5)(q-7)/64\ , \\
&\qquad\qquad (q \equiv 1 \bmod 4)\ .
\end{aligned} \tag{7.103}$$

Adding together (7.102) and the three equations in (7.103), and recalling (7.99), we get

$$N_{12}(q) + 3N_{24}(q) = (q-2)(q-5)^2/16 \qquad (q \equiv 1 \bmod 4)$$

as claimed in Lemma 7.20 (ii).

All of (7.99)–(7.103) are false for $q \equiv 3 \bmod 4$; nevertheless, the proof for $q \equiv 3 \bmod 4$ is quite similar. To prove Lemma 7.20 for q even, we have a much simpler task: we merely count in two ways the ordered cross-ratio triples (λ, μ, ν) with $0, 1, \mu, \mu^{-1}, \nu, \nu^{-1}$ all distinct. This completes the proof of Lemmas 7.19, 7.20.

Lemmas 7.17–7.20 allow us to write down the numbers $N_t(q)$ explicitly; we fail to do this because the formulas fall into six separate groups: two groups for q even, two for $q \equiv 3 \bmod 4$, and two for $q \equiv 1 \bmod 4$. (Compare Lemmas 7.18, 7.20, for example.) The complexity would be even greater for the numbers $N_{t,d}(q)$, q odd; at best we could expect a division into cases modulo 24.—Note, incidentally, that although Lemmas 7.17–7.20 together with equations such as (7.98) and (7.99) through (7.103) are not enough to yield the numbers $N_{t,d}$ and $N'_{t,d}, N^*_{t,d}$ for a general prime-power q, they suffice with little additional computation for q small. In Appendix III, we give a table for the first few values of q.

The following theorem gives a crude summary of our results:

Theorem 7.21. *The total number of equivalence classes of non-linear triples* (7.39) *of disjoint circles of* $IP(q)$ *under the group*

G of Theorem 7.5 *is asymptotic to* $q^3/48$. *More explicitly, the numbers* $N_t(q)$ *are asymptotic to* $q/2$, $q^2/4$ *and* $q^3/48$ *respectively for* $t = 4, 12, 24$; *and (when q is odd)* $N_6(q)$ *is asymptotic to* $q/4$.

Proof. This is immediate from Lemmas 7.17 through 7.20, in the context of the long discussion of nonlinear triples which began right after the proof of Theorem 7.15.

The reader need not fear that we shall now emback upon a thorough discussion of nonlinear quadruples of disjoint circles of $IP(q)$—for one thing, we do not know the theory in any detail. We shall make only a few remarks. First of all, such quadruples exist for $q > 5$, and they fall into two broad types. For the one type, there is one linear subtriple and three nonlinear subtriples. For the other type, there are four nonlinear subtriples. In addition, the common orthogonal circles of the subtriples can form interesting configurations. I have examined (but not classified) the quadruples for $q = 7$. Janet McDonald, in examining some of the subregular spreads for $q = 8$, has found, in particular, a maximal quintuple of disjoint circles of $IP(8)$ having 10 nonlinear subtriples and 10 distinct pairs of commmon conjugate points.

We end our study of nonlinear sets of disjoint circles by returning to Lemma 7.8, taking K to be a complete linear set, and asking whether we can choose a nonlinear set L to satisfy (iia), one or both cases of (iib), or some variation thereof. First we need a condition for tangency of circles. The lemma which follows becomes familiar indeed when we take $P = \infty$ and think of C, C' as two circles of the real Euclidean affine plane A with distinct centers Q, Q'; then D becomes the line joining the centers.

Lemma 7.22. *Let* P, Q, Q', R *be four distinct points of* $IP(q)$. *Let* D *be the circle containing* P, Q, Q'. *Let* C *be the circle through* R *with* P, Q *as conjugate points, and let* C' *be the circle through* R *with* P, Q' *as conjugate points. Then a necessary and sufficient condition that* C, C' *be tangent is that* R *lie on* D.

Proof. We may assume without loss of generality that

$$P = \infty, \quad Q = 0, \quad Q' = s, \quad R = 1 \qquad (s \neq 0, 1)$$

where s is in $PG(q^2)$. Then the inversion for C has form

$$x \to x^{-q}$$

and the equation for C is

$$C: x^{q+1} = 1 . \tag{7.104}$$

The inversion for C' has form $x \to h(x)$ where

$$h(x) = (\lambda x^q + \alpha)/(\beta x^q - \lambda^q)$$

and $\alpha, \beta \in GF(q)$, $\lambda \in GF(q^2)$. Since, by hypothesis, $h(\infty) = s$, we may assume $\beta = 1$, $\lambda = s$. Since, also $h(1) = 1$, then

$$1 - s^q = s + \alpha ; \qquad \alpha + s^{q+1} = 1 - s - s^q + s^{q+1} = (1 - s)^{q+1} .$$

Therefore

$$h(x) - s = (1 - s)^{q+1}/(x - s)^q ,$$

and the equation for C' is

$$C' : \ (x - s)^{q+1} = (1 - s)^{q+1} . \tag{7.105}$$

By construction, C, C' have $R = 1$ as a common point. However, C, C' also have s^{1-q} as a common point, as we may verify from (7.104), (7.105). Hence C, C' are tangent if and only if $s^{1-q} = 1$; that is, if and only if s is in $GF(q)$. On the other hand, the circle through $P = \infty$, $Q = 0$, $R = 1$ is

$$C_0 = \infty \cup GF(q) .$$

Thus R will be on D if and only if $D = C_0$; hence, if and only if s is in $GF(q)$. This completes the proof of Lemma 7.22.

Lemma 7.23. *Assume $q > 4$. Let P, Q be two distinct points of $IP(q)$, and let $K = K(P, Q)$ be the complete linear set of $q - 1$ disjoint circles of $IP(q)$ with P, Q as conjugate points. Let C' be a circle not in K, and suppose that C' contains i of the points P, Q, is tangent or secant to t or s, respectively, of the circles in K, and is disjoint from d of the circles in K. Then, according as q is even or odd, the following possibilities are exhaustive. All can occur:*

I. *q even.*

(I. 1) $i = 2$, $t = q - 1$, $s = 0$, $d = 0$.

(I. 2) $i = 1$, $t = 0$, $s = q/2$, $d = (q - 2)/2$.

(I. 3) $i = 0$, $t = 1$, $s = q/2$, $d = (q - 4)/2$.

II. q *odd*.

(II. 1) $i = 2$, $t = 0$, $s = (q-1)/2$, $d = (q-1)/2$.

(II. 2) $i = 1$, $t = 1$, $s = (q-1)/2$, $d = (q-3)/2$.

(II. 3) $i = 0$, $t = 0$, $s = (q+1)/2$, $d = (q-3)/2$.

(II. 4) $i = 0$, $t = 2$, $s = (q-1)/2$, $d = (q-5)/2$.

Remark. Note that, in each of the above cases for which $d \geq 2$, C' together with the d circles of K disjoint from C' forms a nonlinear set of $d+1$ disjoint circles of $IP(q)$. For $q \geq 8$, only (I. 1) can fail to yield such sets.

Proof. First we treat the case $i = 2$, so that C' is a circle through P and Q. It will be convenient to use the notation of the proof of Theorem 7.5. Then the $q-1$ circles of K correspond to the $q-1$ inversions h_i given by (7.29), and the $q+1$ circles through P, Q correspond to the $q+1$ inversions H_j given by (7.30). By Lemma 7.12, a circle of K and a circle through P, Q must be tangent if q is even (this disposes of (I. 1)) and secant or disjoint if q is odd. We want to prove (II. 1) in a strong form, namely: *For q odd, a circle with inversion h_i is secant to a circle with inversion H_j if and only if $i + j$ is even.* To see this, consider an element x of $GF(q^2)$. The equation

$$h_i(x) = H_j(x)$$

is easily seen to be equivalent to

$$x^2 = \omega^i U^j . \tag{7.106}$$

Hence (compare Lemma 7.4) (7.106) has solutions in $GF(q^2)$ for all i, j, and, for x satisfying (7.106), either $x, h_i(x)$ is the common pair of conjugate points of the circles or $x = h_i(x)$ is a common point of the circles. We may write $q = 1 + 2n$ for a positive integer n. By the choice of ω, U,

$$\omega^n = -1 = U^{n+1} .$$

Hence, if x satisfies (7.106), then

$$x^{q+1} = (x^2)^{n+1} = (\omega^i U^j)^{n+1} = (-1)^{i+j}\omega^i$$

or

$$x = (-1)^{i+j} h_i(x) .$$

This proves the italicized statement and disposes of (II. 1).

Henceforth we may assume that $i = 0$ or 1. Thus we may assume without loss of generality that C' does not contain P. In this case the conjugate, Q', of P with respect to C' is distinct from P, Q. Now let us turn matters around, choose any point Q' distinct from P, Q, and allow C' to range over the circles in the complete linear set $K' = K(P, Q')$. Then let D be the circle containing P, Q, Q'. Thus, if q is even, every circle C of K and every circle C' of K' is tangent to D (by (I. 1)). In this case, if a circle C' of K' has in common with D a point R, then, by Lemma 7.22, if $R \neq Q$, C' is tangent to a unique circle C of K (namely, that through R) whereas, if $R = Q$, C' is tangent to no circle of K. This explains (I. 2), (I. 3) as far as i, t are concerned. Next let q be odd. Then every circle C of K (and every circle C' of K') is secant to D or disjoint from D; and both cases can occur. If we choose C' to contain Q, then C' meets D in a second point R, distinct from Q, Q', and (by Lemma 7.22), C' is tangent to a unique circle C of K, namely, that through R. If we choose C' to meet D in distinct points R_1, R_2, distinct from Q, Q', then C' is tangent to two circles of K, one through R_1 and one through R_2. If we choose C' to be disjoint from D, then C' cannot be tangent to a circle in K. This explains (I. 2) through (I. 4) as far as i, t are concerned. To complete the proof we merely note the equations

$$t + s + d = q - 1\,, \qquad i + t + 2s = q + 1\,.$$

The first holds because K has $q - 1$ circles; the second, because neither P nor Q is on a circle of K, but each of the remaining $q + 1 - i$ points of C' is on exactly one circle of K. This proves Lemma 7.23.

We note that (II. 1) yields a nonlinear set of $(q + 1)/2$ disjoint circles with a linear subset of $(q - 1)/2$ circles and that (I. 2) yields a nonlinear set of $q/2$ disjoint circles with a linear subset of $(q - 2)/2$ circles. Thus we have shown the existence of case (iia) of Lemma 7.8 and of one of the alternatives in (iib) of Lemma 7.8—for K a complete linear set. The nonlinear disjoint sets of $d + 1$ disjoint circles arising from (I. 3), (II. 2), (II. 3) or (II. 4) do not fit the operative hypotheses of Lemma 7.8—and they also violate the conclusions of that lemma.

Now we are ready to return to the connection between subregular spreads of $PG(3, q)$ and sets of disjoint circles of $IP(q)$.

8. CLASSIFICATION OF THE ANDRÉ PLANES AND THE SUBREGULAR PLANES

It is time to put our various conclusions together. In Section 3 we explained Bose's construction which makes correspond to a spread $\mathcal{S}$ of $PG(3, q)$ a translation plane $\pi(\mathcal{S})$ of order q^2. Subject to a conjecture, $U(q)$, valid when q is a prime, we showed that translation planes $\pi(\mathcal{S})$, $\pi(\mathcal{S}')$ are isomorphic if and only if there exists a collineation of $PG(3, q)$ mapping the spread $\mathcal{S}$ upon the spread $\mathcal{S}'$. In Section 4 we introduced the concept of a regular spread and stated theorems (whose proofs are to be found in Appendix I) showing that the projective linear group $PL(\Sigma)$ of $\Sigma = PG(3, q)$ is transitive on regular spreads. We also explained that the lines and reguli of a regular spread $\mathcal{S}$ constitute the points and circles, respectively, of the inversive plane $IP(q)$, studied in Section 7, and that the subgroup of $PL(\Sigma)$ which maps $\mathcal{S}$ upon itself induces a group of collineations of $IP(q)$ called G in Theorem 7.5. In Section 6 we introduced the concept of a subregular spread of index k and showed that, for $0 \leq k \leq (q-1)/2$, a subregular spread of index k is obtained from a regular spread $\mathcal{S}$ by reversing a unique set of k disjoint reguli of $\mathcal{S}$. Consequently, for $0 \leq k \leq (q-1)/2$, the classification of subregular spreads of index k under $PL(\Sigma)$ is equivalent to the classification of sets of k disjoint circles of $IP(q)$ under the group G of Theorem 7.5.

It is a fact (to be discussed in detail in Appendix II) that the sub-class of the so-called André planes which have representations $\pi(\mathcal{S})$ is identical with the class of planes $\pi(\mathcal{S})$ for which the spread $\mathcal{S}$ is subregular and is obtained from a regular spread by reversing a linear set of disjoint reguli. In particular, if q is a prime, every André plane of order q^2 is so obtained but, if q is a prime-power other than a prime, one obtains only those André planes of order q^2 which are, in a well-defined sense, " 2-dimensional ".

As a consequence we may imbed the class of "2-dimensional" André planes of order q^2 in a larger class of *subregular* planes, $\pi(\mathcal{S})$, namely those planes for which $\mathcal{S}$ is subregular, and we may assign to each such plane an *index*, k, namely the index of subregularity of $\mathcal{S}$.

To classify the André planes, we may use (vii) of Theorem 7.5. Given any prime-power q, we consider the subsets of the set of integers mod $q-1$, and form equivalence classes under the group of substitutions

(8.1) $$i \to si + t \bmod q - 1$$

where $s = \pm 1$. If A is a subset (possibly empty) of the integers $\bmod q - 1$ and if A' is the complementary subset, the equivalence class of A and the equivalence class of A' correspond to isomorphic André planes. (Compare, for example, Lemma 6.2, with special reference to (vi) of that lemma.) Hence we need consider only those subsets with k elements, where $0 \leq k \leq q - 1$; however, in the case that $k = (q-1)/2$, we must combine the equivalence class of a k-element subset A with that of the complementary k-element subset A'. With this modification understood, then, for $0 \leq k \leq (q-1)/2$, each equivalence class of k-element subsets corresponds to an equivalence class of subregular spreads of index k under the projective linear group $PL(\Sigma)$ of $\Sigma = PG(3, q)$. If q is a prime, all collineations of Σ are in $PL(\Sigma)$; in this case, the process just described completely classifies the André planes of order q^2. If

(8.2) $$q = p^e, \quad p \text{ prime}, \quad e > 1,$$

the $PL(\Sigma)$ has index e in the group, $PSL(\Sigma)$, of all collineations of Σ. In this case (even assuming the conjecture $U(q)$) the equivalence classes need to be enlarged by use of the "multiplier"

(8.3) $$i \to pi \bmod q - 1 .$$

(For proof of this last fact, see Appendix II; specifically, Lemma 9.5.)

We give a brief table showing the number of non-isomorphic "2-dimensional" André planes of order q^2 for small q. A star (*) indicates a prime-power for which we have cut down the number of classes using (8.3). The star also serves as a reminder that there may be other "higher-dimensional" André planes of order q^2. The planes of indices 0 and 1 are the Desarguesian plane and the Hall plane, respectively.

(8.4)

q	Total	By Index:	0	1	2	3	4	5
3	2		1	1				
4*	2		1	1				
5	4		1	1	2			
7	8		1	1	3	3		
8*	5		1	1	1	2		
9*	14		1	1	3	4	5	
11	44		1	1	5	8	16	13

Our investigations in Section 7 show that, in addition to the André planes of order q^2 there are many non-André subregular planes. For example, by (7.68), there are 3 non-André subregular planes of order 7^2, index 3, in addition to the 3 André planes of order 7^2, index 3. (Also see the table in Appendix III.) Moreover, for q a large prime, it is easy to see that the number of André planes of order q^2, index 3 is asymptotic to $q^2/12$ whereas (by Theorem 7.21) the number of non-André subregular planes of order q^2, index 3 is asymptotic to $q^3/48$. Although we have not worked out the classification of non-André subregular planes of index exceeding 3, some (but by no means all) of the possible types are indicated by Lemma 7.23.

9. APPENDIX I

Our main purpose in this section is to prove the theorems stated in Section 4. We may regard $PG(3, q)$ as a 4-dimensional vector space V over $GF(q)$, points, lines and planes of $PG(3, q)$ being vector subspaces of V over $GF(q)$ of dimensions 1, 2, 3 respectively. We have used a lemma such as the following in Section 7.

Lemma 9.1. *Let P, Q, R be three distinct points of a line L of $PG(3, q)$ and let $\mathbf{e}$ be any basis element for P. Then there exists a unique basis element $\mathbf{e}'$ for Q such that $\mathbf{e} + \mathbf{e}'$ is a basis element for R.*

We shall omit the (very simple) proof. The following extension of Lemma 9.1 is appropriate:

Lemma 9.2. *Given an ordered double-triple* (4.1) *and any basis element $\mathbf{e}_1$ for the point $A \cap A'$, then vectors $\mathbf{e}_2, \mathbf{e}_1', \mathbf{e}_2'$ are uniquely determined in V such that*

$$(9.1)\quad \begin{aligned} &A \cap A' = \{\mathbf{e}_1\}, && A \cap B' = \{\mathbf{e}_2\}, \\ & && A \cap C' = \{\mathbf{e}_1 + \mathbf{e}_2\} \\ &B \cap A' = \{\mathbf{e}_1'\}, && B \cap B' = \{\mathbf{e}_2'\}, \\ & && B \cap C' = \{\mathbf{e}_1' + \mathbf{e}_2'\} \\ &C \cap A' = \{\mathbf{e}_1 + \mathbf{e}_1'\}, && C \cap B' = \{\mathbf{e}_2 + \mathbf{e}_2'\}, \\ & && C \cap C' = \{\mathbf{e}_1 + \mathbf{e}_2 + \mathbf{e}_1' + \mathbf{e}_2'\}\ . \end{aligned}$$

Proof. By applying Lemma 9.1 to A we get the first row of (9.1) for a unique vector $\boldsymbol{e}_2$. Then, by applying Lemma 9.1 to A' we get the first column of (9.1) for a unique vector $\boldsymbol{e}_1'$. Then, by applying Lemma 9.1 to B, we get the second row of (9.1) for a unique vector $\boldsymbol{e}_2'$. By considering the second and third columns of (9.1) we see that

$$C \cap B' = \{x\boldsymbol{e}_2 + \boldsymbol{e}_2'\}\ , \qquad C \cap C' = \{y(\boldsymbol{e}_1 + \boldsymbol{e}_2) + (\boldsymbol{e}_1' + \boldsymbol{e}_2')\}$$

for some x, y in $GF(q)$. Finally, by considering the last row of (9.1), we recognize that $x = y = 1$.

In (9.1), the vectors $\boldsymbol{e}_1, \boldsymbol{e}_2, \boldsymbol{e}_1', \boldsymbol{e}_2'$ form a basis for V over $GF(q)$. Also

$$\text{(9.2)} \quad \begin{aligned} A &= \{\boldsymbol{e}_1, \boldsymbol{e}_2\}\ , \quad B = \{\boldsymbol{e}_1', \boldsymbol{e}_2'\}\ , \quad C = \{\boldsymbol{e}_1 + \boldsymbol{e}_1', \boldsymbol{e}_2 + \boldsymbol{e}_2'\}\ , \\ A' &= \{\boldsymbol{e}_1, \boldsymbol{e}_1'\}\ , \quad B' = \{\boldsymbol{e}_2, \boldsymbol{e}_2'\}\ , \quad C' = \{\boldsymbol{e}_1 + \boldsymbol{e}_2, \boldsymbol{e}_1' + \boldsymbol{e}_2'\}\ . \end{aligned}$$

Conversely, if $\boldsymbol{e}_1, \boldsymbol{e}_2, \boldsymbol{e}_1', \boldsymbol{e}_2'$ is any basis of V over F, the six lines defined by (9.2) yield an ordered double-triple (4.1) and satisfy (9.1). If, in Lemma 9.2, the vector $\boldsymbol{e}_1$ is replaced by $k\boldsymbol{e}_1$, where k is a non-zero element of $GF(q)$, then by uniqueness, the vectors $\boldsymbol{e}_2, \boldsymbol{e}_1', \boldsymbol{e}_2'$ must also be multiplied by k.

It should now be clear that the projective linear group $PL(\Sigma)$, where $\Sigma = PG(3, q)$, is strictly transitive on the ordered double-triples (4.1). This proves Theorem 4.1.

Let us consider the ordered double-triple (4.1), and hence the basis $\boldsymbol{e}_1, \boldsymbol{e}_2, \boldsymbol{e}_1', \boldsymbol{e}_2'$ of V, as given. Then we define lines as follows:

$$\text{(9.3)} \quad \begin{aligned} &L(\infty) = \{\boldsymbol{e}_1, \boldsymbol{e}_2\}\ , \\ &\qquad L(X) = \{x_{11}\boldsymbol{e}_1 + x_{12}\boldsymbol{e}_2 + \boldsymbol{e}_1', x_{21}\boldsymbol{e}_1 + x_{22}\boldsymbol{e}_2 + \boldsymbol{e}_2'\}\ , \\ &L'(\infty) = \{\boldsymbol{e}_1, \boldsymbol{e}_1'\}\ , \\ &\qquad L'(X) = \{x_{11}\boldsymbol{e}_1 + x_{12}\boldsymbol{e}_1' + \boldsymbol{e}_2, x_{21}\boldsymbol{e}_1 + x_{22}\boldsymbol{e}_1' + \boldsymbol{e}_2'\}\ , \end{aligned}$$

for every 2 by 2 matrix $\boldsymbol{X} = (x_{ij})$ with elements in $GF(q)$. In particular, interchange of $\boldsymbol{e}_2, \boldsymbol{e}_1'$ interchanges $L(\infty)$ with $L'(\infty)$ and $L(X)$ with $L'(X)$. Also

$$\text{(9.4)} \quad \begin{aligned} L(\infty) &= A\ , \qquad L(\boldsymbol{0}) = B\ , \qquad L(\boldsymbol{I}) = C\ , \\ L'(\infty) &= A', \qquad L'(\boldsymbol{0}) = B', \qquad L'(\boldsymbol{I}) = C', \end{aligned}$$

where $\boldsymbol{0}$ and $\boldsymbol{I}$ denote the zero matrix and the identity matrix, respectively.

We may verify that every line skew to $A = L(\infty)$ has form $L(X)$ for a unique matrix X, and conversely. In addition, if X, Y are distinct matrices, lines $L(X), L(Y)$ are skew if $X - Y$ is nonsingular and meet in a point if $X - Y$ has rank 1. Similarly for the lines $L'(X)$ skew to $A' = L'(\infty)$. Moreover, the transversals to A, B, C are $L'(\infty)$ and the lines $L'(kI)$; and the transversals to A', B', C' are the $L(\infty)$ and the lines $L(kI)$; where k ranges over $GF(q)$. Thus

$$(9.5)\qquad \begin{aligned} \mathcal{R} &= \mathcal{R}(A, B, C) = L(\infty) \cup \{L(kI) \mid k \in GF(q)\} ;\\ \mathcal{R}' &= \mathcal{R}(A', B', C') = L'(\infty) \cup \{L'(kI) \mid k \in GF(q)\} . \end{aligned}$$

Every point P of $PG(3, q)$ has the form

$$(9.6)\qquad P = \{x_1 e_1 + x_2 e_2 + x_1' e_1' + x_2 e_2'\}$$

where x_1, x_2, x_1', x_2' are elements of $GF(q)$, not all zero. Thus the doubly-ruled quadric

$$(9.7)\qquad Q = Q(A, B, C) = Q(A, B', C') = Q(\mathcal{R}) = Q(\mathcal{R}')$$

may be described as the set consisting of those points (9.6) such that

$$(9.8)\qquad Q:\ x_1 x_2' - x_1' x_2 = 0 .$$

The polarity, ρ, defined by Q sends a point (9.6), not necessarily on Q, into the *polar plane* consisting of all points

$$Q = \{y_1 e_1 + y_2 e_2 + y_1' e_1' + y_2' e_2'\}$$

such that

$$(9.9)\qquad x_1 y_2' + y_1 x_2' - x_1' y_2 - y_1' x_2 = 0 .$$

Since ρ sends the line joining two distinct points R, S into the line of intersection of the (distinct) polar planes $R\rho$, Sρ, we see readily that

$$(9.10)\qquad \begin{aligned} L(\infty)\rho &= L(\infty) , & L(X)\rho &= L(X^*) ,\\ L'(\infty)\rho &= L'(\infty) , & L'(X)\rho &= L'(X^*) \end{aligned}$$

where

$$(9.11)\qquad X^* = (x_{11} + x_{22})I - X = \begin{pmatrix} x_{22} & -x_{12} \\ -x_{21} & x_{11} \end{pmatrix}$$

is the adjoint of the matrix X. We recall that

$$(9.12) \qquad (X+Y)^* = X^* + Y^*\,, \qquad (XY)^* = Y^*X^*$$

for all 2 by 2 matrices X, Y.

First let q be odd. Then the point (9.6) lies on its polar plane (9.9) if and only if (9.6) satisfies (9.8). From this it follows that the polarity ρ uniquely determines the doubly-ruled quadric Q. In addition, from (9.10), (9.11), a line $L(X)$ (or a line $L'(X)$) is self-conjugate if and only is X is a scalar matrix kI and hence $L(X)$ (or $L'(X)$) is a ruling of Q. We may verify, as well, that no line meeting both of $L(\infty)$, $L'(\infty)$ is self-conjugate.

Next let q be even. Then every point (9.6) lies on its polar plane (9.9). In addition, if P is a point of Σ and L is a line of Σ through P, it is easily seen that $L\rho = L$ precisely when L lies in $P\rho$. Thus the self-conjugate lines of Q are the rulings and the ordinary tangents of Q. Nor does ρ determine Q. Indeed, if a_1, a_2, a_1', a_2', b_1, b_2, b_1', b_2' are elements of $GF(q)$ such that

$$a_1b_2' + b_1a_2' = a_2b_1' + b_2a_1' = k \neq 0\,,$$

then the equation

$$(a_1x_1 + a_2'x_2')(b_1x_1 + b_2'x_2') + (a_2x_2 + a_1'x_1')(b_2x_2 + b_1'x_1') = 0\,,$$

which may be written as

$$a_1b_1x_1^2 + a_2b_2x_2^2 + a_1'b_1'(x_1')^2 + a_2'b_2'(x_2')^2 + k(x_1x_2' + x_2x_1') = 0\,,$$

defines a doubly-ruled quadric with the same polarity ρ as the quadric Q given by (9.8). Hence, for q even, we cannot expect to distinguish between doubly-ruled quadrics on the basis of their polarities alone. Now we return to our general considerations.

A line $L(X)$ is skew to the regulus $\mathfrak{R}$ (given by (9.5)) if and only if $X - kI$ is nonsingular for every k in $GF(q)$; that is, if and only if the matrix X is irreducible. If U is irreducible,

$$(9.13) \qquad \mathcal{S}(U) = L(\infty) \cup \{L(aI + bU) \mid a, b \in GF(q)\} \qquad (U \text{ irreducible})$$

is a set of $q^2 + 1$ skew lines of $PG(3, q)$ and hence is a spread of $PG(3, q)$. Clearly $\mathcal{S}(U)$ contains $\mathcal{R}$ and $L(U)$; we may write

$$(9.13') \qquad \mathcal{S}(U) = \mathcal{R} \cup \{L(aI + bU)\} \mid a, b \in GF(q); b \neq 0\} ,$$

We draw from [3] the fact that $\mathcal{S}(U)$ is a regular spread, the unique regular spread containing $\mathcal{R}$ and $L(U)$.

If U is irreducible, the q^2 matrices $aI + bU$ form an algebra over $GF(q)$ isomorphic to the field $GF(q^2)$. In particular, the mapping $X \to X^q$ is an automorphism of this algebra over $GF(q)$ which coincides with the mapping $X \to X^*$. That is:

$$(9.14) \qquad X = aI + bU, \quad U \text{ irreducible} \Longrightarrow X^* = X^q . \qquad (a, b \in GF(q))$$

By (9.10), (9.11), (9.13), (9.14),

$$(9.15) \qquad \mathcal{S}(U)\rho = \mathcal{S}(U^*) = \mathcal{S}(U^q) \qquad (U \text{ irreducible}) .$$

In particular, ρ partitions the lines of $\mathcal{S}(U)$ skew to $\mathcal{R}$ into $(q^2 - q)/2$ pairs of conjugate nonsecants of $Q = Q(\mathcal{R})$. This is enough for the proof of Theorem 4.3.

If U is irreducible, so that $L(U)$ is skew to $\mathcal{R}$, then $L(U)$ is skew to the opposite regulus $\mathcal{R}'$ also. Hence

$$(9.16) \qquad L'(V) = L(U)$$

for a unique irreducible matrix V. Also, the spread

$$(9.17) \qquad \begin{aligned} \mathcal{S}'(V) &= L'(\infty) \cup \{L'(aI + bV) \mid a, b \in GF(q)\} \\ &= \mathcal{R}' \cup \{L'(aI + bV) \mid a, b \in GF(q); b \neq 0\} \end{aligned}$$

is the unique regular spread containing $\mathcal{R}'$ and $L'(V) = L(U)$. Using the relation (9.16) we may verify that

$$(9.18) \qquad L'(aI + bV) = L(M(a, b)UM(a, b)^{-1}) , \qquad b \neq 0 ,$$

where

$$(9.19) \qquad M(a, b) = \begin{pmatrix} 1 & 0 \\ -a & b \end{pmatrix}, \qquad b \neq 0 .$$

Hence, from (9.16), (9.17), (9.18),

$$(9.20) \qquad \mathcal{S}'(V) = \mathcal{R}' \cup \{L(M(a, b)UM(a, b)^{-1}) \mid a, b \in GF(q); b \neq 0\} .$$

As (a, b) ranges over the $q(q - 1)$ ordered pairs of elements a, b of $GF(q)$ with $b \neq 0$, it is clear from (9.20) that

$$M(a, b)UM(a, b)^{-1}$$

must range over $q(q-1)$ distinct matrices similar to U. *These are all the matrices similar to* U. Indeed, the general linear group $GL(2, q)$ has order $(q^2-1)(q^2-q)$; the centralizer of U in $GL(2, q)$ consists of the group of matrices $aI + bU$, $b \neq 0$, of order $q^2 - 1$; and hence the number of matrices similar to U is $q^2 - q$. As a consequence, $\mathcal{S}(U)$ *and* $\mathcal{S}'(V)$ *have exactly two common lines, namely* $L(U)$ *and* $L(U)\rho = L(U^*)$.

Let ϕ be a nonsingular linear transformation of the vector space V over $GF(q)$ which maps each of the lines $A = L(\infty)$, $B = L(0)$, $C = L(I)$ upon itself. Then, as is easily seen, ϕ must affect the basis e_1, e_2, e_1', e_2' according to

$$\text{(9.21)} \quad \begin{aligned} e_1\phi &= r_{11}e_1 + r_{12}e_2\,, & e_1'\phi &= r_{11}e_1' + r_{12}e_2'\,, \\ e_2\phi &= r_{21}e_1 + r_{22}e_2\,, & e_2'\phi &= r_{21}e_1' + r_{22}e_2'\,, \end{aligned}$$

where

$$\text{(9.22)} \qquad R = (r_{ij})$$

is a nonsingular matrix over $GF(q)$. On the other hand, if R is nonsingular, the unique linear transformation ϕ defined by (9.21) fixes not only A, B, C but every line of the regulus $\mathcal{R} = \mathcal{R}(A, B, C)$. More generally,

$$\text{(9.23)} \qquad L(\infty)\phi = L(\infty)\,, \qquad L(X)\phi = L(R^{-1}XR)$$

for every matrix X over $GF(q)$. Since ϕ fixes the lines of $\mathcal{R}$, then ϕ must map the opposite regulus, $\mathcal{R}'$, upon itself. Since $L(X)\phi = L(R^{-1}XR)$, then, by (9.20), ϕ maps $\mathcal{S}'(V)$ upon itself. That is: *if a linear transformation* ϕ *maps each line of the regulus* $\mathcal{R}$ *upon itself, then* ϕ *maps every regular spread containing the opposite regulus,* $\mathcal{R}'$, *upon itself.* On the other hand, $\mathcal{S}(U)\phi = \mathcal{S}(U)$ if and only if either (i) $R^{-1}UR = U$ or (ii) $R^{-1}UR = U^*$. Of course, (i) holds if and only if

$$R = aI + bU$$

for a, b elements of $GF(q)$, not both zero. As a special example of (ii), consider the linear transformation θ defined by

$$\text{(9.24a)} \quad \begin{aligned} e_1\theta &= e_1 & e_1'\theta &= e_1' \\ e_2\theta &= -re_1 - e_2 & e_2'\theta &= -re_1' - e_2' \end{aligned}$$

where

$$u_{12}r = u_{11} - u_{22} \,. \tag{9.24b}$$

(Note that u_{12} is nonzero, since $\boldsymbol{U}$ is irreducible.) Clearly $\theta^2 = \boldsymbol{I}$. We may verify readily that

$$L(a\boldsymbol{I} + b\boldsymbol{U})\theta = L(a\boldsymbol{I} + b\boldsymbol{U}^*) = L(a\boldsymbol{I} + b\boldsymbol{U}^q) \tag{9.25}$$

for all a, b in $GF(q)$. Hence *θ induces an involution of $PG(3, q)$ which induces ρ on $\mathcal{S}(\boldsymbol{U})$*. Let us note, finally, that ϕ, given by (9.21), will fix every line of the opposite regulus $\mathcal{R}'$ precisely when ϕ induces the identity collineation of $PG(3, q)$.

Similarly, a nonsingular linear transformation ψ of the vector space V over $GF(q)$ will fix every line of the opposite regulus $\mathcal{R}' = \mathcal{R}(A', B', C')$ if and only if

$$\begin{aligned} \boldsymbol{e}_1\psi &= s_{11}\boldsymbol{e}_1 + s_{12}\boldsymbol{e}_1' \,, & \boldsymbol{e}_2\psi &= s_{11}\boldsymbol{e}_2 + s_{12}\boldsymbol{e}_2' \,, \\ \boldsymbol{e}_1'\psi &= s_{21}\boldsymbol{e}_1 + s_{22}\boldsymbol{e}_1' \,, & \boldsymbol{e}_2'\psi &= s_{21}\boldsymbol{e}_2 + s_{22}\boldsymbol{e}_2' \end{aligned} \tag{9.26}$$

where $\boldsymbol{S} = (s_{ij})$ is a nonsingular matrix over $GF(q)$. Such a ψ maps $\mathcal{R}$ upon itself and maps every regular spread containing $\mathcal{R}$ upon itself but (in general) permutes the regular spreads containing $\mathcal{R}'$. If ϕ, ψ are given by (9.21), (9.26) then

$$\phi\psi = \psi\phi \,, \tag{9.27}$$

as we see by checking the effect of each product on the basis elements.

Next let us note that if τ is the unique linear transformation defined by

$$\boldsymbol{e}_1\tau = \boldsymbol{e}_1 \,, \quad \boldsymbol{e}_2\tau = \boldsymbol{e}_1' \,, \quad \boldsymbol{e}_1'\tau = \boldsymbol{e}_2 \,, \quad \boldsymbol{e}_2'\tau = \boldsymbol{e}_2' \,, \tag{9.28}$$

then τ induces an involution of $PG(3, q)$ which interchanges $\mathcal{R}, \mathcal{R}'$.

Now we are ready for the proof of Theorem 4.2. The K_2 of that theorem is the subgroup of $PL(\Sigma)$ induced by the transformations (9.21), and K_1 is induced by the transformations (9.26). By (9.27), $K_1K_2 = K_2K_1$ and, as previously remarked, $K_1 \cap K_2$ consists of the identity collineation. Hence $K_1K_2 = K_1 \otimes K_2$. Next consider a linear transformation α which maps $Q = Q(\mathcal{R}) = Q(\mathcal{R}')$ upon itself. Then either (i) α interchanges $\mathcal{R}$ and $\mathcal{R}'$ or (ii) α maps each of $\mathcal{R}, \mathcal{R}'$ upon itself. If α satisfies (i), then $\alpha\tau$ satisfies (ii), where τ is given by (9.28).

If α satisfies (ii), and if $e_1\alpha = v$, we can find a ψ of form (9.26) such that $\{v\psi\}$ is on $A = L(\infty)$ and then a ϕ of form (9.21) such that $v\psi\phi = e_1$. Hence, if $\beta = \alpha\psi\phi$, then β satisfies (ii) and $e_1\beta = e_1$. Since β satisfies (ii) and since $e_1\beta = e_1$, then β must map $L(\infty)$ upon itself. Hence we can find ϕ_1 of form (9.21) such that $\gamma = \beta\phi_1$ satisfies (ii) and fixes every point of $L(\infty)$. As a consequence, γ must fix every line of $\mathcal{R}'$. Thus, if α satisfies (ii), α is in $K_0 = K_1 \otimes K_2$. It is now clear that K_0 is the set of all collineations in $PL(\Sigma)$ which map each of $\mathcal{R}, \mathcal{R}'$ upon itself, and that K_0 has index 2 in the group, K, of all collineations in $PL(\Sigma)$ which map Q upon itself. This should suffice for the proof of Theorem 4.2. We have previously proved Theorem 4.3.

For the proof of Theorem 4.4 we note the following: Given a regular spread, $\mathcal{S}$, and an ordered set A_1, B_1, C_1 of three distinct (hence skew) lines of $\mathcal{S}$, we can certainly find a collineation in $PL(\Sigma)$ mapping A_1, B_1, C_1 upon A, B, C respectively, and hence mapping $\mathcal{S}$ upon one of the regular spreads containing $\mathcal{R} = \mathcal{R}(A, B, C)$. If $\boldsymbol{U}$ is any preassigned irreducible 2 by 2 matrix over $GF(q)$, such a spread must contain some conjugate, say $\boldsymbol{S}^{-1}\boldsymbol{U}\boldsymbol{S}$, of $\boldsymbol{U}$. Thus, by applying a transformation of form (9.21) with $\boldsymbol{R} = \boldsymbol{S}^{-1}$, we carry the spread into $\mathcal{S}(\boldsymbol{U})$. Therefore, as claimed in Theorem 4.4, $PL(\Sigma)$ is transitive on the regular spreads of $\Sigma = PG(3, q)$.

For the proof of Theorem 4.5 we shall require some additional notation. Consider the spread $\mathcal{S} = \mathcal{S}(\boldsymbol{U})$ defined by (9.13) or (9.13′). In this connection let

$$F = \{a\boldsymbol{I} + b\boldsymbol{U} \mid a, b \in GF(q)\} \qquad (\boldsymbol{U} \text{ irreducible}) \tag{9.29}$$

so that F is a field $GF(q^2)$. Consider the set, $G_0 = G_0(\mathcal{S})$, of all linear transformations of ϕ of V over $GF(q)$ having the form

$$\begin{aligned} e_1\phi &= a_{11}e_1 + a_{12}e_2 + b_{11}e_1' + b_{12}e_2' \\ e_2\phi &= a_{21}e_1 + a_{22}e_2 + b_{21}e_1' + b_{22}e_2' \\ e_1'\phi &= c_{11}e_1 + c_{12}e_2 + d_{11}e_1' + d_{12}e_2' \\ e_2'\phi &= c_{21}e_1 + c_{22}e_2 + d_{21}e_1' + d_{22}e_2' \end{aligned} \tag{9.30a}$$

where the 2 by 2 matrices $\boldsymbol{A} = (a_{ij})$, $\boldsymbol{B} = (b_{ij})$, $\boldsymbol{C} = (c_{ij})$, $\boldsymbol{D} = (d_{ij})$ satisfy

$$\boldsymbol{A}, \boldsymbol{B}, \boldsymbol{C}, \boldsymbol{D} \in F\,, \tag{9.30b}$$

(9.30c) $$AD - BC \neq 0 .$$

Lemma 9.3. *The set $G_0 = G_0(\mathcal{S})$, where $\mathcal{S} = \mathcal{S}(\boldsymbol{U})$, is a subgroup of $PL(\Sigma)$ with the following properties:*

(i) *G_0 maps $\mathcal{S} = \mathcal{S}(\boldsymbol{U})$ upon itself. Indeed, if ϕ is given by (9.30) then*

(9.31) $$L(\boldsymbol{X})\phi = L((\boldsymbol{B}\boldsymbol{X} + \boldsymbol{D})^{-1}(\boldsymbol{A}\boldsymbol{X} + \boldsymbol{C})) , \quad \text{for every } \boldsymbol{X} \in \infty \cup F .$$

(ii) *G_0 is strictly transitive on the ordered quadruples*

(9.32) $$(L_1, L_2, L_3;\ T)$$

consisting of three distinct lines L_1, L_2, L_3 of $\mathcal{S} = \mathcal{S}(\boldsymbol{U})$ and a transversal, T, to L_1, L_2, L_3.

(iii) *If $G = G(\mathcal{S})$ is the subgroup of $PL(\Sigma)$ mapping $\mathcal{S} = \mathcal{S}(\boldsymbol{U})$ upon itself, then*

(9.33) $$G = G_0 \cup \theta G_0$$

where θ is the linear transformation of V over $GF(q)$ defined by (9.24).

Proof. If ϕ is given by (9.30), the matrix of ϕ, relative to the ordered basis $\boldsymbol{e}_1, \boldsymbol{e}_2, \boldsymbol{e}_1', \boldsymbol{e}_2'$, is the block matrix

$$\begin{pmatrix} \boldsymbol{A} & \boldsymbol{B} \\ \boldsymbol{C} & \boldsymbol{D} \end{pmatrix} ,$$

which may be considered as a 2 by 2 matrix over the field F. If ϕ^* is the linear transformation with block matrix

$$\begin{pmatrix} \boldsymbol{D} & -\boldsymbol{B} \\ -\boldsymbol{C} & \boldsymbol{A} \end{pmatrix}$$

then $\phi\phi^*$ has block matrix

$$\begin{pmatrix} \boldsymbol{E} & 0 \\ 0 & \boldsymbol{E} \end{pmatrix}$$

where $\boldsymbol{E} = \boldsymbol{AD} - \boldsymbol{BC} \neq 0$. Since $\boldsymbol{E} \neq 0$, then $\boldsymbol{E}$ is nonsingular, and hence ϕ, ϕ^* are nonsingular. (Note that ϕ^* could be regarded as a "relative adjoint" of ϕ.) It is equally easy to see that G_0 is closed under multiplication.

If ϕ is given by (9.30) then

(9.34) $$L(\infty)\phi = \{a_{11}e_1 + a_{12}e_2 + b_{11}e_1' + b_{12}e_2', \quad a_{21}e_1 + a_{22}e_2 + b_{21}e_1' + b_{22}e_2'\}$$

and, for any 2 by 2 matrix X over $GF(q)$,

(9.35a) $$L(X)\phi = \{y_{11}e_1 + y_{12}e_2 + z_{11}e_1' + z_{12}e_2', \quad y_{21}e_1 + y_{22}e_2 + z_{21}e_1' + z_{22}e_2'\}$$

where

(9.35b) $$Y = XA + C\,, \qquad Z = XB + D\,.$$

Since B is in F, either $B = 0$ or B is nonsingular. In the first case, A is nonsingular and hence $L(\infty)\phi = L(\infty)$; in the second case, $L(\infty)\phi = L(B^{-1}A)$. If $B = 0$ then $Z = D$ for every matrix X, and D is nonsingular, whence $L(X)\phi = L(Z^{-1}Y)$. If $B \neq 0$, and if we restrict X to F, then Z is nonsingular except when $X = -B^{-1}D$; in this exceptional case $BY = BC - AD \neq 0$, so Y is nonsingular and $L(X)\phi = L(\infty)$. If $B \neq 0$ and if X is in F, $X \neq -B^{-1}D$, then $L(X)\phi = L(Z^{-1}Y)$. Hence we have (9.31) for all X in $\infty \cup F$ provided we interpret (9.31) in terms of (9.34) for $X = \infty$ and in terms of (9.35) for X in F. This is enough for the proof of (i).

If we want ϕ to fix each of $L(\infty), L(0), L(I)$, we must take $B = C = 0$ and $D = A$. Then ϕ has the form (9.21) where $R = A$; hence $L(X)\phi = L(A^{-1}XA)$ for every matrix X and $L(X)\phi = L(X)$ for X in F. By choice of A in F we can map the point $\{e_1\}$ into any point of $L(\infty)$—else the matrix U would not be irreducible—and hence we can map the ordered quadruple

(9.36) $$(L(\infty), L(0), L(I); L'(\infty))$$

into

$$(L(\infty), L(0), L(I); T)$$

where T is any transversal to $L(\infty), L(0), L(I)$. If, however, we want ϕ to fix the ordered quadruple (9.36), then ϕ must be both of form (9.21) and of form (9.26) and hence must fix evey point of $\Sigma = PG(3, q)$.

To complete the proof of (ii) of Lemma 9.3 it will be necessary to show only that G_0 is transitive on ordered triples of distinct lines of $\mathcal{S} = \mathcal{S}(U)$. We see this as follows: if we take ϕ of form (9.30) with $B = C = 0$ and $D = I$, then ϕ fixes $L(\infty), L(0)$ and maps $L(I)$ upon $L(A)$; and we may choose A

to be any non-zero element of F. If we take $\boldsymbol{B}=\mathbf{0}$, $\boldsymbol{A}=\boldsymbol{D}=\boldsymbol{I}$, then ϕ fixes $L(\infty)$ and maps $L(\mathbf{0})$ upon $L(\boldsymbol{C})$; and we may choose $\boldsymbol{C}$ arbitrarily in F. If we take $\boldsymbol{A}=\boldsymbol{D}=\mathbf{0}$, $\boldsymbol{B}=\boldsymbol{C}=\boldsymbol{I}$, then ϕ interchanges $L(\infty)$ and $L(\mathbf{0})$. This completes the proof of (ii).

In view of (ii) we need consider, for the proof of (iii), only an element ϕ of $PL(\Sigma)$ which fixes the ordered quadruple (9.36) and maps $\mathcal{S}=\mathcal{S}(\boldsymbol{U})$ upon itself. Since ϕ fixes $L(\infty)$, $L(\mathbf{0})$, $L(\boldsymbol{I})$, ϕ must have form (9.21). Since ϕ also fixes $L'(\infty)$, $r_{12}=0$. Then $r_{11}\neq 0$ and we may assume without loss of generality that $r_{11}=1$. A necessary and sufficient condition that ϕ map $\mathcal{S}(\boldsymbol{U})$ upon itself is that either (a) $\boldsymbol{R}^{-1}\boldsymbol{U}\boldsymbol{R}=\boldsymbol{U}$ or (b) $\boldsymbol{R}^{-1}\boldsymbol{U}\boldsymbol{R}=\boldsymbol{U}^*$. If (a) holds, then $\boldsymbol{R}$ is in F and (because of the form of the first row of $\boldsymbol{R}$) $\boldsymbol{R}=\boldsymbol{I}$. Similarly, if (b) holds, a simple calculation shows that

$$\boldsymbol{R}=\begin{pmatrix}1 & 0\\ -r & -1\end{pmatrix}$$

where r satisfies (9.24b). That is: in case (a), $\phi=\boldsymbol{I}$; and, in case (b), ϕ is the transformation θ defined by (9.24). This proves (iii) and completes the proof of Lemma 9.3.

Now we may apply Lemma 9.3 to the proof of Theorem 4.5. For (i) of Theorem 4.5, we note that $N=N(\mathcal{S})$ is the set of all ϕ of form (9.30) with $\boldsymbol{B}=\boldsymbol{C}=\mathbf{0}$ and $\boldsymbol{D}=\boldsymbol{A}$. We have already noted that N is transitive on the points of $L(\infty)$. Indeed, $\boldsymbol{A}$ may range over the q^2-1 non-zero elements of F; the $q-1$ scalar elements $a\boldsymbol{I}$, $a\neq 0$, yield transformations inducing the identity on Σ; and hence N induces a (cyclic) group of order $q+1$ on the $q+1$ points of $L(\infty)$. Hence N is strictly transitive on the points of $L(\infty)$. If L_1, L_2 are any two distinct lines of $\mathcal{S}$, distinct from $L(\infty)$, each point, P_1, of L_1 lies on a unique transversal, T, to $L(\infty)$ and L_2; and P_1 ranges over the points of L, as the point $L(\infty)\cap T$ ranges over the points of $L(\infty)$. Hence N is also strictly transitive on the points of L_1. This proves Theorem 4.5 (i).

Theorem 4.5 (ii) follows from Lemma 9.3. The θ of Theorem 4.5 (iii) is the transformation defined by (9.24). For (iv) of Theorem 4.5 we may regard the projective line $PG(1, q^2)$ as the set $\infty\cup F$ where F is given by (9.29). Then the isomorphism α maps $L(\infty)$ upon ∞ and $L(\boldsymbol{X})$ upon $\boldsymbol{X}$ for each $\boldsymbol{X}$ in F. The rest of Theorem 4.5 (iv) should be sufficiently clear, especially in view of Section 7.

Next we turn to Theorem 4.6. Let $\mathcal{R}, \mathcal{R}'$ be a pair of opposite reguli of $PG(3, q)$, and let $Q = Q(\mathcal{R}) = Q(\mathcal{R}')$. By Theorem 4.3 and Corollary, there are precisely $(q^2 - q)/2$ regular spreads containing $\mathcal{R}$; every nonsecant of Q is in exactly one of these spreads; and each of the spreads contains exactly $q^2 - q$ nonsecants of Q, forming $(q^2 - q)/2$ pairs of conjugate nonsecants. Similarly for the regular spreads containing $\mathcal{R}'$. If $\mathcal{S}, \mathcal{S}'$ are regular spreads containing $\mathcal{R}, \mathcal{R}'$ respectively then either (i) $\mathcal{S}, \mathcal{S}'$ have no common lines or (ii) $\mathcal{S} = \mathcal{S}(\boldsymbol{U})$, $\mathcal{S}' = \mathcal{S}'(\boldsymbol{V})$ where $L'(\boldsymbol{V}) = L(\boldsymbol{U})$. In case (ii), $\mathcal{S}, \mathcal{S}'$ have exactly two common lines, namely $L(\boldsymbol{U}), L(\boldsymbol{U}^q)$. By simple counting, we see that case (i) cannot occur. Hence we may assume (ii) or, more explicitly, that $\mathcal{S} = \mathcal{S}(\boldsymbol{U})$ is given by (9.13) or (9.13′) and that $\mathcal{S}' = \mathcal{S}'(\boldsymbol{V})$ is given by (9.16), (9.17) and hence by (9.20). By Theorem 4.5 (i), the group $N = N(\mathcal{S})$ is cyclic of order $q + 1$; similarly, $N' = N(\mathcal{S}')$ is cyclic of order $q + 1$. Furthermore, in the notation of Theorem 4.2, N, N' are subgroups of K_1, K_2 respectively. Consequently, $NN' = N \otimes N'$ is abelian of order $(q + 1)^2$. This proves Theorem 4.6 (i). We have already noted the truth of (ii), with $L = L(\boldsymbol{U})$, $L' = L(\boldsymbol{U})\rho = L(\boldsymbol{U}^*)$.

By Theorem 4.5 (i), N fixes every line of $\mathcal{S}$ and is strictly transitive on the points of each line of $\mathcal{S}$. Similarly, N' fixes each line of $\mathcal{S}'$ and is strictly transitive on the points of each line of $\mathcal{S}'$. If P is a point of $L = L(\boldsymbol{U})$, a line common to $\mathcal{S}$ and $\mathcal{S}'$, the orbit of P under NN' is, clearly, the $q + 1$ points of L. Similarly, the points of $L' = L(\boldsymbol{U}^*)$ form a single orbit under NN'. Next let P be a point of $PG(3, q)$ which is neither on L nor on L'. Since $\mathcal{S}'$ is a spread, there exists a unique line, L_1', of $\mathcal{S}'$ containing P. By hypothesis, L_1' is distinct from L, L', so L_1' is not in $\mathcal{S}$. Since $\mathcal{S}$ is a regular spread, the $q + 1$ lines of $\mathcal{S}$ which meet L_1 form a regulus, $\mathcal{R}_1$. Let S be a point of $\Sigma = PG(3, q)$ lying on a line, M_1, of $\mathcal{R}_1$, and suppose that M_1 meets L_1' in the point Q. There exists a collineation in N' which maps P upon Q and a collineation in N which maps Q upon S. Consequently, the orbit of P under $NN' = N \otimes N'$ contains the $(q + 1)^2$ distinct points S lying on the lines of $\mathcal{R}_1$. Since NN' has order $(q + 1)^2$, the points of $\mathcal{R}_1$ constitute the complete orbit of P. Again, if L_1 is the unique line of $\mathcal{S}$ through P and if $\mathcal{R}_2$ is the regulus consisting of all lines of $\mathcal{S}'$ which meet L_1, then the complete orbit of P under NN' is the points of $\mathcal{R}_2$. Thus $\mathcal{R}_1, \mathcal{R}_2$ are reguli of Σ having no common lines but covering the same points. We must conclude that $\mathcal{R}_2$ is the opposite regulus, $\mathcal{R}_1'$, of $\mathcal{R}_1$. Hence the orbit of P under NN' is the doubly-ruled quadric $Q_1 = Q(\mathcal{R}_1) = Q(\mathcal{R}_1')$, with one regulus, $\mathcal{R}_1$, in $\mathcal{S}$ and the other regulus, $\mathcal{R}_1'$, in $\mathcal{S}'$. In this way we see that the

$$(q^2+1)(q+1) - 2(q+1) = (q^2-1)(q+1)$$

points of Σ which are neither in $L = L(\boldsymbol{U})$ nor in $L' = L(\boldsymbol{U}^*)$ are partitioned into $q-1$ disjoint doubly-ruled quadrics Q_i $(i = 1, 2, \ldots, q-1)$, each of which is an orbit under NN' and has one regulus $\mathcal{R}_i$ in $\mathcal{S}$ and the opposite regulus, $\mathcal{R}_i'$, in $\mathcal{S}'$. This proves (iii) and (v) of Theorem 4.6. We see (iv) as follows: among the orbit-quadrics Q_i is the quadric $Q = Q(\mathcal{R}) = Q(\mathcal{R}')$. We know that, if ρ is the polarity with respect to Q, then $L\rho = L'$. But, equally, for every $i = 1, 2, \ldots, q-1$, $\mathcal{S}$ is the unique regular spread containing $\mathcal{R}_i$ and L; $\mathcal{S}'$ is the unique regular spread containing $\mathcal{R}_i'$ and L; and L, L' are the only common lines of $\mathcal{S}, \mathcal{S}'$. Hence L, L' must be conjugate nonsecants of Q_i for every i. This proves (iv). Since (vi), (vii) are obvious from the foregoing, the proof of Theorem 4.6 is now complete.

For the proof of Theorem 4.7 we need the following:

Lemma 9.4. *Let $\mathcal{R}_1, \mathcal{R}_2$ be two distinct reguli of $\mathcal{S} = \mathcal{S}(\boldsymbol{U})$ and, for $i = 1, 2$, let ρ_i be the polarity defined by the doubly-ruled quadric $Q_i = Q(\mathcal{R}_i)$. Then* (a) *$\rho_1\rho_2$ is in $G_0 = G_0(\mathcal{S})$ and* (b) *$\rho_1\rho_2$ fixes at most two lines of $\mathcal{S}$.*

Proof. By Lemma 9.3, for $i = 1, 2$, there exists ϕ_i in G_0 such that $\mathcal{R}\phi_i = \mathcal{R}_i$, where $\mathcal{R}$ is the regulus (9.5). Thus $\rho_i = \phi_i^{-1}\rho\phi_i$; and hence $\phi = \rho_1\rho_2$ is given by

$$\phi = \rho_1\rho_2 = \phi_1^{-1}\rho\phi_1\phi_2^{-1}\rho\phi_2 \tag{9.37}$$

where ρ is the polarity with respect to $Q = Q(\mathcal{R})$. Since ρ and the ϕ_i map $\mathcal{S}$ upon $\mathcal{S}$, certainly ϕ maps $\mathcal{S}$ upon $\mathcal{S}$. From the definition of ρ (see (9.6) through (9.9)) and the fact that the ϕ_i are in $PL(\Sigma)$, it is clear that ϕ is in $PL(\Sigma)$. Hence ϕ is in $G = G(\mathcal{S})$. Since ρ interchanges $L(\boldsymbol{U}), L(\boldsymbol{U}^*)$ and since the ϕ_i fix both of $L(\boldsymbol{U}), L(\boldsymbol{U}^*)$, then ϕ fixes $L(\boldsymbol{U})$. Hence ϕ is in G_0 rather than in the other coset, θG_0. This proves (a) of Lemma 9.4. In view of (a) and Lemma 9.3, especially (ii) of that lemma, if ϕ fixes more than two lines of $\mathcal{S}$, then ϕ fixes every line of $\mathcal{S}$. However, since $\mathcal{R}_1, \mathcal{R}_2$ are distinct, there exists at least one line L in $\mathcal{R}_1$ which is not in $\mathcal{R}_2$. Since L, $\mathcal{R}_2$ are in $\mathcal{S}$, L is skew to $\mathcal{R}_2$ and hence L is a nonsecant of the quadric Q_2. Consequently, $L, L\rho_2$ are distinct (in fact, skew) lines. On the other hand, since L is in $\mathcal{R}_1$, then $L\rho_1 = L$. Hence $L\phi = L\rho_2 \neq L$. Thus ϕ does not fix L and, consequently, ϕ fixes at most two lines of $\mathcal{S}$. This proves Lemma 9.4.

Now we turn to Theorem 4.7. We may assume without loss of generality that $\mathcal{S} = \mathcal{S}(\boldsymbol{U})$. In the proof of Theorem 4.6 we showed that $\mathcal{S}$ could be partitioned into $q - 1$ disjoint reguli $\mathcal{R}_i$ such that the lines $L(\boldsymbol{U}), L(\boldsymbol{U}^*)$ were conjugate nonsecants of each of the doubly ruled quadrics $Q_i = Q(\mathcal{R}_i)$. By Lemma 6.3, the same could be done with $L(\boldsymbol{U}), L(\boldsymbol{U}^*)$ replaced by any two distinct lines A, B of $\mathcal{S}$. The only question which remains is that of uniqueness. Now suppose that A, B, C are three distinct lines of $\mathcal{S}$ and that $\mathcal{R}_1, \mathcal{R}_2$ are two distinct reguli of $\mathcal{S}$ containing C such that A, B are conjugate nonsecants with respect to each of the doubly-ruled quadrics $Q_i = Q(\mathcal{R}_i)$, $i = 1, 2$. If ρ_i is the polarity with respect to Q_i, then $\rho_1\rho_2$ fixes each of the three lines A, B, C of $\mathcal{S}$, in contradiction to Lemma 9.4. This proves uniqueness and completes the proof of Theorem 4.7. Now we have proved all of the theorems stated in Section 4.

In Section 8, when classifying the André planes for the case that $q = p^e$, p prime, $e > 1$, we enlarged the equivalence classes by using the "multiplier" p. That this should be done follows from our concluding lemma:

Lemma 9.5. *Assume that $q = p^e$, p prime, $e > 1$, and let $\mathcal{S}(\boldsymbol{U})$ be the regular spread of $PG(3, q)$ given by (9.13). Then there exists a (semi-linear) collineation σ of $PG(3, q)$ such that*

$$(9.38)\qquad L(\infty)\sigma = L(\infty)\,, \qquad L(a\boldsymbol{I} + b\boldsymbol{U})\sigma = L((a\boldsymbol{I} + b\boldsymbol{U})^p)$$

for all a, b in $GF(q)$. In particular, σ maps $\mathcal{S}(\boldsymbol{U})$ upon itself.

Proof. Let α be the semi-linear transformation of V over $GF(q)$ defined by

$$(x_1\boldsymbol{e}_1 + x_2\boldsymbol{e}_2 + x_1'\boldsymbol{e}_1' + x_2'\boldsymbol{e}_2')\alpha = x_1^p\boldsymbol{e}_1 + x_2^p\boldsymbol{e}_2 + (x_1')^p\boldsymbol{e}_1' + (x_2')^p\boldsymbol{e}_2'$$

for all x_1, x_2, x_1', x_2' in $GF(q)$. We see readily that

$$L(\infty)\alpha = L(\infty)\,, \qquad L(\boldsymbol{X})\alpha = L(\boldsymbol{X}^{(p)})$$

for every matrix $\boldsymbol{X}$ over $GF(q)$, where $\boldsymbol{X}^{(p)}$ is the result of replacing each element of X by its pth power. In particular, if

$$X = aI + bU \tag{9.39}$$

for a, b in $GF(q)$, then

$$X^{(p)} = a^pI + b^pU^{(p)}\,, \qquad X^p = a^pI + b^pU^p\,,$$

where, of course, X^p denotes the pth power of the matrix X. The matrices $U^{(p)}$, U^p are both irreducible and have the same characteristic equation. Hence there exists at least one matrix S over $GF(q)$ such that $S^{-1}U^{(p)}S = U^p$. Consequently, $S^{-1}X^{(p)}S = X^p$ for every X of form (9.39). Therefore, if ϕ is the linear transformation (9.21) with $R = S$, the semi-linear transformation $\sigma = \alpha\phi$ will satisfy (9.38). This completes the proof of Lemma 9.5.

10. APPENDIX II

We shall show the connection between the André planes and subregular spreads. Then we shall briefly discuss a spread —associated with the names of Segré, Tits, Lüneburg and Suzuki—which is far from being subregular.

First we define a Veblen-Wedderburn system in the manner of André, specializing to the case of interest for the present paper. Let q be an arbitrary prime-power, and let T be the automorphism of $GF(q^2)$ such that

$$xT = x^q\,, \qquad xT^2 = x \tag{10.1}$$

for all x in $GF(q)$. Also define the relative norm $N(x)$:

$$N(x) = x \cdot xT = x^{q+1} \qquad x \in GF(q)\,. \tag{10.2}$$

Now let f be any function from $GF(q)$ to the integers 0, 1 such that

$$f(0) = f(1) = 0\,. \tag{10.3}$$

Define a new multiplication, $*$, on $GF(q^2)$ by

$$x * y = xT^{f(N(y))} \cdot y\,, \qquad x, y \in GF(q)\,. \tag{10.4}$$

Then the system $(GF(q^2), +, *)$ is a right Veblen-Wedderburn system which forms a coordinate ring for an André plane. The plane is "2-dimensional" because of the fact that

$$c * x = cx \,, \qquad c * (x + y) = c * x + c * y \,, \qquad c * (x * y) = (c * x) * y \tag{10.5}$$

for all c in $GF(q)$ and x, y in $GF(q^2)$.

For each y in $GF(q)$, let $\boldsymbol{R}(y)$, $\boldsymbol{M}(y)$ denote right multiplication by y in the field $GF(q^2)$ and in the corresponding Veblen-Wedderburn system, respectively. Thus

$$x\boldsymbol{R}(y) = xy \,, \qquad \boldsymbol{M}(y) = \boldsymbol{T}^{f(N(y))}\boldsymbol{R}(y) \tag{10.6}$$

for all x, y in $GF(q^2)$. We may suppose that a basis of $GF(q^2)$ over $GF(q)$ has been chosen, so that $\boldsymbol{T}$, the $\boldsymbol{R}(y)$ and the $\boldsymbol{M}(y)$ can be considered as 2 by 2 matrices over $GF(q)$. Then, in the sense of Section 9, let

$$\begin{aligned} \mathcal{S} &= L(\infty) \cup \{L(\boldsymbol{R}(y)) \mid y \in GF(q)\} \\ \mathcal{S}_* &= L(\infty) \cup \{L(\boldsymbol{M}(y)) \mid y \in GF(q)\} \end{aligned} \tag{10.7}$$

be sets of lines of $PG(3, q)$. Then $\mathcal{S}$ is a regular spread. Indeed, $\mathcal{S} = \mathcal{S}(\boldsymbol{U})$ in the sense of Section 9 where $\boldsymbol{U} = \boldsymbol{R}(y)$ for any y in $GF(q)$ which is not in $GF(q)$. Also $\boldsymbol{R}(x)$ has determinant $N(x)$ for every x in $GF(q^2)$. Hence, for each non-zero k in $GF(q)$, the set

$$\mathcal{R}(k) = \{L(\boldsymbol{R}(y)) \mid N(y) = k\}$$

is a regulus of $\mathcal{S}$, and $L(\infty), L(\boldsymbol{0})$ are conjugate nonsecants of the doubly-ruled quadric $Q(\mathcal{R}(k))$. Hence the $q - 1$ reguli $\mathcal{R}(k)$ form a complete linear set of disjoint reguli of $\mathcal{S}$. We draw from [3] the fact that the set

$$\mathcal{R}(k)' = \{L(\boldsymbol{TR}(y)) \mid N(y) = k\}$$

is the opposite regulus to $\mathcal{R}(k)$. Now we define

$$\mathcal{R}_*(k) = \{L(\boldsymbol{M}(y)) \mid N(y) = k\}$$

for each non-zero k in $GF(q)$, and note that

$$\mathcal{R}_*(k) = \mathcal{R}(k) \quad \text{or} \quad \mathcal{R}(k)'$$

according as $f(k) = 0$ or 1. Hence $\mathcal{S}_*$ is subregular of linear type. This completes our discussion of the André planes.

Next we specialize to the case

$$q = 2^{2t-1}\,, \qquad t \geq 2\,, \tag{10.8}$$

and define

$$s = 2^t\,. \tag{10.9}$$

Thus the mapping $x \to x^s$ is an automorphism of $GF(q)$ such that

$$x^{ss} = x^2\,, \qquad \text{for every} \quad x \in GF(q)\,. \tag{10.10}$$

We define the following family of matrices:

$$\boldsymbol{R}(x, y) = \begin{pmatrix} x & y \\ y & x^{s+1} + y^s \end{pmatrix}, \qquad \text{for every} \quad x, y \in GF(q)\,, \tag{10.11}$$

and verify readily that

$$\boldsymbol{R}(k^2x, k^{s+2}y) = \boldsymbol{D}(k)\boldsymbol{R}(x, y)\boldsymbol{D}(k) \tag{10.12}$$

$$\boldsymbol{R}(x + a, y + a^{s/2}x + b) = \boldsymbol{R}(a, b) + \boldsymbol{M}(a)\boldsymbol{R}(x, y)\boldsymbol{M}(a)^T \tag{10.13}$$

for all x, y, k, a, b in $GF(q)$, where

$$\boldsymbol{D}(k) = \begin{pmatrix} k & 0 \\ 0 & k^{s+1} \end{pmatrix}, \qquad \boldsymbol{M}(a) = \begin{pmatrix} 1 & 0 \\ a^{s/2} & 1 \end{pmatrix}. \tag{10.14}$$

If

$$z = \det \boldsymbol{R}(x, y) = x(x^{s+1} + y^s) + y^2\,, \tag{10.15}$$

we find that

$$(x^{s+1} + y^s)^{s+1} + zy^s = xz^s\,. \tag{10.16}$$

Note that, for w in $GF(q)$, $w^{(s+1)(s-1)} = w$. Hence, if $z = 0$, (10.16) yields $x^{s+1} + y^s = 0$, then (10.15) yields $y = 0$, and thus, finally, $x = 0$. In other words

$$\det \boldsymbol{R}(x, y) = 0 \Longleftrightarrow (x, y) = (0, 0)\,. \tag{10.17}$$

Now it is easy to check, with the help of (10.16), that

$$\boldsymbol{R}((x^{s+1} + y^s)z^{-1}, yz^{-1}) = \boldsymbol{R}(x, y)^{-1} \qquad \text{if} \quad (x, y) \neq (0, 0)\,, \tag{10.18}$$

where z is given by (10.18). On replacing y by $y + a^{s/2}x$ in (10.13) and using (10.17), we see, in view of the form of $\boldsymbol{M}(a)$, that

(10.19) $\boldsymbol{R}(x+a, y+b) - \boldsymbol{R}(a, b)$ is nonsingular for

$$(x, y) \neq (0, 0)\ .$$

Now we are ready to discuss the spread of Segré, Tits, Lüneburg, Suzuki. With q of form (10.8) we suppose (in the context of Section 9) that $\boldsymbol{e}_1, \boldsymbol{e}_2, \boldsymbol{e}_1', \boldsymbol{e}_2'$ is a basis of V over $GF(q)$ and that $\mathcal{S}$ is the set of lines of $PG(3, q)$ consisting of $L(\infty)$ and lines $L(\boldsymbol{X})$ where $\boldsymbol{X}$ ranges over the q^2 matrices $\boldsymbol{R}(x, y)$ defined by (10.11). By (10.19), $\mathcal{S}$ is a spread. Next, as a convenient notation, if ϕ is a collineation of $PG(3, q)$ which sends a line $L(\boldsymbol{X})$ into a line $L(\boldsymbol{Y})$, let us write $\boldsymbol{X}^\phi = \boldsymbol{Y}$. Thus, if ϕ has form (9.30a), where $\boldsymbol{A}, \boldsymbol{B}, \boldsymbol{C}, \boldsymbol{D}$ are over $GF(q)$ but not necessarily in the field F defined by (9.29), the proof of Lemma 9.3 indicates that

$$\boldsymbol{X}^\phi = (\boldsymbol{X}\boldsymbol{B} + \boldsymbol{D})^{-1}(\boldsymbol{X}\boldsymbol{A} + \boldsymbol{C}) \tag{10.20}$$

provided the right hand side has a meaningful interpretation. With this understanding, we specify the following collineations of $PG(3, q)$ in terms of the matrices $\boldsymbol{A}, \boldsymbol{B}, \boldsymbol{C}, \boldsymbol{D}$:

$$\eta(k):\quad \boldsymbol{B} = \boldsymbol{C} = \boldsymbol{0},\quad \boldsymbol{D} = \boldsymbol{D}(k)^{-1},\quad \boldsymbol{A} = \boldsymbol{D}(k) \qquad (k \neq 0)\ ;$$

$$\tau(a, b):\quad \boldsymbol{B} = \boldsymbol{0},\quad \boldsymbol{D} = \boldsymbol{M}(a)^{-1},\quad \boldsymbol{A} = \boldsymbol{M}(a)^T,\quad \boldsymbol{C} = \boldsymbol{M}(a)^{-1}\boldsymbol{R}(a, b)\ ; \tag{10.21}$$

$$\omega:\quad \boldsymbol{B} = \boldsymbol{C} = \boldsymbol{I},\quad \boldsymbol{D} = \boldsymbol{A} = \boldsymbol{0}\ .$$

Then, by (10.12), (10.13),

$$\infty^{\eta(k)} = \infty,\quad 0^{\eta(k)} = 0;\quad \boldsymbol{R}(x, y)^{\eta(k)} = \boldsymbol{R}(k^2x, k^{s+2}y), \qquad (k \neq 0)\ , \tag{10.22}$$

$$\infty^{\tau(a,b)} = \infty,\quad 0^{\tau(a,b)} = \boldsymbol{R}(a, b);\quad \boldsymbol{R}(x, y)^{\tau(a,b)} = \boldsymbol{R}(x + a, y + a^{s/2}x + b)\ , \tag{10.23}$$

and, by (10.18),

$$\infty^\omega = 0,\quad 0^\omega = \infty;\quad \boldsymbol{R}(x, y)^\omega = \boldsymbol{R}((x^{s+1} + y^s)z^{-1}, yz^{-1}) \quad \text{if}\ \ (x, y) \neq (0, 0)\ , \tag{10.24}$$

where z is given by (10.15).

Now let G be the group of all collineations in $PL(\Sigma)$ which map $\mathcal{S}$ upon $\mathcal{S}$. Certainly G contains ω, the $\tau(a, b)$ and the $\eta(k)$. In addition:

(i) G is transitive on the ordered pairs of distinct lines of $\mathcal{S}$.

(ii) G is generated by ω, the $\tau(a, b)$ and the $\eta(k)$.

(iii) G is a simple group of order $(q^2 + 1)q^2(q - 1)$, namely the Suzuki group of that order.

Certainly (i) is clear from (10.24), (10.25). To prove (ii) we need only consider a ϕ in $PL(\Sigma)$ which fixes $L(\infty)$ and $L(0)$; thus (10.20) holds with $\boldsymbol{B} = \boldsymbol{C} = 0$, whence $\boldsymbol{X}^\phi$ is linear in the elements of $\boldsymbol{X}$. Then, if we insist that ϕ maps $\mathcal{S}$ upon $\mathcal{S}$, a slightly tedious computation shows that $\phi = \eta(k)$ for some k. Assuming (i), (ii), the order of G is clearly correct in (iii), but we shall say nothing further about (iii). For a much more elaborate account of the geometry of the Suzuki group, see [4].

We shall end our account by showing that $\mathcal{S}$ is not subregular. In fact, $\mathcal{S}$ contains no regulus whatever. In view of (i), we need only consider a regulus, $\mathcal{R}$, of $PG(3, q)$ which contains $L(\infty)$, $L(0)$ and some third line, $L(\boldsymbol{X})$, of $\mathcal{S}$. We draw from [3] the fact that $\mathcal{R}$ consists of $L(\infty)$ and the q lines $L(k\boldsymbol{X})$ where k ranges over $GF(q)$. Now

$$\boldsymbol{X} = \boldsymbol{R}(x, y)\,, \qquad (x, y) \neq (0, 0)$$

for some x, y in $GF(q)$, and

$$k\boldsymbol{X} = \boldsymbol{R}(kx, ky) + \begin{pmatrix} 0 & w \\ 0 & 0 \end{pmatrix}$$

where $w = x^{s+1}k^{s+1} + y^s k^s + (x^{s+1} + y^s)k$. Thus $\mathcal{S}$ will contain $\mathcal{R}$ if and only if $w = 0$ for every k in $GF(q)$. We may regard w as a polynomial in k of degree at most $s + 1$. Since $q = 2^{t-1}s \geq 2s > s + 1$, we cannot have $w = 0$ for every k in $GF(q)$ unless $x = y = 0$, contrary to assumption. Thus $\mathcal{S}$ contains no reguli whatsoever.

11. APPENDIX III

In the table which follows, nonlinear triples of disjoint circles of $IP(q)$ are classified into orbits under the group of all collineations of $IP(q)$. In the notation of Section 7, the table gives first a summary showing the number of orbits with $d = 0, 1, 2$ or 3 and then a breakdown into various types. (The numbers for q even are listed under $d = 0$.) When q is a prime, we are classifying with respect to the group G of Theorem 7.5.

When $q = p^e$, p prime, $e > 1$ (in the table, when $q = 8, 9$ or 16) we have, in essence, enlarged G by using the field automorphism $x \to x^p$. This is appropriate in view of Lemma 9.5. To each of the orbits enumerated in the table corresponds an isomorphism class of translation planes of order q^2 and "dimension" two which are subregular of index 3 and of nonlinear type. See Section 8.

The latter part of the summary suggests that for q a large prime, the numbers of orbits corresponding to $d = 0, 1, 2, 3$ are approximately proportional to $1, 2, 2, 1$ respectively. This would mean that the respective numbers are approximately $q^3/288$, $q^3/144$, $q^3/144$, $q^3/288$ for q a large prime.

SUMMARY

					{1}	{6}		{4}		{12}′			
q	$d=0$	$d=1$	$d=2$	$d=3$	$d=0$	$d=0$	$d=2$	$d=0$	$d=3$	$d=0$	$d=1$	$d=2$	$d=3$
5	1	0	0	0	1								
7	0	1	1	1			1		1				
8*	2	—	—	—	—	—	—	1	—	—	—	—	—
9*	2	1	1	1	1	1			1				
11	2	4	8	4			2	1	3		1		1
13	3	10	11	6	1	2		1	2			2	
16*	19	—	—	—	—	—	—	2	—	—	—	—	—
17	11	26	28	14	1	3		2	4			2	
19	18	38	42	21			4	4	3		2		2

	{12}*				{24}′				{24}*			
q	$d=0$	$d=1$	$d=2$	$d=3$	$d=0$	$d=1$	$d=2$	$d=3$	$d=0$	$d=1$	$d=2$	$d=3$
5												
7		1										
8*	1	—	—	—	—	—	—	—	—	—	—	—
9*		1	1				1					
11	1	3	2								4	
13	1	6	4	4		2	1			2	4	
16*	10	—	—	—	—	—	—	—	7	—	—	—
17	4	12	10	8		6	2		1	8	14	2
19	8	14	12	11		2	6	2	6	20	20	3

References

1. André, J. "Über nicht-Desarguessche Ebenen mit transitiver Translations-Gruppe." *Math. Zeit.* **60** (1954), 156–186.
2. Bruck, R. H. and Bose, R. C. "The Construction of Translation Planes from Projective Spaces." *J. Algebra* **1** (1964), 85–102.
3. Bruck, R. H. and Bose, R. C. "Linear Representations of Projective Planes in Projective Spaces." *J. Algebra* **4** (1966), 117–172.
4. Lüneburg, H. "Die Suzukigruppen und ihre Geometrien." *Lecture Notes in Mathematics*, **10**. Springer-Verlag, Berlin-Göttingen-Heidelberg-New York, 1965.
5. Tits, J. "Ovoides et groupes de Suzuki." *Arch. Math.* **13** (1962), 187–198.

CHAPTER 28

Cyclical Generation of Linear Subspaces in Finite Geometries

C. RADHAKRISHNA RAO, *Indian Statistical Institute, Calcutta, India*

1. INTRODUCTION

This paper describes methods by which all linear spaces of any given dimension in a finite projective or Euclidean geometry can be cyclically generated starting from an initial set of such linear spaces. It also establishes the existence of *spreads* in finite geometries and provides rules for cyclical generation of disjoint spreads.

A spread S as defined by Bruck and Bose [2] is a collection of linear subspaces in a projective geometry Σ such that each point of Σ is contained in one and only one member of S. That is, S is a collection of skew subspaces which contain all the points of the geometry. More generally, we may define a *μ-fold spread* S as a collection of linear subspaces of a given dimension such that each point of Σ is contained in exactly μ members of S. We shall show that such spreads exist.

We will also be interested in disjoint spreads, which together contain all linear spaces of a given dimension. Such disjoint μ-fold spreads provide resolvable designs for balanced incomplete block experiments, i.e., an experiment which can be laid down in groups of adjacent blocks such that each group contains exactly μ replications of the varieties. The special case of $\mu = 1$, where each group of blocks contains a single replication of the varieties, is well known. Examples are given where the least value of μ which is combinatorially feasible is greater than one. Methods of generating all the disjoint spreads cyclically starting from one spread, and also of generating a spread starting from one member of it, are discussed in a few important cases.

My interest in this work arose from two fundamental theorems by Singer [11] and Bose [4] on difference sets, which essentially follow from a cyclical generation of linear subspaces of a geometry described in this paper. Extensions and generalizations of their results on difference sets are contained in the earlier papers of the author [6], [7], [8]. Examples of spreads and the generation of disjoint spreads are also discussed in [8]. Applications of these results in compact listing of solutions to balanced incomplete designs and in the construction of confounded factorial designs by punched card technique are given by the author in [9] and [10].

2. LINEAR SUBSPACES IN PROJECTIVE GEOMETRY

2.1. Analytical representations of *PG*(*t*, *m*)

A point of $PG(t, m)$ is represented by a vector of $(t + 1)$ components

$$(x_0, x_1, \ldots, x_t) \tag{2.1.1}$$

where x_i are elements of $GF(m)$, a Galois field of m elements, and not all x_i in (2.1.1) are simultaneously zero. Two vectors $(x_0, x_1, \ldots, x_t)$ and $(y_0, y_1, \ldots, y_t)$ represent the same point if there exists a non-zero element $\alpha \in GF(m)$ such that $x_i = \alpha y_i$, $i = 0, 1, \ldots, t$. All points which satisfy a set of $(t - d)$ independent linear equations

$$a_{i0}x_0 + a_{i1}x_1 + \ldots + a_{it}x_t = 0\,, \qquad i = 1, \ldots, t - d\,, \tag{2.1.2}$$

are said to constitute a linear subspace of d dimensions, which may be simply called a *d-flat*.

An alternative representation which is exploited in this and previous papers is as follows. A point of $PG(t, m)$ is represented by a non-zero element of $GF(m^{t+1})$ with the convention that two elements ξ and $b\xi$ of $GF(m^{t+1})$ represent the same point if $b \in GF(m)$. A d-flat is defined by the set of points

$$(a_0\xi_0 + a_1\xi_1 + \dots + a_d\xi_d) \tag{2.1.3}$$

where $a_0, a_1, \dots, a_d$ run independently over the elements of $GF(m)$ and are not simultaneously zero, and $\xi_0, \xi_1, \dots, \xi_d$ are fixed elements of $GF(m^{t+1})$ which are not linearly dependent (i.e., there exists no linear combination of $\xi_0, \xi_1, \dots, \xi_d$ with coefficients in $GF(m)$ such that its value is the zero element). The points $\xi_0, \xi_1, \dots, \xi_d$, which lie on the d-flat (2.1.3), are called the *defining points* of the d-flat (2.1.3).

If x is a primitive element of $GF(m^{t+1})$, then the points of $PG(t, m)$ can be represented by the first v powers of x:

$$x^0, x^1, \dots, x^{v-1}, \tag{2.1.4}$$

where $v = (m^{t+1} - 1)/(m - 1)$ or simply by integers

$$0, 1, \dots, v. \tag{2.1.5}$$

The points of a d-flat are the distinct integers $i \pmod v$ satisfying the relation

$$x^i = a_0x^{\alpha_0} + a_1x^{\alpha_1} + \dots + a_dx^{\alpha_d}, \tag{2.1.6}$$

where $x^{\alpha_0}, x^{\alpha_1}, \dots, x^{\alpha_d}$ are fixed elements not satisfying any linear relation and $a_0, a_1, \dots, a_d$ run over the elements of $GF(m)$.

The correspondence between the representations of a point as a $(t + 1)$-vector of elements belonging to $GF(m)$ and as an element of $GF(m^{t+1})$ or as an integer is brought out by the power cycle of the elements of $GF(m^{t+1})$ in terms of a primitive element x which satisfies an irreducible equation of the $(t + 1)$-th degree

$$x^{t+1} - (a_0x^t + \dots + a_t) = 0. \tag{2.1.7}$$

The left hand side of (2.1.7) is known as the minimum function and any element of $GF(m^{t+1})$ can be represented as a power of x or as a polynominal of degree less than $(t+1)$,

$$x^k = b_0 x^t + \ldots + b_t \,, \tag{2.1.8}$$

which is the residue class of x^k with respect to the minimum function (2.1.7). The equation (2.1.8) provides the unique correspondence

$$(b_0, \ldots, b_t) \leftrightarrow x^k \leftrightarrow k \,. \tag{2.1.9}$$

The correspondence between the vector and power cycle representation of the elements of $GF(m^t)$ is given in the appendix for various values of m and t. Representation of the elements as powers of x is useful in multiplication and as polynomials in x is useful in addition. Methods of obtaining minimum functions are found in [1] and [5].

2.2. Cyclical generation of linear spaces in $PG(t, m)$

The key theorem leading to cyclical generation of linear spaces is the following.

Theorem 1. *Let the points on a given d-flat be represented by the integers*

$$(c_0, c_1, \ldots, c_{k-1}) \,, \qquad k = (m^{d+1} - 1)/(m-1) \,. \tag{2.2.1}$$

Then the integers

$$(c_0 + i, c_1 + i, \ldots, c_{k-1} + i) \bmod v \,, \tag{2.2.2}$$

$$v = (m^{t+1} - 1)/(m-1) \,,$$

for any given i also define some d-flat.

Proof. Let $x^{c_0}, \ldots, x^{c_d}$ be the independent points defining the d-flat (2.2.1). To prove the result it is enough to show that $x^{c_0+i}, \ldots, x^{c_d+i}$ are independent. Suppose they are not. Then there exist elements $b_0, \ldots, b_d$ belonging to $GF(m)$ such that

$$x^{c_d+i} = b_0 x^{c_0+i} + \ldots + b_{d-1} x^{c_{d-1}+i} \,. \tag{2.2.3}$$

Dividing both sides of (2.2.3) by x^i we find that $x^{c_0}, \ldots, x^{c_d}$ are dependent, which is a contradiction.

Theorem 1 shows that starting with an initial d-flat represented by a set of integers we can generate other d-flats by adding successively integers $1, 2, \ldots$ to each element of the d-flat and reducing the result mod v. Not all d-flats so generated need be distinct. The smallest integer θ such that at the θ-stage the initial d-flat is reproduced is called the cycle of the initial d-flat. The maximum value of θ is, of course, v.

Theorem 2. *All the d-flats of PG(t, m) can be generated cyclically from a set of initial d-flats which may be of different cycles.*

We start with an initial flat and generate cyclically all the distinct flats it provides. Then we choose another initial flat which has not already occurred and generate some more d-flats and so on. We give a few examples.

Consider the lines in $PG(3, 2)$. There are $v = 15$ points and 35 lines with three points on each line. The points are represented by the elements $x^0, \ldots, x^{14}$ of $GF(2^4)$ or by integers $0, \ldots, 14$. All of them can be generated from the initial lines (0, 6, 8), (0, 11, 14) of cycle 15, with the equations $(a_0x^0 + a_1x^6)$ and $(a_0x^0 + a_1x^{11})$, and the line (0, 5, 10) of cycle 5 with the equation $(a_0x^0 + a_1x^5)$.

0, 6, 8	0, 11, 14	0, 5, 10
1, 7, 9	1, 12, 0	1, 6, 11
2, 8, 10	2, 13, 1	2, 7, 12
3, 9, 11	3, 14, 2	3, 8, 13
4, 10, 12	4, 0, 3	4, 9, 14
5, 11, 13	5, 1, 4	
6, 12, 14	6, 2, 5	
7, 13, 0	7, 3, 6	
8, 14, 1	8, 4, 7	
9, 0, 2	9, 5, 8	
10, 1, 3	10, 6, 9	
11, 2, 4	11, 7, 10	
12, 3, 5	12, 8, 11	
13, 4, 6	13, 9, 12	
14, 5, 7	14, 10, 13	

In $PG(4, 2)$, there are 31 points and 155 lines. It may be seen that all the lines can be generated from the initial lines

(2.2.4) (0, 1, 18), (0, 2, 5), (0, 4, 10),

$$(0,\ 7,\ 22),\quad (0,\ 8,\ 20)\,,$$

each with the full cycle of 31.

It is seen that each initial line in (2.2.4) gives rise to 31 lines which constitute a 3-fold spread in $PG(4, 2)$. The different initial lines give rise to disjoint spreads covering all the lines of the geometry. The compact representation (2.2.4) provides us with a resolvable balanced incomplete block design (*BIBD*) in 5 groups of blocks, each group containing 3 replications of varieties. Thus, whenever a representation of d-flats of $PG(m, t)$ is available in terms of some initial flats of full cycle $v = (m^{t+1} - 1)/(m - 1)$, we have a resolvable *BIBD*.

It may also be observed that starting from one of the initial lines, say (0, 1, 18), we could have obtained three other initial lines by multiplying successively by 2 and reducing the integers mod 31. There is, however, no general theory which gives a simple procedure for obtaining the different initial flats.

In practice, it is not necessary to generate all the d-flats from a chosen initial flat before selecting another initial flat. Let the first initial flat be represented by the integers

$$(c_0, c_1, \ldots, c_k) \tag{2.2.5}$$

and let S_1 be the set of all differences $(c_i - c_j) \bmod v$, $i \neq j$, arising out of integers in (2.2.5). The next initial flat, if necessary, is chosen in such a way that the set of differences S_2 arising from it is not the same as S_1. Similarly a third initial d-flat is chosen such that the set of differences S_3 is not the same as S_1 or S_2 and so on.

Theorem 3. *The points of a d-flat of cycle θ can be written in the form*

$$\begin{array}{llll} c_0, & c_0 + \theta, & \ldots, & c_0 + \overline{r-1}\,\theta \\ c_1, & c_1 + \theta, & \ldots, & c_1 + \overline{r-1}\,\theta \\ \cdots & \cdots & & \\ c_q, & c_q + \theta, & \ldots, & c_q + \overline{r-1}\,\theta \end{array} \tag{2.2.6}$$

where $c_i - c_j \neq 0 \bmod \theta$, $v = r\theta$ *and* r *is of the form* $(m^{t+1} - 1)/(m - 1)$.

A necessary condition for the existence of an initial flat of cycle $\theta < v$ is that v and the number of points on a d-flat are not relatively prime.

Proof. Consider any flat of cycle $\theta < v$. Since any initial flat is reproduced at the vth stage of cyclical generation it follows that v must be a multiple of θ, say $v = r\theta$. Further if x^{c_0} is a point on an initial d-flat of cycle θ, then $x^{c_0+k\theta}$ for a given k is a point on the flat obtained at the $k\theta$ stage of cyclical generation. Since the flat obtained at the $k\theta$ stage is the same as the initial flat, we find that $x^{c_0+k\theta}$ is also a point on the initial flat.

Now consider the set A of points

(2.2.7) $$x^{c_0}, x^{c_0+\theta}, \ldots, x^{c_0+\overline{r-1}\theta}$$

and let the subset B

(2.2.8) $$x^{c_0}, x^{c_0+\theta}, \ldots, x^{c_0+i\theta}$$

be independent. Since, by choice, $x^{c_0+\overline{i+1}\theta}$ is dependent on B it follows that $x^{c_0+j\theta}$, $j \geq (i+1)$, are all dependent on B. Now let x^{b_1} be a point on the flat of cycle θ and independent of B and x^c be a point dependent on B. Since x^{b_1} lies on a flat of cycle θ, $x^{b_1+k\theta}$ for any integer k also lies on it. Further

(2.2.9) $$x^c = a_0x^{c_0} + \ldots + a_ix^{c_0+i\theta}$$

for some $a_0, \ldots, a_i$. Multiplying both sides of (2.2.9) by $x^{b_1-c_0}$,

(2.2.10) $$x^{b_1+c-c_0} = a_0x^{b_1} + \ldots, + a_ix^{b_1+i\theta}$$

which means that $x^{b_1+c-c_0}$ also lies on the same d-flat. Similarly, if x^{b_2} is a point independent of B, then $x^{b_2+c-c_0}$ also lies on the same d-flat. Let the equation of the d-flat be, by a suitable choice of points $x^{b_1}, \ldots, x^{b_{d-i}}$,

(2.2.11) $$(a_0x^{c_0} + \ldots + a_ix^{c_0+i\theta} + a_{i+1}x^{b_1} + \ldots + a_dx^{b_{d-i}})\,.$$

Multiplying by x^{c-c_0} we obtain the d-flat

(2.2.12) $$(a_0x^c + \ldots + a_1x^{c+i\theta} + a_{i+1}x^{b_1+c-c_0} + \ldots + a_dx^{b_{d-i}+c-c_0})$$

which must be the same d-flat as (2.2.11) since the points defining the d-flat (2.2.12) are all on the d-flat (2.2.11). Hence the initial flat has cycle $c - c_0$ which must be a multiple of θ, i.e., $c = c_0 + \lambda\theta$. Thus all points on the i-flat defined by the set B are of the form $x^{c_0+\lambda\theta}$. But the number of such distinct points is r. Hence $r = (m^{i+1} - 1)/(m - 1)$, the number of points on an i-flat. This proves the first part of the theorem.

The second part of the theorem follows since the number of points on the d-flat (2.2.6) is a multiple of r and we have already shown that $v = \theta r$.

As a consequence of Theorem 3, we have Theorem 4 relating to the generation of $(t - 1)$-flats in $PG(t, m)$.

Theorem 4. *Every $(t - 1)$-flat in $PG(t, m)$ is of full cycle v and all the $(t - 1)$-flats can be cyclically generated from a single $(t - 1)$-flat with points on it represented by integers.*

Proof. The number of points on a $(t - 1)$-flat is $(m^t - 1)/(m - 1)$ and by Theorem 3, a necessary condition for a partial cycle is that $v = (m^{t+1} - 1)/(m - 1)$ and $(m^t - 1)/(m - 1)$ have a common factor of the form $(m^{i+1} - 1)/(m - 1)$. It is easily seen that this is not possible. Hence every $(t - 1)$-flat is of cycle v. But the number of $(t - 1)$-flats is also v, which proves Theorem 4.

Thus, to find all the $(t - 1)$-flats of $PG(t, m)$, it is only necessary to write down the points on a suitably chosen $(t - 1)$-flat. Considering the points of $PG(t, m)$ as vectors of $(t - 1)$ elements belonging to $GF(m)$, as in (2.1.1), we observe that all vectors with the ith element zero define the $(t - 1)$-flat with the equation $x_i = 0$. To obtain the integers representing the points we go through the power cycle of $GF(m^{t+1})$ and write down the indices of the primitive element less than $v = (m^{t+1} - 1)/(m - 1)$ corresponding to the vectors with the ith element zero (choosing any fixed value for i).

For example, consider $PG(4, 2)$ with 31 3-flats and 15 points on each. The value of v is 31. Going through the power cycle of $GF(2^5)$ we find that the following integers

(2.2.13) 0, 1, 2, 3, 4, 7, 11, 14, 15, 16, 22, 23, 26, 28, 29

correspond to vectors with the first element zero. All the 31 3-flats are obtained by adding successively $0, 1, \ldots, 30$ to all the integers of (2.2.13) and reducing mod 31.

2.3. Spreads in $PG(t, m)$

As stated in the introduction, a μ-fold spread S is a collection of linear subspaces of a given dimension such that each point of the geometry occurs in exactly μ subspaces of S. Naturally all the subspaces of a given dimension in $PG(t, m)$ constitute a spread with the maximum value of μ. Our interest lies in constructing a spread with the minimum possible value of μ. We prove the following theorem which establishes the existence of spreads.

Theorem 5. *The necessary and sufficient condition for the existence of a d-flat of cycle $\theta < v$ is that $\phi(t, 0, m)$ and $\phi(d, 0, m)$ have a factor of the type $r = \phi(i, 0, m)$, i.e., $(t + 1)$ and $(d + 1)$ have a common factor $i + 1$, $i > 0$. Otherwise every d-flat is of full cycle v.*

Proof. The necessity is already established in Theorem 3. To prove the sufficiency, let $\theta = \phi(t, 0, m)/\phi(i, 0, m)$. Then the points

$$0, \theta, \ldots, \overline{r-1}\,\theta, \tag{2.3.1}$$

where $r = \phi(i, 0, m)$ lie on an i-flat. Consider the set S_1 of points $0, \theta, \ldots, i\theta$ which are independent. Take a point b_1 independent of S_1 and consider the set S_2 of points

$$b_1, b_1 + \theta, \ldots, b_1 + i\theta. \tag{2.3.2}$$

It may be seen that the points in $S_1 \cup S_2$ are all independent. Taking a point b_2 independent of $S_1 \cup S_2$ we consider the set S_3 of points $b_2, b_2 + \theta, \ldots, b_2 + i\theta$. The points in $S_1 \cup S_2 \cup S_3$ are all independent. We can continue this process till the set S_{q+1}, where $(q + 1) = (d + 1)/(i + 1)$, is obtained. The totality of $(d + 1)$ independent points in $S_1 \cup S_2 \cup \ldots, \cup S_{q+1}$ generate a d-flat.

If c is any point on this flat, then it is easily seen that $c + \theta, \ldots, c + (r - 1)\theta$ are also points on it. Therefore the d-flat so constructed has cycle θ.

It is easy to see that by a special choice of $b_1, \ldots, b_q$ we can ensure that the d-flat constructed as above has no subcycle $< \theta$. As an example we can have $b_1 = 1, \ldots, b_q = q$.

Theorem 6. *Let $(t + 1)$ and $(d + 1)$ have $(i + 1)$ as a common* *factor. Then a μ-fold spread of d-flats in $PG(t, m)$ exists, where $\mu = \phi(d, 0, m)/\phi(i, 0, m)$, which assumes the value unity when $\phi(d, 0, m)$ divides $\phi(t, 0, m)$.*

Under the conditions of Theorem 6, it is shown in Theorem 5 that there exists a d-flat of cycle $\theta = \phi(t, 0, m)/\phi(i, 0, m)$, whose points can be represented as

$$
\begin{array}{llll}
c_0, & c_0 + \theta, & \ldots, & c_0 + \overline{r-1}\,\theta \\
 & \ldots\ldots\ldots\ldots & & \\
c_{\mu-1}, & c_{\mu-1} + \theta, & \ldots, & c_{\mu-1} + \overline{r-1}\,\theta
\end{array}
\tag{2.3.3}
$$

where $r = \phi(i, 0, m)$ and $r\mu = \phi(d, 0, m)$. Starting from the d-flat (2.3.3), we can generate θ others by adding successively the integers $0, 1, \ldots, (\theta - 1)$ and reducing mod v. The θ d-flats so obtained obviously constitute a μ-fold spread.

For example, there is a 1-fold spread of $(t-1)$-spaces in $PG(2t-1, m)$ since $\phi(2t-1, 0, m)$ is divisible by $\phi(t-1, 0, m)$, for which a non-analytical proof is given by Bruck and Bose [2]. Another general example of 1-fold spread is that of $(k-1)$ spaces in $PG(kt-1, m)$.

An example of an (m^2+1)-fold spread is that of 3-spaces in $PG(4k+1, m)$ and so on. We shall show how to construct such a 5-fold spread of 3-spaces in $PG(5, 2)$, corresponding to $k = 1$ and $m = 2$, using the method described in Theorem 5. We find that $\phi(5, 0, 2) = 63$ and $\phi(3, 0, 2) = 15$ have the common factor $3 = \phi(1, 0, 2)$. The value of $\theta = 63/3 = 21$ and $r = 3$. Let us consider the 3-space passing through the points x^0, x^{21}, x^1, x^{22}, where $x^i \in GF(2^6)$. The points generated by the linear combinations

$$a_1x^0 + a_2x^{21} + a_3x^1 + a_4x^{22}, \tag{2.3.4}$$

where $a_i \in GF(2)$, are

$$
\begin{array}{r}
0,\ 21,\ 42 \\
1,\ 22,\ 43 \\
4,\ 25,\ 46 \\
14,\ 35,\ 56 \\
16,\ 37,\ 38
\end{array}
\tag{2.3.5}
$$

recording only the indices (integers). The 3-space (2.3.5) has a cycle of 21, and the 21 sets generated from it by adding $0, 1, \ldots, 20$ and reducing the integcrs mod 63 provide a 5-fold spread. Other examples may be given following exactly the method described in Theorem 5.

2.4. Generation of disjoint spreads

It may be noted that the cyclical generation of flats from an initial flat involves the transformation of integers

$$0 \to 1 \to 2 \to \ldots \to v \to 0 \tag{2.4.1}$$

by which points on one flat are transformed to points on another flat. The transformation is indeed linear and non-singular, thus preserving the dimension of the flat.

We consider another transformation of points with a cycle shorter than v in the special case of $PG(t, 2)$. Let y be a primitive element of $GF(2^t)$. Then a point in $PG(t, 2)$ can be represented by a pair of coordinates (ξ, η), where ξ takes values in $GF(2)$ and η takes the values $(0), y^0, \ldots, y^{q-1}$ of $GF(2^t)$, where $q = 2^t - 1$, y is a primitive element of $GF(2^t)$, (0) is the zero element of $GF(2^t)$ and the combination $\xi = 0$, $\eta = (0)$ is not considered. The total number of points is thus

$$2(2^t) - 1 = 2^{t+1} - 1\,. \tag{2.4.2}$$

It is seen from the representation of $y^i \in GF(2^t)$ as a vector of t elements belonging to $GF(2)$, that for given (ξ, η), which may be considered as a vector of $(t+1)$ elements, there corresponds an element x^α of $GF(2^{t+1})$. The correspondence can be written

$$(\xi, \eta) \leftrightarrow \alpha\,, \tag{2.4.2}$$

where α is an integer. Now we consider the cyclical transformation of points

$$(\xi, y^0) \to (\xi, y^1) \to \ldots \to (\xi, y^{q-1}) \to (\xi, y^0) \tag{2.4.4}$$

for a fixed ξ. In terms of integers representing points of $PG(t, 2)$, the cyclical transformation (2.4.4) can be written

$$\alpha_0 \to \alpha_1 \to \ldots \to \alpha_{q-1} \to \alpha_0 \tag{2.4.5}$$

where $(\xi, y^i) \leftrightarrow \alpha_i$, $i = 0, 1, \ldots, q - 1$. The cycle of the transformation (2.4.5) is of order $q = 2^t - 1$. A different choice of ξ in (2.4.4) provides a cyclical transformation of a disjoint set of integers (from 1 to $v = 2^{t+1} - 1$). Thus there are two cyclical transformations each of length $(2^t - 1)$. Now the point $(1, (0))$, where 1 is the non-null element of $GF(2)$, (0) is the null element of $GF(2^t)$, is not covered by the above transformations, and we shall take this point to be invariant under the transformation.

Theorem 7. *Under the transformations* (2.4.5) *of points, a μ-fold spread goes over to a μ-fold spread.*

It can be shown that the transformation (2.4.5) is linear and nonsingular. Hence a subspace of a given dimension transforms to a subspace of the same dimension. Since a μ-fold spread contains each integer from 1 to v exactly μ times and the transformation is one to one, it follows that among the transformed subspaces of a spread the same frequency of each integer holds.

By repeating the transformation we obtain a spread each time and at the qth stage the initial spread is reproduced. But the q spreads so generated may not be disjoint. They will be disjoint if no member of the initial spread can be obtained from another of its members by repeated transformation of the type (2.4.4).

We shall illustrate by determining a spread of lines in $PG(3, 2)$ and deriving all the disjoint spreads. There are 15 points and 35 lines. A spread exists by Theorem 6, since 15 and 3, the number of points on a line, have $3 = 2^2 - 1$ as a common factor. The value of θ defined in Theorem 6 is $15/3 = 5$ and to generate an initial spread we consider the initial line passing through the points x^0, x^5, where $x^0, x^5 \in GF(2^4)$. The three points on the line, recording only the indices of x, are

$$(0,\ 5,\ 10)\,. \tag{2.4.6}$$

By adding 0, 1, 2, 3, 4 we generate the initial spread

$$(0,\ 5,\ 10),\ (1,\ 6,\ 11),\ (2,\ 7,\ 12),\ (3,\ 8,\ 13),\ (4,\ 9,\ 14)\,. \tag{2.4.7}$$

To generate other spreads we need a transformation of the type (2.4.5) which is constructed as follows by first considering the power cycle of $GF(2^3)$. The transformation mentioned in (2.4.4) for $\xi = 0$ is

$$(0, y^0) \to (0, y^1) \to \ldots \to (0, y^6)\,, \tag{2.4.8}$$

where $y^0, \ldots, y^6$ are the non-null elements of $GF(2^3)$. Using the power cycle of $GF(2^3)$ given in the appendix, we write the vector equivalents of y^i, obtaining the transformation of vectors

$$\begin{aligned}(0001) \to (0010) \to (0100) \to (0101) \to (0111)\\ \to (0011) \to (0110)\,.\end{aligned} \tag{2.4.9}$$

Using the power cycle of 2^4, we write down the indices corresponding to the vectors (2.4.9)

$$0 \to 1 \to 2 \to 9 \to 7 \to 12 \to 13 \to 0\,, \tag{2.4.10}$$

thus providing a transformation of integers of cycle 7. Similarly, considering 1 as the first coordinate in (2.4.8), we obtain the transformation

$$4 \to 10 \to 14 \to 11 \to 6 \to 5 \to 8 \to 4\,. \tag{2.4.11}$$

The number 3 corresponding to (1, (0)) transforms to itself. Starting with the spread (2.4.7), we generate the others using the transformations (2.4.9) and (2.4.10).

0, 5, 10	1, 6, 11	2, 7, 12	3, 8, 13	4, 9, 14
1, 8, 14	2, 5, 6	9, 12, 13	3, 4, 0	10, 7, 11
2, 4, 11	9, 8, 5	7, 13, 0	3, 10, 1	14, 12, 6
9, 10, 6	7, 4, 8	12, 0, 1	3, 14, 2	11, 13, 5
7, 14, 5	12, 10, 4	13, 1, 2	3, 11, 9	6, 0, 8
12, 11, 8	13, 14, 10	0, 2, 9	3, 6, 7	5, 1, 4
13, 6, 4	0, 11, 14	1, 9, 7	3, 5, 12	8, 2, 10

The 7 disjoint spreads provide all the lines of the geometry. It is, however, not proved that a transformation of the type (2.4.5) does always provide the maximum number of disjoint spreads.

3. LINEAR SUBSPACES IN EUCLIDEAN GEOMETRY

3.1. Analytical representation of $EG(t, m)$

The points of $EG(t, m)$, Euclidean geometry of t dimensions, can be represented by vectors

$$(3.1.1) \qquad (b_1, \ldots, b_t), \qquad b_i \in GF(m).$$

Two vectors $(b_1, \ldots, b_t)$ and $(c_1, \ldots, c_t)$ represent the same point if and only if $b_i = c_i$ for all i. All points which satisfy a set of $(t - d)$ consistent linear equations

$$(3.1.2) \qquad a_{i0} + a_{i1}b_1 + \ldots + a_{it}b_t = 0, \qquad i = 1, 2, \ldots, t - d,$$

are said to constitute a d-flat.

As in the case of $PG(t, m)$, an alternative analytical representation is available for $EG(t, m)$. A point is represented by one and only one element of $GF(m^t)$. If $\xi_0, \xi_1, \ldots, \xi_d$ are $(d + 1)$ linearly independent points, then the totality of $(m^d - 1)$ points

$$(3.1.3) \qquad (a_0\xi_0 + \ldots + a_d\xi_d), \quad a_i \in GF(m), \quad \Sigma a_i = 1,$$

define a d-flat (passing through the points $\xi_0, \ldots, \xi_d$). If x is a primitive element of $GF(m^t)$, then

$$(3.1.4) \qquad x^k \equiv b_t x^{t-1} + b_{t-1}x^{t-2} + \ldots + b_1 x^0,$$

which provides the correspondence

$$(3.1.5) \qquad x^k \leftrightarrow (b_1, \ldots, b_t),$$

showing that the two analytical representations considered are identical.

As in the case of $PG(t, m)$, we represent the points by the integers $0, 1, \ldots, m^t - 2$ corresponding to $x^0, \ldots, x^{m^t-2}$, the powers of the primitive element and a symbol ∞ corresponding to the null element (0) of $GF(m^t)$. The method of cyclical generation of d-flats consists in applying the transformation

$$(3.1.6) \qquad 0 \to 1 \to 2 \to \ldots \to m^t - 2 \to 0$$

$$\infty \to \infty$$

successively starting with an initial d-flat, whose points are given in terms of the integers from 0 to $m^t - 2$ and the symbol ∞.

It is easily shown from definitions that the number of d-flats in $PG(t, m)$ is $\phi(t, d, m)$ and that in $EG(t, m)$ is $\phi(t, d, m) - \phi(t - 1, d, m)$, where

$$\phi(x, y, z) = \frac{(z^{x+1} - 1) \ldots (z^{x-y+1} - 1)}{(z - 1) \ldots (z^{y+1} - 1)}. \tag{3.1.7}$$

3.2. Generation of subspaces in $EG(t, m)$

Theorem 8. *In $EG(t, m)$, the cycle of any d-flat not passing through the point* (0)*, the null element of $GF(m^t)$, is $v' = m^t - 1$, and any d-flat passing through the point* (0) *has a cycle $\theta' \leq \theta = (m^t - 1)/(m - 1)$.*

Proof. The first part of the theorem easily follows, for if the result is not true, then it would imply that $(m^t - 1)$ and m^d have a common factor, which is impossible.

To prove the second part, we observe that if x^c is a point on a flat passing through (0), then $x^{c+i\theta}$ also lies on it, which implies that θ is a multiple of the cycle of the flat. The maximum value of the cycle is thus θ.

The number b of d-flats not passing through (0) can be generated from b/v' initial flats. The d-flats through the point (0) may have different cycles, depending on the common factors of d and t.

3.3. Spreads in $EG(t, m)$

There is no difficulty in establishing the existence of spreads in $EG(t, m)$. Their cyclical generation, however, poses an interesting problem. We shall consider the special case of spreads of $(t - 1)$-flats in $EG(t, m)$.

Applying Theorem 8 we find that all the $(t - 1)$-flats can be generated from any one $(t - 1)$-flat not passing through (0) which has the cycle v' and any $(t - 1)$-flat passing through (0) which has the cycle $\theta = (m^t - 1)/(m - 1)$. Consider a $(t - 1)$-flat not passing through (0), with the points represented by the integers

$$d_1, d_2, \ldots, d_k. \tag{3.3.1}$$

It is easily shown that the $(t-1)$-flats

$$(3.3.2)\qquad \begin{array}{llll} c_1+d_1, & c_1+d_2, & \ldots, & c_1+d_k \\ c_2+d_1, & c_2+d_2, & \ldots, & c_2+d_k \end{array}$$

do not have a common point if $c_1 - c_2 \neq 0 \bmod \theta$. Then it follows that the $(m-1)$ $(t-1)$-flats

(3.3.3)

$$\begin{array}{llll} d_1+i, & d_2+i, & \ldots, & d_k+i \\ d_1+i+\theta, & d_2+i+\theta, & \ldots, & d_k+i+\theta \\ \cdots & & & \\ d_1+i+\overline{m-2}\,\theta, & d_2+i+\overline{m-2}\,\theta, & \ldots, & d_k+i+\overline{m-2}\,\theta \end{array}$$

are all skew for any given i. To the set (3.3.3) we add another $(t-1)$-flat consisting of the rest of the integers from 0 to $(v'-1)$ and the symbol ∞.

The m $(t-1)$-flats so constructed constitute a spread of $(t-1)$-flats in $EG(t, m)$. By letting $i = 0, 1, \ldots, (\theta-1)$, we obtain θ disjoint spreads covering all the $(t-1)$-flats of $EG(t, m)$. The whole process of generating a spread from a given $(t-1)$-flat represented by the integers

$$(3.3.4)\qquad d_1, d_2, \ldots, d_k\,,$$

and generating disjoint spreads may be written as

$$(3.3.5)\qquad PC(\theta)[(d_1, d_2, \ldots, d_k)S(\theta) + R] \bmod (m^t - 1)\,.$$

In the notation within the square brackets, $S(\theta)$ denotes cyclical generation in steps of θ, i.e., other flats are generated by adding $\theta, 2\theta, \ldots, (m-2)\theta$ to the integers of the initial flat $(d_1, \ldots, d_k)$ and R stands for the rest of the points constituting the mth $(t-1)$-flat. Thus $(d_1, d_2, \ldots, d_k)S(\theta) + R$ is a compact representation of one spread. The symbol $PC(\theta)$ stands for a partial cycle of θ, i.e., by adding successively $1, 2, \ldots, (\theta-1)$ to the integers of the initial spread, other disjoint spreads are obtained. The element ∞ which occurs in R remains invariant under the addition of integers.

The method of obtaining the integers representing an initial $(t-1)$-flat is again simple. We note that the equation $x_i = \alpha \neq 0$, where x_i denotes the ith coordinate of a point in

$EG(t, m)$ represented as a vector and α is a given element of $GF(m)$. Then the integers $d_1, \ldots, d_k$ representing the flat $x_i = \alpha$, are the powers of x, the primitive element of $GF(m^t)$ such that the ith component in their vector representation (3.1.1) has the given value α. These can be easily written by going through the power cycle of the elements of $GF(m^t)$ given in the appendix.

A general method of obtaining disjoint spreads of subspaces of any given dimension in $EG(m, t)$ using punched card techniques is described in a previous paper [9]. For this purpose, we first establish a correspondence between a d-flat in $PG(m, t-1)$ and a spread of $(t-d-1)$-flats in $EG(m, t)$. The spreads associated with two different d-flats in $PG(m, t-1)$ are disjoint. Then considering all the d-flats in $PG(m, t-1)$, we obtain the maximum number of disjoint spreads covering all the $(t-d-1)$ flats in $EG(m, t)$. The cyclical representation of d-flats in $PG(m, t)$ is used in this connection. For details the reader is referred to [9].

Appendix

Power cycle of Galois fields $GF(m^t)$

(Correspondence between the index, power of the primitive element, and vector representations)

Vec.	Ind.	Vec.	Ind.	Vec.	Ind.	Vec.	Ind.
$GF(2^2)$		$GF(2^4)$		$GF(2^5)$		$GF(2^5)$	
01	0	1000	3	10000	4	01110	23
10	1	1001	4	01001	5	11100	24
11	2	1011	5	10010	6	10001	25
		1111	6	01101	7	01011	26
		0111	7	11010	8	10110	27
$GF(2^3)$		1110	8	11101	9	00101	28
001	0	0101	9	10011	10	01010	29
010	1	1010	10	01111	11	10100	30
100	2	1101	11	11110	12		
101	3	0011	12	10101	13		
111	4	0110	13	00011	14	$GF(2^6)$	
011	5	1100	14	00110	15	000001	0
110	6			01100	16	000010	1
				11000	17	000100	2
		$GF(2^5)$		11001	18	001000	3
$GF(2^4)$		00001	0	11011	19	010000	4
0001	0	00010	1	11111	20	100000	5
0010	1	00100	2	10111	21	100001	6
0100	2	01000	3	00111	22	100011	7

Vec.	Ind.	Vec.	Ind.	Vec.	Ind.	Vec.	Ind.
$GF(2^6)$		$GF(2^6)$		$GF(2^7)$		$GF(2^7)$	
100111	8	001100	60	1001110	46	0101110	98
101111	9	011000	61	0011111	47	1011100	99
111111	10	110000	62	0111110	48	0111011	100
011111	11			1111100	49	1110110	101
111110	12			1111011	50	1101111	102
011101	13	$GF(2^7)$		1110101	51	1011101	103
111010	14	0000001	0	1101001	52	0111001	104
010101	15	0000010	1	1010001	53	1110010	105
101010	16	0000100	2	0100001	54	1001101	106
110101	17	0001000	3	1000010	55	1100111	107
001011	18	0010000	4	0000111	56	0011001	108
010110	19	0100000	5	0001110	57	0110010	109
101100	20	1000000	6	0011100	58	1100100	110
111001	21	0000011	7	0111000	59	1001011	111
010011	22	0000110	8	1110000	60	0010101	112
100110	23	0001100	9	1100011	61	0101010	113
101101	24	0011000	10	1000101	62	1010100	114
111011	25	0110000	11	0001001	63	0101011	115
010111	26	1100000	12	0010010	64	1010110	116
101110	27	1000011	13	0100100	65	0101111	117
111101	28	0000101	14	1001000	66	1011110	118
011011	29	0001010	15	0010011	67	0111111	119
110110	30	0010100	16	0100110	68	1111110	120
001101	31	0101000	17	1001100	69	1111111	121
011010	32	1010000	18	0011011	70	1111101	122
110100	33	0100011	19	0110110	71	1111001	123
001001	34	1000110	20	1101100	72	1110001	124
010010	35	0001111	21	1011011	73	1100001	125
100100	36	0011110	22	0110101	74	1000001	126
101001	37	0111100	23	1101010	75		
110011	38	1111000	24	1010111	76		
000111	39	1110011	25	0101101	77	$GF(3^2)$	
001110	40	1100101	26	1011010	78	01	0
011100	41	1001001	27	0110111	79	10	1
111000	42	0010001	28	1101110	80	21	2
010001	43	0100010	29	1011111	81	22	3
100010	44	1000100	30	0111101	82	02	4
100101	45	0001011	31	1111010	83	20	5
101011	46	0010110	32	1110111	84	12	6
110111	47	0101100	33	1101101	85	11	7
001111	48	1011000	34	1011001	86		
011110	49	0110011	35	0110001	87		
111100	50	1100110	36	1100010	88	$GF(3^3)$	
011001	51	1001111	37	1000111	89	001	0
110010	52	0011101	38	0001101	90	010	1
000101	53	0111010	39	0011010	91	100	2
001010	54	1110100	40	0110100	92	102	3
010100	55	1101011	41	1101000	93	122	4
101000	56	1010101	42	1010011	94	022	5
110001	57	0101001	43	0100101	95	220	6
000011	58	1010010	44	1001010	96	101	7
000110	59	0100111	45	0010111	97	112	8

Vec.	Ind.	Vec.	Ind.	Vec.	Ind.	Vec.	Ind.
$GF(3^3)$		$GF(3^4)$		$GF(5^2)$		$GF(5^3)$	
222	9	1022	32	10	1	414	26
121	10	2221	33	43	2	242	27
012	11	0212	34	42	3	221	28
120	12	2120	35	32	4	011	29
002	13	2202	36	44	5	110	30
020	14	0022	37	02	6	003	31
200	15	0220	38	20	7	030	32
201	16	2200	39	31	8	300	33
211	17	0002	40	34	9	204	34
011	18	0020	41	14	10	341	35
110	19	0200	42	33	11	114	36
202	20	2000	43	04	12	043	37
221	21	1002	44	40	13	430	38
111	22	2021	45	12	14	402	39
212	23	1212	46	13	15	122	40
021	24	1121	47	23	16	123	41
210	25	0211	48	11	17	133	42
		2110	49	03	18	233	43
		2102	50	30	19	131	44
$GF(3^4)$		2022	51	24	20	213	45
0001	0	1222	52	21	21	431	46
0010	1	1221	53	41	22	412	47
0100	2	1211	54	22	23	222	48
1000	3	1111	55			021	49
2001	4	0111	56			210	50
1012	5	1110	57	$GF(5^3)$		401	51
2121	6	0101	58	001	0	112	52
2212	7	1010	59	010	1	023	53
0122	8	2101	60	100	2	230	54
1220	9	2012	61	403	3	101	55
1201	10	1122	62	132	4	413	56
1011	11	0221	63	223	5	232	57
2111	12	2210	64	031	6	121	58
2112	13	0102	65	310	7	113	59
2122	14	1020	66	304	8	033	60
2222	15	2201	67	244	9	330	61
0222	16	0012	68	241	10	004	62
2220	17	0120	69	211	11	040	63
0202	18	1200	70	411	12	400	64
2020	19	1001	71	212	13	102	65
1202	20	2011	72	421	14	423	66
1021	21	1112	73	312	15	332	67
2211	22	0121	74	324	16	024	68
0112	23	1210	75	444	17	240	69
1120	24	1101	76	042	18	201	70
0201	25	0011	77	420	19	311	71
2010	26	0110	78	302	20	314	72
1102	27	1100	79	224	21	344	73
0021	28			041	22	144	74
0210	29			410	23	343	75
2100	30	$GF(5^2)$		202	24	134	76
2002	31	01	0	321	25	243	77

Vec.	Ind.	Vec.	Ind.	Vec.	Ind.	Vec.	Ind.
$GF(5^3)$		$GF(7^2)$		$GF(11^2)$		$GF(11^2)$	
231	78	53	3	29	4	85	56
111	79	56	4	78	5	8*	57
013	80	16	5	16	6	2*	58
130	81	54	6	54	7	88	59
203	82	66	7	*9	8	0*	60
331	83	03	8	*7	9	*0	61
014	84	30	9	87	10	17	62
140	85	45	10	**	11	64	63
303	86	12	11	07	12	92	64
234	87	14	12	70	13	43	65
141	88	34	13	46	14	*5	66
313	89	15	14	25	15	67	67
334	90	44	15	38	16	12	68
044	91	02	16	51	17	14	69
440	92	20	17	79	18	34	70
002	93	51	18	26	19	11	71
020	94	36	19	48	20	04	72
200	95	35	20	45	21	40	73
301	96	25	21	15	22	75	74
214	97	31	22	44	23	96	75
441	98	55	23	05	24	83	76
012	99	06	24	50	25	6*	77
120	100	60	25	69	26	42	78
103	101	13	26	32	27	95	79
433	102	24	27	*1	28	73	80
432	103	21	28	27	29	76	81
422	104	61	29	58	30	*6	82
322	105	23	30	39	31	77	83
424	106	11	31	61	32	06	84
342	107	04	32	62	33	60	85
124	108	40	33	72	34	52	86
143	109	32	34	66	35	89	87
333	110	65	35	02	36	1*	88
034	111	63	36	20	37	94	89
340	112	43	37	98	38	63	90
104	113	02	38	*3	39	82	91
443	114	33	39	47	40	5*	92
032	115	05	40	35	41	59	93
020	116	50	41	21	42	49	94
404	117	26	42	*8	43	55	95
142	118	41	43	97	44	09	96
323	119	42	44	93	45	90	97
434	120	52	45	53	46	23	98
442	121	46	46	99	47	18	99
022	122	22	47	03	48	74	100
220	123			30	49	86	101
				81	50	9*	102
		$GF(11^2)$		4*	51	13	103
$GF(7^2)$		01	0	65	52	24	104
01	0	10	1	*2	53	28	105
10	1	*4	2	37	54	68	106
64	2	57	3	41	55	22	107

Vec.	Ind.	Vec.	Ind.	Vec.	Ind.	Vec.	Ind.
$GF(11^2)$		$GF(11^2)$		$GF(11^2)$		$GF(11^2)$	
08	108	71	111	84	114	31	117
80	109	56	112	7*	115	91	118
3*	110	19	113	36	116	33	119

* represents the integer 10

References

1. Alanen, J. D. and Knuth, D. E. "Tables of Finite Fields," *Sankhyā*, **26** (1964), 305–326.
2. Bruck, R. H. and Bose, R. C. "The Construction of Translation Planes from Projective Spaces," *J. Algebra*, **1** (1964), 85–102.
3. Bose, R. C. "On the Construction of Balanced Incomplete Block Designs," *Annals of Eugenics*, **9** (1939), 353–399.
4. Bose, R. C. "The Affine Analogue of Singer's Theorem," *J. Ind. Math. Soc.*, **6** (1942), 1–15.
5. Bose, R. C., Chowla, S. and Rao, C. R. "On the Integral Order (mod p) of the Quadratics $x^2 + 2x + b$, with Applications to the Construction of Minimum Functions for $GF(p^2)$, and to some Number Theory Results," *Bull. Calcutta Math. Soc.*, **36** (1964), 153–174.
6. Rao, C. R. "Extension of the Difference Theorems of Singer and Bose," *Science and Culture*, **10** (1944), 57.
7. Rao, C. R. "Finite Geometries and Certain Derived Results in Theory of Numbers," *Proc. Nat. Inst. Sci. India*, **11** (1945), 136–149.
8. Rao, C. R. "Difference Sets and Combinatorial Arrangements Derivable from Finite Geometries," *Proc. Nat. Inst. Sci. India*, **12** (1946), 123–135.
9. Rao, C. R. "A Simplified Approach to Factorial Experiments and the Punched Card Technique in the Construction and Analysis of Designs," *Bull. Inst. Internat. Statist.*, **33** (1951), 1–28.
10. Rao, C. R. "A Study of BIB Designs with Replications 11 to 15," *Sankhyā, Ser. A*, **23** (1961), 117–127.
11. Singer, J. "A Theorem in Finite Projective Geometry and Some Applications to Number Theory," *Trans. Amer. Math. Soc.*, **43** (1938), 377–385.

Part V

GRAPHS

CHAPTER 29

Some Classes of Perfect Graphs

C. BERGE, *International Computation Center, Italy*

1. INTRODUCTION

Given an unoriented graph G, a set S of vertices is said to be *stable* (or "internally stable") if no two vertices in S are adjacent; a set C of vertices is a *clique* if every pair of vertices in C is adjacent. We shall denote by $\gamma(G)$ the *chromatic number* of G, that is, the smallest number of colors which can be used to color the vertices of G in such a way that no two adjacent vertices have the same color. $\omega(G)$ will denote the *clique number* of G, that is, the maximum number of vertices which can constitute a clique in G. $\alpha(G)$ will denote the *stability number* of G, that is, the maximum number of vertices which constitute a stable set in G. $\theta(G)$ will denote the *partition number* of G, that is, the smallest number of disjoint cliques which cover all the vertices of G.

It is obvious that $\alpha(G) \leq \theta(G)$, because if k cliques partition the set of all vertices, a stable set S has not more than one element in each of the cliques; the graphs G for which $\alpha(G) = \theta(G)$ seem to have a great importance in several fields, and C. Shannon has noticed that only such a graph can be used in communication theory as a perfect channel (see [1], p. 38). On the other hand, it is obvious that $\omega(G) \leq \gamma(G)$, because if k vertices constitute a maximum clique, we need at least k colors to color the vertices of G. As the determination of $\gamma(G)$ is of great importance in many combinatorial problems (four-color conjecture, triads and statistical completion of block designs, etc.), several authors have tried to determine when we have $\gamma(G) = \omega(G)$.

2. TWO BASIC LEMMAS

Recall a few definitions. Given a graph G, let us denote by X the set of all its vertices, and by U the set of all its edges. We shall assume that G is finite, that is, $|X| < \infty$, and that a pair of vertices cannot be joined by more than one edge; in addition, we assume that G has no loop, (that is, an edge having its two extremities at the same vertex). A *subgraph* of G is a graph defined by a subset $A \subset X$ and all the edges of G having their two extremities in A.

A graph G is said to be *γ-perfect* if every subgraph $A \subset X$ satisfies $\gamma(A) = \omega(A)$;

a graph G is said to be *α-perfect* if every subgraph $A \subset X$ satisfies $\alpha(A) = \theta(A)$;

a graph is *perfect* if it is both γ-perfect and α-perfect.

If G is connected, a subset $A \subset X$ is said to be an *articulation set* if, by removing it, we create several connected components $C_1, C_2, \ldots$; the subgraphs $A \cup C_1$, $A \cup C_2$, ... are the *leaves* relative to A. We have two basic results:

Lemma 1. *If a connected graph G possesses an articulation set A which is a clique, and if each of its leaves is γ-perfect, then G is γ-perfect.*

Proof. If $\omega(G) = k$, then there exists a k-clique in at least one of the leaves relative to A. If G' is this leaf, it satisfies $\gamma(G') = \omega(G') = k$; thus it can be colored with k colors. If we color each leaf with k colors, and if we give the same name to the $|A|$ different colors used in clique A, we obtain a coloration in k colors for the entire graph G. Then, $\gamma(G) \leq k$. On the other hand, we have $\gamma(G) \geq \omega(G) = k$. Thus, we have finally

$$\gamma(G) = k = \omega(G)\,.$$

Since this equality holds also for any subgraph, the graph G is γ-perfect.

Lemma 2. *If a connected graph G possesses an articulation set A which is a clique, and if each of its leaves is α-perfect, then G is α-perfect.*

Proof. Let A be an articulation set which is a clique and assume first that there are only two connected components C and C' attached to A. Put

$$\alpha(A \cup C) = k\,, \qquad \alpha(A \cup C') = k'\,.$$

Case 1. *There exists a maximum stable set S of $A \cup C$ which does not meet A, and a maximum stable set S' of $A \cup C'$ which does not meet A.*

Thus, $\alpha(C) = k$, and $\alpha(C') = k'$. $A \cup C$ can be partitioned into $\alpha(A \cup C) = k$ cliques, and C' can be partitioned into $\alpha(C') = k'$ cliques; therefore $A \cup C \cup C'$ can be partitioned into $k + k'$ cliques. On the other hand, $S \cup S'$ is a stable set, and $|S \cup S'| = k + k'$. Hence,

$$k + k' \leq \alpha(G) \leq \theta(G) \leq k + k'\,.$$

We thus have $\alpha(G) = \theta(G)$.

Case 2. *There exists a maximum stable set S of $A \cup C$ which does not meet A, and every maximum stable set of $A \cup C'$ meets A.*

Then $\alpha(C) = k$, $\alpha(C') = k' - 1$, $A \cup C$ can be partitioned into k cliques, and C' can be partitioned into $k' - 1$ cliques, so $A \cup C \cup C'$ can be partitioned into $k + k' - 1$ cliques. On the other hand, if S' is a maximum stable set of C', $S \cup S'$ is a stable set, and $|S \cup S'| = k + k' - 1$. Hence,

$$k + k' - 1 \leq \alpha(G) \leq \theta(G) \leq k + k' - 1$$

Thus, we have $\alpha(G) = \theta(G)$.

Case 3. *Every maximum stable set S of $A \cup C$ meets A, every maximum stable set S' of $A \cup C'$ meets A, and there exists an S and an S' such that $S \cap S' \neq \phi$.*

In this case, $\alpha(C) = k - 1$, and $\alpha(C') = k' - 1$. Thus, C can be partitioned into $k - 1$ cliques, C' can be partitioned into $k' - 1$ cliques, and $C \cup C' \cup A$ can be partitioned into $k + k' - 1$ cliques. As $S \cap S'$ is a set of one element $a \in A$, the set $S \cup S'$ is stable, and $|S \cup S'| = k + k' - 1$. Hence,

$$k + k' - 1 \leq \alpha(G) \leq \theta(G) \leq k + k' - 1\,.$$

Hence, $\alpha(G) = \theta(G)$.

Case 4. *Every maximum stable set S of $A \cup C$ meets A, every maximum stable set S' of $A \cup C'$ meets A, and we cannot have $S \cap S' \neq \phi$.*

In this case, $(S \cup S') - A$ is a stable set, and

$$|(S \cup S') - A| = k + k' - 2\,.$$

Let $\bar{C}$ be the union of C and of all elements in A which belong to a maximum stable set of $A \cup C'$. We have $\alpha(\bar{C}) = k - 1$, and we can partition $\bar{C}$ into $k - 1$ cliques. Let $\bar{C}'$ be the union of C' and $(A - \bar{C})$. We have $\alpha(\bar{C}') = k' - 1$, and we can partition $\bar{C}'$ into $k' - 1$ cliques. Hence,

$$k + k' - 2 \leq \alpha(G) \leq \theta(G) \leq k + k' - 2\,.$$

Hence, $\alpha(G) = \theta(G)$.

Thus, if there are only two leaves, we have always $\alpha(G) = \theta(G)$. If this is always true with less than p leaves, it will be true for a graph G with p leaves because we can consider as one leaf the union of distinct leaves. Thus the result is proved.

3. COMPARABILITY GRAPHS

A graph G is said to be a *comparability graph* if it is possible to orientate each edge in such a way that the relation " there is an oriented edge going from vertex a to vertex b," or in short, $a > b$, is a strict order. That is,

(1) $a > b$, $b > c$ implies $a > c$

(2) $a > b$ implies not $b > a$.

A subgraph of a comparability graph is obviously a comparability graph; a characterization of such graphs has been given independently by P. Gilmore and A. Hoffman [6] and by A. Ghouila-Houri [4].

Example. *Bipartite graph.* A graph is said to be *bipartite* if it has at least one edge and if it does not contain any odd cycle. It is well known that for such a graph G, one has $\gamma(G) = 2$. If we color the vertices with two colors, and if we direct each edge from the first color to the second color, the conditions (1) and (2) are trivially true, so the graph G is a comparability graph. The converse is not true; the graph of Figure 1 is a comparability graph, but not bipartite. The reader can easily check that the graph of Figure 2 is not a comparability graph.

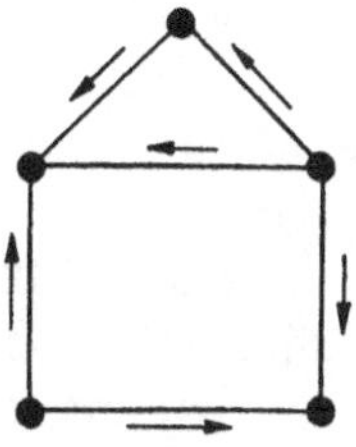

Figure 1

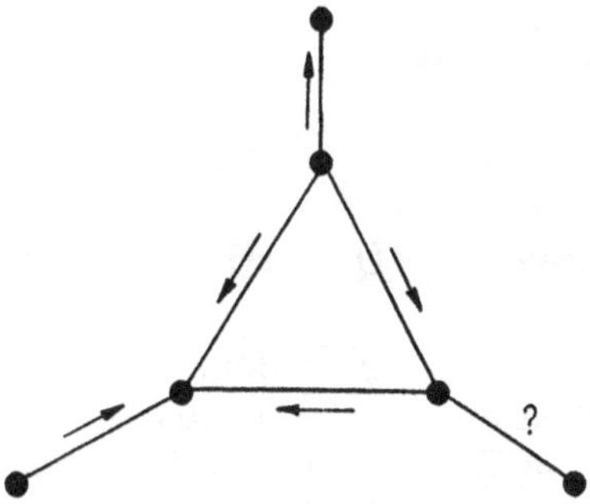

Figure 2

For a bipartite graph G, one has

$$\omega(G) = 2 = \gamma(G)\,.$$

On the other hand, it is known that one has also $\alpha(G) = \theta(G)$. This is equivalent to a very famous theorem of König. Thus, a bipartite graph is *perfect*.

Theorem 1. *A comparability graph is γ-perfect.*

Proof. Consider a graph G with an orientation $>$ which satisfies (1) and (2). Thus, the graph G has no circuits, and, as G is finite, to each vertex x we can assign a finite number $f(x)$ representing the length of the longest path issuing from x. If $\max_x f(x) = k - 1$, there exists a path with k vertices. There exists no clique with more than k vertices because such a clique would contain a path passing through each of its vertices (Theorem of Rédei), and the longest path contains only k vertices. Thus, we have

$$\omega(G) = k\,.$$

On the other hand, consider k colors $0, 1, \ldots, k-1$, and assign color $f(x)$ to vertex x. Two adjacent vertices x and y cannot have the same color because if the edge $[x, y]$ is oriented from x to y, we have $f(x) > f(y)$; therefore

$$\gamma(G) \leq k\,.$$

As we have always

$$\gamma(G) \geq \omega(G) = k\,,$$

we have finally

$$\gamma(G) = k = \omega(G)\,.$$

Theorem 2. (Dilworth[1]). *A comparability graph is α-perfect.*

Proof. A very famous theorem of Dilworth states: "If the orientation of the edges of a graph G satisfies (1) and (2), then $\alpha(G)$ is equal to the smallest number of disjoint paths which cover all the vertices". But to each path of G corresponds a clique, and to each clique corresponds a path; therefore, we have $\alpha(G) = \theta(G)$.

4. TRIANGULATED GRAPHS

A graph is said to be *triangulated* if every cycle of length greater than three possesses a chord, that is, an edge joining two non-consecutive vertices of the cycle. Triangulated graphs arise in many contexts.

Example 1. *Interval graphs.* Consider on a line a finite family of intervals, and draw a graph whose vertices represent the intervals, two vertices being joined if the two corresponding intervals intersect; such a graph is called an *interval graph.*

[1] Ann. Math. (2), **51** (1950), 161-166.

The problem of characterizing a graph representing a family of intervals was first put by G. Hajos [8] as follows: In a university, each student has to go once a day to the library; at the end of the day we ask each of them whom he has met, and we draw a graph G whose vertices represent the students, two vertices being joined if the two corresponding students met at the library. For each student, we have corresponding an interval of time during which he stayed at the library, and G is the representing graph of this family of intervals. Hajos [9], later gave an algorithm to locate the intervals and P. C. Gilmore and A. J. Hoffman [6] gave a complete characterization of interval graphs. See also Lekkerkerker and Boland [11].

Interval graphs arise also in psychology as follows: Consider on a line a finite number of points $P_1, P_2, \dots, P_n$, and an infinite family of intervals Ω; two points, P_i and P_j are said to be *indistinguishable* if there exists in Ω an interval which contains both P_i and P_j.

A problem which has been considered in psychosociology is to characterize the graphs of indistinguishable pairs of points. In fact, such a graph G is the representing graph of a family of intervals $I_1, I_2, \dots, I_n$. Interval I_i corresponding to point P_i is defined as follows: I_i is the intersection of interval $[P_i, +\infty]$ with the union of all the intervals of Ω which contain P_i.

Consider two points P_i and P_j, with $P_i < P_j$. If P_i and P_j are indistinguishable, one has $P_j \in I_i$; therefore $I_i \cap I_j \neq \phi$. Conversely, if $I_i \cap I_j \neq \phi$, one has $P_j \in I_i$; therefore P_i and P_j are indistinguishable.

Thus, each graph of indistinguishable pairs of points is a representing graph of a family of intervals (and each representing graph of a family of intervals is also a graph of indistinguishable pairs of points).

Example 2. A graph G is said to be a *cactus* if it is connected and does not possess any cycle of length greater than 3. Cacti are considered in physics and are obviously triangulated.

Example 3. Given a graph G, its *adjoint* G^* is a graph whose vertices represent the edges of G, two vertices being joined if they represent adjacent edges of G. *The graph G^*, adjoint of a cactus G, is triangulated.* Assume that there exists in G^* a cycle $(u_1^*, u_2^*, \dots, u_k^*)$ without a chord, with $k > 3$; it corresponds to a cycle of G with edges $u_1, u_2, \ldots, u_k$. This contradicts the fact that G is a cactus.

We must remark that "triangulated" and "comparability" graphs are two independent properties. The graph pictured in Figure 1 is a comparability graph, but it is not a triangulated graph; on the other hand, the graph of Figure 2 is triangulated, but is not a comparability graph.

Theorem 3. *If a triangulated graph G is connected and is not a clique, it contains an articulation set which is a clique.*

Proof. As G is not a clique, there exists at least two non-adjacent vertices, so there exists at least one articulation set. Let A be a minimal articulation set, whose removal creates several connected components $C, C', C'', \ldots$. Every element a of A is joined by an edge to every component because, if not, $A - \{a\}$ would be also an articulation set, and A would not be a minimal articulation set. Consider in A two distinct element a_1 and a_2. There exists a chain

$$\mu = [a_1, c_1, c_2, \ldots, c_p, a_2], \qquad c_1, c_2, \ldots, c_p \in C,$$

and consider μ a shortest chain of that kind. There exists also a chain

$$\mu' = [a_2, c_1', c_2', \ldots, c_q', a_1], \qquad c_1', c_2', \ldots, c_q' \in C',$$

and consider μ' a shortest chain of that kind. The cycle

$$\mu + \mu' = [a_1, c_1, c_2, \ldots, c_p, a_2, c_1', c_2', \ldots, c_q', a_1]$$

does not possess any of the following chords:

$$\left.\begin{array}{l} [a_1, c_i] \text{ with } i \neq 1, \\ [c_i, c_j] \text{ with } i \neq j \pm 1, \\ [a_2, c_i] \text{ with } i \neq p, \end{array}\right\} \text{because } \mu \text{ would not be a shortest chain.}$$

$[c_i, c_j']$, because C and C' are two disjoint connected components.

$$\left.\begin{array}{ll}[a_2, c'_j] \text{ with } j \neq 1, \\ [c'_i, c'_j] \text{ with } i \neq j, \\ [a_1, c'_j] \text{ with } j \neq q,\end{array}\right\} \text{because } \mu' \text{ would not be a shortest chain.}$$

As the graph G is triangulated, the cycle $\mu + \mu'$ (whose length is at least 4) possesses a chord, and this chord is necessarily $[a_1, a_2]$.

Thus, every pair of vertices $a_1, a_2 \in A$ is joined, and therefore A is a clique.

Corollary 1. *A triangulated graph G is γ-perfect.*

Proof. Notice first that every subgraph of G is also a triangulated graph; because if G' contained a cycle without a chord, G would also contain a cycle without a chord.

Our statement is obviously true for triangulated graphs with 1 or 2 vertices; we assume that it is true for graphs with less than n vertices, and prove it for a given graph G with n vertices.

We can assume that G is neither a disconnected graph nor a clique (because the statement would be proved). So, by Theorem 3, G possesses an articulation set A which is a clique, and every subgraph G' of any one of its leaves satisfies

$$\gamma(G') = \omega(G');$$

hence, by Lemma 1 (Section 1), we have

$$\gamma(G) = \omega(G).$$

Corollary 2. (Hajnal—Suranyi's theorem). *A triangulated graph is α-perfect.*

The proof is exactly as above, using Lemma 2 instead of Lemma 1.

5. UNIMODULAR GRAPHS

Given a graph G with vertices $x_1, x_2, \ldots, x_n$, let $C_1, C_2, \ldots, C_m$ be its maximal cliques. A matrix $\boldsymbol{M} = (m^i_j)$, with n columns and m rows, is said to be the *clique-incidence* matrix of G if

$$m_j^i = \begin{cases} 0 & \text{if} \quad x_j \notin C_i \\ 1 & \text{if} \quad x_j \in C_i . \end{cases}$$

Definition 1. *A graph G is said to be unimodular if its clique-incidence matrix $\boldsymbol{M}$ is totally unimodular (that is, if every square submatrix of $\boldsymbol{M}$ has a determinant equal to* 0, + 1 *or* − 1).

Obviously, if G is unimodular, a subgraph of G has a clique-incidence matrix which is a submatrix of $\boldsymbol{M}$; hence this subgraph is also unimodular. An alternate definition is:

Definition 2. *For $A \subset X$, a clique C of a graph G is said to be even in A if $|C \cap A|$ is even; G is said to be unimodular if every non-empty subgraph A contains two disjoint sets A_1 and A_2 (not both empty) such that any maximal clique C, even in A, satisfies $|C \cap A_1| = |C \cap A_2|$.*

The equivalence of the two definitions is a particular case of a result of A. Ghouila-Houri [5] and P. Camion [3].

Example 1. A bipartite graph is unimodular, because it is well known that the edge-incidence matrix of a bipartite graph is totally unimodular.

Example 2. The adjoint G^* of a bipartite graph G is unimodular, because the clique-incidence matrix of G^* is nothing else than the transpose $\boldsymbol{M}^*$ of the edge-incidence matrix $\boldsymbol{M}$ of G.

Example 3. An interval graph is unimodular. We have seen that an interval graph G can represent a family of points on a line, two points P_i and P_j being joined if and only if there exists an interval $\omega \in \Omega$ which covers both of them (see Example 1, Section 3). A set A of vertices can be ordered by the natural order of the points of the line that they represent: call A_1 the points whose order is even, and A_2 the points whose order is uneven. A maximal clique C corresponds to an interval of points on the line. Thus, if C is even in A, we have $|C \cap A_1| = |C \cap A_2|$, and, by Definition 2, the graph G is unimodular.

Remark. The concept of a unimodular graph is independent of the one of a triangulated graph, or of a comparability graph. For instance, the triangle inscribed in a hexagon (Figure 3) is triangulated but is not unimodular.

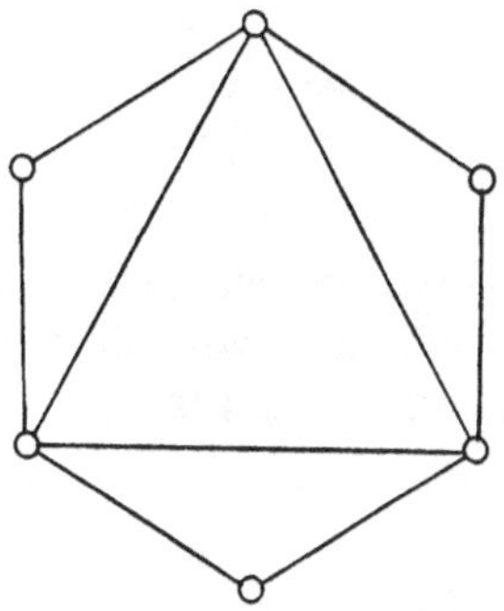

Figure 3

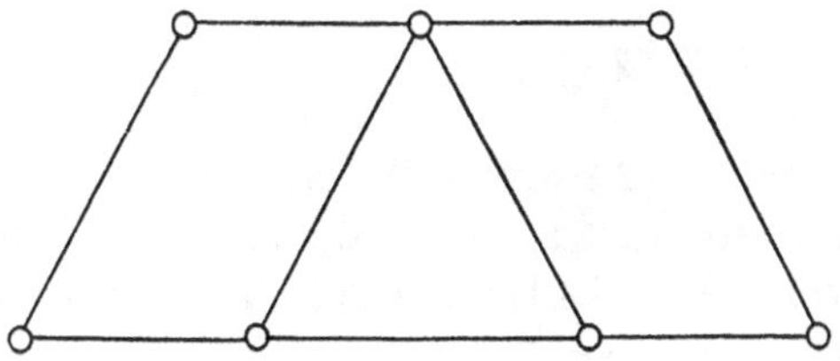

Figure 4

The graph of Figure 4 is unimodular, but it is not a triangulated graph, nor a comparability graph (since it contains an odd cycle without a triangular chord).

Theorem 4. *A unimodular graph G is α-perfect.*

Proof. Actually, this statement is already given in different forms by different authors (see Hoffman [10]). All we need is to prove that $\alpha(G) = \theta(G)$.

We shall define a stable set S by a vector

$$\boldsymbol{\alpha} = (\alpha_1, \alpha_2, \ldots, \alpha_n)\,,$$

with $\alpha_j = 1$ if $x_j \in S$, $\alpha_j = 0$ if $x_j \notin S$. A maximum stable set is given by a linear program in integers:

(1) $\boldsymbol{\alpha} \geq 0$,

(2) $\boldsymbol{\alpha} \leq \mathbf{1} = (1, 1, \ldots, 1)$,

(3) $\boldsymbol{M\alpha} \leq \mathbf{1}$,

(4) maximize $\sum_{j=1}^{n} \alpha_j$.

Condition (2) can be deleted, because it is contained in (3). The dual linear program is:

(1′) $\boldsymbol{\lambda} = (\lambda_1, \lambda_2, \ldots, \lambda_m) \geq 0,$

(2′) $\boldsymbol{M}^* \boldsymbol{\lambda} \geq \mathbf{1},$

(3′) minimize $\sum_{i=1}^{m} \lambda_i$

We can also add

(4′) $\boldsymbol{\lambda} \leq \mathbf{1},$

because (2′) and (3′) imply (4′). In other words, we are trying to find a minimal family of cliques which cover all the vertices. As the matrix $\boldsymbol{M}$ is totally unimodular, by the duality theorem, we have:

$$\alpha(G) = \max \sum_{j=1}^{n} \alpha_j = \min \sum_{i=1}^{m} \lambda_i = \theta(G) .$$

Theorem 5. *A unimodular graph G is γ-perfect.*

Proof. All we need is to prove that $\gamma(G) = \omega(G)$. Let k be the largest number of elements in a clique. Consider the clique-incidence matrix $\boldsymbol{M}$, and find a vector $\boldsymbol{\alpha} = (\alpha_1, \alpha_2, \ldots, \alpha_n)$ which satisfies

$(0, 0, \ldots, 0) \leq \boldsymbol{\alpha} \leq (1, 1, \ldots, 1),$

$\langle \boldsymbol{M}^i, \boldsymbol{\alpha} \rangle = \mathbf{1}$, if clique C_i contains exactly k vertices, and

$\langle \boldsymbol{M}^i, \boldsymbol{\alpha} \rangle \leq \mathbf{1}$, if clique C_i contains less than k vertices.

$\boldsymbol{\alpha} = (1/k, 1/k, \ldots, 1/k)$ is a point which satisfies these inequalities, hence they are consistent. But by the unimodular property of the matrix $\boldsymbol{M}$, there is an integral solution

$$\boldsymbol{\alpha}^1 = (\alpha_1^1, \alpha_2^1, \ldots, \alpha_n^1) .$$

Consider the set S_1 of all vertices x_i such that $\alpha_i^1 = 1$: it is a stable set, and it meets all the cliques with k elements. Color with a first color the vertices of S_1. The subgraph obtained by deleting S_1 is also unimodular and its largest clique contains $k - 1$ elements. Color with a second color a stable set S_2 which meets all the cliques with $k - 1$ elements, etc..

By such a process, we can color all the vertices of G with k colors; hence $\gamma(G) = k = \omega(G)$.

This theorem contains the fundamental theorem of bipartite graphs, namely that the chromatic class of a bipartite graph is equal to the maximum degree of its vertices (see, for example, Berge [1], p. 95). It shows that this statement still holds for multigraphs (when there may be more than one edge connecting the same two vertices).

Acknowledgements

We are indebted to A. J. Hoffman and P. Gilmore for suggestions and helpful discussions. We are indebted to M. H. McAndrew for the proof of Theorem 5, which is somewhat shorter than our original version.

References

1. Berge, C. *The Theory of Graphs and its Applications*, Methuen-Wiley, 1961.
2. Berge, C. "Farbung von Graphen, deren samtliche bzw. deren ungerade Kreise starr sind," (Zusammenfassung), Wiss. Z. Martin-Luther-Univ. Halle-Wittenberg, 1961.
3. Camion, P. "Matrices totalement unimodulaires et problèmes combinatoires," Thèse, Université libre de Bruxelles, 1963.
4. Ghouila-Houri, A. "Caractérisation des graphes non orientés dont on peut orienter les arêtes de manière à obtenir le graphe d'une relation d'ordre," *C. R. Acad. Sci., Paris*, **254** (1962), 1370.
5. Ghouila-Houri, A. "Caractérisation des matrices totalement unimodulaires," *C. R. Acad. Sci. Paris*, **254** (1962), 1192.
6. Gilmore, P. C. and Hoffman, A. J. "A Characterization of Comparability Graphs and of Interval Graphs," unpublished.
7. Hajnal, A. and Suranyi, T. "Uber die Auflosung von Graphen vollstandige Teilgraphen," *Ann. Univ. Sci. Budapest. Eötvös Sect. Math.*, **1** (1958), 113.
8. Hajos, G. "Uber eine Art von Graphen," *Int. Math. Nachrichten*, **11** (1957).
9. Hajos, G. Talk at Budapest, Oct. 1959 (unpublished).
10. Hoffman, A. J. "Some Recent Applications of the Theory of Linear Inequalities to Combinatorial Analysis," *Proc. Symp. Applied Math.*, **10** (1960), 113.
11. Lekkerkerker, C. G. and Boland, J. C. "Representation of a Finite Graph by a Set of Intervals on the Real Line," *Fund. Math.*, **51** (1962), 45.

12. Shannon, C. E. " The Zero-Error Capacity of a Noisy Channel," *IRE Trans.*, **IT-3** (1956), 3.

Discussion on Professor Berge's Paper

PROFESSOR M. E. WATKINS: Professor Berge has taken three very large classes of graphs and shown them to be both α-perfect and γ-perfect. The proofs for α-perfectness and γ-perfectness, however, are totally independent. On the other hand, if one considers graphs that are not perfect, the simplest examples are the circuits Z of length $2n+1$ where $n \geq 2$. Thus $\alpha(Z) = n$, $\theta(Z) = n+1$, $\gamma(Z) = 3$ and $\omega(Z) = 2$. That is, they are neither α-nor γ-perfect.

One is thus led to inquire, are any general classes of graphs known which are α-perfect but not γ-perfect or which are γ-perfect but not α-perfect? It would appear that the concomitance of α- and γ-perfectness in graphs has some underlying theoretical basis which is worthy of investigation.

If a graph G has n vertices, let $\bar{G}$ denote its complement in the complete graph K_n. Since $\alpha(G) = \omega(\bar{G})$ and $\gamma(G) = \theta(\bar{G})$ and since only section subgraphs are considered, it seems likely that G is perfect if and only if $\bar{G}$ is perfect. Our investigation would be reduced to determining when, if at all, one can have G α-perfect (or γ-perfect) without having $\bar{G}$ be α-perfect (or γ-perfect).

α- and γ-perfectness are induced hereditary properties, (see S. Hedetniemi, " On Hereditary Properties of Graphs," unpublished), and perhaps this problem can be approached by some of the techniques of this paper.

CHAPTER 30

A Geometrical Version of the Four Color Problem[1]

W. T. TUTTE, *University of Waterloo, Canada*

1. INTRODUCTION

By a *four-coloring* of a map on the sphere we mean a coloring of its regions in four colors so that no two regions with a common border have the same color. (A finite number of isolated points do not constitute a border.) The Four Color Problem is the problem of deciding whether or not every map on the sphere has a four-coloring.

The Four Color Problem seems at first to be a problem of of 2-dimensional topology, but it can be replaced in more than one way by a problem of pure graph theory.

We can for example represent a map by a graph whose vertices correspond to regions. Two vertices are joined by an edge if and only if the two regions have a common border. Evidently no edge joins a vertex to itself, that is, the graph is without loops. We are thus led to define a four-coloring of any graph as a coloring of its vertices in four colors so that no two vertices of the same color are joined by an edge. We

[1] This paper is presented as an introduction to a longer and more technical one entitled "On the algebraic theory of graph colorings" which appeared recently in the *Journal of Combinatorial Theory* (1, 15-50). The longer paper develops a suggestion of O. Veblen for converting the Four Color Problem into a problem about finite projective geometries.

note that no graph with a loop has a four-coloring.

We may ask for a non-trivial sufficient condition for a graph G without loops to have a four-coloring. Hadwiger has suggested the following conjecture: if G has no four-coloring then it can be transformed into a complete 5-graph by deleting some edges and contracting others to single vertices. If this conjecture could be proved the Four Color Problem would be disposed of, since the complete 5-graph cannot be drawn on the sphere with no edges crossing.

It is well known that the Four Color Problem can be reduced to the case in which exactly three regions of the map meet at each vertex. Such a map is defined by a graph, drawn in the sphere, in which exactly three edges meet at each vertex, and which has no isthmus. (An isthmus of a graph is an edge whose deletion disconnects that component of the graph in which it lies.) P. G. Tait showed that the map would have a four-coloring if and only if the edges of the graph could be colored in three colors so that just one edge of each color was incident with each vertex.

We refer to a graph having three edges at each vertex, loops being counted twice, as *trivalent*, and we define a *Tait coloring* of a trivalent graph, planar or non-planar, as a coloring of its edges in three colors so that one edge of each color is incident with each vertex. It is not difficult to show that a trivalent graph with an isthmus has no Tait coloring. We note that a trivalent graph with a loop necessarily has an isthmus incident with the same vertex.

We may replace the Four Color Problem by that of finding a sufficient condition for a given trivalent graph G to have a Tait coloring. There seems to be no widely recognized conjecture about such a condition.

It is known however that the Petersen graph has no Tait coloring. This graph is formed from two pentagons $a_1a_2a_3a_4a_5$ and $b_1b_3b_5b_2b_4$ by making the five joins a_ib_i. Perhaps we shall not soon be refuted if we conjecture that any trivalent graph with no isthmus and no Tait coloring can be reduced to a Petersen graph by deleting some edges and contracting others to single vertices.

2. GEOMETRICAL REPRESENTATIONS

Let G be any graph. If S is a set of vertices of G we

define the *coboundary* δS of S as the set of all edges having one end in S and one not in S. If U is a set of edges we define its *boundary* ∂U as the set of all vertices v of G such that the number of members of U incident with v, loops being ignored, is odd. If ∂U is null we say U is a *cycle* of G.

Two sets of edges of G will be called *orthogonal* if the number of members of their intersection is even. It can be shown that a set of edges is a cycle (coboundary) if and only if it is orthogonal to every coboundary (cycle).

The set of cycles and the set of coboundaries of G are groups with respect to the operation of addition of sets modulo 2.

We shall say that a set of edges of G is *linearly dependent* with respect to cycles (coboundaries) if and only if some subset of it is a non-null cycle (coboundary) of G.

Consider a mapping F of the set $E(G)$ of edges of G onto a set S of points of a projective geometry $PG(q, 2)$ over $GF(2)$. Given a subset E_1 of k edges of G we form the subset FE_1 of S and say that FE_1 also has k points. Thus each point is counted as many times as it has corresponding edges. We call F a *direct embedding* of G in $PG(q, 2)$ if, for each $E_1 \subseteq E(G)$, FE_1 is a linearly dependent set of points if and only if E_1 is linearly dependent with respect to cycles. The set FE_1 is counted as linearly dependent if it includes repeated points. A direct embedding of G is possible only if G has no loop, for each loop is the only member of some cycle, whereas no point of $PG(q, 2)$, taken by itself and counted once only, forms a linearly dependent set. If G is loopless it is always possible to construct a direct embedding.

Let F be a direct embedding of G in which S spans $PG(q, 2)$. Then it can be shown that the $(q-1)$-spaces in $PG(q, 2)$ are in $1-1$ correspondence with the non-null coboundaries of G, by the following rule. If f is a non-null coboundary and Σ_f is the corresponding $(q-1)$-space, then a point of S belongs to Σ_f if and only if its corresponding edges are not in f. We note that two distinct edges corresponding to the same point of S must constitute a cycle. Hence by the orthogonality of cycles and coboundaries they are either both in f or both not in f.

In an analogous way we can define a *dual embedding* of G in terms of linear dependence of edges with respect to coboundaries. Such an embedding can be constructed whenever G has no isthmus.

Now let us attempt to describe the four-colorings and Tait

colorings of G, if any, in terms of direct and dual embeddings. In the case of a four-coloring we take the four colors to be the four 2-vectors over $GF(2)$. Let U_1 be the set of vertices for which the first component of the color-vector is 1, and U_2 the set for which the second component is 1. A given edge must join vertices of different colors and must therefore appear either in ∂U_1 or in ∂U_2. Any four-coloring therefore is associated with a pair $\{f, g\}$ of coboundaries such that each edge appears in at least one. Conversely, given such a pair $\{f, g\}$ we can construct a corresponding four-coloring. If G is trivalent and has a Tait coloring let the three colors involved be α, β and γ. It is easily verified that the edges colored α and β constitute a cycle, as do those colored β and γ. Hence, the Tait coloring is associated with a pair $\{f, g\}$ of cycles such that each edge belongs to at least one of them. Conversely, given such a pair of cycles and using trivalency, we can construct a Tait coloring. We can use the characterization in terms of pairs of cycles to extend the notion of a Tait coloring to non-trivalent graphs.

Let $\{f, g\}$ be a pair of coboundaries of G such that each edge appears in at least one. Consider a direct embedding of G in $PG(q, 2)$ for which the point-set S spans the entire geometry. Consider the spaces Σ_f and Σ_g, taking for example $\Sigma_f = PG(q, 2)$ if f is null. Their intersection contains a $(q-2)$-space which contains no point of S. Conversely, given such a $(q-2)$-space we can represent it as the intersection of two distinct $(q-1)$-spaces, and these correspond to a pair of coboundaries associated with a four-coloring. The same reasoning applies to pairs of cycles and the dual embedding of G.

Instead of asking which graphs are not 4-colorable, or which have no Tait coloring, we may merge these two problems into a single geometrical one. This geometrical problem asks for the classification of those sets of points in $PG(q, 2)$ that meet every $(q-2)$-space.

3. BLOCKS

Let k be a positive integer. We use the term "k-block" to mean a set B of points in $PG(q, 2)$ whose dimension is $\geq k$ and which includes at least one point from each subspace of dimension $q - k$. Evidently a k-block in $PG(q, 2)$ is a k-block in any subspace of $PG(q, 2)$ that contains it.

Our problem is the classification of the 2-blocks. As an

example, we may take the plane, which has exactly 7 points. We call this the Fano block.

We try to reduce our problem to that of classifying some special kinds of 2-blocks. For example, we may define a minimal k-block as one that contains no smaller k-block, and we may then attempt to classify the minimal 2-blocks. It is clear that the Fano block is minimal.

A more promising procedure is suggested by Hadwiger's Conjecture. Given a graph G let us delete the edges of a set U and consider the subgraph G_1 of G defined by the remaining edges and their incident vertices. We write $E(G_1) = V$. It is easily verified that the cycles of G_1 are those cycles of G that are contained in V, whereas the coboundaries of G_1 are the intersections with V of the coboundaries of G. It follows that given a direct embedding of G in $PG(q, 2)$ we can obtain one of G_1 by retaining only those points of S that correspond to members of V. If G_1 has no isthmus we can obtain a dual embedding of it, given one of G, as follows. We write Σ for the subspace generated by the points of S corresponding to members of U. Let Π be a subspace of $PG(q, 2)$ which does not meet Σ and which has the maximum dimension consistent with this property. We project the points of V from Σ onto Π. That is for each point P we construct the subspace generated by P and Σ, first noting that P is not in Σ since G_1 has no isthmus. This meets Π in a single point P', and we call P' the projection of P. It is found that the points P' corresponding to the points of S not in Σ constitute a dual embedding of G_1 in Π.

We can also consider the effect on G of contracting each edge of U to a single vertex. Let the resulting graph be G_2. It is now found that the coboundaries of G_2 are those coboundaries of G that are contained in V, and that the cycles of G_2 are the intersections with V of the cycles of G. We may therefore obtain embeddings of G_2 just as for G_1, except that we delete points of S for the dual embedding and use the projection process for the direct one. For the projection process G_2 must have no loop.

In this connection it is convenient to define a *tangential* k-block as one that cannot be converted into a smaller k-block by the process of projection described above. More precisely we put the definition as follows. Given a subset T of the k-block S we define a *tangent* of T, with respect to S, as a $(q-k)$-subspace of $PG(q, 2)$ that meets S only in the points of T and points linearly dependent on them. We say S is a *tangential* k-block if every non-null subset of S, of dimension not exceeding $q - k$, has a tangent.

The Fano block is trivially tangential if $q = 2$. Any block which is tangential in $PG(q, 2)$ is easily seen to be tangential in any subspace of $PG(q, 2)$ that contains it.

The following theorem is helpful.

Theorem 3.1. *Every tangential k-block is minimal.*

Proof. Suppose a k-block B is tangential but not minimal. Then there exists $P \in B$ such that $B - \{P\}$ is a k-block. But this is impossible, for $\{P\}$ has a tangent and this tangent is a $(q - k)$-space not meeting $B - \{P\}$.

Let us now propose the problem of classifying the tangential 2-blocks. Success with this should enable us to dispose of the Four Color Problem. For suppose S is a 2-block occurring as a direct embedding of some graph G which has no four-coloring. We can convert S into a tangential 2-block S' by a sequence of operations of projection (by Theorem 3.1). Then S' is a direct embedding of graph G' obtained from G by contracting some edges to points, and which is still not four-colorable. If the Hadwiger Conjecture is true G' will always be a complete 5-graph, save for possible multiple joins. Thus a complete list of tangential 2-blocks would enable us to check Hadwiger's Conjecture for four-colorings, assuming that we could determine for each one whether or not it was a direct embedding of a graph.

It seems then that we are justified in offering the problem of the classification of the tangential 2-blocks as a geometrical version of the Four Color Problem.

4. THE TANGENTIAL 2-BLOCKS

So far the tangential 2-blocks have been determined as far as the 5-dimensional case.

The only 2-dimensional tangential 2-block is evidently the Fano block.

It can be shown that the only minimal 2-block of three dimensions is obtained from $PG(3, 2)$ by deleting five points in general position. Evidently each line in $PG(3, 2)$ meets one of the 10 remaining points, and so these points constitute a 2-block B. The five chosen points determine 10 lines. As each point of B lies on one of these it follows that B is tangential, and therefore minimal. On the other hand if a 2-block B is given no three points of its complement are collinear. Moreover each plane contains a point of the complement if B is minimal, for then B contains no Fano block. These results ensure that the complement contains 5 points in general position, and no other point.

The tangential 3-dimensional 2-block just described consists of 10 points lying in threes on 10 lines in a Desargues configuration. We therefore refer to it as the *Desargues block.*

It is found that there is no 4-dimensional tangential 2-block and only one 5-dimensional one. The 5-dimensional tangential 2-block can be regarded as a dual embedding of the Petersen graph. Its existence is associated with the fact that the Petersen graph has no Tait coloring. We refer to this 2-block as the *Petersen block.*

It is not known if any tangential 2-blocks other than the Fano, Desargues, and Petersen block exist. Even the 6-dimensional case has not yet been analyzed.

Discussion on Professor Tutte's Paper

PROFESSOR M. E. WATKINS: It has been conjectured in the paper we have just heard that if a trivalent graph G without isthmus has no Tait coloring, then some subgraph of G can be mapped onto the Petersen graph by a connected homomorphism. It is worth noting that the converse of this conjecture is false.

In fact, let us consider a special class of graphs $P(n, k)$ which, for want of an original name, we shall call *pinwheels.* Here n and k are integers with $n \geq 2$ and $1 \leq k \leq [n/2]$. The vertices of $P(n, k)$ are

$$\{x_0, x_1, \ldots, x_{n-1}, y_0, y_1, \ldots, y_{n-1}\}$$

and if subscripts are read as residues modulo n, the edges of $P(n, k)$ are all those of the form $[x_i, x_{i+1}]$, $[x_i, y_i]$, and $[y_i, y_{i+k}]$.

Thus $P(5, 2)$ is the Petersen graph. The graph $P(10, 4)$ can be reduced to a Petersen graph by removing a certain 5 of the edges and contracting certain others to single vertices, but $P(10, 4)$ does have a Tait coloring, as is shown in the accompanying figure.

I am led to offer the following

Conjecture. *Every pinwheel $P(n, k)$ has a Tait coloring except for the Petersen graph $P(5, 2)$.*

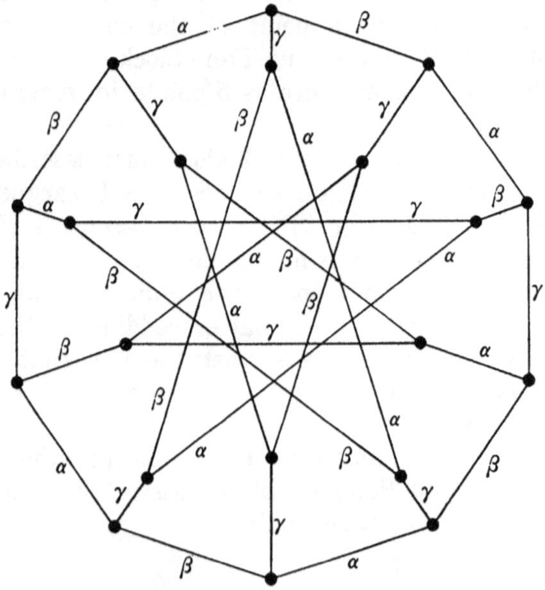

My remaining remarks are attempts to bring to the surface some of the concepts underlying this rather deep paper.

1. If G is planar, it has dual graph G^*. The correspondence between the cycles (coboundaries) of G and the coboundaries (cycles) of G^* is evident. However, what notion underlies that of a dual embedding, particularly if G is not a planar graph?

2. If one takes the points of the Desargues block as vertices of a graph and joins any two by an edge if and only if they do not lie on a common line in the block, then the resulting graph is the Petersen graph, whose dual embedding is associated with the Petersen block. Is this "duality" between the 3- and 5-dimensional cases more than a coincidence?

CHAPTER 31

Edge Colorings in Bipartite Graphs

JON FOLKMAN AND D. R. FULKERSON[1],
The Rand Corporation

1. INTRODUCTION

The following edge-coloring problem can be posed for any finite graph G. Given a finite sequence of positive integers $p_1, p_2, \ldots, p_l$, when can the edges of G be colored (edges on the same vertex having distinct colors) with l colors in such a way that precisely p_i edges of G have color i, $i = 1, 2, \ldots, l$? In this general form, the problem is no doubt extremely difficult. Even for the case of bipartite graphs, where one might reasonably expect major simplification to occur, very little seems to be known. The question in this case, rephrased in terms of (0, 1)-matrices, becomes: When can a (0, 1)-matrix A be written as a sum

$$(1.1) \qquad A = P_1 + P_2 + \ldots + P_l,$$

where each P_i is a permutation matrix of size p_i, that is, P_i

[1] This research is sponsored by the United States Air Force under Project RAND—Contract No. AF 49(638)-1700—monitored by the Directorate of Operational Requirements and Development Plans, Deputy Chief of Staff, Research and Development, Hq. USAF.

has at most one 1 in each row and column and contains p_i 1's? Two special cases of this problem have been examined in [1, 4]. In [1] it is shown that if A has τ 1's and maximum row or column sum k, if p is an integer in the interval $1 \leq p \leq [\tau/k]$, and if r and q are the unique integers such that $r \geq 0$, $0 \leq q < p$, $\tau = (k+r)p + q$, then A can be written as a sum of $k + r$ permutation matrices of size p and one permutation matrix of size q. The main result of [4] is a combinatorial duality formula for the maximum number of permutation matrices of size m contained in an m-by-n (0, 1)-matrix A, where $m \leq n$. Denoting this maximum number by $h(A)$, it is shown in [4] that

$$(1.2) \qquad h(A) = \min_{B \subset A} \left[\frac{\tau(B)}{s(B)} \right].$$

Here B is an e-by-f submatrix of A, $\tau(B)$ denotes the number of 1's contained in B, $s(B) = e + f - n$, and the minimum is taken over all B such that $s(B) > 0$. The content of formula (1.2) can be rephrased as follows. Let

$$(1.3) \qquad p_1 = p_2 = \ldots = p_h = m, \qquad p_{h+1} = p_{h+2} = \ldots = p_l = 1.$$

Then a necessary and sufficient condition for a decomposition (1.1) is that for any e-by-f submatrix B of A, $e = 0, 1\ldots, m$, $f = 0, 1, \ldots, n$, we have

$$(1.4) \qquad \tau(B) \geq \sum_{j=(m-e)+(n-f)+1}^{\infty} p_j^* .$$

In (1.4) p_j^* denotes the number of integers p_i that are greater than or equal to j, that is, the p-sequence and the p^*-sequence are conjugate partitions of the integer $p_1 + p_2 + \ldots + p_l$. It is also understood that equality holds in (1.4) for $B = A$.

Although conditions (1.4) are necessary for a decomposition (1.1) in which P_i contains p_i 1's, it can be seen from examples that they are not in general sufficient. We shall prove in Secs. 2 and 3, however, that these conditions are sufficient if

$$(1.5) \qquad p_1 = p_2 = \ldots = p_h = p, \qquad p_{h+1} = p_{h+2} = \ldots = p_l = q,$$

thereby generalizing the main result of [4]. Our attempts to find necessary and sufficient conditions for arbitrary p_i have not been successful. What little information we have on the general case is presented in Sec. 4.

There is a kind of coloring problem involving matroids for which conditions analogous to (1.4) are known to be both necessary and sufficient. A matroid $M = (E, F)$ is a finite set E of elements and a family F of subsets of E, called *independent* sets, such that (1) every subset of an independent set is independent, and (2) for every set $X \subset E$, all maximal independent subsets of X have the same cardinality, called the *rank* $r(X)$ of X. It is known [2] that the elements E of matroid M can be partitioned into independent sets $I_1, I_2, \ldots, I_l$ of respective sizes $p_1, p_2, \ldots, p_l$ if and only if, for every $X \subset E$,

$$|X| \geq \sum_{j=r(\bar{X})+1}^{\infty} p_j^* . \tag{1.6}$$

Here $\bar{X} = E - X$, $|X|$ denotes the cardinality of set X, and equality is assumed to hold for $X = E$. Thus the coloring problem is one in which a set of elements can have the same color if they form an independent subset of E, and we are asked to color elements of E in such a way that p_i elements have color i, $i = 1, 2, \ldots, l$. If we say that a set of edges in the bipartite graph G having edge set E (a set of 1's in the (0, 1)-matrix $\boldsymbol{A}$) is "independent" if it forms a *matching*, that is, if no two edges of the set are on the same vertex (no two 1's lie in the same row or column), the family F of "independent" sets thus defined satisfies axiom (1) for matroids, but does not satisfy axiom (2). However, an analog of (1.6) for the resulting coloring problem would be

$$|X| \geq \sum_{j=\rho(\bar{X})+1}^{\infty} p_j^* , \qquad \text{all } X \subset E , \tag{1.7}$$

where $\rho(\bar{X})$ denotes the maximum size of a matching in set $\bar{X}$ of edges. Conditions (1.7), which appear on the surface to be stronger than (1.4), are again necessary, but not sufficient, for the desired coloring. Indeed, using the König theorem on maximum matchings in bipartite graphs, it can be shown that (1.4) and (1.7) are actually equivalent systems of inequalities. Thus the matroid result mentioned above extends in only a limited way to the non-matroidal situation we are concerned with here.

2. A DECOMPOSION THEOREM FOR BIPARTITE GRAPHS

Let $G = [M, N; E]$ be a bipartite graph with edge set E and vertex "parts" $M = \{1, 2, \ldots, m\}$, $N = \{1, 2, \ldots, n\}$. Thus each edge of G joins a vertex of M to a vertex of N. Throughout this and the following section we shall need to single out subsets of E of the following kind—all the edges of G that join vertices in $X \subset M$ to vertices in $Y \subset N$. We denote such a set of edges by (X, Y). Thus $E = (M, N)$.

Let $G' = [M, N; E']$ and $G'' = [M, N; E'']$ be subgraphs of G and suppose that E', E'' is a partition of E, empty sets not being excluded. We then write $G = G' + G''$, and say that G is the *sum* of its subgraphs G', G''. (If we let $\boldsymbol{A}$, $\boldsymbol{A}'$, $\boldsymbol{A}''$ be the m-by-n adjacency matrices for G, G', G'', respectively, we have $\boldsymbol{A} = \boldsymbol{A}' + \boldsymbol{A}''$.) Let the degree in graph G of vertex $i \in M$ be denoted by r_i, that of vertex $j \in N$ by s_j. Similarly, let r'_j, s'_j, and r''_j, s''_j denote degrees in G' and G'', respectively. The question we raise and answer in this section is the following. For each $i \in M$ and $j \in N$ let ρ'_i, σ'_j and ρ''_j, σ''_j be specified nonnegative integers. We also specify two nonnegative integers τ', τ''. When can we write $G = G' + G''$, where the degrees in G' satisfy $r'_i \leq \rho'_i$, $s'_j \leq \sigma'_j$, the degrees in G'' satisfy $r''_i \leq \rho''_i$, $s''_j \leq \sigma''_j$, and G' has τ' edges, G'' has τ'' edges? As we shall see in the next section, the answer to this question leads to a solution of the edge-coloring problem for bipartite graphs if the integers $p_1, p_2, \ldots, p_l$ satisfy (1.5). Moreover, the method of proof, which uses basic results of network flow theory [3], provides an efficient edge-coloring algorithm in this special case.

Theorem 2.1. *Let $G = [M, N; E]$ be a bipartite graph having degrees r_i, $i \in M$, and s_j, $j \in N$, and $\tau = |E|$ edges. Further, let ρ'_i, σ'_j, ρ''_i, σ''_j, τ', τ'' be specified nonnegative integers satisfying*

$$\tau = \tau' + \tau''; \tag{2.1}$$

$$r_i \leq \rho'_i + \rho''_i, \qquad i \in M; \tag{2.2}$$

$$s_j \leq \sigma'_j + \sigma''_j, \qquad j \in N. \tag{2.3}$$

Then

$$G = G' + G'', \tag{2.4}$$

where $G'(G'')$ has $\tau'(\tau'')$ edges and degrees $r'_i \leq \rho'_i$ for $i \in M$ ($r''_i \leq \rho''_i$ for $i \in M$) and $s'_j \leq \sigma'_j$ for $j \in N$ ($s''_j \leq \sigma''_j$ for $j \in N$) if and only if, for each $X \subset M$, $Y \subset N$, we have

$$\tau' - \sum_{i \in \bar{X}} \rho'_i - \sum_{j \in \bar{Y}} \sigma'_j \leq |(X, Y)|, \tag{2.5}$$

$$\tau'' - \sum_{i \in \bar{X}} \rho''_i - \sum_{j \in \bar{Y}} \sigma''_j \leq |(X, Y)|, \tag{2.6}$$

$$|(X, Y)| \leq \sum_{i \in X} \rho'_i + \sum_{j \in Y} \sigma''_j, \tag{2.7}$$

$$|(X, Y)| \leq \sum_{i \in X} \rho''_i + \sum_{j \in Y} \sigma'_j. \tag{2.8}$$

Here $\bar{X} = M - X$, $\bar{Y} = N - Y$, and $|(X, Y)|$ denotes the number of edges joining X to Y in G.

Proof. The necessity of each of the conditions (2.5)–(2.8) can easily be verified directly. Suppose (2.4) holds with G' and G'' as specified, and let t' denote the number of edges joining X to Y in G', t'' the number of edges joining X to Y in G''. Then

$$\tau' \leq \sum_{i \in \bar{X}} r'_i + \sum_{j \in \bar{Y}} s'_j + t' \leq \sum_{i \in \bar{X}} \rho'_i + \sum_{j \in \bar{Y}} \sigma'_j + |(X, Y)|,$$

$$|(X, Y)| = t' + t'' \leq \sum_{i \in X} r'_i + \sum_{j \in Y} s''_j \leq \sum_{i \in X} \rho'_i + \sum_{j \in Y} \sigma''_j,$$

verifying (2.5) and (2.7). Similarly for (2.6) and (2.8).

The sufficiency of conditions (2.5)–(2.8) can be established using known results about flows in networks. We begin by imbedding the graph G in an appropriate flow network G^* as follows. The vertex set V of G^* consists of $x_1, x_2, \ldots, x_m$, corresponding to part M of G, and $y_1, y_2, \ldots, y_n$, corresponding to part N of G, plus two additional vertices that we label a and b. The directed edges of G^* are those ordered pairs (x_i, y_j) that correspond to the edges of G, plus the ordered pairs (a, x_i) for $i \in M$, (y_j, b) for $j \in N$, and (b, a). We now impose nonnegative integral lower bounds $l(x, y)$ and capacities (upper bounds) $c(x, y)$ on the amount of flow $f(x, y)$ in edge (x, y) of G^* as follows:

$$\begin{aligned} &l(a, x_i) = \max(0, r_i - \rho''_i), \qquad c(a, x_i) = \rho'_i, \\ &l(x_i, y_j) = 0, \qquad c(x_i, y_j) = 1, \\ &l(y_j, b) = \max(0, s_j - \sigma''_j), \qquad c(y_j, b) = \sigma'_j, \\ &l(b, a) = \tau', \qquad c(b, a) = \tau', \end{aligned} \tag{2.9}$$

Then an integral-valued flow f satisfying Kirchoff's conservation law at all vertices of G^* and also satisfying the prescribed bounds $l(x, y) \leq f(x, y) \leq c(x, y)$ for all edges (x, y) of G^* picks out subgraphs G', G'' of G satisfying the requirements of the theorem by putting edge (i, j) of G in G' or G'' according as $f(x_i, y_j) = 1$ or $f(x_i, y_j) = 0$. It therefore suffices to show that (2.5)–(2.8) imply the existence of such a flow.

It is known [5] that such a flow exists if, for every subset $Z \subset V$ of the vertices of G^*, the sum of the lower bounds on edges from Z to $\bar{Z} = V - Z$ is less than or equal to the sum of the capacities on edges from $\bar{Z}$ to Z:

$$\sum_{\substack{x \in Z \\ y \in \bar{Z}}} l(x, y) \leq \sum_{\substack{x \in \bar{Z} \\ y \in Z}} c(x, y) . \tag{2.10}$$

We shall show that (2.5)–(2.8) imply (2.10).

Let $Z \subset V$. First assume $a \in Z$, $b \in \bar{Z}$. Let $\bar{S}$ denote the subset of indices $i \in M$ such that $x_i \in \bar{Z}$ and T the subset of indices $j \in N$ such that $y_j \in Z$. Then (2.10) may be written as

$$\begin{aligned} &\sum_{i \in \bar{S}} \max(0, r_i - \rho_i'') + \sum_{j \in T} \max(0, s_j - \sigma_j'') \\ &\leq \tau' + |(\bar{S}, T)| . \end{aligned} \tag{2.11}$$

Let $\bar{X} \subset \bar{S}$ be that subset of $\bar{S}$ on which $r_i - \rho_i'' > 0$, and let $\bar{Y} \subset T$ be that subset of T on which $s_j - \sigma_j'' > 0$. Then (2.11) becomes

$$\begin{aligned} &\tau + |(\bar{S}, T)| - \sum_{i \in \bar{X}} r_i - \sum_{j \in \bar{Y}} s_j \\ &\geq \tau'' - \sum_{i \in \bar{X}} \rho_i'' - \sum_{j \in \bar{Y}} \sigma_j'' . \end{aligned} \tag{2.12}$$

But the left-hand side of (2.12) is at least $|(X, Y)|$, where $X = M - \bar{X}$, $Y = N - \bar{Y}$, and hence (2.6) implies (2.12).

Next suppose $a \in Z$, $b \in Z$. Let $\bar{S}$ denote the subset of indices $i \in M$ such that $x_i \in \bar{Z}$, let $\bar{Y}$ denote the subset of indices $j \in N$ such that $y_j \in Z$, and let $Y = N - \bar{Y}$. Then (2.10) may be written as

$$\sum_{i \in \bar{S}} \max(0, r_i - \rho_i'') \leq |(\bar{S}, \bar{Y})| + \sum_{j \in Y} \sigma_j' . \tag{2.13}$$

Let $X \subset \bar{S}$ be that subset of $\bar{S}$ on which $r_i - \rho_i'' > 0$, and $\bar{X} = M - X$. Then (2.13) becomes

(2.14)
$$\sum_{i \in X} r_i - |(\bar{S}, \bar{Y})| \le \sum_{i \in X} \rho_i'' + \sum_{j \in Y} \sigma_j' .$$

Since left-hand side of (2.14) is at most $|(X, Y)|$, we see that (2.8) implies (2.14).

The remaining two cases, $a \in \bar{Z}$, $b \in Z$ and $a \in \bar{Z}$, $b \in \bar{Z}$, can be dealt with similarly. In the first case, (2.5) implies (2.10); in the second, (2.7) implies (2.10).

This completes the proof of Theorem 2.1.

In Corollaries 2.2 and 2.3 below, we specialize the primed parameter occurring in (2.1)–(2.3) in order to note simplifications that occur in the existence conditions (2.5)–(2.8). We shall be particularly interested in the case of constant bounds on degrees in G' and in G''.

Corollary 2.2. *Let* $\rho_i' = k'$ *and* $\rho_i'' = k''$, *all* $i \in M$, *and let* $\sigma_j' = k'$, $\sigma_j'' = k''$, *all* $j \in N$. *Then there is a decomposition* (2.4) *if and only if the inequalities*

(2.5a)
$$\tau' - k'(|\bar{X}| + |\bar{Y}|) \le |(X, Y)| ,$$

(2.6a)
$$\tau'' - k''(|\bar{X}| + |\bar{Y}|) \le |(X, Y)|$$

hold for all $X \subset M$, $Y \subset N$.

Proof. It suffices to show that (2.7) and (2.8) hold automatically. Suppose that (2.7) fails for some $X \subset M$, $Y \subset N$:

(2.15)
$$|(X, Y)| > k'|X| + k''|Y| .$$

Let $k = \max(r_1, \ldots, r_m, s_1, \ldots, s_n)$ be the maximum degree in G. Thus $k \le k' + k''$ by (2.2) and (2.3). Then

(2.16)
$$-|(X, Y)| \ge -k|X| ,$$

(2.17)
$$-|(X, Y)| \ge -k|Y| .$$

Adding (2.15) and (2.16), and (2.15) and (2.17), yields

$$0 > (k' - k)|X| + k''|Y| \ge k''(|Y| - |X|) ,$$

$$0 > k'|X| + (k'' - k)|Y| \ge k'(|X| - |Y|) ,$$

a contradiction. Hence (2.7) holds. Similarly for (2.8).

If we further specialize parameters by taking $k' = 1$, $k'' = k - 1$, where k is the maximum degree in G, a generalization of a result due to Dulmage and Mendelsohn [1] is obtained. In this case there always exists a value of τ' that produces a decomposition (2.4). Indeed, it is shown in [1] that if the bipartite graph G has τ edges and maximum degree k, there exists a matching in G of size $\tau' = [\tau/k]$ that "hits" all vertices of degree k; that is, each vertex of degree k is incident with some edge of the matching. Corollary 2.3 below describes the full range of values of τ' for which such a matching exists.

Corollary 2.3. *Let $k > 0$ be the maximum degree in the bipartite graph $G = [M, N; E]$ having $\tau = |E|$ edges, and let*

$$\max_{\substack{X \subset M \\ Y \subset N}} \{\tau - |(X, Y)| - (k-1)(|\bar{X}| + |\bar{Y}|)\} = \sigma, \tag{2.18}$$

$$\min_{\substack{X \subset M \\ Y \subset N}} \{|(X, Y)| + |\bar{X}| + |\bar{Y}|\} = \rho. \tag{2.19}$$

Then, for each τ' in the interval

$$\sigma \leq \tau' \leq \rho, \tag{2.20}$$

there exists a decomposition $G = G' + G''$ where G' is a matching of size τ' and G'' has maximum degree $k-1$. In particular, the integer $\tau' = [\tau/k]$ satisfies (2.20).

Proof. We show first that the interval (2.20) is nonempty. Let $\tau' = [\tau/k]$. If $\tau' > \rho$, then there are $X \subset M$, $Y \subset N$ such that

$$\tau > k\{|(X, Y)| + |\bar{X}| + |\bar{Y}|\},$$

a contradiction. Hence $\tau' \leq \rho$. Next suppose $\tau' < \sigma$. Then there are $X \subset M$, $Y \subset N$ such that

$$\tau - [\tau/k] > |(X, Y)| + (k-1)(|\bar{X}| + |\bar{Y}|)$$

and hence

$$\tau(k-1) > k|(X, Y)| + k(k-1)(|\bar{X}| + |\bar{Y}|),$$

again a contradiction. Hence $\sigma \leq [\tau/k] \leq \rho$. Corollary 2.3 now follows from Corollary 2.2.

The nonnegative integers σ and ρ defined in (2.18) and (2.19) can be described in other ways. It is easy to see that the minimum in (2.19) occurs for X, Y such that (X, Y) is empty, and hence ρ is the minimum number of vertices that cover all edges in G. By the König theorem, this is equal to the size of a maximum matching in G (frequently called the *term rank* of G). The integer σ can also be described in terms of certain matchings in G. If we let $S \subset M$, $T \subset N$ be the vertices of maximum degree k in G, and let $\rho(X, Y)$ denote the size of a maximum matching in the subgraph of G having edge set (X, Y), then it can be shown that

$$\sigma = \rho(S, N) + \rho(M, T) - \rho(S, T) . \tag{2.21}$$

We conclude this section with some further discussion of the existence conditions (2.5)–(2.8) of Theorem 2.1. Our first comment concerns (2.7), (2.8). We noted in Corollary 2.2 that these conditions could be dispensed with in the case of constant bounds on degrees in G' and G''. In the general case, however, it can be seen from examples that these conditions are essential. Our second comment concerns interpretations of conditions (2.5)–(2.8), viewed individually. Suppose we know, for example, that (2.5) holds. What does this say about G, if anything? It is not difficult to see (by taking $\rho_i'' = \sigma_j'' = \infty$ in Theorem 2.1, for instance) that inequalities (2.5) are equivalent to the existence of a decomposition $G = G' + G''$ where $r_i' \le \rho_i'$ for $i \in M$, $s_j' \le \sigma_j'$ for $j \in N$, and G' has τ' edges. Similarly for (2.6). Conditions (2.7), on the other hand, are equivalent to the existence of a decomposition $G = G' + G''$, where $r_i' \le \rho_i'$ for $i \in M$ and $s_j'' \le \sigma_j''$ for $j \in N$, and similarly for (2.8). For example, consider (2.7). Let ρ_i' and σ_j'' be specified nonnegative integers, and take $\rho_i'' = \sigma_j' = \infty$. Then (2.8) clearly holds. Define

$$\tau' = \min_{X \subset M} \{|(X, N)| + \sum_{i \in \bar{X}} \rho_i'\} = |(X_0, N)| + \sum_{i \in \bar{X}_0} \rho_i' , \tag{2.22}$$

so that (2.5) is valid. Suppose that (2.6) were violated for some $X \subset M$, $Y \subset N$. Then clearly $X = M$, and hence

$$\tau' < \tau - |(M, Y)| - \sum_{j \in \bar{Y}} \sigma_j'' = |(M, \bar{Y})| - \sum_{j \in \bar{Y}} \sigma_j'' .$$

By (2.22) we have

$$|(X_0, N)| + \sum_{i \in \bar{X}_0} \rho_i' < |(M, \bar{Y})| - \sum_{j \in \bar{Y}} \sigma_j''$$

$$\sum_{i \in \bar{X}_0} \rho_i' + \sum_{j \in \bar{Y}} \sigma_j'' < |(M, \bar{Y})| - |(X_0, N)| \le |(\bar{X}_0, \bar{Y})| ,$$

contradicting (2.7). Hence (2.7) implies $G = G' + G''$, where $r'_i \le \rho'_i$, $s''_j \le \sigma''_j$, and G' has the number of edges given by (2.22). To sum up, Theorem 2.1 can be viewed as saying that if G can be decomposed in four different ways, each of which satisfies a certain subset of requirements in the theorem, there will be a single decomposition of G satisfying all requirements in the theorem.

We state this result for the situation of Corollary 2.2 explicitly.

Corollary 2.4. *Let the bipartite graph G have τ edges and maximum degree k. Suppose $\tau = \tau' + \tau''$ for nonnegative integers τ', τ'' and let k', k'' be nonnegative integers satisfying $k \le k' + k''$. If G has a subgraph H' having τ' edges and degrees not exceeding k', and also a subgraph H'' having τ'' edges and degrees not exceeding k'', then $G = G' + G''$ where $G'(G'')$ has $\tau'(\tau'')$ edges and degrees not exceeding $k'(k'')$.*

3. EDGE COLORINGS

In this section we assume that (1.5) holds and apply the results of Section 2 to the edge-coloring problem described in Section 1.

We say that a sequence of positive integers $p_1, p_2, \ldots, p_l$ is *color-feasible* in graph G if there is an edge coloring of G in which precisely p_i edges have color i, $i = 1, 2, \ldots, l$.

Theorem 3.1. *Assume*

$$p_1 = \ldots = p_h = p, \qquad p_{h+1} = \ldots = p_l = q. \tag{3.1}$$

The sequence (3.1) *is color-feasible in the bipartite graph $G = [M, N; E]$ if and only if the inequalities*

$$|(X, Y)| \ge \sum_{j=|\bar{X}|+|\bar{Y}|+1}^{\infty} p_j^* \tag{3.2}$$

hold for all $X \subset M$, $Y \subset N$. Here the p^*-sequence is the conjugate of (3.1), *and equality is assumed to hold in* (3.2) *for $X = M$, $Y = N$.*

Proof. Suppose the sequence $p_1, p_2, \ldots, p_l$ is color-feasible in G. The number of edges in $E - (X, Y)$ having color i is at most $|\bar{X}| + |\bar{Y}|$; hence the number of edges in (X, Y) having color i is at least $\max(0, p_i - |\bar{X}| - |\bar{Y}|)$. Summing over i yields

$$|(X, Y)| \geq \sum_{i=1}^{l} \max(0, p_i - |\bar{X}| - |\bar{Y}|) = \sum_{j=|\bar{X}|+|\bar{Y}|+1}^{\infty} p_j^* .$$

Now assume (3.1) and (3.2). We shall first apply Corollary 2.2 with

$$h = k', \qquad l - h = k'', \qquad ph = \tau', \qquad q(l-h) = \tau'' .$$

To this end, we note that by (3.2), with $X = M$, $Y = N$, we have $\tau' + \tau'' = |(M, N)| = \tau$. Also taking $X = M - \{i\}$, $Y = N$ in (3.2) yields

$$|(M - \{i\}, N)| \geq \sum_{j=2}^{\infty} p_j^* ,$$

and hence

$$r_i = |(i, N)| \leq p_1^* = l , \qquad i \in M .$$

Similarly, $s_j \leq l$ for $j \in N$. Thus the maximum degree k in G satisfies

$$k \leq k' + k'' = l .$$

We now check that (3.2) implies the existence conditions (2.5a) and (2.6a) of Corollary 2.2. Let X, Y be arbitrary subsets of M, N, respectively. Then

$$|(X, Y)| \geq \sum_{j=|\bar{X}|+|\bar{Y}|+1}^{\infty} p_j^* \geq ph - h(|\bar{X}| + |\bar{Y}|) , \tag{3.3}$$

verifying (2.5a). Similarly,

$$|(X, Y)| \geq \sum_{j=|\bar{X}|+|\bar{Y}|+1}^{\infty} p_j^* \geq q(l-h) - (l-h)(|\bar{X}| + |\bar{Y}|) . \tag{3.4}$$

It follows from Corollary 2.2 that $G = G' + G''$ where G' has ph edges and degrees not exceeding h, and G'' has $q(l-h)$ edges and degrees not exceeding $l - h$.

We can further decompose G', and also G'', by use of the following lemma.

Lemma 3.2. *Let G be a bipartite graph having τ edges and maximum degree k. Then G decomposes into a sum of h matchings, each of size p, if and only if $\tau = ph$ and $k \leq h$.*

The necessity of these conditions is obvious. The sufficiency is a consequence of the Dulmage-Mendelsohn theorem described in Section 1. Sufficiency can also be established by induction on h, using Corollary 2.3, as follows. The case $h = 1$ is trivial. Assume the validity of the lemma for $h - 1$ and consider h. If $k < h$, then $[\tau/k] \geq [\tau/h] = p$. By Corollary 2.3, G has a matching G_1 of size p. Moreover, the graph $G - G_1$ obtained by deleting edges of G_1 from G has $p(h - 1)$ edges and the assumption $k < h$ implies that $G - G_1$ has maximum degree less than or equal to $h - 1$. On the other hand, if $k = h$, Corollary 2.3 implies that G has a matching G_1 of size p that hits all vertices of degree h. Thus again $G - G_1$ has $p(h - 1)$ edges and maximum degree $h - 1$. The lemma now follows from the induction assumption.

We return to the decomposition $G = G' + G''$ reached prior to the statement of Lemma 3.2 in the proof of Theorem 3.1. It follows from the lemma that G' decomposes into a sum of h matchings, each of size p, and that G'' decomposes into a sum of $l - h$ matchings, each of size q. This completes the proof of Theorem 3.1.

By taking $p_{h+1} = \ldots = p_l = 1$ in (3.1), we obtain the following corollary.

Corollary 3.3. *Let $h(G, p)$ denote the maximum number of disjoint matchings, each of size p, contained in the bipartite graph $G = [M, N; E]$. Then*

$$h(G, p) = \min_{\substack{X \subset M \\ Y \subset N}} \left[\frac{|(X, Y)|}{p - |\bar{X}| - |\bar{Y}|} \right], \tag{3.5}$$

the minimum in (3.5) *being taken over all X, Y such that $p - |\bar{X}| - |\bar{Y}| > 0$.*

This result is a direct generalization of formula (1.2) for the case $p = |M| = m \leq n$.

4. SOME REMARKS ON THE GENERAL CASE

It can be seen from examples that conditions (3.2) are not

sufficient for the sequence $p_1, p_2, \ldots, p_l$ to be color-feasible in a bipartite graph if this sequence contains three or more distinct positive integers. For instance, consider the tree in Figure 4.1 having the adjacency matrix shown there. This graph

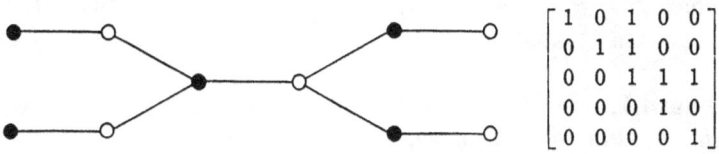

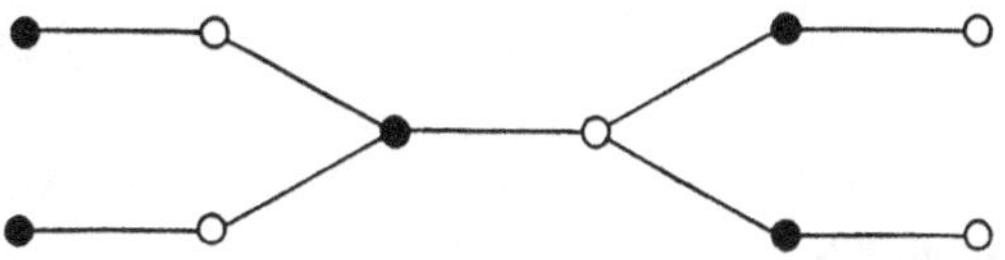

$$\begin{bmatrix} 1 & 0 & 1 & 0 & 0 \\ 0 & 1 & 1 & 0 & 0 \\ 0 & 0 & 1 & 1 & 1 \\ 0 & 0 & 0 & 1 & 0 \\ 0 & 0 & 0 & 0 & 1 \end{bmatrix}$$

Figure 4.1.

satisfies (3.2) for the sequence 5, 3, 1, but this sequence is not color-feasible. Each of the sequences 5, 2, 2 and 4, 4, 1 satisfies (3.2) and is color-feasible.

Let $P = (p_1, p_2, \ldots, p_n)$ and $Q = (q_1, q_2, \ldots, q_n)$ be two sequences of nonnegative integers. The sequence P is said to *majorize* Q, written $P \succ Q$, provided that with subscripts renumbered in accordance with

$$p_1 \geq p_2 \geq \ldots \geq p_n\,, \qquad q_1 \geq q_2 \geq \ldots \geq q_n\,, \tag{4.1}$$

we have

$$\sum_{i=i}^{e} p_i \geq \sum_{i=1}^{e} q_i \qquad e = 1, 2, \ldots, n-1\,, \tag{4.2}$$

$$\sum_{i=1}^{n} p_i = \sum_{i=1}^{n} q_i\,. \tag{4.3}$$

If $P \succ Q$ and $Q \succ P$, we say that P and Q are *equivalent* and write $P \approx Q$. Thus $P \approx Q$ means that with the numbering selected in (4.1), we have $p_i = q_i$, $i = 1, 2, \ldots, n$. The majorization relation can be viewed as a partial order on the set of all n-lists (unordered n-tuples) of nonnegative integers satisfying (4.3).

It follows from Theorem 3.1 that if P and Q each contains at most two distinct positive integers, and if P is color-feasible and $P \succ Q$, then Q is also color-feasible. This is in fact generally so, without any special assumptions on P and Q. To prove this, we first make a definition and establish a lemma. Let $P = (p_1, p_2, \ldots, p_n)$, and suppose $p_i > p_j$. Then the sequence P' obtained from P by defining

$$p'_i = p_i - 1\,,$$
$$(4.4) \qquad p'_j = p_j + 1\,,$$
$$p'_k = p_k\,, \qquad k \neq i, j\,,$$

satisfies $P \succ P'$. Moreover, if $p_i \geq p_j + 2$, we have $P \not\approx P'$. We call the transformation (4.4) a *transfer* from i to j on P.

Lemma 4.1. *If $P \succ Q$, then P can be transformed into Q by a finite sequence of transfers.*

Proof. Select the numbering so that P and Q are monotone decreasing, and suppose $P \not\approx Q$. Let l be the last integer in the interval $1 \leq l \leq n - 1$ for which strict inequality holds in (4.2). Then $p_{l+1} < q_{l+1}$. There are integers in the interval $1 \leq i \leq l$ for which $p_i > q_i$. Let k be the last such. Thus $p_k > q_k \geq q_{l+1} > p_{l+1}$. Let P' be obtained from P by a transfer from k to $l + 1$. Then $P \succ P' \succ Q$ and $P \not\approx P'$. Repetition of this process establishes the lemma.

Using Lemma 4.1, it is easy to prove Theorem 4.2, below. In Theorem 4.2 it is unnecessary to suppose the graph G to be bipartite.

Theorem 4.2. *Let G be an arbitrary graph and suppose that $P = (p_1, p_2, \ldots, p_n)$ is color-feasible in G. If $P \succ Q = (q_1, q_2, \ldots, q_n)$, then Q is also color-feasible in G.*

Proof. By Lemma 4.1, it is enough to prove that if P' is obtained from P by a transfer from i to j, then P' is also color-feasible. Let $G = G_1 + G_2 + \ldots + G_n$, where G_i is a matching of size p_i, $i = 1, 2, \ldots, n$. Consider the matchings G_i, G_j, where $p_i > p_j$. Each connected component of the graph $G_i + G_j$ is either an even circuit or a chain, with edges alternately in G_i and G_j. Moreover, since $p_i > p_j$, at least one component must be an odd chain having first and last edges in G_i. Let G'_i and G'_j be obtained from G_i and G_j by interchanging G_i-edges and G_j-edges in this chain. This produces a coloring of G in which p'_i edges have color i.

In view of Theorem 4.2, one possible approach to the general edge-coloring problem might be in the direction of attempting to characterize those lists of nonnegative integers that are color-feasible and maximal in the sense of majorization, that is, are not majorized by any other color-feasible list. (In the example of Fig. 4.1, for instance, (5, 2, 2) and (4, 4, 1) are the only two such lists.) However, even for the case of bipartite graphs, we don't know how to construct one such list, let alone all of them. The main information we have in this direction is contained in Theorem 4.3, below, the proof of which is modeled on König's proof that the edges of a bipartite graph having maximum degree k can always be colored with k colors [6].

Theorem 4.3. *Let G be a bipartite graph having maximum degree k. Then each maximal color-feasible list for G contains exactly k positive members.*

Proof. Let $p_1, \ldots, p_r$ be a maximal color-feasible list for G and assume that $p_1 \geq p_2 \geq \ldots \geq p_r > 0$. Since some vertex of G has k edges incident to it, we must have $r \geq k$. Suppose that $r > k$. Let $G = G_1 + \ldots + G_r$ be a decomposition of G into matchings such that G_i has p_i edges. Let (u, v) be an edge in G_{k+1}. Now u is incident to at most $k - 1$ edges of G other than (u, v). Hence, there is an s with $1 \leq s \leq k$ such that u is incident to no edge of G_s. Similarly, there is a t with $1 \leq t \leq k$ such that v is incident to no edge of G_t. Interchanging u and v if necessary, we may assume that $s \leq t$.

We will now construct a decomposition $G = G'_1 + \ldots + G'_r$ of G into matchings with the following properties:

$$(4.5)\quad \begin{cases} G'_i = G_i & \text{for } i \neq s, t\,, \\ \text{number of edges of } G'_s \geq \text{number of edges of } G_s\,, \\ \text{neither } u \text{ nor } v \text{ is incident to an edge of } G'_t\,. \end{cases}$$

If u is not incident to any edge of G_t, take $G'_i = G_i$ for all i. If u is incident to some edge of G_t, let U be the component of $G_s + G_t$ containing u. Since u is incident to no edge of G_s, U is a chain with u as an endpoint. Suppose v is in U. Then v would have to be the other endpoint of U. Furthermore, the number of edges in U would have to be even, since the edges are alternately in G_t and G_s, with the first edge in G_t and the last one in G_s. Hence, U together with the edge (u, v) would form a circuit in G of odd length. This is impossible because G is bipartite. Therefore v is not in U. Now let $G'_i = G_i$ for $i \neq s, t$, and let G'_s and G'_t be obtained from G_s and G_t by interchanging G_s-edges and G_t-edges in the chain U. Conditions (4.5) now hold.

Since neither u nor v is incident to an edge of G'_t, we may define a decomposition $G = G''_1 + \ldots + G''_r$ of G into matchings by setting

$$(4.6) \qquad \begin{cases} G''_t = G'_t \cup \{(u, v)\}, \\ G''_{k+1} = G'_{k+1} - \{(u, v)\}, \\ G''_i = G'_i \qquad \text{for } i \neq t,\ k+1. \end{cases}$$

Let q_i be the number of edges in G''_i. Then $q_1, q_2, \ldots, q_r$ is color-feasible for G. Furthermore, by (4.5) and (4.6) we have

$$\begin{aligned} q_i &= p_i \qquad \text{for } i \neq s, t, k+1, \\ q_s &= p_s + e, \\ q_t &= p_t - e + 1, \\ q_{k+1} &= p_{k+1} - 1, \end{aligned}$$

where $e \geq 0$ is the excess of the number of edges in G'_s over the number of edges in G_s. Since $s \leq t < k + 1$ it now follows that

$$\sum_{i=1}^{j} q_i \geq \sum_{i=1}^{j} p_i \qquad \text{for } 1 \leq j \leq r,$$

$$\sum_{i=1}^{k} q_i > \sum_{i=1}^{k} p_i.$$

Since the list P is in monotone decreasing order, this implies that $Q \succ P$ and $Q \not\approx P$, contradicting the assumption that P was a maximal color-feasible list. Hence $r = k$.

References

1. Dulmage, A. L., and N. S. Mendelsohn, "Some Graphical Properties of Matrices with Nonnegative Entries," to appear in *Aequationes Mathematicae.*
2. Edmonds, Jack, and D. R. Fulkerson, "Transversals and Matroid Partition," *J. Res. N.B.S.*, **69 B** (1965), 147–153.
3. Ford, L. R., Jr., and D. R. Fulkerson, *Flows in Networks*, Princeton University Press, Princeton N.J., 1962.
4. Fulkerson, D. R., "The Maximum Number of Disjoint Permutations Contained in a Matrix of Zeros and Ones," *Canad. J. Math.*, **16** (1964), 729–735.

5. Hoffman, A. J., "Some Recent Applications of the Theory of Linear Inequalities to Extremal Combinatorial Analysis," *Proc. Symposia in Appl. Math.*, **10** (1960), 113–129.
6. König, D., "Uber Graphen und ihre Anwendung auf Determinantentheorie und Mengenlehre," *Math. Ann.*, **77** (1916), 453–465.

CHAPTER 32

The Eigenvalues of the Adjacency Matrix of a Graph

A. J. HOFFMAN[1], *IBM Watson Research Center*

1. INTRODUCTION

We treat ordinary graphs (i.e., finite, undirected, at most one edge joining a pair of vertices, and no edge joining a vertex to itself). If G is a graph on w vertices, we denote by $A = A(G)$ the adjacency matrix of G; namely,

$$A = (a_{ij}) = \begin{cases} 1 & \text{if vertex } i \text{ and vertex } j \text{ are adjacent (joined by an edge)} \\ 0 & \text{otherwise} \end{cases}$$

A is a symmetric matrix of 0's and 1's, with diagonal 0, and has real eigenvalues $\lambda_1 \geq \ldots \geq \lambda_w$, and we shall call these numbers the spectrum of G. Clearly, the spectrum of G yields some information about G; the question is, how much?

This article will survey the work of the past eight years on this question by the author and his friends. It will be apparent from the survey that the results achieved so far barely scratch the surface of what appears to be a rich area of

[1] This research was supported in part by the Office of Naval Research under Contract No. Nonr 3775(00).

investigation. We hope that the survey also indicates what have been the two principal charms of the subject: the first is that one meets in the course of the work some very pretty graphs. The second is that it is amusing to use the basic theorems of the theory of real symmetric matrices in an off-beat combinatorial context, and we shall mention these theorems as they are used.

It is also worth remarking that, although the focus of interest here is on graphs, related questions arise in group theory, elliptic geometry, experimental designs, telephony, mechanics, etc. [17].

We describe the results under three headings: girth graphs, line graphs and imbedding.

2. GIRTH GRAPHS

The *girth* of a graph is the length of the smallest circuit that the graph contains. Let G be a regular graph of valency v, girth g. What is the smallest number of vertices that G can contain? We first derive a trivial inequality, and pose (but do not solve) the question of finding all graphs for which this inequality holds as an equation.

To derive the trivial inequality, observe that if we build the graph starting from a single vertex, it must stay tree-like for several "levels" until the level becomes approximately half the girth. If the girth is odd, say $2k + 1$, then we have, at the 0th level, one vertex; at the first level, v vertices; at the second level, $v(v-1)$ vertices; ...; at the kth level, $v(v-1)^{k-1}$ vertices. Thus the number of vertices w satisfies

$$w \geq 1 + v + v(v-1) + \ldots + v(v-1)^{k-1}. \tag{2.1}$$

Now assume that (2.1) holds as an equality. It is shown in [15] that in case $k = 3$, this can only happen if $v = 2$ (i.e., if G is a simple heptagon), and in case $k = 2$, the only possibilities are $v = 2, 3, 7, 57$ (whether a graph exists for $v = 57$ is unknown, in the other cases the graphs exist and are unique: respectively the pentagon, Petersen graph and an unnamed graph of 50 vertices). Nothing is known about other values of k.

In case $k = 3$, it is easy to show that the adjacency matrix A satisfies the equation

$$A^3 + A^2 - 2(v-1)A - (v-1)I = J, \tag{2.2}$$

where J is the matrix every entry of which is unity. Consequently, if λ is any eigenvalue of A other than v, (2.2) implies λ is a root of the polynominal

$$x^3 + x^2 - 2(v-1)x - (v-1) = 0. \tag{2.3}$$

But (2.3) is irreducible, hence all eigenvalues of A other than v itself occur with equal multiplicity. It is then easy to infer from $Tr\ A = 0$ that $v = 2$.

In case $k = 2$, A satisfies the equation

$$A^2 + A - (v-1)I = J, \tag{2.4}$$

and using the fact that the multiplicity of an eigenvalue is a positive integer, one deduces $v = 2, 3, 7$ or 57. The case $v = 7$ involves some detailed calculations on the dimension of eigenspaces in order to prove the graph unique.

We now turn to the case g even, say $g = 2k$ [21]. Then using reasoning similar to that employed in the case g odd, one sees that

$$w \geq 2(1 + n + \ldots + n^{k-1}), \tag{2.5}$$

where $n = v - 1$. If $n = 1$, the simple $2k$-gon yields equality in (2.5), so assume $n > 1$. If $k = 2$, the complete bipartite graph $K_{n+1,n+1}$ satisfies (2.5) as an equality. If $k = 3$ or $k = 4$, the graphs one obtains are in 1-1 correspondence with projective 2 and 3-space with $n + 1$ points on a line. For $k = 6$, graphs have been found, but not a complete description of all of them. (See [1]). For $k = 5, 7, 8, 9, \ldots$, there are none for which $n > 1$. The argument in brief is that one can (as in (2.2) and (2.4)) find a polynomial $P(x)$ of degree k such that $P(A) = J$. It is known (see [11]) that such an equation holds for $A = A(G)$ if and only if G is connected and regular. Further, in that case, the unique polynomial of least degree (called the polynomial of the graph) for which the equation holds is given by

$$P(x) = w\frac{\prod (x - \alpha_i)}{\prod (d - \alpha_i)},$$

where $d > \alpha_1 > \ldots > \alpha_t$ are the distinct eigenvalues of A.

Since in the case under discussion G has diameter k and $P(x)$ has degree k, $P(x)$ is the polynomial of the graph, hence the distinct eigenvalues of A can be calculated. Next, it is shown that if two eigenvalues of A have different squares, they have different multiplicities. But for $k = 5, 6, 7, \ldots$, $n > 1$, A has eigenvalues which are algebraic conjugates (hence have the same multiplicity) but have different squares.

In very recent work, Alan Gewirtz has considered graphs of girth $2k$, with the additional stipulation that a fixed number t of paths join any pair of vertices at distance k from each other. He has shown that these graphs are regular, found several necessary conditions for their existence, and constructed several beautiful graphs fulfilling all conditions.

3. LINE GRAPHS

If G is a graph, the line graph $L(G)$ of G (also called the interchange graph of G) is the graph whose vertices are the edges of G, with two vertices of $L(G)$ adjacent if and only if the corresponding edges of G have exactly one common vertex. A striking characterization of line graphs has been recently given [18], but there remains the question of trying to characterize line graphs by spectral properties. Let us denote by $\lambda(A)$ the algebraically least eigenvalue of a symmetric matrix A, and by $\lambda(G)$ the least eigenvalue of $A(G)$. Then it is easy to show that $\lambda(L(G)) = -2$ for any G with more edges than vertices. The number -2 is rather large for the least eigenvalue of the adjacency matrix of a graph, and all the research on characterizing line graphs by spectral properties has depended heavily on exploiting this fact.

In a series of papers ([9], [10], [12], [13], [14]), the following general situation was studied: a particular family of regular, connected line graphs was examined, the distinct eigenvalues in the spectrum of each graph in the family calculated, and the question raised of whether any regular connected graph with the same distinct eigenvalues was a member of the family. In the case of $L(K_n)$, the line graph of the complete graph on n vertices, the answer was yes except for $n = 8$. In the case of the line graph of a finite projective plane (regarded as a bipartite graph) and the line graph of the affine plane, the answer was yes, with no exceptions; in the case of the line graph of a balanced symmetric incomplete block design (v, k, λ), the answer was yes unless $v = 4$, $k = 3$, $\lambda = 2$. (See [2], [5], [6], [7], [8], [19], [20] for related work not highlighting the eigenvalue concept.)

The basic idea in these papers was to reconstruct the vertices of the original graphs by finding cliques in the putative line graphs. The process of constructing these cliques is facilitated by proving that the graph in question cannot contain a claw of order 3 (a claw of order k is a graph consisting of $k + 1$ vertices, of which one is adjacent to each of the others, but no two of the others are adjacent). It turns out that a regular graph G whose valency is at least 13 cannot contain a claw of order 3. The matrix theory arguments on which this depends are the well-known minimum characterization of $\lambda(A)$ and the fact that, for a symmetric matrix, different eigenvalues have corresponding eigenvectors orthogonal. Of course, settling the details of all cases and finding the exceptions requires detailed use of all the spectral information provided, such as the polynomial of the graph. But most of the information in these papers is summarized by the statement that, in the cases considered, when G was a regular connected graph of sufficiently high valency with $\lambda(G) = -2$, then G was a line graph. It is attractive to conjecture that this is true in general, but that cannot be so. If H_{2n} is the graph formed by deleting a 1-factor from the complete graph on $2n$ vertices, then $\lambda(H_{2n}) = -2$, but H_{2n} is not a line graph if $n > 3$. However, Ray-Chaudhuri and the author have proven that H_{2n} is the only exception for sufficiently high valency: namely, if G is a regular connected graph of valency > 16, $\lambda(G) = -2$, then G is a line graph or $G = H_{2n}$ for some n. An example due to J. J. Seidel shows that 16 cannot be reduced to 15 in the above theorem.

Let $v(G)$ be the minimum valency of the vertices of a graph G (not assumed regular). Ray-Chaudhuri has proved that, if $v(G) > 43$, $\lambda(G) = -2$, and if, for any pair of adjacent vertices i and j, there are at least two vertices adjacent to i but not to j, then G is a line graph.

4. IMBEDDING

Let G be a graph, and suppose $G \subset H$. From matrix theory, it is clear that $\lambda(G) \geq \lambda(H)$, but how close can $\lambda(H)$ come to $\lambda(G)$, when $v(H)$ is arbitarily large? To study this question, we define

$$\mu(G) = \lim_{v\to\infty} \sup_{\substack{G\subset H \\ v(H)\geq v}} \lambda(H) .$$

A. M. Ostrowski and the author have given a formula for $\mu(G)$. Let Γ be the set of $(0, 1)$ matrices C each of which has w rows, has the property that every row sum is positive, but loses that property if any column is deleted. If $A = A(G)$, then

$$\mu(G) = \max_{C \in \Gamma} \lambda(A - CC^T) . \tag{4.1}$$

The proof of (4.1) uses, from graph theory, Ramsey's theorem and a Ramsey-like theorem for bipartite graphs due to Kovari, Sos and Turan; from matrix theory, it uses the minimum characterization of $\lambda(M)$, where M is any symmetric matrix. Some inequalities follow naturally from (4.1), such as

$$\mu(G) \geq \lambda(G) - 1$$

$$\mu(G) \geq -\text{(maximum valency of the vertices of } G) .$$

One can also prove without much difficulty that $\mu(P) = -2$, if P is any simple polygon. The most interesting use of (4.1) is to find $\mu(K)$, when K is a claw of order k. The most attractive manner of stating the answer is in an inverse way. Define $k(G)$ to be the largest order of a claw contained in G. Then, for $\alpha \geq 1$,

$$\lim_{v\to\infty} \max_{\substack{\lambda(G)\geq -\alpha \\ v(G)\geq v}} k(G) = [\alpha]^2 + [(\alpha - 1)(\alpha - [\alpha])] .$$

References

1. Benson, C. T. "Minimal Regular Graphs of Girths Eight and Twelve," *Can. J. Math.*, **18** (1966), 1091–1094.
2. Bose, R. C. "Strongly Regular Graphs, Partial Geometries and Partially Balanced Designs," *Pacific J. Math.*, **13** (1963), 389–419.
3. Bose, R. C. "A Theorem on Regular Graphs," *Notices*, 1966.
4. Bose, R. C. and Laskar, Renu "A Characterization of Tetrahedral Graphs," *Notices*, 1966.
5. Bruck, R. H. "Finite Nets II, Uniqueness and Embedding," *Pacific J. Math.*, **13** (1963), 421–457.

6. Chang, L. C. "The Uniqueness and Nonuniqueness of the Triangular Association Schemes," *Science Record* **3** (1959), 604–613.
7. Chang, L. C. "Association Schemes of Partially Balanced Designs with Parameters $n = 28$, $n_1 = 12$, $n_2 = 15$ and $p_{11}^2 = 4$," *Science Record*, **4** (1960), 12–18.
8. Connor, W. S. "The Uniqueness of the Triangular Association Scheme," *Ann. Math. Statist.*, **29** (1958), 262–266.
9. Hoffman, A. J. "On the Uniqueness of the Triangular Association Scheme," *Ann. Math. Statist.*, **31** (1960), 492–497.
10. Hoffman, A. J. "On the Exceptional Case in a Characterization of the Arcs of a Complete Graph," *IBM J. Res. Develop.*, **4** (1960), 497–504.
11. Hoffman, A. J. "On the Polynomial of a Graph," *Amer. Math. Monthly*, **70** (1963), 30–36.
12. Hoffman, A. J. "On the Line Graph of a Projective Plane," *Proc. Amer. Math. Soc.*, **16** (1965), 297–302.
13. Hoffman, A. J. and Ray-Chaudhuri, D. K. "On the Line Graph of a Finite Affine Plane," *Canad. J. Math.*, **17** (1965), 687–694.
14. Hoffman, A. J. and Ray-Chaudhuri, D. K. "On the Line Graph of a Symmetric Balanced Incomplete Block Design," *Trans. Amer. Math. Soc.*, **116** (1965), 238–252.
15. Hoffman, A. J. and Singleton, R. R. "On Moore Graphs with Diameters 2 and 3," *IBM J. Res. and Develop.*, **4** (1960), 497–504.
16. Laskar, Renu "A Characterization of Cubic Lattice Graphs," *Amer. Math. Soc. Notices*, 1966.
17. Lint, J. H. van and Seidel, J. J. "Equilateral Point Sets in Elliptic Geometry," *Proc. Koninklijke Akad. van Wetenschappen, Ser, A, Mathematical Sciences*, **69** (1966), 335–348.
18. Rooij, A. C. M. van and Wilf, H. S. "The Interchange Graph of a Finite Graph," *Acta Math. Acad. Sci. Hung.*, **16** (1965), 263–270.
19. Shrikhande, S. S. "On a Characterization of the Triangular Association Scheme," *Ann. Math. Statist.*, **30** (1959), 39–47.
20. Shrikhande, S. S. "The Uniqueness of the L_2 Association Scheme," *Ann. Math. Statist.*, **30** (1959), 781–798.
21. Singleton, Robert "On Minimal Graphs of Maximum Even Girth," *J. Combinatorial Theory*, **1** (1966), 306–332.

CHAPTER 33

Labelling to Obtain a Maximum Matching

M. L. BALINSKI[1], *The City University of New York*

1. INTRODUCTION

This paper describes a very simple labelling algorithm for solving the maximum matching problem on a graph. In contrast with the approach [4] which depends on "shrinking odd cycles" and, hence, on consideration of a hierarchy of reduced graphs this method permits all work to proceed on the given graph through use of purely "local" information stored only on the vertices of the graph. To contrast these approaches and enable the discussion to be self-sufficient Section 2 reviews basic theorems and the shrinking approach in a simple manner before proceeding to labelling. The success of labelling techniques in the closely related network flow problems together with the practical importance of integer programs defined on a graph [1], [2] has motivated this development which is now being extended to the general integer programming problem on a graph.

2. THE PROBLEM

A graph $G = \{V, E\}$ is taken to be a finite set of vertices

[1] This work was supported by the Army Research Office, Durham, under contract No. DA-31-124-ARO(D)-366.

V together with a set of distinct edges E which are unordered pairs of distinct vertices. A *matching* M of the graph is a subset of the edges E with the property that no two edges of M are incident at a vertex. The *maximum matching problem* is to find a matching having a maximum number of edges.

Given a graph G and a matching M an *alternating path* is a simple path (no vertices in common) whose successive edges alternately belong to and do not belong to M. An *augmenting path* is an alternating path connecting a pair of *exposed vertices*, vertices which are not incident to an edge of M. It is obvious that if a matching M admits an augmenting path, M cannot be maximum, since a simple reversal of assignment of edges in the augmenting path to M results in a new matching having one more edge than M. Not so obvious is the reverse statement which permits

Theorem 1. (Berge [3] for matching; Norman-Rabin [5] for covering). *A matching M is maximum if and only if G admits no augmenting paths relative to M.*

Proof. Suppose that M is a matching which admits no augmenting path but that $M^* \neq M$ is a maximum matching. Consider the subgraph G' of G containing all vertices of G and edges e of G which satisfy: $e \in M$ and $e \notin M^*$ or $e \notin M$ and $e \in M^*$. This is the set of edges where M and M^* differ.

Consider any connected component of the subgraph. Such a component can only be either a simple path or a cycle, for otherwise, the component would have a vertex with three incident edges. But this would mean that either at least two edges of M or at least two edges of M^* are incident, contradicting the fact that M and M^* are matchings. If the component is a simple path P we distinguish three cases: (i) the extreme edges of P are both in M and not in M^*; or (ii) both in M^* and not in M; or (iii) one is in M and not M^*, the other in M^* and not M.

In case (i) M^* admits an augmenting path in the component, contradicting its maximality. In case (ii) M admits an augmenting path, contradicting the hypothesis. In case (iii) the extreme vertices have no incident edges belonging to M or M^* other than those in the component itself. Therefore the component has an even number of edges and M^* may be changed by taking every edge $e \in M^*$ of the component out of M^*, and putting every edge $e \notin M^*$ of the component into M^*. This does not change the cardinality and hence gives a new M^* which is a maximum matching. But the change leads to a new G' containing fewer edges. We call such a change a *redefinition* of M^*.

If the component is a cycle, the cycle must contain an even number of edges. This permits the same redefinition of M^*. Thus, in a finite number of such redefinitions we must have $|M^*| = |M|$, completing the proof.

Our augument yields

Corollary. *If M_1 and M_2 are both maximum matchings of G one may be obtained from the other through a finite number of redefinitions.*

Thus to solve the matching problem a method is needed to find augmenting paths or to show that none exist. Edmonds [4] correctly pointed out that the theorem itself does not indicate an obvious algorithm. For to have a "good" algorithm it is necessary to show that the number of steps required to obtain a solution is less than exponential in the size of the graph, i.e., the algorithm must be better than sheer exhaustive search and its difficulty should increase only algebraically with the size of the graph. Edmonds proposed a "good" method which is based on the idea of "shrinking" odd cycles of edges into pseudonodes and working on the reduced graph in searching for augmenting paths. Given a simple odd cycle of edges together with its vertices B in $G = (V, E)$, the *reduced graph* G/B, said to be obtained from G by *shrinking* B, is the graph consisting of vertices v_i in G but not in B, and a (pseudo) vertex v_B; and consisting of edges (v_i, v_j) for v_i, v_j in G but not B if $(v_i, v_j) \in E$, and edges (v_B, v_j) for $v_j \notin B$ if $(v_i, v_j) \in E$ for some $v_i \in B$.

Lemma 1 ([4]). *Let B be a simple odd cycle of $2k+1$ edges together with its vertices. Then, if M_1 is a matching in G/B there exists a maximum matching M_B of B, $|M_B| = k$, such that $M = M_1 \cup M_B$ is a matching for G.*

Proof. Since M_1 is a matching of G/B only one edge of M_1 is incident to v_B hence to some v_b in B. Thus, in B, choose for M_B the k unique edges which are not pairwise incident at a vertex and leave v_b exposed. The lemma implies that if an augmenting path can be found in G/B relative to some M_1 then an augmenting path obtains in G relative to the corresponding $M = M_1 \cup M_B$.

Theorem 2 (Edmonds [4]). *Let M be a matching leaving at least two vertices of G exposed, B a simple odd cycle of $2k+1$ edges together with its vertices, and $M_B \subset M$, $|M_B| = k$, a maximum set of matching edges in B. Suppose v_b, the unique vertex not incident to an edge of M_B in B, is either exposed or connected to an exposed vertex v_0 along an alternating path beginning with an edge in M. Then M is a maximum matching for G if and only if $M \cap (G/B)$ is a maximum matching for G/B.*

Proof. If M is a maximum matching and $M_1 = M \cap (G/B)$ is not, then the latter admits an augmenting path which, by Lemma 1, implies M does as well. This is a contradiction by Theorem 1.

Suppose, then, that M_1 is a maximum matching but that M is not. Then there exists an augmenting path P in G relative to M connecting exposed vertices v_1 and v_2 which must include an edge of B. Let $P_i \subset P$ be the part of the path first joining v_i to a vertex of B $(i = 1, 2)$, and let P_0 be the alternating path joining v_b to v_0. If either P_1 or P_2 first hits a node of B along an edge in M, hence in M_1, then $(P_1 \cup P_2) \cap (G/B)$ is an augmenting path, contradicting the maximality of M_1. Thus we may assume P_1 and P_2 both first hit a vertex of B along an edge not in M, hence hit v_B along an edge not in M_1. If P_0 is distinct from either P_1 or P_2 the same contradiction results. So, suppose that P_0 first hits P_1 (and not P_2) from v_B to v_1 in G/B at v_a. Then, either the cycle of edges formed by P_0 and P_1 on G/B between v_B and v_a is even or it is odd.

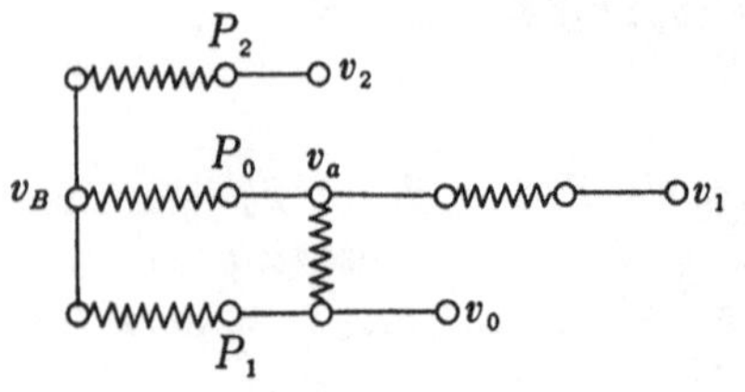

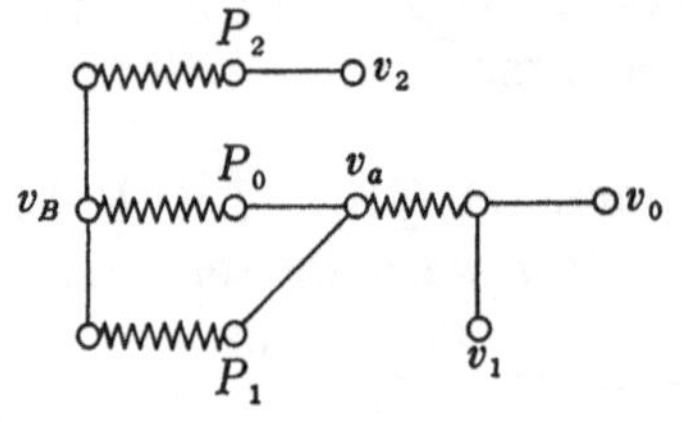

If the cycle is even then define P_2^1 to be the alternating path going from v_2 to v_B along P_2 and v_B to v_a along P_0; define P_1^1 to be the alternating path going from v_1 to v_a along P_1; and P_0^1 to the alternating path going from v_0 to v_a along P_0. The paths P_0^1, P_1^1, and P_2^1 satisfy the same assumptions as P_0, P_1, P_2

did (where v_a takes the role of v_B), but the length of the path P_0^1 is strictly less than that of P_0. Repeat. In a finite number of redefinitions either P_0^k is distinct from P_1^k or P_2^k, a contradiction; or P_0^k is identical to P_1^k or P_2^k implying it is distinct from one or the other, again a contradiction; or, an odd cycle D is formed. In this last case shrink D to obtain the graph $G_2 = (G/B)/D$. In this graph the edge $P_1 \cap G_2$ incident to v_D in G_2 belongs to $M_1 \cap G_2$. Therefore $(P_1 \cup P_2) \cap G_2$ is an augmenting path in G_2 which implies, by Lemma 1, that $G_1 = G/B$ admits an augmenting path as well. This is a contradiction and establishes the theorem.

These observations lead to the following algorithm for finding augmenting paths or showing none exists [4]. Given a graph G and some matching M in which at least two nodes of G are exposed (otherwise no augmenting path is to be found) call one exposed node, say v_0, even. Then, iteratively define nodes to be *even* or *odd* by the rules below. We will say that a node which has been defined odd or even is *paired*. A node defined to be even (odd) has an alternating path joining it to v_0 with first edge in M (not in M).

1. If vertex v is even, w unpaired, w exposed, $(v, w) \in G$ an augmenting path from w to v_0 exists.

2. If vertex v is even, w unpaired, w not exposed, $(v, w) \in \tilde{M}$ $(= E - M)$, define w to be odd.

3. If vertex v is even, w is even, $(v, w) \in \tilde{M}$, or if v is odd, w is odd, and $(v, w) \in M$ an odd cycle B of $2k + 1$ edges containing k matching edges and the vertices v and w exists. Shrink B, to obtain a reduced graph, and define v_B to be even.

4. If vertex v is odd, w unpaired, $(v, w) \in M$, define w to be even.

At every stage the (reduced) graph of paired vertices is a tree with every even (odd) vertex v connected to v_0 along an alternating path beginning with an edge in M (not in M). These rules must result either in an augmenting path in some reduced graph which can be used to determine a larger matching M^1 in G, or in a reduced graph $G_k = G/B_1/\ldots/B_k$ where every even vertex is connected only to odd vertices. Suppose the subgraph of G corresponding to paired vertices in G_k together with all edges joining them is J, and that the matching edges M in J are M_J. M_J is a maximum matching for J. Eliminate from G all edges having exactly one end in J to obtain two disjoint subgraphs J and $G - J$. Use the algorithm only on $G - J$, and repeat. If no augmenting paths can be found in

$G - J$ the matching M is a maximum matching by virtue of

Theorem 3 [4]. *$M = M_1 \cup M_J$ is a maximum matching of G if and only if M_J and M_1 are maximum matchings for J and $G - J$, respectively.*

Proof. Clearly, if either M_J or M_1 is not a maximum matching then $M = M_1 \cup M_J$ is not. So suppose M_J and M_1 are maximum matchings but M is not. Then there exists an augmenting path P in G which must use one or more edges (v, w) with $v \in J$ and $w \in G - J$. Suppose only one such edge is in P. Then v_0 must be one of the exposed nodes of P. Therefore there must exist an alternating path beginning at v and terminating at v_0 in J, and the node corresponding to or including v in G_k must be odd, since otherwise w would be paired. $P \cap G_k$ is alternating in $J \cap G_k$, contains an even number of edges $2k$ and $k + 1$ nodes defined to be even since v_0 is even and every edge of $M_J \cap G_k$ has exactly one even incident node. This implies two even nodes are incident in G_k, a contradiction. If there is more than one edge $(v, w) \in P$, $v \in J$, $w \in G - J$, the same argument applies, completing the proof.

It should be noted why this is a "good" algorithm. If n is the number of edges in G then given any matching M the algorithm requires looking at each edge at most once to either prove M is a maximum or locate an augmenting path which permits improving M. Thus it is clear that in at most n^2 looks the problem must be solved, since at most n looks locates an augmenting path, if any exists.

The difficulty with this approach, however, is that the "memory" requirement in its implementation can be excessive. It would appear that shrinking odd circuits to obtain reduced graphs, and reduced-reduced graphs, etc., together with the necessity of expanding back to the original G makes it extremely difficult to keep track of and organize the information necessary to implement the algorithm in a completely prescribed manner, e.g., as a computer program. Building upon the philosophical discussion of Edmonds in which he discusses the importance of a "good" algorithm in terms of the number of steps necessary to find a solution, it seems important as well to consider the memory requirement necessary to carry out such a step. Users of modern day computing machinery are happy to testify to the fact that memory and speed, or number of steps, compete. Unfortunately, there seems to be no theory by which the nature of this competition can be analyzed. Nevertheless, motivated by this consideration, and by the success of labelling techniques in the closely related network flow problem, a labelling approach is developed below for solving the maximum matching problem. This approach is purely local in nature; that is, requires only information stored at vertices of the original graph G, and completely does away with shrinking odd cycles and reduced graphs.

3. LABELLING

Given a graph G and a matching M, let v_0 be an exposed vertex, and assume at least one other exposed vertex exists. Then we assign labels to vertices, according to stated rules, having the following meanings. If an arbitrary vertex x carries the label

(1) $[x', -]$, then there exists an alternating path, denoted $p_1(x)$, of labelled vertices beginning at x with edge $(x, x') \notin M$ and ending at v_0;

(2) $[-, x'']$, then there exists an alternating path, denoted $p_2(x)$, of labelled vertices beginning at x with edge $(x, x'') \in M$ and ending at v_0;

(3) $[x', x'']$, then both of the above simultaneously.

In the latter case x is said to be *doubly labelled*; in the former cases, *singly labelled*. The alternating path or paths $p_i(x)$, where i indicates that the path begins with edge (x, y), y the ith component of x's label, are defined inductively by "backtracking" from vertex to vertex as indicated by alternate components of the labels. We use the symbols $-$, $+$, 0 as components of the labels to mean, respectively, the component is empty, is nonempty, and is either empty or not. Finally, every labelled vertex x carries an *exponent*, $e(x)$, which is a nonnegative integer used in determining order in the application of the labelling rules.

Rule 0. *Label an exposed vertex* v_0 *with* $[v_0, v_0]$ *and let* $e(v_0) = 0$.

Then, at any stage, use any one of the following rules given a labelled vertex v^* (the "origin") with label $[0, +]$ and $e(v^*) = \max\{e(x): x$ labelled $[0, +]$ and admitting the application of at least one rule$\}$. If no such v^* exists, the current matching M gives a maximum matching M_J for the subgraph J consisting of all labelled vertices of G together with edges connecting them. Confine subsequent labelling to the subgraph $G - J$, and store or remember M_J.

Rule 1 ("breakthrough"). *There exists a vertex w, unlabelled, $(v, w) \notin M$ and w is exposed. This means an augmenting path has been found between exposed vertices w and v_0. Reverse the assignment of edges to M in $p_2(v^*)$ and adjoin (v^*, w) to M, to obtain a new matching. Erase all labels and exponents and return to rule* 0.

Rule 2 ("tandem label"). *There exist vertices w_1 and w_2, unlabelled, $(v^*, w_1) \notin M$ and $(w_1, w_2) \in M$. Label w_1 with $[v^*, -]$, w_2 with $[-, w_1]$ and let $e(w_1) = e(v^*) + 1$, $e(w_2) = e(v^*) + 2$.*

Rule 3 ("double label"). *There exists a vertex w with label* $[0, +]$, *$(v^*, w) \notin M$, $e(v^*) \geqq e(w)$. v^* and certain vertices on $p_2(v^*)$ should receive a second label since an alternating path from v^* to w via $p_2(w)$ to v_0 exists. Let $v^* = v_1, \ldots, v_k, v_{k+1}$ be successive vertices on $p_2(v^*)$ with $e(v_j) > e(w)$ for $j \leqq k$ and $e(v_{k+1}) \leqq e(w)$. Then give second labels to $v^* = v_1, \ldots, v_k$ (changing any labels if some are already doubly labelled) with v^* receiving w in the first component, v_2 receiving v in the second, v_3 receiving v_2 in the first, etc. Redefine exponents by setting $e(v_j) = e(w)$ for $1 \leqq j \leqq k$.*

These are the only possibilities we need consider. The other logical possibilities either cannot occur or lead to no additional or potential augmenting paths. The cases are these: (1) There exists a node w labelled with $[+, -]$ where $(v^*, w) \notin M$, which simply indicates the existence of more than one alternating path from w to v_0 beginning with an edge not in M. (2) There exists nodes w_1 unlabelled, w_2 labelled, $(v^*, w_1) \notin M$, $(w_1, w_2) \in M$. But this is impossible by the rules. For if w_2 has a label $[0, +]$ then this label must be $[0, w_1]$, since M is a matching, implying w_1 is labelled after all. So, suppose w_2 has a label $[+, -]$. This is impossible for no rule assigns a label of form $[+, -]$ without giving its neighbor along an edge of M a label of form $[-, +]$. This completes the description of the algorithm.

This algorithm is good in the sense of Edmonds. Let m be the number of vertices of G, n the number of edges of G. Usually one would expect $n > m$ for otherwise, if G is a connected graph, the problem is trivial. If we count the number of times a label is assigned (including changed) as a step, it takes at most m^2 steps before either an augmenting path is located or no further labelling is possible. For each vertex can be labelled at most m times, and there are m vertices. Since there are at most m possible origins for labelling, m^3 is as upper bound on the number of labellings.

4. STRUCTURE OF LABELLED GRAPHS AND VALIDITY OF ALGORITHM

The simplicity of the algorithm is not matched by a simplicity of structure. It seems that to achieve a simpler structure in labelled graphs it is necessary to make more stringent demands in the labelling process.

In order to be able to describe the pertinent structure of labels we introduce a number of terms. A vertex x is *usable* if it carries a label $[0, +]$ and either has an unlabelled neighbor or a neighbor $\bar{x}$ with label $[0, +]$ and $e(x) > e(\bar{x})$. A *branch* of a vertex x, denoted $B(x)$, is the set of vertices $B(x) = \{y : y$ labelled, $x \in p_1(y)$ or $x \in p_2(y)\}$. We take $x \in B(x)$.

Lemma 1. (i) *The exponents along any defined alternating path $p_i(x)$ are nonincreasing.*

(ii) *The exponents of two successive vertices along any $p_i(x)$ differ by exactly* 1 *if both are singly labelled, and differ by at most* 1 *along an edge of M.*

(iii) *Any vertex with label* $[0, +]$ *has an even exponent.*

Proofs are immediate by observing rules 2 and 3 preserve all properties.

Lemma 2. *If a node x is doubly labelled* $[+, +]$, *then the node $\bar{x}$ incident to x along $(x, \bar{x}) \in M$ is also doubly labelled.*

Double labelling can occur only by labelling according to rule 3, and the parity of exponents is sufficient to establish this fact.

Lemma 3. *Suppose v carries the label* $[-, +]$ *and $z \in B(v)$ is* *usable. Then every node receiving a label after v contains v in its paths unless v becomes doubly labelled.*

Proof. This simply says that a branch grown by labelling from a singly labelled v $[-, +]$ is grown until no further growth is possible before a distinct branch is considered. To prove the lemma assume it true at some stage of labelling. We show it must still be true after one labelling step. Suppose $z \in B(v)$ is usable and v has label $[-, +]$. Then $e(y) \geqq e(v)$ for $y \in B(v)$ and the origin of labelling $v^* \in B(v)$. For, if $v^* \notin B(v)$, $e(v^*) \geqq e(v)$ and hence v should not have been labelled when it was since it was labelled from a node with exponent $e(v) - 2$.

Suppose rule 2 is used from v^* to label w_1 and w_2 previously not labelled. Then, clearly, v belongs to the paths of w_1 and w_2.

Suppose rule 3 is used from v^* due to a neighbor w having label $[0, +]$, $e(v^*) > e(w)$. If $e(w) < e(v)$ then, since $v \in p_2(v^*)$, v would be doubly labelled. If $e(w) \geqq e(v)$ and $w \in B(v)$ then, since $v \in p_2(w)$, v would belong to the path of any newly labelled node. If $e(w) \geqq e(v)$ and $w \notin B(v)$ then, again, v should not have been labelled when it was.

Lemma 4. *Suppose v carries the label $[-, +]$, $e(v) \leqq e(u)$ and $B(u) \cap B(v) = \phi$ at some stage of labelling. Then if $y \in B(u)$ and $z \in B(v)$ have labels $[0, +]$ they do not both have unlabelled neighbors.*

Proof. Either v received the label $[-, +]$ before y, or y received the label $[0, +]$ before v was labelled. So, if the lemma is false, then in the first case v belongs to the paths of y, contradicting $B(u) \cap B(v) = \phi$; in the second case v should not have been labelled when it was since it was labelled from a node having exponent $e(v) - 2 < e(y)$.

Theorem 4. *Suppose v carries the label $[-, +]$, u is labelled, and $e(v) \leqq e(u)$ at some stage of labelling. Then, if $B(u) \cap B(v) = \phi$, and $y \in B(u)$ and $z \in B(v)$ have labels $[0, +]$*

(i) *y (respectively, z) usable implies $z(y)$ has no neighbor in $B(u)$ (in $B(v)$); and*

(ii) *y and z are not neighbors.*

However, if $B(u) \cap B(v) \neq \phi$ then

(iii) *v belongs to all the paths of u.*

It should be pointed out that it is not possible to strengthen (i) and (ii) to say that $y(z)$ usable implies $z(y)$ is not usable; that is, a branch admitting of no growth at one stage of labelling may admit growth at a later stage. The following example illustrates this fact. Vertices v_i are indexed in the order in

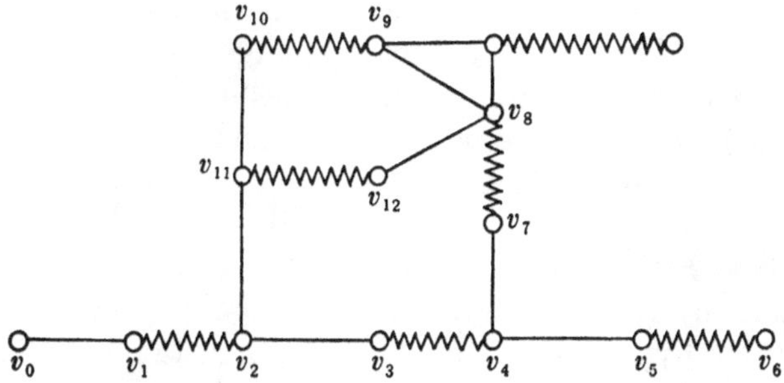

which they are labelled and vertices without names are not labelled. Suppose v_0 through v_6 are labelled. $B(v_6) = v_6$ cannot be grown. Labelling continues from v_4 to through v_{12}, double labelling occurs due to v_{12} and v_8, then due to v_{11} and v_2. At this point $e(v_9) = 8$, $e(v_6) = 6$ yet both branches have usable vertices with $B(v_9) \cap B(v_6) = \phi$. The same example shows it is even impossible to assert that $y(z)$ usable implies $z(y)$ has no unlabelled neighbor. Also, the asymmetry in v and u is necessary as shown by the following example. For (iii) $B(v_2) \cap B(v_3) \neq \phi$ yet v_2 is on only one path of v_3. For (i) and (ii), $B(v_3) \cap B(v_6) = \phi$, yet both branches have vertices labelled $[0, +]$

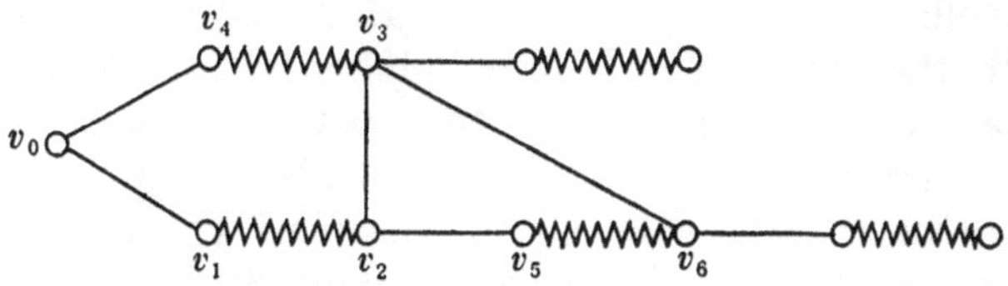

having neighbors in the other branch (and unlabelled neighbors).

To prove the theorem we will use induction; namely, we show that if the structure described holds before use of labelling rules 2 or 3 then the same structure obtains after the labelling. However, as a preliminary, we establish

Lemma 5. *Suppose* $e(v) \leq e(u)$, v *carries the label* $[-, +]$, *and* u *is labelled immediately after a labelling by rule* 3. *Assume the structure of Theorem* 4 *obtains before this labelling. Then*

$B(u) \cap B(v) = \phi$ holds after the labelling if and only if it holds before.

Proof. Suppose $B(u) \cap B(v) = \phi$ before but $B(u) \cap B(v) \neq \phi$ after. Then there must exist a vertex $x \in B(u) \cap B(v)$ after labelling while before either (a) $x \notin B(v)$ or (b) $x \notin B(u)$ and $x \in B(v)$.

(a) There must exist a node y belonging to a path of x, with $v \notin p(y)$ whose label becomes changed by rule 3. Hence, $y \in p_2(v^*)$, where v^* is the origin of labelling, and either $v \in p_2(w)$ or $v \in p_2(v^*)$ between the vertices v^* and y, where w is the neighbor of v^* prompting the labelling. In the first instance, $e(v) \leqq e(w) < e(y)$ and $v^* \in B(y)$, $w \in B(v)$. But by Theorem 4 (ii) implies $B(v) \cap B(y) \neq \phi$, since v^* and w are labelled with $[0, +]$, and by (iii) this implies v belongs to the paths of y, contrary to hypothesis. In the second instance v would have to be doubly labelled by rule 3.

(b) There must be a node y in a path of x, $u \notin p(y)$ whose label becomes changed by rule 3. Thus, $y \in p_2(v^*)$ and $u \in p_2(w)$ or $u \in p_2(v^*)$ between v^* and y. In the first instance, $e(v) \leqq e(u) \leqq e(w) < e(y)$ and $x \in B(v) \cap B(y)$ implying v belongs to both paths of y and hence $v \in p_2(v^*)$. But this contradicts (ii) since $v^* \in B(v)$ and $w \in B(u)$ are neighbors carrying labels $[0, +]$. In the second instance we know (by (iii)) that v belongs to the paths of x. So, if $e(v) \leqq e(y)$, then v belongs to the paths of y hence to a path of u, a contradiction. Otherwise, $e(v) > e(y)$. But, then, after labelling $e(u) = e(w) < e(y) < e(v)$, again a contradiction.

The converse is easily established. Suppose $B(u) \cap B(v) \neq \phi$ before labelling but $B(u) \cap B(v) = \phi$ after labelling. By (iii) v belongs to the paths of u before but not after. This can only mean that a node y belonging to a path of u and containing v in its paths came to be doubly labelled. But this implies $y \in p_2(v^*)$ whence $v^* \in B(v)$ and is usable. By Lemma 3 this means all newly defined paths contain v. This establishes Lemma 5.

We turn to Theorem 4 and treat each part separately. It is trivial to verify that labelling rule 2 preserves the properties, so we consider only use of rule 3 and recall Lemma 5 applies.

(i) Suppose $v^* \in B(v)$ before applying rule 3. Then we show that if $y \in B(u)$ carries a label $[0, +]$ then it can have no neighbor in $B(v)$. For otherwise, before labelling either (a) $y \notin B(u)$ or (b) $y \in B(u)$ and has label $[+, -]$ or (c) $y \in B(u)$, has label $[0, +]$, but has no neighbor in $B(v)$. In case (a) this implies $w \in B(u)$ or $v^* \in B(u)$, both contradicting $B(v) \cap B(u) = \phi$. In case (b) $y \in p_2(v^*)$, hence v is in the paths of y, again contradicting $B(v) \cap B(u) = \phi$. Finally, in case (c) some neighbor t of y comes to belong to $B(v)$ due to double labelling a vertex $s \in p_2(v^*)$ belonging to a path of t. $e(s) \geqq e(v)$, since otherwise, v would be doubly labelled after rule 3. But then $v^* \in B(s)$ is usable, $t \in B(s)$, and $y \in B(u)$, with y labelled $[0, +]$, contradicting our inductive hypothesis (i).

Suppose $v^* \in B(u)$ before applying rule 3. We show that if $z \in B(w)$ carries a label $[0, +]$ then it can have no neighbor in $B(u)$ after labelling. For otherwise, before labelling either (a) $z \notin B(v)$ or (b) $z \in B(v)$ and has label $[+, -]$ or (c) $z \in B(v)$, has label $[0, +]$, but has no neighbor in $B(u)$. In case (a) either $w \in B(v)$ or $v^* \in B(v)$, contradicting $B(v) \cap B(u) = \phi$. In case (b) v belongs to the path of z, z to the $p_2(v^*)$, implying $v^* \in B(v)$, also a contradiction. In case (c) some neighbor t of z comes to belong to $B(u)$ due to double labelling some $s \in p_2(v^*)$. This means that either $w \in B(u)$ or $u \in p_2(v^*)$. If $w \in B(u)$ then $e(u) \leqq e(w) < e(s)$. If $u \in p_2(v^*)$, then the new exponent of s, which is strictly less than its old exponent $e(s)$, equals the new exponent of u which is assumed greater or equal to $e(v)$. So, $e(s) \geqq e(v)$ in either case. Consider $B(s)$ and $B(v)$. $v^* \in B(s)$, and is usable, $t \in B(s)$, $z \in B(v)$ with label $[0, +]$, and t and z are neighbors. This contradicts our hypothesis (i) before labelling.

So suppose $v^* \notin B(v)$ and $v^* \notin B(u)$. If $B(v)$ grows then either $w \in B(v)$ or $v \in p_2(v^*)$, i.e., $v^* \in B(v)$. The latter is a contradiction. So is the former since then $w \in B(v)$ and $v^* \in B(v^*)$ are neighbors with $e(v) \leqq e(w) < e(v^*)$ so that v belongs to the paths of v^*. So $B(v)$, by Lemma 3, has no usable nodes. So, if $y \in B(u)$ is usable then $z \in B(v)$ having label $[0, +]$ can have a neighbor $t \in B(u)$ after labelling only if $t \notin B(u)$ before. A contradiction is derived in precisely the same manner as case (c) immediately above. This establishes (i).

(ii) The arguments for (i) apply directly to (ii) except for minor modifications for the case $v^* \notin B(v)$ and $v^* \notin B(u)$. In this case $B(v)$ cannot, as shown above, grow. So the only question concerns a $z \in B(v)$ with label $[0, +]$ gaining a neighbor $t \in B(u)$ having label $[0, +]$. If $t \notin B(u)$ before labelling the same argument used above applies. So suppose $t \in B(u)$ and has the label $[+, -]$ before labelling but t has label $[+, +]$ after labelling. Then $t \in p_2(v^*)$, that is, $v^* \in B(t)$ or $v^* \in B(u)$, a contradiction.

(iii) Since $B(v) \cap B(u) \neq \phi$ after labelling, then $B(v) \cap B(u) \neq \phi$ before labelling by rule 3. Hence v belonged to the paths of u before labelling. If rule 3 caused this to change then some node s would become relabelled, s in a path of u and v in the paths of s before labelling. But this means $v^* \in B(v)$ and all new paths include v. This completes the proof.

The validity of the algorithm is now quickly established.

Theorem 5. *Labels assigned by the algorithm have the meanings prescribed in* (1), (2), (3).

Proof. It is obvious that rule 2 preserves the correctness of the labels' meanings. The only difficulty which could conceivably arise is that in the application of rule 3 labels are assigned which define an alternating path which is not simple. But this cannot occur. For changes in paths occur only because some v_j, just doubly labelled is assigned the new path $v_{j-1}, \ldots, v_1 = v^*$, $p_2(w)$ (see definition of rule 3). The path is simple since $p_2(v^*)$ is by induction; and $p_2(w)$ is simple by induction; and $p_2(w)$ cannot contain a vertex of $B(v_j)$ for $1 \leqq j \leqq k$ since before use of rule 3 $e(v_j) > e(w) \geqq e(x)$ for $x \in p_2(w)$.

It is interesting to note that the labels of $v_1, \ldots, v_k$ must be changed as well as their exponents even if they were already labelled before a use of rule 3 for otherwise simplicity could be destroyed.

Suppose that labelling originates at v_0 and that a vertex v either has an alternating path beginning with an edge in M but has no second label, or an alternating path beginning with an edge not in M but has no first label, or both. Then we will say that labelling from v_0 is *not complete*. Otherwise, it is *complete*.

Theorem 6. *If labelling originates at exposed node v_0 and terminates with no breakthrough, then labelling from v_0 is complete.*

Proof. Suppose labelling terminates but is not complete. If this is due to a vertex v being unlabelled then there must also exist a pair of singly labelled vertices which should be doubly labelled. For, let $v = v_h, \ldots, v_0$ designate the alternating path p from v to v_0 with $(v_{2i}, v_{2i+1}) \notin M$ and $(v_{2i-1}, v_{2i}) \in M$, and let u be the labelled vertex of the path with lowest index having an unlabelled neighbor along p with higher index. u must be a vertex v_{2i} and have a label $[+, -]$, and v_{2i-1} carries the label

$[-, v_{2l}]$. Yet both of these should be doubly labelled. So we consider this possibility and show a contradiction results.

Let v_{2k} be the vertex of lowest even index on the path with label $[+, -]$, so that v_{2k-1} has label $[-, v_{2k}]$, $k \leq l$. Consider v_{2k-2}. It has, by hypothesis, a label of form $[-, +]$ or $[+, +]$ (it clearly must have a label). Suppose it is $[-, +]$. Then, since $e(v_{2k-1}) = e(v_{2k-2})$ we conclude $v_{2k-1} \in p_2(v_{2k-2})$ and $v_{2k-2} \in p_2(v_{2k-1})$, a contradiction. So v_{2k-2} must have the label $[+, +]$ and $v_{2k-2} \in B(v_{2k-1})$. Assume, inductively, that v_{2i} for $i = k-1, k-2, \ldots, j+1$ is doubly labelled, $e(v_{2i}) = e(v_{2k-1})$ and $v_{2i} \in B(v_{2k-1})$, and consider v_{2j}. v_{2j+1} is doubly labelled, and $e(v_{2j}) = e(v_{2j+1}) = e(v_{2k-1})$. If v_{2j} is singly labelled, then $v_{2k-1} \in p_2(v_{2j})$ and $v_{2j} \in p_2(v_{2k-1})$, a contradiction. So v_{2j} has label $[+, +]$ and $v_{2j} \in B(v_{2k-1})$. Therefore, $v_0 \in B(v_{2k-1})$. But this means $0 = e(v_0) = e(v_{2k-1}) = e(v_{2k}) + 1$ or $e(v_{2k}) = -1$, a contradiction. This proves the theorem.

Corollary. *Suppose labelling originates at exposed node v_0 and terminates with no breakthrough. Let $J(v_0)$ be the set of nodes labelled from v_0. Then there exists no augmenting path in the graph G which uses nodes of $J(v_0)$ (whatever is the matching M not incident to $J(v_0)$).*

Proof. This corollary justifies the restriction of subsequent labelling to the subgraph $G - J(v_0)$ consisting of nodes not in $J(v_0)$ and edges not incident to $J(v_0)$. Clearly no augmenting path exists which includes v_0. So an augmenting path p as described in the corollary would have to connect two exposed nodes not in $J(v_0)$. Further, since labelling from v_0 terminated, the only nodes of $J(v_0)$ connected to nodes of $G - J(v_0)$ have labels $[+, -]$. Therefore p would have to connect at least two such nodes, say u_1 and u_2 by an alternating path in $J(v_0)$ whose first and last edges in $J(v_0)$ are in M. But, then, either u_1 or u_2 or both should be doubly labelled, contradicting Theorem 6.

5. REMARKS

The labelling algorithm given above represents one step in obtaining such methods for solving general integer programs on a graph and describing the associated structure of labels and the role of alternating paths. The motivations for carrying out this program are manifold. The class of problems has important practical applications [1] [2]; it is closely related to network flow problems, and an understanding of this class would throw light on the structure of more general integer programming problems. We take as the formulation for such problems

(4) *maximize* $\sum c_j x_j$ *when* $\sum \alpha_j x_j \mathrm{R} \alpha_0$, $x_j \geqq 0$, x_j *integer-valued where* α_j, $j \neq 0$, *is a column of* 0's *and* 1's *containing at most two* 1's, α_0 *is a column of nonnegative integers, and R represents relations which are either* $\geqq$, *or* $\leqq$ *or* $=$.

The simple graph matching problem treated in this paper is of form (4) with all R representing $\leqq$, α_0 a column of 1's, $c_j = 1$ and α_j containing exactly two 1's for all $j \neq 0$. The matrix $A = (\alpha_1, \ldots, \alpha_n)$ is the node-arc incidence matrix of the graph G. The network flow problems are of form (4) except that each α_j either contains one non-zero entry which can be ± 1, or two with one $+1$ and one -1. Note that the minors of A in (4) are all of form $\pm 2^k$, for some integer $k \geqq 0$.

In conclusion, we indicate one generalization to a restricted problem (4) which can itself be generalized further in various directions depending on the form of (4) which is considered (these are being developed in a separate paper). Suppose we consider a *weighted graph* G, a graph G in which every edge e is assigned a positive weight $w(e) > 0$. Problem: find a *maximum weighted matching* on G, that is, find a matching the sum of whose weights is a maximum. This is a problem (4) with $c_j > 0$ the weights, and remaining data as prescribed above for the simple matching problem.

Given a matching M in a weighted graph G define an *augmenting path* P relative to M to be an alternating path or alternating simple cycle having no edge of M incident to only one vertex of P and with the property $w_M(P) = \{\sum w(e) - \sum w(d) : e \in P \cap \tilde{M},\ d \in P \cap M\} > 0$.

Theorem 7. *M is a maximum weighted matching if and only if M admits no augmenting path.*

If M admits an augmenting path it is not of maximum weight.

Suppose that M is a matching which admits no augmenting path but that $M^* \neq M$ is a maximum weighted matching. Consider, as above, the subgraph G' of G containing all vertices of G and edges e of G which satisfy $e \in M$ and $e \notin M^*$ or $e \notin M$ and $e \in M^*$. Look at a connected component H of G'. H can only be a simple path or a simple cycle consisting of an even number of edges and having no edge of M or of M^* incident at only one node. In either case since M admits no augmenting path, $w_M(H) \leqq 0$. But since M^* is a maximum matching it can admit no augmenting path so that $0 \geqq w_{M^*}(H) = -w_M(H)$ or $w_{M^*}(H) = w_M(H) = 0$, proving that the weights of M and M^* must be identical.

References

1. Balinski, M. L. "Integer Programming: Methods, Uses, Computation," *Management Sci.*, **12** (1965), 253–313.
2. Balinski, M. L. and Quandt, R. E. "On an Integer Program for a Delivery Problem," *Operations Res.*, **12** (1964), 300–304.
3. Berge, Claude. "Two Theorems in Graph Theory," *Proc. Nat. Acad. Sci. U.S.A.*, **43** (1957), 842–844.
4. Edmonds, Jack. "Paths, Trees and Flowers," *Canad. J. Math.*, **17** (1965), 449–467.
5. Norman, R. Z. and Rabin, M. O. "An Algorithm for the Minimum Cover of a Graph," *Proc. Amer. Math. Soc.*, **10** (1959), 315–319.

Discussion on Professor Balinski's Paper

PROFESSOR B. B. BHATTACHARYYA: The labelling algorithm for the maximum matching problem may be viewed as a special method for solving a certain class of integer-programming problems. Linear programmers have been successful in solving transportation-model-type integer programming problems without any special difficulty, as the extreme points of the convex set of feasible solutions of the associated problem, with integral restrictions removed, are integer-valued. But a genuine difficulty may arise in solving the maximum matching problem in a non-bipartite graph as we may get fractional optimal solutions by removing integral restrictions. This difficulty was met by Edmonds by the device of "shrinking blossoms." It would be of help to understand the labelling algorithm if we set up a correspondence between the two algorithms.

Edmonds starts by rooting an alternating tree in a matching and partitions the vertices into outer and inner vertices.

The vertices labelled (0, +) in the labelling algorithm seem to be outer vertices of the rooted alternating tree. The break-through rule corresponds to the situation where the planted tree is augmenting. The tandem label rule enlarges the already existing tree while non-existence of a vertex v^* under Rule 0 indicates that the tree is hungarian. The "double label rule" deals with a blossoming tree, and this rule provides a method of keeping the matching in the blossom in view; and hence it becomes possible to avoid the problem of resurrecting shrinking blossoms at the end of the calculations.

It seems plausible that one should be able to devise a labelling algorithm for the maximum matching problem by solving the associated minimum cover problem by using the concept of "reducing path" introduced by Norman and Rabin. In this procedure labelling should start naturally from a vertex which has more than one edge incident on it.

It has been noted by Edmonds that the maximum matching problem may be solved as an ordinary linear programming problem, by substituting for the zero-one conditions the additional constraints that the variables are non-negative and that for any set R of $2K + 1$ vertices, the sum of the variables that correspond to the edges with both end points in R is not greater than K. It may be of interest to study the dual of this problem to see whether it leads to some simple algorithm.

The possibility of extending the algorithm to the weighted maximum matching problem is of great interest. This would be very helpful to an integer programmer as otherwise he has to use more laborious cutting plane techniques to deal with the situation.

www.ingramcontent.com/pod-product-compliance
Lightning Source LLC
Chambersburg PA
CBHW021022210326
41598CB00016B/887